PRACTICAL MATLAB®
BASICS FOR ENGINEERS

Handbook of Practical MATLAB® for Engineers

Practical MATLAB® Basics for Engineers

Practical MATLAB® Applications for Engineers

PRACTICAL MATLAB® FOR ENGINEERS

PRACTICAL MATLAB®
BASICS FOR ENGINEERS

Misza Kalechman

Professor of Electrical and Telecommunication Engineering Technology

New York City College of Technology

City University of New York (CUNY)

 CRC Press
Taylor & Francis Group
Boca Raton   London   New York

CRC Press is an imprint of the
Taylor & Francis Group, an **informa** business

CRC Press
Taylor & Francis Group
6000 Broken Sound Parkway NW, Suite 300
Boca Raton, FL 33487-2742

© 2009 by Taylor & Francis Group, LLC
CRC Press is an imprint of Taylor & Francis Group, an Informa business

No claim to original U.S. Government works
Printed in the United States of America on acid-free paper
10 9 8 7 6 5 4 3 2 1

International Standard Book Number-13: 978-1-4200-4774-5 (Softcover)

Library of Congress Cataloging-in-Publication Data

Kalechman, Misza.
    Practical MATLAB basics for engineers / Misza Kalechman.
        p. cm.
    Includes bibliographical references and index.
    ISBN 978-1-4200-4774-5 (alk. paper)
    1. Electric engineering--Mathematics. 2. MATLAB. I. Title.

TK153.K18 2007
620.001'51--dc22   2008000268

**Visit the Taylor & Francis Web site at
http://www.taylorandfrancis.com**

**and the CRC Press Web site at
http://www.crcpress.com**

Contents

Preface

Practical MATLAB® Basics for Engineers is a simple, easy-to-read, introductory book of the basic mathematical concepts and principles, using the MATLAB® language to illustrate and evaluate numerical expressions and data visualization of large classes of functions and problems, written for beginners with no previous knowledge of MATLAB. MATLAB is a registered trademark of The MathWorks, Inc. For product information, please contact

The MathWorks, Inc.
3 Apple Hill Drive
Natick, MA 01760-2098 USA
Tel: 508 647 7000
Fax: 508-647-7001
E-mail: info@mathworks.com
Web: www.mathworks.com

Once the mathematical concepts are introduced and understood by the reader, MAT-LAB is then used in *Practical MATLAB® Applications for Engineers* in the analysis and synthesis of engineering and technology problems, for the case of continuous and discrete time systems.

MATLAB is a powerful, comprehensive, user-friendly, and interactive software package that is gaining acceptance as the ideal computational choice for scientists and engineers and is becoming an industrial standard, used to solve a wide range of problems in other diverse areas such as economics, business, technology, engineering, science, and education.

The reason that MATLAB has replaced other technical computational languages is that MATLAB is based on simple and easy-to-use programming tools, graphic facilities, built-in functions, and an extensive number of toolboxes.

Each chapter of this book is self-contained, in the sense that a serious attempt was made to provide the reader with all the theoretical concepts required to fully understand each chapter's material using simple numerical examples as well as direct language.

The idea is that with a relatively smaller set of functions, the reader can begin to write programs. Each chapter contains in addition a number of worked-out examples, systematically solved and chosen to illustrate general types of solutions to classes of problems often encountered in industry and academia.

The only thing that this book requires from the reader is an open and logical mind, basic skills, common sense, and academic maturity equivalent to those in the first year of college in science, technology, engineering, or a senior at a technical high school.

In summary, an effort has been made to accomplish the following goals and objectives:

- To provide reasonable proficiency in a relatively short time
- To be practical
- To introduce concepts in a compact, simple, and direct way
- To teach core skills that will aid the reader in the classroom and careers
- To be easy to read and understand, friendly, and interesting
- To provide many numerical and worked-out examples
- To be self-contained with little or no outside assistance

- To be organized by topics and complexity
- To be a valuable resource to
  - The engineering and technology student
  - The professional engineering student (preparing for the PE license)
  - The technical consultant
  - The practicing engineer

Author

Misza Kalechman is a professor of electrical and telecommunication engineering technology at New York City College of Technology, part of the City University of New York.

Mr. Kalechman graduated from the Academy of Aeronautics (New York), Polytechnic University (BSEE), Columbia University (MSEE), and Universidad Central de Venezuela (UCV; electrical engineering).

Mr. Kalechman was associated with a number of South American universities where he taught undergraduate and graduate courses in electrical, industrial, telecommunication, and computer engineering; and was involved with applied research projects, designs of laboratories for diverse systems, and installations of equipment.

He is one of the founders of the Polytechnic of Caracas (Ministry of Higher Education, Venezuela), where he taught and served as its first chair of the Department of System Engineering. He also taught at New York Institute of Technology (NYIT); Escofa (officers telecommunication school of the Venezuelan armed forces); and at the following South American universities: Universidad Central de Venezuela, Universidad Metropolitana, Universidad Catolica Andres Bello, Universidad the Los Andes, and Colegio Universitario de Cabimas.

He has also worked as a full-time senior project engineer (telecom/computers) at the research oil laboratories at Petroleos de Venezuela (PDVSA) Intevep and various refineries for many years, where he was involved in major projects. He also served as a consultant and project engineer for a number of private industries and government agencies.

Mr. Kalechman is a licensed professional engineer of the State of New York and has written *Practical MATLAB for Beginners* (Pearson), *Laboratorio de Ingenieria Electrica* (Alpi-Rad-Tronics), and a number of other publications.

1

Trends, the Industry, and MATLAB®

Unless you try to do something beyond what you have already mastered, you will never grow.*

Ralph Waldo Emerson

1.1 Introduction

In this chapter, a general look is taken at the computer, field of computing, skills associated with computer programming and computer languages, problem solving and algorithms, as well as economic shifts produced by the changes in technology that are having an impact on the world around us, job market, and of course our lives.

Obviously everyone has their own opinion about the world around them. This opinion is shaped by background, education, values, and above all by experiences.

We don't see things as they are, we see them as we are.*

Anais Nin

The main objective of this book is to attempt to see things as they are. Some see technology as a "graying industry," but others see it as opportunities especially when computing technology focuses less on the tools of technology and more on how technology is used in the search for scientific breakthroughs, the development of new products and services, or the way work is done.

Presently, it is universally accepted that computers are an essential tool of the educational process in the technologies, humanities, sciences and engineering, as well as industries and business. The computer has changed our lives: the way we study, work, and do business.

Bill Gates, the cofounder and chairman of Microsoft, summarized his view of the computing field by saying

> We are on the threshold of extraordinary advances in computing that will affect not only the sciences but also how we work and our culture.
> We need to get the brightest people working on those opportunities.

The meaning of computing has also changed over the last decades. Let us analyze some of the changes and trends.

---

* O'Brien, M.J. and Lary, S., *Profit from Experience*, Bard & Stephen, Austin, TX, 1995.

There was a time, not long ago, when the word *computer* was a job description associated with special people, with strong analytical minds, who performed tedious mathematical calculations for huge military and engineering projects.

With the passing of time, *computers* evolved and became more associated with machine languages, compilers, and tables of numbers.

Computers today are known as machines that perform symbolic computations, animation, graphics, interactive calculations, and act as an intelligent communication device, replacing in many instances the plain old telephone (pot).

The modern computer is based on the original model developed by John von Neumann back in 1952. He recognized that the real power of the computer is based on simple logical operations, binary in nature, which executed one instruction at a time in strict serial order at fantastic speeds. Today's computers can perform multiprocessing or parallel computations, and information can be received from a number of sources such as other computers or communication devices or systems through the Internet, or the World Wide Web.

A few words about the Web. The Web is a medium that has the potential to provide universal access to information for almost everyone, independent of boundaries, cultures, and locations. The Web is the most important part of the Internet. The Internet is a worldwide network of computers, owned and supervised by no particular entity or agency or more directly stated by no one. The Internet was originally developed by the U.S. Department of Defense, in 1969, under the project name of Advanced Research Project Agency Network (Arpanet) whose main research objective was to keep the U.S. military sides communicated in the event of a nuclear war. Its first test and practical application was to serve as a communication medium among nuclear physicists located in dispersed and distant geographic locations, employing a variety of communication systems and devices. This first test was performed by the European Particle Laboratory, part of a larger organization known as European Organization for Nuclear Research (CERN). From the early days, in March 1989 (led by Tim Berners-Lee, an Oxford graduate student) engineers recognized the importance of finding a simple and efficient solution to the communication problem of large, geographically extended organizations.

The same needs exist in private and government organizations, such as banks, hospitals, insurance and investment corporations, airline and oil companies, as well as government agencies such as law enforcement, military, education, and health.

The communication and information revolution of the last decades of the twentieth century was centered on the computer and the Internet. This revolution started in the early 1950s with the development of the solid-state transistor and will probably continue well into the twenty-first century.

As the devices and technologies improved over the last half-century (1960–2008), so did productivity, quality of life, and industrial competitiveness, creating new jobs and economic opportunities.

Understanding today's technologies is the basis for learning tomorrow's technologies, applications, and business opportunities. Computing is almost an infinitely malleable and universal tool. Software can be programmed to do all manner of tasks and is continuously being improved. So, computing is more like biology; it evolves unlike traditional industrial technologies such as steam, electricity, and the internal combustion engine. For example, deoxyribonucleic acid (DNA) codes that contain the secrets of life and evolution can be explored and simulated using computer codes.

Disciplines as diverse as weather forecasting, oil exploration, drug research and marketing, drug side effects, and chemical analysis rely heavily on computers and computer simulation. Even the entertainment industry (sound and video) and modern automobiles are largely controlled and monitored by a network of microprocessors and software. Today's automobile is commonly referred to as a computer on wheels.

The computer and the network it is connected to is as powerful as the software it uses. This book deals with one such software package named MATLAB that is gaining acceptance in the scientific and business communities.

The Matrix Laboratory package referred to as MATLAB was originally designed to serve as the interactive link to the numerical computation libraries LINPACK and EISPACK that were used by engineers and scientists when they were dealing with sets of equations.

Today, MATLAB is a computer language designed for technical computing, mathematical analysis, and system simulation. It is interactive in nature and is specifically designed to solve problems in the engineering fields, sciences, and business applications, and appears to be evolving as the preferred tool in the processes of engineering analysis and synthesis.

The MATLAB software was originally developed at the University of New Mexico and Stanford University in the late 1970s. By 1984, a company was established named as Matwork by Jack Little and Cleve Moler with the clear objective of commercializing MATLAB. Over a million engineers and scientists use MATLAB today in well over 3000 universities worldwide and it is considered a standard tool in education, business, and industry.

The basic element in MATLAB is the matrix, and unlike other computer languages it does not have to be dimensioned or declared.

MATLAB's original objective was to be the tool to solve mathematical problems in linear algebra, numerical analysis, and optimization; but it quickly evolved as the preferred tool for data analysis, statistics, signal processing, control systems, economics, weather forecast, and many other applications. Over the years, MATLAB evolved creating an extended library of specialized built-in functions that are used to generate among other things two-dimensional (2-D) and 3-D graphics and animation and offers numerous supplemental packages called toolboxes that provide additional software power in special areas of interest such as

- Curve fitting
- Optimization
- Signal processing
- Image processing
- Filter design
- Neural network design
- Control systems
- Statistics

Why is MATLAB becoming the standard in industry, education, and business? The answer is that the MATLAB environment is user-friendly and the objective of the software is to spend time in learning the physical and mathematical principles of a problem and not about the software. The term *friendly* is used in the following sense: the MATLAB software executes one instruction at a time. By analyzing the partial results and based on these results, new instructions can be executed that interact with the existing information already stored in the computer memory, without the formal compiling required by other competing high-level computer languages.

This interactive environment between the machine and the user is particularly important in the solution of problems in which the information at one point of the process may be the guide to the next step in the solution of a particular problem. This computation environment is probably the one that a new engineer, technologist, or technician is most likely to encounter in tomorrow's industries.

1.2   The Job Market

Today, the key to economic growth and economic survival of regions and nations is to have an adequate number of well-trained engineers, technologists, and technicians to support the society's industrial and commercial infrastructure.

To identify technical areas of growth that may impact the job market, some of the present global economic conditions and trends are identified and discussed first.

In 2004, the total U.S. job market exceeded 131 million, with a huge service sector, which now employs more than 80% of America's workers. The U.S. economy needs to add 2–3 million jobs annually, just to keep unemployment at a reasonable healthy level.

An estimated 35–40% of the new jobs are in the electronic-telecommunication-computer area, and nearly 3.5 million are employed as information technology professionals (2004). The U.S. government is a big employer and can add large numbers of jobs to the market depending on political (security, terrorism, etc.) and global conditions (agreements, wars, intervention, conflicts, disasters, etc.).

In 2003, the (U.S.) federal government employed 1.9 million civilian workers, 1.5 million in the military, and 800,000 in the postal service, which brought the total number employed by the federal government to 4.2 million, equal to 3% of the total (U.S.) job market. Government policies such as taxes, interest rates, trade agreements, economic indexes (such as consumer and confidence), and foreign competition may also have an effect on the economy and of course the job market.

1.3   Market and Labor Trends

Some market and labor trends are summarized below:

a. The general economic and job conditions, according to the U.S. Department of Labor, is that more than 1 million jobs of the 1.2 million jobs created in the period 1999–2004 are part-time or temporary (*The New York Times*, October 10, 2004).

   It can be safely stated that job trends are driven by part-time and temporary employment. The main reason for this is probably the cost of labor benefits usually paid to full-time employees.

   In 2003, there were 25 million part-time workers in the United States and from this figure, only 4.8 million had some kind of benefits.

   The trends indicate that part-time jobs would represent approximately 20% of the overall job market in the United States.

b. Today's market trends can be summarized by a simple sentence—*do more with less*, which means that the use of technology (computerized and intelligent systems) will increase, whereas union jobs and job security in general will be on the decline.

c. The Fortune 500 American companies have been downsizing and outsourcing for the past 30 years. Meanwhile, small and midsize firms have been growing much more rapidly. The result is that the labor force must be much more flexible and able to adjust to rapid changes.

d. Clearly, the U.S. economy is moving the job market away from industries that export or compete with imports, especially manufacturing, to industries that are insulated from foreign competition, such as housing and health. Since 2000, almost 3 million jobs in the manufacturing areas were lost, whereas membership in the National Association of Realtors has risen 50%.

e. In the technologies, for example, the leap from copper to optical fiber (from 1998 to 2003) eliminated 15.5% of the cable jobs. But the new fiber jobs paid 26% more than those in the cable industry and employment grew at 22.6%, according to the Economic Policy Institute (a Washington-based research center), whereas the total number of telecommunication workers represented by unions has fallen 23% since 2000 (Bureau of Labor Statistics).

f. After years of encouraging workers to take early retirement as a way to cut jobs, a growing number of American companies are hunting for older workers because they have lower turnover rate and in many cases better job performance.

Some statistics may illustrate this point—in the 65–69 age group, about one-third of men and almost one-fourth of women were working in 2004.

In activities like nursing where statistics are available, the following occurred: In 2002–2003, hospitals raised pay scales and hired 130,000 nurses over the age of 50, which makes up more than 70% of the 185,000 hired in these years.

g. The only way that labor can squeeze out more efficiency is by evolving, which means that people have to learn more than one job in their career even if they stay with the same company.

h. Statistical data supports the economic expert's finding that a new worker (a recent graduate) should expect not just four or five job changes over a lifetime, but four or five different careers over a lifetime.

i. The job market trend indicates increases in
   • Self-employment
   • Home office and online jobs
   • Contract work
   • Temporary or contingent work
   • Consulting

j. Job market trends also indicate decreases in services and technology, in the form we are accustomed to. The reasons for the decline are partially due to shifts in the technologies and trade, which is addressed later in this section. Clearly, the job market rewards people that possess individual talent. Higher education pays off because it provides technical knowledge and filters out people who have organizational skills, discipline, self-motivation, and social adeptness.

k. Furthermore, trade and technology are rapidly transforming the service economy, as we traditionally know it. The United States as well as the global economy is in a transition period and it will surely adjust over time to the new realities, creating new sources of work that will employ new workers with new skills and talents.

l. Statistical data and economic studies indicate that foreign competition and outsourcing (from China, India, etc.) are having a growing impact on the U.S. global economy and will surely affect the job markets in the coming decades.

m. According to the Kaiser Foundation "globalization of manufacturing means that more manufacturing and service related industries are outsourced." Obviously, the reason for outsourcing and moving abroad is not just to find lower wages and keep operating costs down, but also to get smart, dedicated workers and in many cases better infrastructure.

   The overseas worker is generally well educated and trained, focused and efficient, and receives generally a lower salary and little or no benefits.

   Why should any employer, anywhere in the world, hire American workers if other people, just as well educated, are available for half the wages or less?

n. No one knows with precision how many jobs are leaving the United States. Government estimates are

   i. 102,000 in 2003

   ii. 143,000 in 2004

   Unless someone abolishes the Internet and global economic integration, it will be hard to stop and reverse this trend.

o. A few words about foreign competition using India as an example. India's service industry posted $12.3 billion in export revenues in the year ending 2004, a 30% rise over the previous year. India's outsourcing industry employed over 800,000 employees and its growth is estimated to be 30–40% per year. General Electric and City Group are some of the American corporations that use India's outsourcing industry. The leading outsourcing companies in India earned as much as two-thirds of their revenues from U.S. customers (*The New York Times*, November 4, 2004). Of course, India is not an isolated case. Identical problems are faced by the U.S. economy from competing countries in all five continents.

p. According to the Bureau of Labor Statistics, outsourcing is responsible for 1.9% of layoffs in the United States. Economic experts predict that the efficiencies due to outsourcing will create more jobs at better wages than the ones destroyed (Brooks, 2007; Lohr, 2007). Over the years, the H-1 visa that allows a person to work in the United States for 3 years and be renewed for an additional 3 years has been used by U.S. companies to recruit the brightest workers from around the world. The current visa cap (2007) is 65,000, which poses a serious challenge to the U.S. job market. Meanwhile, the outsourcing market is estimated to be in the order of $386 billion in 2007 and growing with high-quality talents from eastern and central Europe like Poland, Hungary, the Czech Republic, and Slovakia with an estimated outsourcing business of $2 billion in 2007 and an expected growth rate of 30% by 2010, compared with 25% for the global market (Tagliabue, 2007).

q. The old line of U.S. companies, the last bastion of fully paid employee benefits are struggling in the global market, and few can afford to pay 100% of worker's health insurance premiums. The number of individual premiums plummeted from 29% in 2000 to 17% in 2004, and family health coverage premiums paid by private companies dropped from 11% in 2000 to 6% in 2004.

r. Some figures about costs of health benefits are provided as follows to give some insight to the magnitude of the problem facing the American manufacturing and service industries. For example, General Motors (GM), the largest private purchaser of health services, spent an estimated $4.8 billion a year with earnings of only $1.2 billion to provide health coverage to all employees (active and 400,000 retirees and dependents). At GM, each U.S. worker has to support 2.5 retirees, adding an average of $2200 to the price of each vehicle ($1625 on health care and $675 on pension), whereas its market share has declined steadily since 1996.

Toyota, with profits of $10.2 billion, which is more than double the combined profit of the big three (GM, Ford, and Daimler-Chrysler), reported that the health care obligations are not large enough to affect in any significant way its profits (*The New York Times*, October 25, 2004).

s. GM, which does set aside money for future retiree benefits, has reported (*The New York Times*, July 25, 2005) that the sum of its health care promised to retirees was $77.47 billion in 2004, which is $9.93 billion up from 2003.

t. GM is not an isolated case. Boeing, which estimated its retiree health and other nonpension obligations at $8.14 billion at the end of 2004, has assets of less than $100 million to cover them.

u. Because of the soaring cost of health care coverage, an estimated 40% of companies with more than 5000 employees no longer offer retiree health benefits.

v. In the 3-year period of 2002–2005, profits at the seven largest companies in the Silicon Valley area, the nation's high technology heartland, increased by an average of 500%, whereas employment has declined.

The increase in profits is dramatic. These actions are driven in part by the automation that Silicon Valley has largely made possible, allowing companies to create more value with fewer workers, keeping a brain trust of creative people, managers, and engineers in the United States, and hiring workers for lower level tasks elsewhere (*The New York Times*, July 3, 2005).

w. An analysis published in the San Jose Mercury News found that the top 100 public companies in the Silicon Valley (Stross, 2006)* region had revenues of $336 billion in 2004, an increase of 14% from the previous year, clearly indicating a high productivity (profits and sales) jobless trend.

1.4 Technical Know-How: Trends and Facts

Some facts and trends about technical knowledge are summarized as follows:

- Human knowledge is doubling every 10 years.
- In the past decade (1995–2005), more scientific knowledge was created than in all human history.
- Computational power based on powerful microprocessors is doubling every 18–24 months.
- A weekend edition of *The New York Times* contains more information than the average person was likely to come across in a lifetime during the seventeenth century in England.

---

* One-third of all venture investment deals went to the San Francisco Bay area. This number has not changed for the past 10 years. The New England region is far behind at 10%.

- According to Daniel Reed, director of the Renaissance Computing Institute (a collaboration of researchers from the University of North Carolina, Duke University, and North Carolina State University), computing has become the third pillar of science, along with theory and experimentation.

- The present educational system was designed in the 1900s for people to do routine work. The present market requires people who can imagine things that have never been thought before (Friedman, 2006).

- More and more routine work can be digitized and automated, including white-collar work.

- Some useful global statistics—59 and 66% of all undergraduates receive degrees in science, technology, and engineering in China and Japan, respectively, whereas it is only 32% in the United States.

- In the present job market, 85% of the jobs in the United States require advanced training or education (Caputo, 2006). Studies show that as much as 85% of measured growth in U.S. per capita income is due to technological changes driven by highly educated well-trained people applying their talents, expertise, and skills in science and technology (Exxon Mobil, 2006).

- U.S. industry is presently spending more on lawsuits than on research and development (R&D). R&D represents the most important source of value creation and investments for a company that is likely to pay dividends in the future. Few other investments can pay off the way R&D can.

- The United States is the world's biggest investor in R&D (34% of the total), but the data are troubling. R&D spending grew for decades until 2002 when it dropped for the first time in 50 years. According to the figures from the National Science Foundation, R&D climbed slightly in 2003, to $281.9 billion, and is estimated to increase to $312.1 billion by 2004 (Bernasek, 2006).

- It seems that the federal government will continue to spend more on developing weapon systems and spacecraft and less on basic and applied research, which is the foundation of the innovative competitive industrial capacity.

- Basic research is the foundation of innovation because it advances scientific knowledge and generates ideas, which the industry can then use to develop products and services. But, basic research is a risky investment in the sense that there is no guarantee that the knowledge gained from research may pay off commercially.

- U.S. companies over the years have developed research sites overseas, raising concerns about how the research benefits will filter back to the United States. Approximately 40% of the American high-tech industry already has an R&D presence in Asia and plans are on to increase this share.

- Federal investment in research as a share of the total economic output is estimated to drop to 0.4% in 2007 from 0.5% in 2006 and may drop even further as large unfunded commitments like Social Security and Medicare come due.

- Estimates indicate that China and India will account for 31% of the world's R&D personnel by the year 2007, up from 19% in 2004.

- It seems that R&D investments have been a declining priority for the last U.S. administrations. In the 1960s, the government accounted for 67% of the total U.S. R&D spending. Presently, the share is approximately 30%, whereas corporate America makes up most of the remaining.

- According to the U.S. Bureau of Labor Statistics, the labor market will experience a shift from *hard hats* to *pencil and paper pushers*. Employment in industries is expected to grow at 6.7% from 2002 to 2012, yet the number of installers and repairers is expected to grow just by 2%. The number of computer-related jobs will jump by 14.5%, whereas sales and retail jobs are expected to increase by 16.5%.

- *Rising Above The Storm* is a report written by some of the best minds in the country recruited from the Academy of Science, National Academy of Engineering, and Institute of Medicine (October 2005) and organized by two U.S. senators, Lamar Alexander and Jeff Bingaman. The explicit objective of the report is to come up with recommendations of how to enhance America's technological base. The report states that, because of globalization, the U.S. worker in virtually every sector must now face competitors who live just a mouse-click away. The report also indicates that the U.S. economic leadership is eroding at a time when many other nations are gathering strength.

- Technology has changed very rapidly in the past 20 years. Economists, educators, and industrial experts predict that technology is expected to change 500 times faster in the next 20 years.

- Three recommendations for success for the coming decades from different schools of thoughts are summarized as follows:

We need to get back to basic blocking and tackling, educating more Americans in the skills needed for the 21st century jobs.

Charles Vest
Former president of Massachusetts Institute of Technology (MIT)

Across many nations, the market increasingly rewards people with high social customer-service skills.

Lawrence Katz
Harvard University

The most important community for an individual will not necessarily be a company, but a looser community of people with similar skills and social connections. Continually building up those skills and connections is what a career is today.

Robert B. Reich
Professor of economic and social policy
Brandeis University
Former Secretary of Labor
Clinton administration

1.5   What Constitutes Essential Knowledge

Let us explore what constitutes the essential attributes for survival and growth in the present competitive and technological driven economy. It is widely recognized that essential knowledgeable skills are

- Reading
- Writing
- Problem solving

which are the basic communication, organization, and technical–logical–mathematical skills required in the modern workplace and for further growth.

The marketable skills in addition to the preceding essentials are

- Information processing
- Management and administration

It is widely recognized by educators, economists, experts, and industrial leaders that the process of learning is more important than the product, which merely entails a collection of facts that happen to be current at a particular time.

Certainly, facts are important in science, engineering, and technology, but far more important is to

- Navigate and access information
- Analyze the information
- Use the information in a creative and meaningful way
- Work and act in a team as a team

It is far more important to find, analyze, and process information and see the big picture than to acquire a skill with a particular technology, the usual definition of computer literacy.

It is far more important to learn methodology than facts.

It is far more important to learn how to learn, which means learning where and how to get information and even more important is to

- Know how to manage information and its complexities
- Master modeling and abstraction
- Think analytically in terms of algorithms
- Implement systematically, step-by-step, any algorithm

The key to employment success will be the ability to process information into useful, practical, and marketable knowledge. Workers will get jobs only if they or their firm offer a unique innovative product or service, which demands a skilled and creative labor force able to conceive, design, manufacture, and market (Friedman, 2006).

Most experts agree that the marketable skills required for the high echelon jobs are

- Problem solving
- Developing algorithms
- Recognizing patterns
- Using simulation and programming
- Being a team worker

In the simplest technical terms,

> Computing is more important than number.

> R.W. Hamming

It is widely recognized and accepted by educators, labor experts, economists, and educational leaders that in the coming decades, the biggest employment gains will be in occupations that rely on

- Unique or specialized skills
- Intelligence
- Imagination
- Creativity

The following quote well defines the knowledge and skills of the successful employee.

> If you have only technical knowledge you are vulnerable. But if you can combine business or scientific knowledge with technical savvy, there are a lot of opportunities; and it's a lot harder to move that kind of work offshore.

> **Professor Thomas W. Malone**
> *Sloan School of Management at the MIT*
> *Author of "The Future of Work" (Harvard Business School Press, 2004)*
> *(The New York Times, August 23, 2005)*

1.6 Technological Trends

There are good reasons to believe that the electronic-telecommunication-computer industry will remain an industry with opportunities in the coming decades (U.S. Department of Labor).

This industry, an industry of industries, central to any modern society indicates strong growth potential. A way that new technology can move ahead is by increasing its focus on the use of technology in specific fields instead of being narrowly fascinated with the tools. This will afford technology with high growth potential in a wider world, beyond the engineers from Silicon Valley.

A summary of current and future technologies and their applications that will impact and may revolutionize the economy and job market in the coming decades are summarized is given as follows:

- Radio tagging technologies (International Business Machines and Hewlett Packard [IBM/HP]) are heavily involved in radio frequency ID (RFID) are predicted to be used in the coming decade by such corporations like Procter & Gamble, Gillette, Boeing, Airbus, and drug and pharmaceuticals companies, as well as libraries and government agencies.
- Smart phone systems with new powerful operating systems (OSs) will provide a number of services besides the old services (television, pictures, sports, games, etc.). In 2005, of the 180 million cell phone subscribers in the United States, the majority of users were teenagers that were practically living on the phones.

  As of 2005, an estimated 76% of teenagers, aged 15–19, and 90% of the people in their early 20s regularly use their phones for text messages, purchasing ring tones and wallpaper for their handsets, playing games, and other personalization services with an estimated contribution of $2.6 billion just to the U.S. economy.

- Nanotechnology is expected to touch every part of the economy in the same way as computers have. The National Science Foundation predicted in 2001 that nanotechnology would contribute $1 trillion to the U.S. economy by 2015. Some U.S. experts even predict that this figure might be low.

- Special-purpose computers and control systems such as robots will affect every sector of the economy. It is estimated that 4.1 million electronic robots are in service by 2008, the time of this publication.

- IBM, Sony, and Toshiba are working on the latest microprocessor chip known as the "Cell." The Cell architecture consists of a network of eight processors, a 5.6 GHz clock that could have a theoretical peak performance of 256 billion mathematical operations per second, which places this chip according to its processing power among the top 500 supercomputers (Markoff, 2005).

- Intel and HP over the last decade (1998–2008) had invested millions of dollars on the Itanium chip that may have an impact on the huge video gaming and digital home entertainment industries.

- The Intel corporation, the world's largest chip maker and the University of California are working on an indium phosphate microprocessor that can switch on and off billions of times a second and transmit data at 100 times the speed of laser-based communication and use laser light rather than wires. Japanese scientists, in a related effort, are pursuing an equivalent result with a different material, the chemical element erbium (Markoff, 2006, 2007). Intel is also developing an 80-processor engine described as the Teraflop chip with computing power that matches the performance speed of the world's fastest supercomputer of just a decade ago. This chip will be available within 5 years (by 2012) and will be used in standard desktops, laptops, and server computers.

- There is no one in the government or medical field who does not consider it crucial and overdue to have electronic records in doctor's offices and hospitals. Health care specialists agree that information technology, if properly used, could help reduce medical errors and costs. Fewer than 10% of American hospitals have computerized clinical systems with electronic patient records and software for tracking their status, treatments, prescriptions, and progress. Only 20–25% of the nation's 650,000 licensed doctors outside the military and the Department of Veterans Affairs are using electronic patient records (*The New York Times*, July 21, 2005).

- A mere 25% of physicians in the United States use *ePOCRATeS®*, a software package which provides updated information on diseases, diagnostics, drugs, billing references, and insurance plans. This package saves an average of 11–30 minutes a day of the doctor's time, typically valued at $250 an hour, at a cost of only $30–$150 a year. The fees are small compared with the physician's time, since a major portion of the services costs are paid by the pharmaceutical companies.

- Silicon Valley's *dot com* era may be giving way to the *watt com* era. The new mission of many Silicon Valley companies is to develop alternative energy, such as wind- and solar power, solar panels, ethanol plants, and hydrogen power cars in a $1 trillion domestic market. For many in Silicon Valley, high tech has given way to clean tech (Richtel, 2007).

- The rise in oil prices (over $108 barrel on March 10, 2008) combine with the rising concern about the environment such as greenhouse gases from oil and coal burning are turning policy makers, environmentalists, scientists, engineers, and economists to alternate cheaper and cleaner energies such as geothermal, solar- and wind power.

- The U.S. Geothermal Energy Associates (GEO) released a report (2007) assessing the progress in the generation of geothermal energy in which the United States, the leader in online geothermal capacity, is expected to double its output in the period 2007–2015 as a result of inacting a federal tax incentive in 2005 by the U.S. congress (Gawell, 2007).

  The solar energy market for silicone-based photovoltaic panels is growing by 42% annually for the last 5 years, and since 2004, the market value of the world solar companies has grown from 1 billion to 71 billion, a 7000% increase (Hodge, 2007).

  Wind power already supplies 1% of America's domestic electrical needs, providing power to 4.5 million homes, with over 1 million homes or 3% of the electrical needs in Texas, the wind capital. A recent study by Emerging Energy Research, a consulting firm in Cambridge, MA, estimates investments of 65 billion in the next 7 years (2008–2015). In European countries such as Denmark, 20% of the electrical power is derived from wind, a goal that the United States want to emulate (Krauss, 2008).

- Propelled by mounting soaring oil costs, climate change and global warming, bio-fuels in the form of ethanol is becoming the leading alternative of the green tech revolution as an alternative source of renewable energy. From 1998–2008 the U.S. quintupled its production of ethanol, and the U.S. Congress is working on incentives for another five-fold increase in the next decade (2008–2018). Overall world wide investments in biofuels increased from $5 billion in 1995 to $38 billion in 2005, and estimates predict $100 billion by 2010 (Grunwald, 2008).

- Medicare, which claims that the lack of electronic records is the biggest impediment to improve health care, is providing the medical doctors, free of charge, a software package called *Vista* (and its new version Vista-Office) to computerize their medical practices beginning in August 2005. *Vista* has been used for over two decades by the Department of Veterans Affairs in 1300 inpatient and outpatient facilities and contains over 10 million records and treats more than 5 million veterans a year.

  Vista presents many problems, the most important one is that it is difficult to install, maintain, and operate.

- The military spends about $12 billion a year in basic- and applied research and advanced technology development in the following areas:

  - Electronic sensors
  - Robotics
  - Artificial intelligence
  - Biotechnology
  - Brain and cognitive science
  - Large-scale modeling and simulation

  These activities are creating a significant number of jobs in private as well as government sectors and have a multiplying effect on the economy.

The global positioning satellite system (GPS), for example, first developed for precision-guided munitions, is essential for cell sites to serve the cell phone industry and has the potential to revolutionize the civil air traffic control system. American companies not only draw heavily on the Pentagon's work, but they have also come to depend on it. America's ability to translate the Pentagon's technology based on commercial achievements is the model of the world.

1.7   Objective of This Book

The objective of this book is to address in a meaningful and practical way, some of the technical issues of the present changing economy, and be a means of providing the reader with some of the skills and knowledge necessary to get a well-paying technical job by mastering an essential tool such as MATLAB and, more important, a number of broad essential technical skills. Hopefully, it allows the reader to hit the ground running.

This book is written specifically to support the independent learner, serve as a textbook in an introductory course in MATLAB (high school or college), or a companion or reference (handbook) in a number of standard college courses.

1.8   Organization

The book *Practical MATLAB® Basics for Engineers* consists of nine chapters intended to be used as a textbook in an undergraduate freshmen or sophomore course that introduces programming and the use of an engineering language, and the book *Practical MATLAB® Applications for Engineers* consists of six chapters dedicated to principles, exercises, and applications geared to the electrical, electronics, computer, telecommunication engineering technologies, or technology in general.

The emphasis of the applications is in using MATLAB to solve types of engineering problems from basic circuit analysis (direct current [DC] and alternating current [AC]) to signal analysis, Laplace, Fourier, Z-transforms, filters (analog and digital), etc.

Each chapter of this book and the book titled *Practical MATLAB® Applications for Engineers* is structured as follows:

- Introduction
- Objectives
- Background
- Examples
- Further analysis
- Application problem

Chapters 2 through 9 of this book are dedicated to

- Basic math concepts such as functions, algebra, geometry, arrays, vectors, matrices, trigonometry, precalculus, and calculus
- The MATLAB language syntax rules, notation, operations, and computational programming

The knowledge gained in these eight chapters is then applied in the chapters of *Practical MATLAB® Applications for Engineers*, where the section titled Questions is omitted, since the assumption is that the reader is more mature and disciplined at this point in the learning process and drill questions are no longer appropriate.

The contents of the chapter's sections are summarized as follows.

Introduction. Each chapter starts with a brief description of the main topics and in some cases a compressed history of the events relevant to the chapter's material is included.

Objectives. Each chapter has a set of objectives that clearly establish the chapter's goals.

Background. Each chapter introduces all the concepts required to fully understand the discussion of the chapter's material in the form of rules. The notation used is R.c.n, where R stands for rule, c for the chapter, and n for the rule, concept, or definition number.

Theorems are stated, and theoretical results are quoted omitting formal proofs. Concepts are introduced using simple and direct language with explanations and examples that are easy to understand and visualize and in many cases can be worked out by hand.

In some cases, MATLAB is also employed in verifying mathematical or physical relations. In this way, the reader can quickly learn, review, and refresh the theory and start using the concepts in the form of MATLAB instructions first and programs later.

Hopefully, with a relatively smaller set of instructions and simple examples, the reader can quickly begin to write programs. The programs presented in this book have been tested under different versions of MATLAB. The view in this matter is best summarized by the following quote:

It is profoundly erroneous truism, repeated by all copy books and by eminent people, when they are making speeches, that we should cultivate the habit of thinking of what we are doing.

The precise opposite is the case.

Civilization advances by extending the number of important operations which we can perform without thinking about them.

Alfred North Whitehead

Examples. Each chapter has a number of worked-out problems with both analytical and MATLAB solutions, when appropriate or possible. The emphasis of each example's solutions is on the development of an approach leading to an algorithm and a corresponding program. The examples are chosen to illustrate general types of solutions to classes of practical problems often encountered in industry or academia.

The programs presented are not necessarily the fastest or shortest, since the primary purpose is to illustrate the logical and systematic approaches to solving broad classes of problems as well as to provide maximum clarity by choosing the most frequently encountered instructions.

Further analysis. Each chapter presents questions about the example problems to drill, review, and stimulate creative thinking. The reader is encouraged to follow the examples by executing the commands as they occur. This book is designed to be used by the reader while working on the computer. A lot of effort has been invested to make this book as easy as possible for the reader to work through without any assistance.

The best way to learn programming is by *doing*. In working out the example problems, the reader can systematically gain experience and incorporate fundamental concepts and practices into practical applications.

Application problems. At the end of each chapter a number of problems are presented. Some of the problems are encouraged to be solved by hand, others are drill problems that may include numerical manipulations, whereas still others are application problems in which the command window as well as M-files (editor window) are used.

M-files are encouraged as solutions for classes of problems where different sets of data can be tested. The M-file concept is presented in Chapter 9, in some depth, but simple file structures are introduced and employed as early as Chapter 2.

It should be emphasized that an attempt was made to provide the reader with all the theoretical concepts required in each chapter. The section titled Background of each chapter provides the reader with most of the fundamental concepts necessary to understand and follow the example problems, as well as to solve the application problems.

Both books are self-contained, and coverage of the fundamental theory and applications is sufficiently broad to make it an ideal companion to a number of college and technical high school level courses.

A serious effort has been made to make both books readable user friendly and the learning process climate a pleasant and less intimidating experience.

It should also be pointed out that these books (*Practical MATLAB® Basics for Engineers* and *Practical MATLAB® Applications for Engineers*) are also for the beginners as well as for the more seasoned or mature engineering reader. The material in the first five chapters of *Practical MATLAB® Basics for Engineers* assumes that the reader has no experience in programming and no mathematical background, except algebra and trigonometry. This makes it ideal for some high schools.

The only thing that these books require from the reader in general is an open and logical mind, basic skills, common sense, and academic maturity equivalent to the first year of college in science, technology, or engineering or being a senior at a secondary school.

The examples in the form of programs are presented with comments, when first introduced, so that the reader can follow the logic steps in the solution of a problem, with emphasis on new or important points.

The material in these books is presented and organized in a way that they can be used in a formal educational environment, but could also be for the self- or independent learner and graduate student who needs to review and refresh MATLAB and its many applications.

Many engineering and technical schools now require a course in MATLAB early in the curriculum. In many schools, MATLAB has replaced the traditional Formula Translator (Fortran), Beginners All-purpose Symbolic Instruction Code (Basic), or Programming Language One (PL/1) programming courses. In some specialized fields such as digital signal processing and linear and control systems, MATLAB is becoming the accepted standard software.

Although designed to serve engineering and technology courses, these books are also appropriate for students in the natural sciences, economics, business, social sciences disciplines, and in general disciplines in which numerical or quantitative methods are used.

The novice would probably run into difficulties when trying to learn MATLAB using the standard available textbooks. Most of the available MATLAB textbooks are either for programmers and assume that the reader is familiar with computers, models, and mathematical algorithms or are designed to be used in advanced engineering applications such as filter design, linear systems, digital signal processing, control systems, and communication.

Practical MATLAB® Basics for Engineers is different; it is written for the true beginner with no background, experience, or training in engineering or science.

In summary, an effort has been made to accomplish the following goals and objectives:

- To allow reasonable proficiency in a relatively short time
- To be practical
- To introduce concepts in a compact, simple, and direct way
- To be easy to read and understand
- To contain many numerical and worked out examples
- To be self-contained with little or no assistance
- To be organized by topics and complexity
- To be a valuable resource to
  - The engineering and technology student
  - The professional engineering student (preparing for the professional engineer [PE] license)
  - The technical consultant
  - The practicing engineer

1.9 What Is a Computer? What Constitutes Hardware? What Constitutes Software?

It is widely accepted that a good programmer should have a basic knowledge of the hardware and software components of a computer system.

A computer is a machine capable of executing a set of instructions called a program, which constitutes a coded version of the solution of a particular problem.

Computers are made up of hardware and software.

The term "computer hardware" refers to anything that can be seen, touched, or felt; usually, the computer itself is represented by three building blocks as shown in Figure 1.1. Typically, the hardware is specified by the manufacturer's model of the central processing unit (CPU) (8, 16, 32, or 64 bits processor; the higher the number, the faster and more powerful it is), memory size, intern clock that represents the speed of operation, and connecting busses.

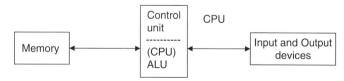

FIGURE 1.1
Simplified diagram of a computer.

A bus is a group of wires that link the building blocks of the computer and are used as the means to deliver or receive information (instructions or data) to and from the components, inside and outside the computer (peripherals). Most computers have three busses—the address-, data-, and control bus. Each one of the busses defines the type of information it is capable of carrying. The bus sizes may affect the memory size, speed of the computer, as well as its complexity and performance.

The execution of the instructions that make up a computer program is done by the CPU in conjunction with the system software stored in read only memory (ROM) (defined later in this section).

The CPU is the engine that controls the execution of the program's instructions and interrupts. An interrupt is a request from a device, which consists of an electrical signal sent to the CPU to stop and defer what it is doing and take care of the requests and then resumes the original task. Some CPUs can work on the solution of multiple tasks, a characteristic commonly referred to as multitasking. The CPU consists of an arithmetic logic unit (ALU), a control unit, a clock, and a central memory. The control unit is responsible to fetch, decode, and execute the program's instructions stored in the memory. The ALU is responsible for all the arithmetic and logic operations in the program.

Computer memories can be classified as central and external. The central memory is the main memory and is semiconductor-based. Semiconductor memories are designated as

- ROM

- Random access memory (read and write) (RAM)

- Erasable programmable read only memory (E-PROM)

The *ROM* is where the resident programs are stored. The ROM is installed by the manufacturer and cannot be erased or changed. The programs in ROM are converted from program instructions to machine language commands. Machine language consists of binary characters (on or off) and is the only (characters) language the CPU understands. The CPU with the help of the software stored in ROM converts machine language to other higher-order languages. The ROM's software is permanently stored in a memory chip and remains unchanged even when the computer is turned off.

The *RAM* is the primary memory in a computer and is used to store data and low-level programming instructions. All the information stored in RAM can be erased at will and new information can be stored in the same (memory) location. The RAM information is destroyed when the computer is turned off.

The *E-PROM* is a programmable ROM, but the information can be erased by exposure to ultraviolet light.

Solid-state memory is often referred to as volatile and nonvolatile depending on whether the information stored is lost when the power is turned off or if the information is retained in the absence of power.

External memory refers to hard- and floppy disks. These elements are also known as magnetic disks and are random access storage devices. Disks are mechanical devices that turn at a constant speed in the 2000–4000 rpm range and are accessed by the read and write heads of a movable arm.

Floppy disks or diskettes are removable storage devices. Floppy disks have diameters of $3\frac{1}{2}$ in. or $5\frac{1}{4}$ in. (and the old 8 in.) and are usually referred by their physical dimension and are becoming progressively absolute. The $3\frac{1}{2}$ in. disk has a capacity of either 720 kB double density (DD), or 1.44 MB high density (HD). The $5\frac{1}{4}$ in. disk has a capacity of either 360 kB (DD), or 1.2 MB (HD).

The input and output devices (peripherals) are devices by which information is fed to or received from the computer. The typical input devices are the keyboard and the mouse, whereas the typical output devices are the monitor and the printer.

Computer software refers to information and as such cannot be seen, touched, or felt. It is what makes the computer possible to operate and make decisions. It is the brain and soul of the machine. It is divided into

- System software
- Application software

The system software consists basically of the OS. The OS is a computer program or a series of programs that supervise the execution of all the other programs and in addition provides the interface between the user's programs and the available hardware. The OS is also responsible for controlling and managing the computer software in an efficient, effective, and user-friendly environment. Two of the most popular OSs are Unix and Windows. Fortunately, they are similar in their design and functionality. In summary, the OS is responsible for

- Managing the work or programs to be executed by the CPU
- Being the machine and user interface (controls also the peripherals)
- Organizing and keeping track of the execution of the user's programs

Application software consists of specialized packages designed to be used for solving specific classes of problems. Application software is brought into the system via the disk drive (RAM). When using application software, the programmer operates the software under the supervision of the OS.

MATLAB is an application software that can run on many computer platforms, using a number of different OSs. Some of the systems are

- Macintosh PC (68020, 68030, 68840, 68882, and up)
- Unix workstations from Sun Microsystems
- HP 9000 series
- IBM (Intel 486+ coprocessor, Pentium, Pentium Pro)
- IRS series 4D
- Digital Equipment Corporation (DEC) RISC
- DEC Alpha
- Virtual Address eXtension (VAX)
- Cray super computers

The programmable software languages are divided into three types.

- Machine language
- Assembly language
- High-level language

Machine language (Silverman and Tukiew, 1988) uses binary digits (ones and zeros) to define operations as well as operands. It is the only language that the CPU understands.

Any instruction, data, or command can always be represented by a string of ones and zeros (on or off) no matter how complex the operation may be, provided that the CPU is designed to execute such an operation.

Assembly language is one step above machine language. Mnemonic codes (memory aids) are used to specify the operations and operands performed by the CPU by converting the long binary sequences representing machine operations into compact hexadecimal codes.

Examples of typical assembly instructions are the addition of the contents of a memory location with the contents of a register or the transfer of information from a memory location to a register. Assembly- and machine languages are referred to as hardware-based languages, and a translator is required to convert machine codes into assembly language codes. This translation is done by a program called the translator or the assembler. Assembly language has a one-to-one relation to machine language and is used mainly when data is input or output directly from electronic devices, processed at the electrical level (bits and bytes), or when data and operations have to be performed at the microprocessor speed set by its internal clock.

A high-level language is several steps higher in sophistication than assembly language. The instructions are more like or resemble English. They closely follow standard mathematical relations. High-level languages must be either compiled or interpreted into machine language for execution. The difference between compiling and interpretation is that an interpreter converts each instruction into machine code and then checks for syntax errors, whereas a compiler performs the conversion and error checking simultaneously.

The programs written by programmers are usually known as source programs. Source codes are translated into object codes or machine executable instructions with absolute memory addresses. A source program therefore may result in the generation of multiple machine language instructions. The most frequently used high-level languages are summarized as follows (Linderburg, 1982):

- *Fortran.* This language was introduced by IBM in 1957 and is one of the first languages widely adopted and used by the scientific community. The main objective was to solve complex mathematical problems. This language is relatively easy to learn, but involves formatting (input as well as output).

- *Formula calculator* (FOCAL). This language consists of simple instructions and was designed to serve the scientific community. It requires little input or output formatting, but the language is harder to learn.

- *Algorithmic language* (ALGOL). This language was developed mainly by John Backus and introduced in 1958 as a universal, multipurpose language.

- *Common business oriented language* (COBOL). This language was introduced in 1958 to basically serve the business community in areas such as accounting and inventories. It is an excellent file handler and uses English-like words and sentences.

- *PL/1.* This language was introduced in 1966 by IBM as a multipurpose language designed for both the scientific and business communities (good for processing both numbers as well as strings).

- *BASIC.* This language was developed at Dartmouth College in the late 1970s and early 1980s and introduced in 1976. The instructions and algebraic equations are English-like and similar to Fortran. It was a popular computer language developed to be used as a teaching tool in colleges and universities. In the 1990s, the language evolved into Visual Basic.

- *Lisp.* This language is a symbolic, tree-structure language used for searches, qualitative decision making, and artificial intelligence applications.

- *A programming language (APL).* This language was developed by Iverson at the IBM Corporation. The main feature is that it consists of operators that can carry out functions requiring dozens of statements in other languages. It is an extremely powerful language that is particularly good for handling vectors and scalars.

- *Pascal (and Modula-2).* This language was developed in 1968, pioneered by Niklaus Wirth, and named after the eighteenth century French mathematician Blaise Pascal. It is a language that is essentially machine-independent and is particularly useful to build data structures.

- *Forth.* This language was designed basically for process control applications by Charles Moore in the late 1960s.

- *Ada.* This language was developed in the early 1980s for the U.S. Department of Defense. It is a modular language. The National Aeronautics and Space Administration (NASA) is one of the main users. Its space shuttle employs over 1 million lines of Ada's programming code.

- *C, C++.* This language was developed by Dennis Richie at the Bell Telephone Laboratories in the early 1970s. The original language combined the properties and features of high- and low-level programming languages. It is a modular language and its main application is in the control of the computer hardware.

- *Simula, comprehensive school mathematics program (CSMP), general purpose simulation system (GPSS), electronics workbench, MicroSim PSpice, laboratory virtual instrumentation engineering workbench (LabVIEW), SIMSCRIPT, graph algorithm and software package (GASP)* are very specialized simulation and control computer languages.

- *Mathematica.* This language was developed by Wolfram Research Inc. and is primarily used by engineers and scientists. Its main applications include numerical, graphical, and art schematic computations.

- *Programmation en logique (Prolog).* This language is based on formal logic and is considered by engineers and scientists as the fifth generation computer language.

- *Mathcad.* This language was developed by MathSoft Inc., Massachusetts, and designed for engineering and scientific computation.

- *RPG* (report program generator) is a language that was developed to generate reports.

- *Java.* This language was developed by Sun Microsystems and introduced during the SunWorld'95 Conference in May 1995. This language is based on an old language and compiling system technique known as University of California, San Diego (UCSD) Pascal. The P-code was developed by Kenneth Bowles in the late 1970s. Java is a network-oriented programming language used to facilitate efficient communications among many diverse electronic terminal devices in a home or business environment. The main purpose of Java is to be the medium employed in sharing information and have a centralized control.

- *Standard generalized markup language (SGML).* This language is used to describe other languages, which in turn is used to describe documents.

- *Hypertext markup language (HTML).* This is a universal, simple language for formatting, embedding of images and graphics, and hypertextual linking, also called hyperlinks of documents. This language is used in Web pages. HTML is defined by SGML and is a language that is independent of the terminal devices.

As the reader can appreciate, there are different types of languages, but fortunately only one language is required to use a computer and every computer knows at least one language.

A brief mention of the economic relation involved between the hardware and software components can be made. This relation changed drastically during the past 30 years. In the 1970s, software developments consumed approximately 20% of the total cost of a project, whereas the cost of hardware was estimated to be approximately 80%.

Currently, the cost of hardware and software are reversed. Eighty percent of the cost of a project is used in software development, upgrading, and maintenance, whereas 20% is used in hardware. During the recent years, hardware costs decreased dramatically, whereas software costs soared. The main reasons are better, cheaper, and faster microprocessors and communication devices that are mass-produced. The software component is currently focused on solving complex and difficult problems by very specialized programs in almost all disciplines, from engineering, biology, medicine, climate forecasting, and mining to business applications. The variety of specialized software applications and the economic impact in man-hours exceed the hardware platform, which in most cases is standard.

1.10   What Is MATLAB®?

MATLAB is an efficient, user-friendly, interactive software package, which is very effective for solving engineering, mathematical, and system problems. Two versions of MATLAB are commercially available—the professional and student. The professional version includes only the standard tool box, and any other tool boxes must be purchased separately. The size of the matrices is limited by the memory constrains and is expensive. The student version of MATLAB includes the basic tool box, Simulink, and symbolic tool box functions. The size of the matrices is large, but limited and inexpensive.

This book uses features of MATLAB from old as well as new versions and professional as well as student versions. Some of the main features are

- Full support of all languages, graphics, and external interfacing.
- In the older versions, the maximum matrix size was limited to 16,384 elements, which was large enough to process 128×128 matrices. In the newer versions, this limit will most likely increase.
- The toolboxes that are included in the standard student packages are signal processing, control systems, and symbolic math.
- No other toolbox can be used with the standard student edition, but it is likely that this requirement may change in newer versions.
- Programs can be externally interfaced to C and Fortran files (called MEX files).
- A math coprocessor is strongly recommended to improve efficiency.
- For any problems encountered when using MATLAB commands, the software online help facility should be used by typing help at the MATLAB prompt ($>>$) (discussed in Section 1.12).

As an additional advice, it is recommended that the reader who purchases MATLAB software should complete and return the registration card as a user, since this will entitle him/her to replace defective compact discs (CDs) at no charge and qualify for discount upgrades.

For any additional information regarding MATLAB and any of its products (toolboxes), contact

MATLAB Work Inc.

24 Prime Park Way, Natick, MA 01760-1500

Phone (508) 647-7000 e-mail: info@mathworks.com

1.11 Conventions Used in This Book

The following table describes the notations used in this book.

Convention	Definition
Times New Roman font	Used to represent MATLAB instructions or data entered by the user, such as the program code
Bold font (Times New Roman)	Used to indicate MATLAB responses usually displayed in the command window
Italic font (Times New Roman)	Used to define MATLAB instructions, commands, functions, ranges, domains, limits, relations, and key words
Angle brackets (< >)	Used to denote a key on the keyboard or an order pair. For example, <enter>, <xo,yo>

The following example indicates the input–output relation of a MATLAB command and its response (on the command window) using the preceding defined convention.

```
>>A=3*2 <enter>  ←——————  input by user

   A  =  ←——————  output by MATLAB, stored in
      6  ←——————  the computer memory
```

1.12 MATLAB® Windows

The assumption here is that the reader is sitting in front of an active computer and MATLAB is installed. To begin MATLAB, click the MATLAB icon on the computer's desktop or select MATLAB from the Start or Program menu. The prompt >> or *EDU>>* is the program prompt indicating that you are in the MATLAB environment (see Figure 1.2).

Each instruction line in the command window begins with a prompt (>> or *EDU>>*), which is automatically inserted by MATLAB. An instruction is executed after pressing the *enter* or *return* key. The designation <enter> means that the enter key was pressed. This action is implicit after a command. The result of the execution of a command appears on the next line. The result can be

- An error message (when an error is committed)
- A MATLAB prompt, meaning that the instruction was executed and MATLAB is waiting for the next command
- A MATLAB output

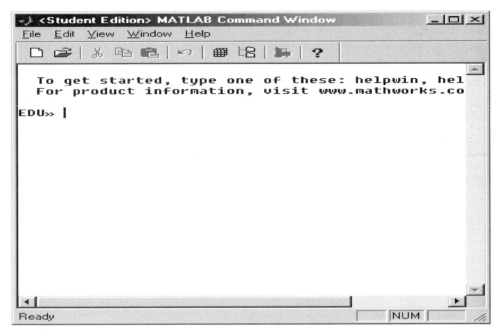

FIGURE 1.2
Command window.

This activity of entering and executing commands is carried out in the main window (called the command window) and is used to enter single line commands only. Besides the main window (command window), there are two more windows of interest that are defined as follows:

- The figure window, which is used to display graphs and plots executed by a program entered at the command window (see Figure 1.3).
- The editor or debugger window is the place where programs are created and modified. These programs can be saved in the form of files (discussed in Chapter 9).

When the user first enters MATLAB, the main program window or *command* window is active (see Figure 1.2). The *edit* window (see Figure 1.4) is used only when a program is created or modified and then stored in a file. The graphic or *figure* window is created when plots are generated as a result of executing a set of instructions, as illustrated in Figure 1.5, where the right window is the command window showing a short, four MATLAB instructions program that creates the plot of *sin*(*x*) versus *x*, shown at the left, in the command window.

There is another window that may be of interest to the user—the history window. It allows the user to see all the previous instructions executed in the command window.

By clicking the mouse at the appropriate (figure) prompt, MATLAB switches from the command window to the graphic window.

To exit MATLAB, click *quit* or *exit* in the File menu or type at the MATLAB prompt (>>) *exit* followed by <enter>. Finally, to abort or terminate a program, press the *Ctrl* and C keys simultaneously at the command window.

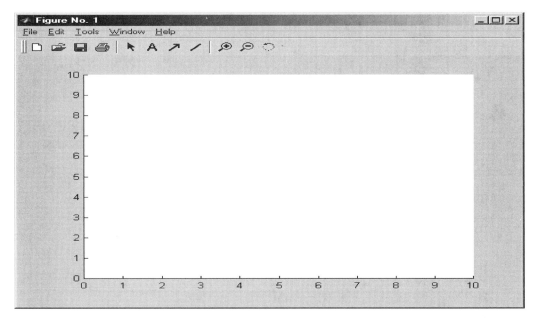

FIGURE 1.3
Graphic window.

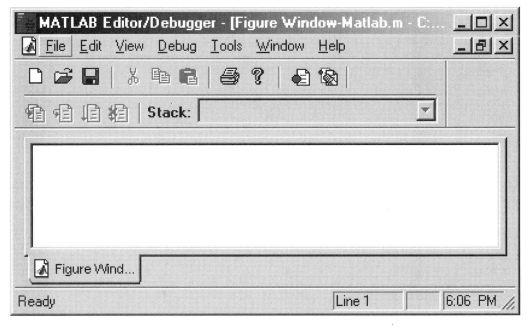

FIGURE 1.4
Edit window.

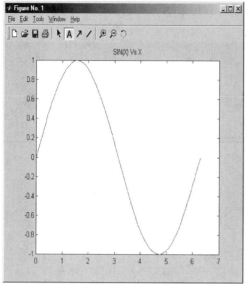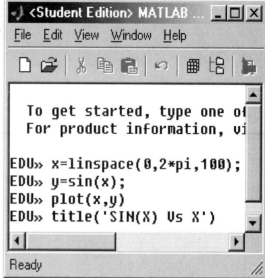

FIGURE 1.5
Figure- and command window.

1.13 A Word about Restrictions on the User's Software

The software that the reader has purchased or using is copyright protected. Copyright means that the author of a package or program has legal exclusive rights to copy, distribute, sell, or modify the software. If you are not the owner of the copyright, it is then illegal to copy, sell, or distribute that software. When copyrighted software is purchased, the user owns only one copy, which can be used by one user, in one workstation.

Software packages can also be licensed, that is, an agreement between the publisher and user(s). Licensed software means that the user is not buying a package, but rather is paying for permission to use a package. Licensed software can be for single users, multiple users (network), concurrent users (more than one copy of the software), and site users (used by any user in a location or organization). If the software is not copyrighted, it is public with no restrictions, that is, it can be copied, distributed, sold, and changed. The only restriction of public domain software is that no one can apply for a copyright on it.

1.14 Help

MATLAB offers a number of help instructions that can be accessed from the command window. A list of the most important help commands is as follows:

- *help*
- *lookfor*
- *whatsnew*

- *info*
- *helpwin*
- *helpdesk*

In addition to the help commands, MATLAB has a demo program. Type *demo* followed by <enter> and MATLAB activates a program that shows many of its features and capabilities and can serve as an introduction to MATLAB or a short tutorial as follows (see Figures 1.6 and 1.7).

```
EDU>> demo
```

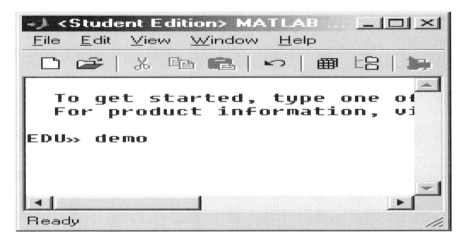

FIGURE 1.6
Command window showing the activation of *demo*.

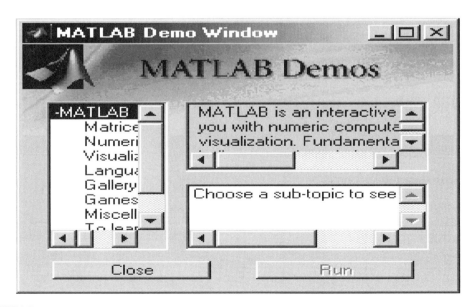

FIGURE 1.7
Demo window.

To use the *help* command, type *help* after the system prompt >> followed by the topic, followed by <enter>. For example, *help demos* <enter> can be used to find out more about the available demos in the system (depends on the MATLAB version). Some of the *demos* may be particularly useful for the beginner in a particular area. For example, in matrix algebra, the following demos may be of interest to the reader:

- *matdemo*, where matrix computation is introduced.
- *rrefmovie,* where the reduced row echelon form is introduced.

Additional examples on the use of the help command are illustrated as follows:

```
EDU>> help log <enter>
```

MATLAB returns the information about the natural logarithm, as shown in Figure 1.8.

If the topic is not known, by typing *help MATLAB* followed by <enter>, MATLAB returns a list of topics (see Figure 1.9), or a complete list of elementary MATLAB functions is displayed by typing *help elfun*.

To use the *lookfor* command, type *lookfor* followed by the topic followed by <enter>, and MATLAB returns a list of MATLAB help topics that contain key words that best describe a file. For example, the *lookfor* command for the case of the *sqrt* function is shown in Figure 1.10. If the instruction *lookfor* is followed by a specific topic, the search for key words is done through all the function files.

The command *whatsnew* displays the information about changes, innovations, and the latest improvements in MATLAB. It can be used with or without arguments, as shown in Figure 1.11. The command info is used in the same way as the *whatsnew* command.

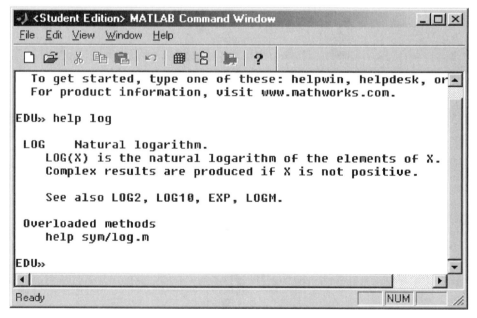

FIGURE 1.8
The help log command.

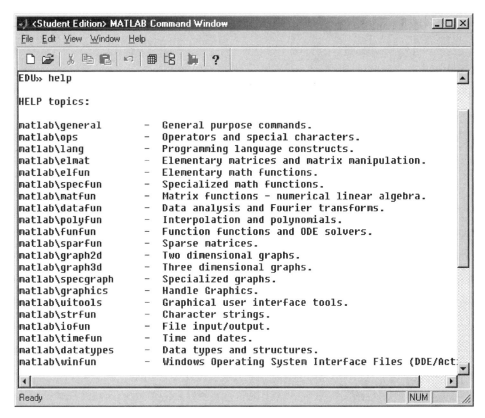

FIGURE 1.9
The help command.

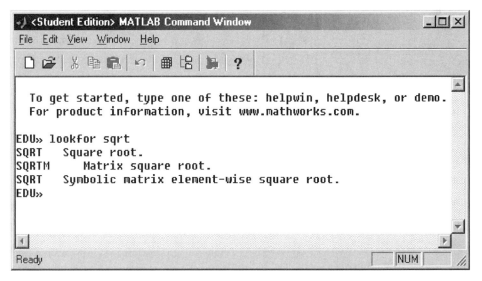

FIGURE 1.10
The *lookfor sqrt* command.

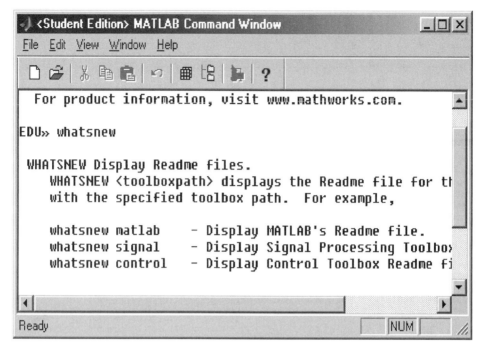

FIGURE 1.11
The *whatsnew* command.

The *helpwin* command is shown in Figure 1.12 and when executed, the user will then be taken for a ride in order to see the many help facilities available online.

Finally, the command *helpdesk* executed at the MATLAB system prompt opens MATLAB's helpdesk in a separate browser.

1.15   The Problem

Science, engineering, and technology students are trained to solve problems, with or without the help of computers.

Let us define what constitutes a problem.

The word problem is derived from the Greek word *problema*, which translates as "something thrown forward." The modern concept of the word problem can be equated to a question mark (?) that requires an answer often in the form of a solution.

The accepted definition of problem is as follows:

> a question raised for inquiry, consideration, or solution

A question becomes a problem when the question is of interest and presents some challenge with no obvious or immediate solution. Problem solving can be described as the process of arriving at solutions to a problem, question, or situation, which involves the use of mathematical, physical, or logical reasoning.

Not all questions constitute problems. A question becomes a problem when, besides posing a challenge, with no apparent answer, understanding, reason, strategies, knowledge,

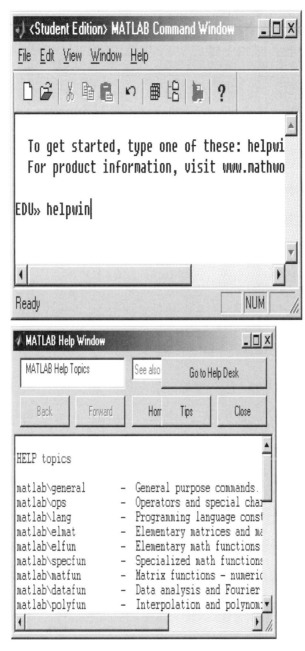

FIGURE 1.12
The *helpwin* command.

skills, and abilities play a role in finding at least one solution. Problems do not always have straightforward solutions. Problems can be classified according to their type of solution as algorithmic* or heuristic.

---

* An algorithm is nothing more than a set of rules for solving a problem in a finite number of steps. The word algorithm is derived from the Latin translation of the greatest Muslim mathematician of the eighth century Al-Khowarazmi. He developed algebra (Arabic Al-gabr), which means equations or restorations.

Algorithmic solutions are solutions that follow an algorithm, and an algorithm is a step-by-step approach leading to the solution of a problem. Computers are ideal machines to implement algorithmic solutions, since computers solve problems by executing sequences of instructions, which in many cases involve a repetitive sequence of actions. This sequential form of an algorithm implementation is convenient when a computer language is used. Algorithmic problems are usually quantitative in nature and require numerical computations to reach a solution. All computer languages when used to solve problems essentially employ algorithms, which are generally very long processes when done by hand by ordinary people, but highly appropriate for machine implementation. Some examples of algorithmic solutions are

- Instructions to get to the nearest bus stop
- Instructions in assembling a barbecue grill
- Instructions to make a phone call
- Instructions to fill a bank deposit form
- Instructions to install computer software

A heuristic solution, however, is a solution that does not follow a step-by-step approach and is mainly based on reasoning built on practice, knowledge, and experience and in many cases the method is trial and error. Heuristic solutions are qualitative in nature, based on human judgment, values, principles, and experience. Some examples of heuristic solutions are

- Buying a car
- Choosing a cell phone provider
- Choosing an investment
- Choosing a college

The following example clearly illustrates two distinct solutions to basically the same problem—the algorithmic- and heuristic approach.

Example 1.1

Assume that a capital of $1000 is available to be invested during a period of 3 years, with the sole objective of obtaining the highest possible return.

Two different investment choices (strategies) are presented and discussed as follows:

- Choice number 1 (algorithmic solution)
- Choice number 2 (heuristic solution)

Choice Number 1

The $1000 is deposited in a fix certificate of deposit (CD) at a local bank that earns a fixed interest of 6% per annum. The net profit or gain after 3 years would then be $191.02.

Table 1.1 traces the growth of the $1000 during the 3-year period.

During the first year, the principal is $1000 and the interest earned by the end of the year is $(1000) * (0.06) = $60. At the end of the first year, the interest is compounded, increasing the principal to (1000 + 60)$ = $1060, and the interest would be

TABLE 1.1

Growth of $1000 Placed at 6% Annually during 3 Years

Year	Principal at Beginning	Interest Earned	Principal at Year's End
1	$1000	$60	$1060
2	$1060	$63.60	$1123.60
3	$1123.60	$67.42	$1191.02

$1060 * (0.06) = 63.60 by the end of the second year. Continuing with this line of thought in a repetitive way, the total amount accumulated after 3 years, interest plus principal would be $1191.02.

Let us build a mathematical algorithm to solve this problem. Let the principal and interest be denoted by the variable P and I, respectively. Then in terms of P and I, the amount at the end of the first year would be given by

$$P + P * I = P * (1 + I)$$

At the end of the second year, the total amount accumulated in the savings account would then be

$$P * (1 + I) + P * (1 + I) * I = P * (1 + I) * (1 + I) = P * (1 + I)^2$$

And at the end of the third year, the total amount in the savings account would be

$$P * (1 + I)^2 + P(1 + I)^2 * I = P * (1 + I)^2 * (1 + I) = P * (1 + I)^3$$

In general, the principal P, after n years at an annual fixed interest rate I, would then grow to (Kurtz, 1985)

$$P * (1 + I)^n$$

It is obvious that this type of solution is algorithmic, deterministic, based on a repetitive approach, and modeled with equations that show the precise steps involved in reaching a unique numerical solution.

Choice Number 2

The capital of $1000 is now invested in stocks. In order to make a profic stock investments involve research and knowledge about the particular stock (company) picked, such as past performance, market trends, economic indicators (such as consumer confidence, employment, inflation), as well as experience and intuition.

With all the research done and using all the available data as well as professional advice and life experience, there is no guarantee that the stock chosen will outperform during the same 3 years (the time allocated), the solution indicated in choice number 1. This type of solution is called heuristic. It is uncertain, unpredictable, and for the present case impossible to estimate with precision the value of the investment after 3 years.

It is obvious that strategies as well as the ability to recognize, formulate, and solve classes of problems can be taught and learned. Either solution, algorithmic or heuristic, involves strategies, knowledge, understanding, abilities, and skills that can be grouped together and are also called heuristics.

Heuristics are general rules or guidelines that help in planning the actions and strategies involved in the problem-solving process. This methodology is generally hard to teach and discuss and is often neglected in the classroom, but it has been practiced by the engineering profession for many years and is commonly referred as the engineering method.

1.16   Problem-Solving Techniques (Heuristics)

Problem-solving techniques are also referred to as heuristics* and are a set of steps or actions followed in the problem-solving process. Basically, these steps are designed to provide answers to the following questions:

- How do we start the process of analyzing a given problem?
- How do we start solving a problem?
- What abilities are required?
- What strategies are followed?

In a sense, heuristics are general guidelines, problem-solving procedures, or rules that may help in the solution of a problem. In particular, the more complex a problem is, the more we need a systematic approach to reach a solution. These rules are as follows:

- Read carefully and clearly identify the purpose and goal of the problem.
- Understand the problem; restate the problem to eliminate ambiguities and clarify its objectives.
- Identify the known information and look for hidden assumptions and consider extreme cases to gain insight into a situation.
- Identify the unknown- and wanted information.
- Restate and simplify the problem in terms of known and unknown information and state any additional assumptions and approximations. Human communication tends to be imprecise and by restating the problem, more than one interpretation may emerge.
- Break the problem, if possible, into smaller, simpler, easy to manage problems.
- Select appropriate notation to identify the known and unknown information, and if beneficial, define intermediate variables.
- Make a graph, figure, or drawing to help visualize the abstract elements of a problem (include a flowchart when appropriate).
- Construct a table. In some cases, a table may indicate a pattern that may lead to a solution or a better understanding of the problem.
- Replace the variables defined in the mathematical relations by their units and check for consistency.
- Construct a physical model when possible.
- Determine which principles, equations, or models best describe the relation that transforms the known information (called inputs) into the unknown (wanted) variables (called outputs).
- Guess a solution and check if indeed the guessed solution makes sense. Use trial and error method.
- State general solutions and systematically list other approaches, exhausting all possibilities, eliminating the impossible but not the improbable.

---

* Do not be confused with heuristic solution.

- Select from all possible solutions the best one. The term "best" should be defined by the problem solver. Best could mean different things to different people—it could mean the shortest, clearest, easiest, simplest, or cheapest solution.

- Once a solution is known, it is appropriate to work backwards. Verify if it is valid and correct. Analyze and test the solution with simple data to see if indeed the solution satisfies the requirements of the problem. Estimate the results and analyze the implications, such as does the solution make mathematical, logical, or physical sense?

- Test the solution using extreme and special cases and search for patterns or symmetries.

- Find alternate solutions and compare them.

1.17 Proofs and Simulations

Proving or verifying relations, equations, equalities, and theorems may also constitute a problem, and a particular methodology is usually followed in the form of hypothesis, proof (synthetic or analytic), and conclusion.

Hypothesis refers to the statements that are sufficient to arrive to a conclusion. Synthetic proof is a proof based on hypothesis as well as other known information such as axioms, hard fact definitions, and other proven or accepted theorems. Analytic proof is a backward proof that starts by exploring a conclusion and analyzing the conditions that satisfy the conclusion.

The formal mathematical proofs are avoided in this chapter, and MATLAB is used as an exhaustive tool to verify relations. The following example illustrates the technique.

Verify the following equality $\sinh(x) = (e^x - e^{-x})/2$ over the range $0 \le x \le 2\pi$.

MATLAB Solution
```
>> x = linspace(0,2*pi,10);
>> y1 = sinh(x);
>> y2 = (exp(x)-exp(-x))./2;
>> disp(' ********R E S U L T S ************')
>> disp('      x           sinh(x)          [exp(x)-exp(-x)]/2 ');
>> disp(' *****************************');
>> disp('     ');
>> [x' y1'   y2']
>> disp(' *****************************');

************R E S U L T S ***************
       x        sinh(x)         [exp(x)-exp(-x)]/2
*****************************************
ans =
       0        0         0
       0.6981  0.7562    0.7562
       1.3963  1.8963    1.8963
       2.0944  3.9987    3.9987
       2.7925  8.1305    8.1305
       3.4907  16.3885   16.3885
       4.1888  32.9639   32.9639
       4.8869  66.2687   66.2687
       5.5851  133.2054  133.2054
       6.2832  267.7449  267.7449
*****************************
```

The preceding program generates three columns. They are

1. The first column represents x
2. The second column represents $y1 = sinh(x)$
3. The third column represents $y2 = (e^x - e^{-x})/2$

By inspection it is easy to observe that $y1 = y2$, for any (arbitrary) value of x.

Simulation can also be very useful in analyzing a given problem and finding a solution using the power of the computer. Simulation means that variables that represent a given (usually physical) system can be created and analyzed under certain conditions over an interval or range of interest, usually time, and quantitative or qualitative relationships among variables can be observed.

1.18 Computer Solutions

A computer does not understand human language, has no feeling, preferences, and intelligence. The only thing a computer is capable of doing is executing instructions extremely fast, without making mistakes or getting tired. For a computer to execute instructions, it is necessary to have an effective communication channel between the user and the machine—a task that is accomplished by means of a computer language. A computer language is similar to the human language in the sense that it follows a set of well-defined syntax and semantic rules with the specific goal of being able to communicate an idea. The syntax rules govern the grammar, format, and punctuation, whereas the semantic rules provide the meaning.

Every computer understands at least one language, which then becomes the user or machine interface. In this book, the interface language is MATLAB.

Problems can be solved using an algorithm or a simulation by means of a computer program. A computer program is a set of instructions written in a computer language that has to be input into a computer following strict and well-defined syntax rules (in general by using a keyboard or mouse). Syntax rules are strict grammar rules and the correct spelling of key words are essential. Any variation of the rules causes errors.

The instructions must follow the strict syntax rules so that the source code can be translated into machine code. If an instruction is incorrect, the computer will give an error message or a wrong answer. An error is called a bug. Errors must be found and corrected, a process called debugging. The instructions must be properly sequenced and all syntax errors must be corrected before a program can produce the desirable result. The instructions or programs when accepted by the computer are first stored in the computer memory and then executed in a sequential order.

All calculations are done by the ALU. The results or partial results are stored in memory, whereas the control unit manages the flow of instructions.

Information is usually stored in the computer memory as a variable.

What is a variable?

A variable is a name that can represent data, numbers, or strings that may change during the execution of a program. MATLAB requires that all variables' names be assigned a value before they are used, except when defined as symbols (see Chapter 7).

For example, the instruction $A = 1.0$ followed by pressing <enter> means that the variable name A is assigned the value 1.0, and the value of 1.0 is stored in memory location A.

If the next instruction of the program is $A = A + 2$ followed by <enter>, then the variable A is assigned a new value of $A = 1 + 2$ and the new value for A becomes 3, and 3 is then stored in the same memory location A. Note that the previous value of 1 in A is lost and cannot be recovered.

The symbol "=" in most computer languages does not mean the algebraic sign equal to but means "is assigned." A constant, on the other hand, takes a specific value during the execution of a program and does not change. An example of a constant is $\pi = 3.1415....$.

It is a good programming practice to name variables and constants according to what they represent. For example, β can be used as an argument of an angle, $R1$ as a resistance, $I1$ as a current, w for angular velocity, and t for time. Variables and constants are connected by operators to implement or model expressions, algorithms, or equations. These expressions are then processed by the computer software following a given hierarchy.

The computer operators or connectors can be

1. Arithmetic $\{+, -, *, /, \wedge\}$
2. Relational $\{= =, >=, <=\}$
3. Logical $\{\sim, \&, |\}$

These operators tell the computer what, how, and when decisions are made, and what type of processing needs to be done. The instructions that make up a program must follow a correct sequence to lead to the correct result, but there may be more than one correct sequence. The shortest correct sequence leading to the correct result constitutes the best or the most efficient program.

Finally, most computer languages have libraries of functions. Functions are instructions that are frequently used in the solution of problems. They are defined by key words by the computer software and perform specific computations such as $sin(x)$, $abs(x)$, $log(x)$, and $max(x)$.

Variables, constants, operators, functions, expressions, and equations are used in the solution of problems, and by applying the heuristics presented in Section 1.9 will hopefully lead the reader to the systematic solution of wide classes of problems.

1.19 The Flowchart

Probably the most important step in the process of problem solving by using a computer is the construction of a flowchart {heuristic (G)}, also called flow diagram.

The flowchart is a graphical or pictorial representation of an algorithm showing the steps involved as well as the interrelations of these steps in the solution of a problem. A flowchart also defines the problem in terms of the known and unknown variables, selection of appropriate notation in defining the variables and constants, and clearly indicates

TABLE 1.2

Flowchart Symbols

Symbol	Meaning
⬭ (ellipse)	Indicates the beginning and end of a program
▭ (rectangle)	Process or instruction box, use for data manipulation, computation, or movement of data
⟶ (arrow)	Flow line indicates the direction or sequence in which data flows or the instructions are executed
◯ (circle)	Connectors are used to show continuation of logic flow
◇ (diamond)	Decision or branching box represents a point in the program where the logic flow will follow one of two paths, depending on the situation (yes or no, or true or false)

the sequence of steps and decisions, program operations, principles, and equations used. Each step in the algorithm process is defined by a box with a specific shape where the shape identifies the action or process. Loops represent repeating sequences of instructions controlled by conditions imposed by the algorithm. Once a flowchart is created, the writing of a corresponding computer program is an easy translation.

This book uses five symbols in the construction of flowchart. They are shown and defined in Table 1.2.

Simple problems may not require the construction of flowcharts, but for complicated programs, flowcharts are useful to organize the sequential steps involved in the logic implementation of an algorithm solution.

Two examples are presented as follows, shown in Figures 1.13 and 1.14, which illustrate the construction of flowcharts for the case of algorithmic, recursive solutions.

Example 1.2

The flowchart presented in Figure 1.13 illustrates the solution of the problem analyzed in Table 1.1 and Section 1.8.

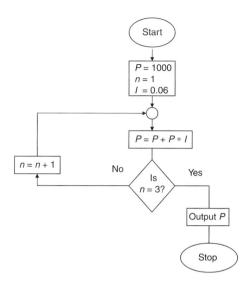

FIGURE 1.13
Flowchart of Example 1.1.

Example 1.3

Construct a flowchart that computes the product and addition of all the integers from 1 to 99. The solution is a recursive algorithm, as illustrated in the flowchart of Figure 1.14.

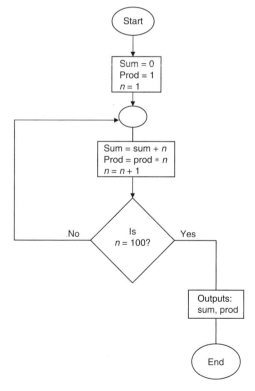

FIGURE 1.14
Flowchart of Example 1.2.

2

Getting Started

In a time of drastic change, the learners inherit the future. The learned usually find themselves equipped to live in a world that no longer exists.

Eric Hoffer

2.1   Introduction

Once MATLAB® is installed, it is time to start using it. MATLAB is best understood by entering the instructions in your computer one at a time and by observing and studying the responses. Learning by doing is probably the most effective way to maximize retention.

I hear and I forget

I see and I remember

I do and I understand

Confucius

In this chapter, MATLAB will be used as a simple calculator, and simple programming techniques are introduced. MATLAB can be used as a powerful calculator that is capable of representing information and performing basic calculations such as addition, subtraction, multiplication, and division. It can also handle trigonometric, logarithmic, and complex numbers. Along with performing basic calculations, MATLAB can also be used to plot and display data in a graphic format, which are topics and subjects of later chapters. To get started, the reader needs to have information about the computer that he/she will be using, such as

1. How to switch on the computer
2. How to access MATLAB* (assuming that MATLAB was installed in the reader's computer earlier)
3. How to quit MATLAB and log off[†]
4. How to access the command window of MATLAB to carry out the first examples

---

* To access MATLAB, click the MATLAB icon with your mouse, or access it by clicking the sequence: Start, Programs, and MATLAB (windows environment). Once you see the MATLAB prompt, given by either character: >> or EDU >>, you are in the command window of MATLAB and you can start entering and executing instructions.
† To exit MATLAB, type *quit* or *exit* at the prompt at the command window.

The assumption in this chapter, as well as in the remaining chapters of this book, is that the reader is seated in front of a computer with access to an active MATLAB command window.

2.2 Objectives

On reading this chapter, the reader should be able to

- Launch MATLAB
- Use MATLAB to perform simple arithmetic operations
- Express constants
- Define, use, and assign values to variables
- Use MATLAB as a calculator
- Express small and large numbers in scientific notation
- Write and execute simple MATLAB commands and functions
- Manage the workspace
- Understand the meaning of punctuation and comments
- Control and manage the output screen
- Interrupt and quit MATLAB
- Use the Edit/Debugger window
- Write simple files

2.3 Background

R.2.1   Constants are real numbers such as 18, 25, or 87 that represent information.*

R.2.2   Numbers can be expressed as positive or negative quantities such as +18, 18, or −18. The plus sign is optional for positive quantities.

R.2.3   Decimal numbers are expressed in the conventional way such as 18.37, +18.37, or −18.37.

R.2.4   Numbers such as 300,000 ($= 3*10^5$) can be expressed in MATLAB using scientific notation as 3e05, or the number −0.003 ($-3*10^{-3}$) can be expressed as −3e−3, whereas 63,000,000 can be expressed as 6.3e7.

R.2.5   MATLAB does not care about spacing (except when working with complex numbers). For example, 4 + 2 = 4 + 2. Algebraic calculations do not require the symbol "=", if the result is not saved.

---

* Recall that real numbers can be rational or irrational. Rational numbers are integers or fractions that can be represented as a decimal number with a finite number of characters, or can be periodic. Irrational numbers cannot be expressed as a fraction, and can only be represented in decimal form by an infinite number of characters. π and $\sqrt{2}$ are examples of irrational numbers.

R.2.6   Commands are entered after the prompt (*EDU >> or >>*), and followed by a carriage return or by pressing the *<enter>* key.

R.2.7   If a mistake is made in entering an instruction, causing an error message, then the whole instruction must be retyped or reentered. No characters can be modified in the command window after the *<enter>* key is pressed. The retyping can be avoided by pressing the ↑ or ↓ keys. That action repeats the last instructions and the error can then be corrected without the need of retyping.
            The error-free instruction can then be executed by pressing the *<enter>* key.

R.2.8   Recall that MATLAB allows using one or several characters to define or assign a variable name. For example, A, AA, ABC, and a5.
            A variable represents data stored in the (RAM) memory of the computer in use.

R.2.9   Variables may represent a scalar, a vector, an array, or a matrix (Chapter 3 deals with arrays, vectors, and matrices).

R.2.10 MATLAB variable names are case sensitive and in general, Aa ≠ aA ≠ AA ≠ aa.
            In some versions of MATLAB, sensitivity can be controlled by using the command *casesen on* or *casesen off*.

R.2.11 Variable names can contain a large number of characters depending on the version, but the characters beyond, let us say the first 31 (character), are ignored. Again, the length of a variable name is likely to change in future MATLAB versions.

R.2.12 MATLAB variable names must start with a letter, followed by any other letter, number, or underscore such as *total_resistance* and *current_1*.

R.2.13 MATLAB's command's structure follows the format

$$variable\_name = expression$$

            where the expression refers to a numerical, mathematical, or logical relation, which is evaluated and the value is then assigned to *variable_name* on the left-hand side of the equal.

R.2.14 MATLAB reserves special variable names to represent a function or a particular constant. The most common reserve variables are listed in Table 2.1. The reader should not create variables using these names.

R.2.15 MATLAB performs calculations based on the last value assigned to a variable.

R.2.16 The command *clear* deletes all the variables defined or used earlier.

R.2.17 The *clear* command can be made selective such as *clear* A, Aa, deleting only the indicated variables: A and Aa; although *clear A** deletes all variable names that start with the character A.

R.2.18 The command *clc* clears the command window, but does not delete the variables defined earlier.

R.2.19 The command *clf* clears the figure window.

R.2.20 The standard MATLAB algebraic symbols and some simple functions are defined in Table 2.2.

R.2.21 Standard mathematical operations are evaluated from left to right, keeping in mind the following rules:

        a. The inside of the innermost set of parentheses in an expression is always evaluated first.

TABLE 2.1

List of Reserved Variable Names

Variable	Description
ans	Temporary variable that stores the most recent answer.
computer	Returns the computer type.
version	MATLAB version.
ver	Returns the information about the license and version of the MATLAB package installed in your computer.
license	License information.
pi	The number $\pi = 3.14159\ldots$
exp(1)	The value of $e = 2.71\ldots$
eps	Represents the accuracy of floating point, the smallest possible positive number with a magnitude of the order of 10^{-10}.
realmin	The smallest real positive number.
realmax	The largest real positive number.
bitmax	The largest positive integer, magnitude of $2^{53} - 1$.
flops	Counts of the floating-point operations. flop(0) starts the count of all algebraic operations such as $+, -, *, /$.
inf	Represents infinity, $(1/0)$.
nan	Not a number, undefined $(0/0)$.
i or j	The value of $\sqrt{-1}$. Denotes the imaginary part of a complex number.
input	Accepts information via keyboard.
date	Represents the current date as a string. For example, 25-Jul-00.
clock	Represents the current date and time as YYMMDDHHMMSS.
beep	Executes a beep sound.
etime (T_f, T_I)	Calculates elapse time in seconds between T_I (initial) and T_f (final). T_I and T_f are in vector form consistins of six elements (year month day hour minute second).
tic, toc	Measures the time between the tic and the toc. The tic starts the stopwatch, and the toc stops the stopwatch and outputs the elapsed time.
cputime	Total time of MATLAB used in seconds.
Pause	Stops executing a program momentarily.
Pause(n)	Stops executing a program during n seconds.

TABLE 2.2

Matlab Operations

Symbol	Operation	Example	Answer
+	Addition	$z = 4 + 2$	$z = 6$
−	Subtraction	$z = 4 - 2$	$z = 2$
/	Right division	$z = 4/2$	$z = 2$
\	Left division	$z = 2\backslash 4$	$z = 2$
*	Multiplication	$z = 4 * 2$	$z = 8$
^	Exponentiation	$z = 4 \wedge 2$	$z = 16$
Functions such as:	square root log2	$z = \text{sqrt}(4)$	$z = 2$
sqrt, log		$z = \log2(4)$	$z = 2$

b. Then the hierarchy of operations is as follows:

   i. Functions such as *sqrt(x)*, *log(x)*, and *exp(x)*

   ii. Exponentiation ($\wedge$)

   iii. Products and division (*, /)

   iv. Addition and subtraction ($+, -$)

R.2.22 For example, evaluate the following mathematical expressions by transforming them into MATLAB (notation) and estimate the values of x and y, respectively, by hand using the operational hierarchy, then use MATLAB, and verify the results obtained by hand.

a. $x = 4 + 2^3 - \sqrt{4^2 - 7}\big/(2^2 - 1) * 4^2 - 1$

b. $y = 1 + (3^2 - 4)^2\big/\left(\sqrt{(2^4 + 9)} + 5\right) * 2^3$

ANALYTICAL Solution

Part (a)

$x = 4+2^3\text{-}\sqrt{4^2\text{-}7}/(2^2\text{-}1) * 4^2\text{-}1$ is converted to MATLAB notation as
$x = 4+2\char`^3\text{-}sqrt(4\char`^2\text{-}7)/(2\char`^2\text{-}1) * 4\char`^2\text{-}1$ applying the hierarchy rules yields
$x = 4+2\char`^3 - sqrt(16\text{-}7)/(4\text{-}1) * 4\char`^2\text{-}1$
$x = 4+2\char`^3\text{-}sqrt(9)/3 * 4\char`^2\text{-}1$
$x = 4+2\char`^3\text{-}3/3 * 4\char`^2\text{-}1$
$x = 4+8\text{-}3/3 * 16\text{-}1$
$x = 4+8\text{-}16\text{-}1$
$x = -5$

Part (b)

$y = 1+(3^2\text{-}4)^2 / (\sqrt{(2^4+9)}+5) * 2^3$ is converted to MATLAB notation as
$y = 1+(3\char`^2\text{-}4)\char`^2 / (sqrt(2\char`^4+9)+5) * 2\char`^3$ applying the hierarchy rules yields
$y = 1+(9\text{-}4)\char`^2 / (sqrt(16+9)+5) * 2\char`^3$
$y = 1+5\char`^2 / (sqrt(25)+5) * 2\char`^3$
$y = 1+5\char`^2 / (5+5) * 2\char`^3$
$y = 1+5\char`^2 / 10 * 2\char`^3$
$y = 1+25 / 10 * 8$
$y = 1+25 / 10 * 8$
$y = 1+20$
$y = 21$

MATLAB Solution
```
.>> % part(a)
>> x = 4+2^3-sqrt(4^2-7)/(2^2-1)*4^2-1

    x =
       -5
```

MATLAB Solution
```
>> % part(b)
>> y = 1+(3^2-4)^2/(sqrt(2^4+9)+5)*2^3

    y =
       21
```

R.2.23 The command *who* lists the variables currently used in the workspace.

R.2.24 The command *whos* lists the variables used with their respective sizes, where the size is the number of elements that make up the variable (Chapter 3 deals with arrays, vectors, and matrices, as well as their sizes).

R.2.25 A comment statement starting with the percentage symbol (%) does not affect any executable MATLAB instruction, and cannot be continued on the next line.

R.2.26 Multiple statements can be placed on one line when they are separated by a comma (,) or semicolon (;).

TABLE 2.3

Precision Formats

MATLAB Instruction	Display	Numerical Output exp(1)
format short	4 decimal digits (default)	2.7183
format long	16 decimal digits	2.71828182845905
format short e	4 decimal digits plus exponent	2.7183e+000
format long e	15 decimal digits plus exponent	2.71828182845904e+000
format bank	2 decimal digits	2.72
format +	+, −, 0 (positive, negative, and zero)	+
format hex	Hexadecimal	4005bf0a8b14576a
format rat	Rational approximation	1457/536
format compact	Suppress extra line-feeds	2.7183
format loose	Puts the extra line-feeds back in	2.7183

R.2.27 A semicolon (;) at the end of an instruction suppresses the echo, whereas a comma does not.

R.2.28 An instruction statement that is long may be continued on the next line if the preceding line ends with an ellipsis, that is, three consecutive dots (…).

R.2.29 MATLAB can be interrupted at any time by pressing the *Ctrl* and *C* keys simultaneously (this action aborts a running MATLAB program).

R.2.30 MATLAB is, in general, case sensitive (mentioned in R.2.10 for the case of variable names), and MATLAB commands and functions always use lowercase characters.

R.2.31 The flow of information can be controlled when many screens of information are available by typing the command *more* at the MATLAB prompt. The output on the screen is then controlled, and one output screen is displayed at a time.

R.2.32 The default data used by MATLAB is of double precision, but the format of the display is defined by setting the format type indicated in Table 2.3.

Table 2.3 uses the display of the value of "e" as an illustration (expressed in MATLAB notation as *exp(1)*).

The formats *long* and *short* use fixed-point notation, whereas all the other formats use floating-point notation, conforming to the Institute of Electrical and Electronics Engineers (IEEE) standard for double-precision arithmetic, discussed later in this section.

R.2.33 The fixed-point representation is used to represent integers, where its magnitude is stored in one place, but the point position is not stored with the number and must be remembered by the programmer. Integers are expressed without the decimal point and are not subject to round-off errors.

R.2.34 The floating-point notation represents an arbitrary decimal number (with a decimal point) that may be stored with or without an exponent, usually a number times 10 raised to a power, emulating scientific notation.

MATLAB represents all the numbers using the floating-point format.

R.2.35 The MATLAB command *isieee* returns a message indicating if the software used by the reader conforms to the IEEE standard. For example, *NaN* (not a number) is the IEEE floating-point standard message for an undefined result such as a number divided by zero.

R.2.36 MATLAB accepts two types of data files: MATLAB and ASCII* files (see Chapter 9 for additional information).

---

* ASCII stands for the American Standard Code Information Interchange, defined in Chapter 3.

R.2.37 MATLAB files are stored in memory in an efficient binary format and can be read and used in any MATLAB environment.

R.2.38 ASCII files are useful if the information stored consists of numbers that may be used in a non-MATLAB environment.

R.2.39 To save data using the ASCII format on a floppy disk, type *a:\myfile.dat variables −*
ASCII, where *a:* is the path to the floppy, *myfile* the file name, *dat* the extension, and *variables* the variable names to be saved.
   For any additional information, refer to Chapter 9 that deals with files and file-related commands.

R.2.40 To read *myfile.dat*, stored in a floppy or in the hard drive (of your computer) into the workspace (command window), type *load myfile*.

R.2.41 To save the content of the workspace in a MATLAB file, type *save mywork*, or click *File*, followed by: *Save Workspace as: mywork*.

R.2.42 The command *load mywork* reads the file: *mywork* into the command window.

R.2.43 MATLAB files with extension *m* are referred to as *Mfiles*.

R.2.44 MATLAB uses two types of M-files: *script* and *function* files. *Script* and *function* files should be saved in the current folder (default), which is available when working in the MATLAB domain. *Script* and *function* files are revisited in Chapter 9, and the concept of accessing a file is redefined as being in the path file search.

R.2.45 The content of a script file may be a program, data, or just a set of instructions that are created using the Edit/Debugger built into MATLAB. The edit window can be accessed in the MS Windows environment by the following sequence of actions; once in the command window: *File→New→M-File*.
   A program can then be typed using the MATLAB word-processor software and keyboard. When typing is completed, select (from the file menu) *File→Save as→*and replace in the dialog box the file name of your choice (starting with a letter) and then (click) *save*. The Editor/Debugger automatically provides the extension *m*, and the variables defined in the script file once called become global or part of the workspace. The file can be called (executed) by typing, while in the command window, the script file's name (with no extension).

R.2.46 *Script* files are used to process a sequence of commands that are stored in the computer memory by typing just one word—the file's name. The file is then executed without any display at the command window; unless the *echo* command is activated, or a command ends with a comma, or a required output is programmed.

R.2.47 A *function* file is created following the same sequence of steps outlined for the script files, but the first line follows the syntax

$$\textit{function [output\_variables]} = \textit{function\_name(input\_variables)}$$

where the *input_variables* are local, which means that their values can be used only within the *function* file.

R.2.48 A function file can be called by typing the following at the command window:

$$\textit{[output\_variables]} = \textit{function\_name(input\_variables)}$$

Function files evaluate and return the *output_variables* given the *input_variables*.

R.2.49 The following example illustrates a script file that is used to solve for the roots of the quadratic equation of the form, $ax^2 + bx + c = 0$:

```
MATLAB Solution
% Script file: rootsquad
% Return the roots of the equation of the form ax^2+bx+c=0
disp('***************************************************') % display
                                                              messages
disp(' This script file solves for the roots of the ');
disp('quadratic equation of the form: ax^2+bx+c=0');
disp('***************************************************')
a = input('***  Enter the coefficient a :');
b = input('*** Enter the coefficient b :'); % the inputs a assigned
                                               to a, b and c
c = input('***  Enter the coefficient c :');
disp('***************************************************')
disp('The roots of the quadratic equation')
disp('of the form: ax^2+bx+c=0, are :')
root1 = (-b+sqrt(b^2-4*a*c))/(2*a)                       % display the
                                                           roots
root2 = (-b-sqrt(b^2-4*a*c))/(2*a)
disp('***************************************************')
```

This file is *Save As … rootsquad* (use the file menu) automatically in the current MATLAB folder.

R.2.50 The preceding script file, *rootsquad* of R.2.49 is tested below for the following quadratic equations:

a. $2x^2 + 3x + 7 = 0$

b. $\pi x^2 + 2\pi x + 3\pi = 0$

c. $log10(32.3)x^2 + sqrt(3^3 + 1.33^{3.3})x + tan(1.112) = 0$

Recall that while at the command window, the script file is called and executed by typing *rootsquad*. The resulting process is shown below:

```
MATLAB Solution
>> rootsquad

***************************************************************
This script file solves for the roots of the
quadratic equation of the form: ax^2+bx+c=0
***************************************************************
***   Enter the coefficient a :2
***   Enter the coefficient b :3
***   Enter the coefficient c :7
***************************************************************
The roots of the quadratic equation
of the form: ax^2+bx+c=0, are :
    root1 =
            -0.7500 + 1.7139i
```

```
                    root2 =
                             -0.7500 - 1.7139i
    *************************************************************

    >> rootsquad

    *************************************************************
    This script file solves for the roots of the
    quadratic equation of the form: ax^2+bx+c=0
    *************************************************************
    ***    Enter the coefficient a : pi
    ***    Enter the coefficient b : 2*pi
    ***    Enter the coefficient c : 3*pi
    *************************************************************
    The roots of the quadratic equation
    of the form: ax^2+bx+c=0, are :
        root1 =
                 -1.0000 + 1.4142i
        root2 =
                 -1.0000 - 1.4142i
    *************************************************************

    >> rootsquad

    *************************************************************
    This script file solves for the roots of the
    quadratic equation of the form: ax^2+ bx+c=0
    *************************************************************
    ***    Enter the coefficient a :log10(32.3)
    ***    Enter the coefficient b :sqrt(3^3+1.33^3.3)
    ***    Enter the coefficient c :tan(1.112)
    *************************************************************
    The roots of the quadratic equation
    of the form: ax^2+bx+c=0, are :
        root1 =
                 -0.4217
        root2 =
                 -3.1810
    *************************************************************
```

R.2.51 Using MATLAB, let us verify the results obtained in the examples of R.2.50 by executing the following instructions:

```
MATLAB Solution
>> % part(a), verify if root1 and root2 (below) satisfy the equation:
     2x^2 + 3x + 7 = 0
>> root1 = -0.7500+1.7139i;
>> root2 = -0.7500-1.7139i;
>> check _ eq _ a _ root1 = 2*root1.^2+3*root1+7

    check _ eq _ a _ root1 =
                         9.3580e-005

>> check _ eq _ a _ root2 = 2*root2.^2+3*root2+7
```

```
    check _ eq _ a _ root2 =
                         9.3580e-005
```

>>% part (b),verify if root1 and root2 (below) satisfy the equation:
πx²+2πx+3π=0
>> root1 = -1.0000+1.4142i;
>> root2 = -1.0000-1.4142i;
>> check _ eq _ b _ root1 = pi*root1.^2+2*pi*root1+3*pi

```
    check _ eq _ b _ root1 =
                         1.2051e-004
```

>> check _ eq _ b _ root2 = pi*root2. ^2 + 2*pi*root2 + 3*pi

```
    check _ eq _ b _ root2 =
                         1.2051e-004
```

>> % part (c) , verify if root1 and root2 (below)
>> % satisfy the equation: log10(32.3)x²+sqrt(3³+1.33³·³) x + tan(1.112)=0
>> root1 = -0.4217;
>> root2 = -3.1810;
>> check _ eq _ 3 _ root1=log10(32.3)*root1.^2+sqrt(3^3+1.33^3.3)*root1+
tan(1.112)

```
    check _ eq _ c _ root1 =
                         2.4166e-005
```

>> check _ eq _ c _ root2=log10(32.3)*root2.^2+sqrt(3^3+1.33^3.3)*root2+
tan(1.112)

```
    check _ eq _ c _ root2 =
                         1.2869e-004
```

Note that the results obtained by the *check_eq_ …* is not exactly zero, but very close to zero, due to the round-off errors and approximations.

R.2.52 Let us now illustrate the use of a function file to solve the same quadratic equation defined in R.2.49 (of the form $ax^2 + bx + c = 0$).

MATLAB Solution
```
function [root1,root2] = func _ quad _ sol(a,b,c)
% function  file : func _ quad _ sol
% Returns the roots of the equation: ax^2 + bx + c = 0
% The outputs are: roots1 and root 2
% The inputs are the coefficients: a, b, and c
disp('************************************************');
disp(' This function file solves for the roots of the '); % display message
disp('quadratic equation of the form: ax^2+bx+c=0'); ); % display message
disp('given the coefficients a, b, and c as inputs'); ); % display message
root1 = (-b+sqrt(b^2-4*a*c))/(2*a);                        % the roots are
                                                             evaluated
root2 = (-b-sqrt(b^2-4*a*c))/(2*a);
disp('*************The roots are:*******************'); % display message
root1                                                      % returns roots
root2
disp('************************************************'); %display message
```
This file is Save As… *func_quad_sol* in the default current MATLAB folder.

R.2.53 The preceding function file, *func_quad_sol*, is tested for the same quadratic equations used for the script file: *rootsquad*, indicated below:

a. $2x^2 + 3x + 7 = 0$

b. $\pi x^2 + 2\pi x + 3\pi = 0$

c. $log10(32.3)x^2 + sqrt(3^3 + 1.33^{3.3})x + tan(1.112) = 0$

Recall that while at the command window, the function file is called by typing *func_quad_sol(a,b,c)*, and the results are shown as follows.

```
MATLAB Solution
>> func _ quad _ sol(2,3,7)

*************************************************************
This function file solves for the roots of the
quadratic equation of the form: ax^2+bx+c=0
given the coefficients a, b, and c as inputs
*************The roots are:*******************************
    root1 =
            -0.7500 + 1.7139i
    root2 =
            -0.7500 - 1.7139i
*************************************************************

>> func _ quad _ sol(pi,2*pi,3*pi)
*************************************************************
This function file solves for the roots of the
quadratic equation of the form: ax^2+bx+c=0
given the coefficients a, b, and c as inputs
*************The roots are:*******************************
    root1 =
            -1.0000 + 1.4142i
    root2 =
            -1.0000 - 1.4142i
*************************************************************

>> func _ quad _ sol(log10(32.3), sqrt(3^3+1.33^3.3),tan(1.112))
*************************************************************
This function file solves for the roots of the
quadratic equation of the form: ax^2+bx+c=0
given the coefficients a, b, and c as inputs
*************The roots are:*******************************
    root1 =
            -0.4217
    root2 =
            -3.1810
*************************************************************
```

R.2.54 Observe that script or function files can be useful in the solution of any problem, specially if the same type of problem is repeatedly encountered. The solution then consists of calling the M-file name. Also observe that script files are profoundly different from function files. Function files take the input variables and return the output variables that must be defined in the first line of the command, and the variables become local. Script files may ask for input variables during the execution of the file and all the variables once defined become global.

R.2.55 It is important that a file or program be written in a way that can be easily understood by any user or reader. A good program is one that is logical in the sequence of instructions, and every instruction has a well-defined and clear objective. Clarity is greatly improved by the appropriate choice of variable names and the use of comments. Comments are nonexecutable statements (R.2.25) that are generally used with the sole objective of improving clarity and readability of the instructions, which make up a program.

R.2.56 Files (script and function), file creation, modification and saving, file commands, file structure and organization, file addresses and search path, and file compiling (parsing) are revisited with details in Chapter 9.

2.4 Examples

Example 2.1

Write a MATLAB program that evaluates the hypotenuse of a right triangle with sides $A = 4$ and $B = 3$, shown in Figure 2.1.

ANALYTICAL Solution

Applying the Pythagorean theorem to the triangle of Figure 2.1, the hypotenuse C is given by

$$C = \sqrt{A^2 + B^2} = \sqrt{3^2 + 4^2} = 5$$

MATLAB Solution
```
>>                      % enter the following sequence of instructions:
>>A = 3;                % length of one side of the right triangle
>>B = 4;                % length of second side of the right triangle
>>C = sqrt(A^2+B^2);    % length of hypotenuse is evaluated
>> Hypotenuse = C;      % displays the result or solution
>> Hypotenuse
   Hypotenuse =
              5
```

Observe that only the command window is used in this example, and the (display) solution is given by the variable *Hypotenuse*.

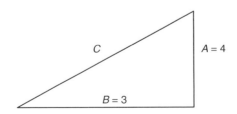

FIGURE 2.1
Right triangle of Example 2.1.

Example 2.2

Write a program that returns the average value giving three arbitrary numbers represented by the variables *A*, *B*, and *C*. Test the program for *A* = 35, *B* = 21, and *C* = 13.

MATLAB Solution
```
Enter the following instructions:*
>> A = input ('Enter the value of the first number: A = ');
```
         Enter the value of the first number: A = 35
```
>> % note that 35 is assigned to A
>> B = input ('Enter the value of the second number: B = ');
```
         Enter the value of the second number: B = 21
```
>> % 21 is assigned B
>> C = input ('Enter the value of the third number: C = ');
```
         Enter the value of the third number: C = 13
```
>> % 13 is assigned C
>> format compact                    % suppress extra line-feed
>> The _ average _ is = (A+B+C)/3    % returns the average
```

   The _ average _ is =
                    23

Note that this program can easily be converted into a script or function file.

Example 2.3

Write a program that returns the balance of a bank account, after a period of $n = 5$ years, where the present value is denoted by the variable *P* and the interest by *I*, for the case of $P = \$7300$, and $I = 2.7\%$.[†]

MATLAB Solution
```
(in the command window)
>>P =7300;
>>I =2.7;
>>n =5;
>>format bank                          % uses two decimal digits
>>Total _ amount = P*(1+I/100)^n       % total amount, capital +
                                         interest
```
   Total _ amount =
                8340.17

Example 2.4

Enter and evaluate the following expressions using MATLAB:

1. $a = e$
2. $b = e^3$

---

[*] The *input* command is defined in Chapter 3, Section 3.3, R.3.2, and also in Table 2.1. Its use was explained in Example 2.1.
[†] See Chapter 1, Section 1.8, for the equations that relate *P*, *I*, and *n*.

3. c = ln(e) + ln(e³)⁽*⁾
4. d = log(e) + log(e³)⁽†⁾
5. e = π
6. f = cos(π/4)
7. $g = e^{(3\sqrt{131})}$‡
8. h = log(5) + log$_e$(5) + log$_2$(5)§

(Exponential and trigonometric functions are revisited in Chapter 4.) At this point, let us gain some experience just by entering the above commands and observing their respective responses.

MATLAB Solution

```
>> % use only the command window
>> format short
>> a = exp(1)

    a =
        2.7183

>> b = exp(3)

    b =
        20.0855

>> c = log(exp(1))+log(exp(3))

    c =
        4

>> d = log10(exp(1))+log10(exp(3))

    d =
        1.7372

>> e = pi

    e =
        3.1416

>> f = cos(pi/4)

    f =
        0.7071

>> g = exp(3*sqrt(131))

    g =
        8.1693e+014

>> h = log10(5)+log(5)+log2(5)

    h =
        4.6303
```

---

* ln(x) is the natural logarithm of x expressed in MATLAB as log(x).
† log(x) is the logarithm base 10 of x expressed in MATLAB notation as log10(x). For additional information about logarithms see Chapter 4.
‡ e^x is expressed in MATLAB as exp(x).
§ log$_2$(x) is expressed in MATLAB as log2(x).

Example 2.5

The main objective of this example is to gain experience by just observing the MAT-LAB responses to some frequently used commands involving constants, and functions defined in this chapter, as well as performing simple numerical calculations, defined by the (%) comments.

```
>> format long      % define in Table 2.3
>> a = pi

   a = 3.14159265358979

>> b = eps          % Matlab's smallest number

   b =
       2.220446049250313e-016

>> flops (0)        % starts the count of algebraic operations
>> c = a+b          % the smallest number that can be added to pi

   c =
       3.14159265358979

>> d = realmax      % Matlab's largest positive real number

   d =
       1.797693134862316e+308

>> e = realmin      % the smallest positive real number

   e =
       2.225073858507201e-308

>> f = bitmax       % exact representation of the largest integer

   f =
       9.007199254740991e+015

>> flops            % counts the number of floating-point operations

   ans =
        2

>> date             % returns the current date

   ans =
        29-Apr-2001

>> format short
>> clock            % returns the current date and time

   ans =
        1.0e+003*
        2.0010    0.0040    0.0290    0.0080    0.0470    0.0551

>> cputime          % returns the time

   ans =
        1.0866e+003

>> computer         % Matlab returns the computer type used

   ans =
        PCWIN
```

```
>> ver              % Matlab's version

***********************************************************************
<Student Edition>MATLAB     Version 5.3.0.62a  (R11)  PCWIN
License Number:0
MATLAB Toolbox              Version 5.3        (R11)  15 January-1999
Symbolic Math Toolbox       Version 2.1        (R11)  11 Septemb-1998
Signal Processing Toolbox   Version 4.2        (R11)  10 July-1998
Control System Toolbox      Version 4.2        (R11)  15 July-1998
***********************************************************************
>> beep             % executes a sound beep, if sound card was (on)
                      installed
>> license

   ans =
        0
```

Example 2.6

Draw a flow chart and write a program that evaluates (see Figure 2.2)

1. The circumference of a circle
2. The area of a circle
3. The volume of a sphere with the following radii: $r1 = 1.5$ and $r2 = 2.5$

ANALYTICAL Solution

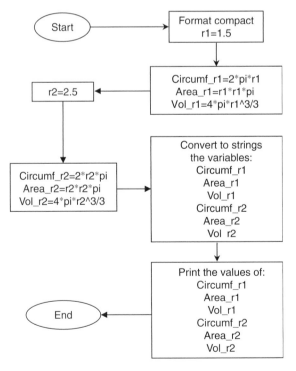

FIGURE 2.2
Flowchart of Example 2.6.

```
MATLAB Solution
>> format compact
>> r1 = 1.5;
>> % evaluate circumf, area and volumes for r1=1.5
>> circumf _ r1 = 2*pi*r1;
>> area _ r1= r1*r1*pi;
>> vol _ r1= 4*pi*r1^3/3;
>> % evaluate circumf, area and volumes for r2=2.5
>> r2 = 2.5;
>> circumf _ r2 = 2*pi*r2;
>> area _ r2 = r2*r2*pi;
>> vol _ r2 = 4*r2*area _ r2/3;
>> % convert numbers to strings
>> cirr1= num2str(circumf _ r1);                    % see footnote (*)
>> arear1= num2str(area _ r1);
>> volr1= num2str(vol _ r1);
>> cirr2= num2str(circumf _ r2);
>> arear2 = num2str(area _ r2);
>> volr2 = num2str(vol _ r2);
>> % returns the results
>> disp('******** results ********')
>> disp('For the circle with radius 1.5')
>> disp(['The circumference is',cirr1])
>> disp(['The area is',arear1])
>> disp(['The volume is',volr1])
>> disp('******************')
>> disp('For the circle with radius 2.5')
>> disp(['The circumference is',cirr2])
>> disp(['The area is', arear2])
>> disp(['The volume is ',volr2])
>> disp('******************')

******** results ********
For the circle with radius 1.5
The circumference is 9.4248
The area is 7.0686
The volume is 14.1372
*********************
For the circle with radius 2.5
The circumference is 15.708
The area is 19.635
The volume is 65.4498
*********************
```

2.5   Further Analysis

Q.2.1   Load and run the program of Example 2.1.

Q.2.2   Run Example 2.1 without the semicolons (;). Comment on your result.

---

* *num2str* converts a number into a sequence of characters (string).

Q.2.3 Modify and rerun Example 2.1 by using the minimum number of lines (not instructions).

Q.2.4 Rerun Example 2.1 with $A = 1/3$ and $B = \pi$.

Q.2.5 Enter and execute the command that returns the variables used in Example 2.1.

Q.2.6 Rerun Q.2.4 by using the commands *format short, format long, format e,* and *format long e.* Compare your results and comment on them.

Q.2.7 Enter, execute and define the command *clear* A.

Q.2.8 Enter and execute the command that returns the variable names and their sizes that are currently in the workplace.

Q.2.9 Enter *clear B.* Comment.

Q.2.10 Repeat Q.2.8. Comment.

Q.2.11 Press *Ctrl* and the C keys simultaneously. Explain its action.

Q.2.12 Convert, save, and run the program of Example 2.1 as a script file. Call this new file Example_2.1.

Q.2.13 Convert, save, and run Example 2.1 as a function file. Define its input and output variables.

Q.2.14 Load and run the program of Example 2.2.

Q.2.15 Modify Example 2.2 such that variables A, B, and C are defined in one line.

Q.2.16 Modify Example 2.2 for the case of five numbers. Use A, B, C, D, and E to define the variables involved.

Q.2.17 Load and run the program of Example 2.3.

Q.2.18 Rerun Example 2.3 for the following cases: $P = \$1000$, $I = 6\%$, and $n = 1, 2$, and 5 years.

Q.2.19 Calculate Q.2.18 by hand and compare the results.

Q.2.20 Load and run the instructions of Example 2.4.

Q.2.21 Using a calculator, evaluate each of the expressions of Example 2.4 and compare them with the answers obtained for Q.2.20.

Q.2.22 Enter and execute the instruction *who,* and record and describe the display.

Q.2.23 Enter and execute the instruction *whos,* and record the display.

Q.2.24 Compare the outputs of Q.2.22 with Q.2.23.

Q.2.25 Enter *clear.*

Q.2.26 Repeat Q.2.22 and record and describe the display.

Q.2.27 Enter *clc* and observe and comment on the results.

Q.2.28 Load and run the program of Example 2.5.

Q.2.29 Use the *tic, tac* commands to measure the execution time employed to run Example 2.5.

Q.2.30 Load and run the program of Example 2.6.

Q.2.31 How many variables are used in Example 2.6? What are the variable types?

Q.2.32 Define the objective of the instruction *format compact.*

Q.2.33 Use the *tic, tac* commands to measure the execution time of Example 2.6.

Q.2.34 Check if the *flop* command is defined by your software. If it is, use it to count the number of operations executed in Example 2.6.

Q.2.35 Define what is a MATLAB function file.

Q.2.36 Define what is a MATLAB script file.

Q.2.37 Discuss what is meant by an M-file.

Q.2.38 What is the syntax of a MATLAB function?

Q.2.39 Give at least three examples of MATLAB functions.

2.6 Application Problems

P.2.1 A $314 coat has a discount of 30%. What is the price of the coat?

P.2.2 What percentage of 60 is 53?

P.2.3 Evaluate the following expressions using the MATLAB arithmetic hierarchy expressed with a minimum number of parenthesis.

    a. $2^5/(2^6 - 1)$

    b. e^4

    c. $\ln(e^4)$

    d. $\log10(e^4)$

    e. $e^{\pi\sqrt{121}}$

    f. $\cos(\pi/4) + \sin^2(\pi/3)$

    g. $\log_e(e^3) + \log_{10}(e)$

    h. $\text{area} = \pi * (\pi/3)^2$

P.2.4 Solve for x for each of the equations given below.

    a. $2^x = 7$

    b. $\ln(x) = 3$

    c. $e^x = 10$

    d. $\sqrt{3^{1/3} \cdot 9^{1/3}} = \left(\frac{1}{3}\right)^x$

    e. $\dfrac{\sqrt{2}}{6^3} = 3^x$

P.2.5 A student gets 9 questions right on a 20-question test.

    a. Write a MATLAB program that returns the percentage of correct answers.

    b. What should the student's grade be if all the questions have the same value?

    c. What should the student's grade be if three questions are worth 5%, two are worth 6%, and the remaining four are worth 7%?

    Table 2.4 indicates the grades as a percentage of correct responses.

TABLE 2.4

Grades as a % of Academic Performance

Percentage	Grade
90–100	A
80–89	B
70–79	C
65–69	D
Less than 65	E

P.2.6   Evaluate the following algebraic expressions by hand (with the help of a calculator) and by using MATLAB. Employ the minimum number of parentheses when using MATLAB, where $A = 1$, $B = 1.5$, $C = 2$, $D = 2.5$, and $E = 3$.

a. $Y = (A + B)^D + \left[\dfrac{D * E}{A + B}\right]^2$

b. $X = \dfrac{((A + B) * C)^2}{D * (E + A)^2}$

c. $Z = \sqrt{B^c + (E^D * (B + D))^2}$

d. $V = B * \left[\dfrac{C^4}{E - D}\right]$

e. $W = 4 * \pi * \sqrt{D}$

f. $R = \dfrac{4^D}{1 - E^B}$

P.2.7   Evaluate

$$Y = \left(1 + \frac{1}{N} + \frac{1}{N^2}\right)^N \quad \text{for } N = 10{,}000 \text{ and } 1000$$

P.2.8   Repeat problem P.2.5 part(a) for the cases of 7, 11, and 15 questions right from the same total of 20, assuming that all the questions are equally important.

P.2.9   A room size is 13.5 by $8^1/_6$ feet. The cost of a square foot of a carpet including installation is $17.15. How much will it cost to carpet the entire room?

P.2.10 Ann earns $8.50/h and she is paid an additional 50% for any time exceeding the first 35 hours weekly. Draw a flow chart and write a program of how much she would earn if she worked 43 and 58 h/week.

P.2.11 Seven cans of soda cost $3.50. How much would a dozen cost?

P.2.12 An investor buys a product for $5635. If the investor wants to make a profit of 18.50%, what should the selling price be?

P.2.13 A product is sold at $730 and its cost is $583. Determine the profit as a percentage of the cost.

P.2.14 A house and its corresponding plot of land were bought for $250,000. The plot costs 2/3 of the price of the house. What is the price of the house?

P.2.15 Using MATLAB, evaluate the following quantities:

a. 5% of 20

b. 5% of 5

c. 100% of 3

d. 150% of 17

e. 10% of 5/8

P.2.16 Write a MATLAB program and draw a clear flowchart that divides $26,500 into four partners A, B, C, and D. The division is made on the following: A receives 3/5 of the amount of B, C receives 1/4 of the amount of A, and D receives 2/3 of the amount assigned to C.

P.2.17 In a given organization, four out of seven workers are men. How many women work for the organization if it employs 580 men?

P.2.18 A student wants to get an average of 91 on his English course. What should be the score on the sixth quiz if he/she has scored 91, 98, 82, 88, and 93 on the earlier five quizzes?

P.2.19 The estimated transmission of a telex costs $0.15 per word. Write a program and draw a flowchart that accepts as inputs the number of words of a message and returns the total cost.

P.2.20 Write a program that returns the perimeter, area of a circle, and the volume of a sphere with a radius of $\pi/4$, with a precision of 16 decimal digits.

P.2.21 Mr. X has agreed to buy a car for $13,800. Registration is $45 and NYC taxes are 8.65%. The car dealer is offering three options:

a. Pay in full with a rebate of $1200.

b. Pay 70% of the amount and the remaining balance at an interest rate of 3.5% for 5 years.

c. Full financing with an interest rate of 4.75% during 3 years with a rebate of $750 at the end.

Which is the best option, if the cost of money is 3.6% a year?

P.2.22 A college has a population of 8000 students, of which 30% are women. The administration wishes to increase enrollment until half the student body are women. How many more women should be enrolled?

P.2.23 A student takes four exams. The exams are worth 10, 20, 30, and 40%, respectively, of the final grade. All exams are graded on the basis of 100. Write a program to determine the student's final grade (use Table 2.4).

P.2.24 A triangle with side lengths a, b, and c is given. The area A is given by Hero's (Balador, 2000) formula as $A = \sqrt{s(s - a)(s - b)(s - c)}$, where $s = (a + b + c)/2$. Write a MATLAB program that returns the area for the following triangles with side lengths:

a. 5, 7, and 9

b. 15, 20, and 32

P.2.25 Three points given in terms of the Cartesian coordinates system define a triangle. Write a program to compute the area of the triangle if the vertex points are given by the following points:

$$P_1(1, 1), P_2(2, 4), \text{ and } P_3(3,2).$$

Note that the distance between two points, $P_1(X_1, Y_1)$ and $P_2(X_2, Y_2)$, in terms of its coordinates is given by

$$\text{Distance} = \sqrt{(X_1 - X_2)^2 + (Y_1 - Y_2)^2}$$

P.2.26 Two Cartesian coordinates points are given by $P_0(X_0, Y_0)$ and $P_1(X_1, Y_1)$, then the slope of the line passing through these points is given by

$$\text{Slope} = \frac{Y_1 - Y_0}{X_1 - X_0}$$

The distance between points P_0 and P_1 is given in P.2.25.

Using MATLAB, compute the slope and distance for the following sets of points:

a. $P_0(5, -1)$; $P_1(3, 2)$

b. $P_0(8, 1)$; $P_1(9, 0)$

c. $P_0(5, 3)$; $P_1(-1, 7)$

P.2.27 Write a MATLAB program that returns the number of seconds in a day, a month, and a year.

P.2.28 Write a program that converts a six-digit time array from hh/mm/ss to minutes.

P.2.29 Write a program that converts the following:

    a. 23°F to °C

    b. 132° to rad

    c. 15 gal to L

    d. 13.5 miles to meters

    e. 13 ft to meters

    f. 3.42 inches to meters

    g. 5.5 ft³ to gal

    h. 6.32 miles to inches

    i. 12.3 qt to ft

    The following relations may be of help:

    1 gal = 3.785 L

    $1°F = (9/5) * °C + 32$

    1 mi = 1609.3 m

    1 ft³ = 23.3168 L

    1 qt = 0.9464 L

    1 ft = 0.3048 m

    1 in. = 0.0254 m

    $1 \text{ rad} = \dfrac{360°}{(2\pi)}$

P.2.30 Three scales: Celsius, Fahrenheit, and kelvin are used by engineers to measure temperature. The conversion formulas are given in Table 2.5.

    Write a MATLAB program that converts

    a. 132.3°C to °F

    b. 273 K to °C

    c. 32 K to °F

    d. The temperature of boiling water expressed in degree Fahrenheit

TABLE 2.5

Temperature Conversion Formulas

Conversion	Formula
°C to °F	$°F = \dfrac{9°C}{5} + 32 = 1.8°C + 32$
°F to °C	$°C = (°F - 32) \cdot \dfrac{5}{9} = \dfrac{°F - 32}{1.8}$
K to °C	$°C = K - 273.15$
°C to K	$K = °C + 273.15$

e. The normal body temperature expressed in degree Fahrenheit and Kelvin

f. The temperature when water freezes expressed in degree Fahrenheit and Kelvin

g. The absolute zero expressed in degree Fahrenheit and degree Celsius*

P.2.31 Information units are usually expressed in bits and bytes, where

$$8 \text{ bits} = 1 \text{ (one) character} = 1 \text{ byte}$$

Estimate the number of characters that can be placed in

a. One sheet of paper

b. One quire

c. One ream

d. Three bundles

e. Five cases

f. Three bales

where one sheet of paper can store 4000 characters (50 lines per sheet * 80 characters per line)

1 quire = 24 sheets

1 ream = 20 quires

1 bundle = 2 reams

1 case = 4 bundles

1 bale = 10 reams

P.2.32 Use Table 2.6 (Foreign Exchange, *The New York Times*, September 8, 2000) to convert the following foreign currencies to U.S. dollars.

a. 132 Danish kroner

b. 128 French francs

c. 2800 Italian liras

d. 205 Jordanian dinars

e. 8521 Venezuelan bolivars

P.2.33 Determine which choice will provide the biggest return in absolute value for an initial investment of $1500, placed.

a. at an annual interest of 5% during 4 years

b. at an annual interest of 4% during 5 years

c. at an annual interest of 4% for the first and second year, 5% for the third year, and 5.5% for the fourth year

P.2.34 The following formula can be used to calculate the monthly payments to repay borrowed money at an annual interest I

$$M = \frac{A \cdot (I/12)(1 - (I/12))}{(1 + I/12)^n - 1}$$

where M is the monthly payments, A the amount borrowed, I the annual interest rate, and n the number of months of the loan.

Determine the monthly payments, if the amount borrowed is $132,500, at an annual interest rate of 7.25% payable in 120, 180, and 240 months.

---

* The absolute zero is defined as 0 K.

TABLE 2.6

Foreign Currencies (9/8/2000)

Currency	In Dollars		In Foreign Currency	
	Fri.	Thu.	Fri.	Thu.
North America/Caribbean				
Canada (Dollar)	0.8770	0.8770	1.4770	1.4771
Dominican Rep (Peso)	0.0548	0.548	18.24	18.24
Mexico (Peso)	0.107550	0.107543	9.2900	9.2900
South America				
Argentina (Peso)	1.0002	1.0002	0.9998	0.9998
Bolivia (Boliviano)	—	—	—	—
Brazil (Real)	0.6485	0.5495	1.8230	1.8200
Chile (Peso)	—	—	—	—
Colombia (Peso)	0.000454	0.000462	2204.50	2210.50
Paraguay (Guarani)	—	—	—	—
Peru (New Sol)	0.2879	0.2880	3.473	3.472
Uruguay (New Peso)	0.0804	0.0804	12.4360	12.4450
Venezuela (Bolivar)	0.0015	0.0015	689.2000	689.25
Asia/Pacific				
Australia (Dollar)	0.5865	0.5685	1.7869	1.7806
China (Yuan)	0.1208	0.1208	6.2790	6.2794
Hong Kong (Dollar)	0.1282	0.1282	7.7984	7.7991
India (Rupee)	0.0219	0.0219	45.600	45.520
Indonesia (Rupiah)	0.000119	0.000119	8395.00	8376.00
Japan (Yen)	0.009415	0.009520	106.21	106.04
Malaysia (Ringgit)	0.2832	0.2832	3.7998	3.7998
New Zealand (Dollar)	0.4183	0.4167	2.3008	2.3998
Pakistan (Rupee)	0.0183	0.0183	54.55	54.56
Philippines (Peso)	0.0219	0.0219	45.58	46.56
Singapore (Dollar)	0.5759	0.5773	1.7386	1.7322
So. Korea (Won)	0.000902	0.000901	1108.50	1110.40
Taiwan (Dollar)	0.0322	0.0322	31.09	31.08
Thailand (Baht)	0.02214	0.02416	41.42	41.39
Vietnam (Dong)	—	—	—	—
Europe				
Britain (Pound)	1.4194	1.4378	0.7045	0.8955
Czech Rep (Koruna)	0.0247	0.0247	40.50	40.50
Denmark (Krone)	0.1170	0.1165	8.5442	8.5850
France (Franc)	0.1324	0.1332	7.5515	7.5073
Italy (Lira)	0.000449	0.000451	2220.08	2216.04
Europe (Euro)	0.06880	0.07390	1.1510	1.1443
Hungary (Forint)	0.0033	0.0039	300.44	300.83
Norway (Krone)	0.1085	0.1084	9.2080	9.2220
Poland (Zloty)	0.2282	0.2283	4.42	4.36
Russia (Ruble)	0.0359	0.0359	27.3800	27.3400
Slovak Rep (Koruna)	0.0204	0.0204	48.91	48.99
Sweden (Krona)	0.1039	0.1046	9.6285	9.5668
Switzerland (Franc)	0.5615	0.5644	1.7809	1.7718
Turkey (Lira)	0.000002	0.000002	681390	669580
Middle East/Africa				
Bahrain (Dinar)	—	—	—	—
Egypt (Pound)	0.2843	0.2843	3.5175	3.5175
Iran (Rial)	—	—	—	—
Israel (Shekel)	0.2480	0.2481	4.0320	4.0300
Jordan (Dinar)	1.4085	1.4085	0.71098	0.70998
Kenya (Shilling)	—	—	—	—

Source: *The New York Times*, Foreign Currency, September 8, 2000.

P.2.35 *The New York Times* reported in an article published in April 21, 1990 that Benjamin Franklin bequeathed $270,000 to the cities of Boston and Philadelphia with the condition that the money could be spent after 200 years of his death. In 1990, Franklin's moneys became available.

Calculate the total amount, in dollars, if the initial capital was invested at the following interest rates:

a. 4% compounded quarterly

b. 4% compounded annually

c. 4% for the first 100 years compounded annually, and 5% for the remaining 100 years compounded quarterly

P.2.36 The present population in the United States is estimated at 289,000,000. Estimate the population in 25 years, if the estimated growth is 2.6% annually (use equation of the population growth provided in P.2.37).

P.2.37 The population in the United States after 1980 can be approximated by the following equation:

$$P(n) = 227e^{0.007n}$$

where n represents the number of years after 1980. Estimate the population in the United States in

a. 2010

b. 2020

c. 2050

P.2.38 Table 2.7 lists the odds of particular events. Determine the odds of

a. Dying in a plane crash wearing glasses

b. Reaching 80 years, without wearing glasses and without a divorce

c. Having high blood pressure or high cholesterol level

d. Dying in a plane or train crash

P.2.39 Write a MATLAB program that estimates your electric bill, if the appliances with the kilowatt consumption are shown in the Table 2.8 and the cost per kilowatt-hour of usage charge by the power provider is $0.13.

Kilowatt hours used (during a month) * cost (per kilowatt hour) = total cost in dollars ($) per month

TABLE 2.7

Statistical Odds

Event	Odds
Dying in a train accident	1 in 10^6
Dying in a plane crash	1 in 10^7
Wearing glasses at some point in life	1 in 2
Having a marriage end in divorce	1 in 2
Someday having a high cholesterol level	1 in 4
Reaching 80 years	1 in 3
Being hurt in a car accident	1 in 75
Someday having a high blood pressure	2 in 5

TABLE 2.8

Electrical Power Consumption

Appliance	Kilowatt
Refrigerator	0.5
Microwave oven	1.0
TV	0.2
Washing machine	0.5
Dryer	4.5
Dishwasher	1.5
Iron	1.25
Toaster	1.3
Air conditioner	1.5

TABLE 2.9

Gasoline Prices and Taxes

Country	Miles	Price per Gallon ($)	Tax ($)	Total Price	Tax Percentage of Total
United Kingdom	165	1.04	3.25	4.29	76
France	260	1.15	2.51	3.66	69
Germany	520	1.13	2.29	3.42	67
Sweden	122	1.33	2.53	3.86	66
Netherlands	133	1.44	2.52	3.96	64
Belgium	82	1.31	2.27	3.58	63
Italy	383	1.34	2.30	3.64	63
Ireland	232	1.27	1.77	3.04	58
United States		1.35	0.39	1.74	22

Note: Estimated prices for the year 2000.

P.2.40 A tourist plans to visit a number of European countries by automobile, and drives the distances indicated in Table 2.9 (prices as of 2000), using a car that yields 22 miles to the gallon. The cost of gasoline is also indicated for each of the countries as well as the taxes paid. How much would the traveler spend on

   a. Gasoline.

   b. Local taxes.

   c. Tax as a percentage of the total.

   d. Tax and total price if he/she would travel the same distances in the United States.

   e. The prices in 2008 are twice as those provided in Table 2.9. Repeat parts a, b, c, and d, if the same distances are traveled in 2008.

   f. Estimate the gasoline prices by 2010 and 2020 for each country in Table 2.9, assuming the same increments as given over the period 2000/2008.

3

Matrices, Arrays, Vectors, and Sets

God made the integers; all else is the work of man.

Leopold Kronecker

3.1 Introduction

The basic element in MATLAB® is the matrix. The name MATLAB stands for **Mat**rix **Lab**oratory, and the language syntax and commands are based on matrix operations and their extensions. Therefore to fully understand the MATLAB language, a summary of basic matrix concepts, definitions, operations, and applications are introduced and discussed in this chapter.

Let us start this discussion by defining what is a matrix. In its simplest form a matrix is a set of numbers or elements arranged in a rectangular grid of horizontal rows and vertical columns. Every row or column in a matrix is also called a vector. An array with m rows and n columns is referred as an m times n matrix, denoted by $m \times n$ indicating the size or the matrix dimension, consisting of a total number of $m * n$ elements. The elements of a matrix are indexed. The purpose of indexing is to make easier the identification process of each element of a matrix by using one or more subscripts.

A constant or scalar can be considered a special matrix consisting of 1 row by 1 column, or a 1×1 matrix. Similarly, a vector may be viewed as a one-dimensional (1-D) array, where one of the indexes is 1, which is usually omitted. A 2-D array is the typical matrix structure encountered in most applications, and requires two indexes: the first identifies the row followed by a comma and the second identifies the column.

Matrices and vectors are frequently enclosed in brackets when used in MATLAB. The elements of a matrix or vector can be real or complex numbers, strings, symbolic expressions (see Chapter 7), or in general any MATLAB function or expression. A vector consisting of only one row or column is commonly referred as a row or column vector (and requires one index for obvious reasons). When $m = n$ (the number of rows is equal to the number of columns), the matrix is referred to as an n-square or simply square matrix. These matrices are frequently used to model physical systems represented by sets of n simultaneous equations, with n unknown variables.

Matrices and vector notation, properties, manipulations, and algebra are introduced and discussed in this chapter, as well as a number of often used special matrices such as

the identity (eye), empty, zeros, ones, magic, rand, randn, diagonal, triangular, symmetric, magic, Hilbert, Hermitian, and *Pascal.*

In MATLAB, all the matrices must be either called or created by the user, but there are different ways of creating a matrix. The simplest way is by typing (entering) each one of the elements of the matrix manually or by calling a built-in function that returns the complete or portion of a desired matrix.

MATLAB is an ideal environment to study and experiment with matrix, and array algebra and manipulations, where relations, properties, and results can easily be verified. What is the reason to study matrices and arrays, and why are matrix concepts important? The simplest answer is that data are frequently supplied and organized in tables or arrays, where the elements can easily be identified by either one or more subscripts, making matrix a natural way to organize, present, or represent data, events, or relations.

Systems of simultaneous linear equations can be expressed using matrix notation and matrix algebra can be used in its solution, a fact that makes possible for ordinary people to easily juggle hundreds of equations simultaneously.

3.2   Objectives

After reading this chapter, the reader should be able to

- Understand the concept of a matrix and array
- Create manually an array, matrix, sequence, or vector
- Create or call special built-in vectors and matrices
- Perform array, vector, and matrix algebra
- Understand matrix manipulation and notation
- Append a row or column vector to a matrix
- Append a matrix to another matrix
- Identify an array or matrix element
- Determine the matrix size, length, and dimension
- Understand the concepts of norm or length, distance, and angle between vectors
- Understand the concepts of inner or dot product
- Understand the concept of cross product
- Understand the concept of orthogonality when applied to vectors and matrices
- Understand the Cauchy–Schwarz inequality
- Solve a set of linear equations using matrix algebra
- Solve a matrix system for its characteristic equation
- Solve a matrix system for its characteristic polynomial
- Solve a matrix system for its eigenvalues and eigenvectors
- Understand the concept of a sparse matrix
- Create sparse matrices
- Explore and create patterns by using the *spy* function
- Use MATLAB commands and techniques to solve a number of matrix, array, and vector problems

3.3 Background

R.3.1 An array organized in terms of m rows and n columns is called a matrix.

R.3.2 In its simplest form a matrix can be created using MATLAB, by typing each element of a row, row by row, with a space or a comma separating the consecutive elements in a row, and semicolons to separate consecutive rows of a matrix.

R.3.3 The elements of a matrix are generally entered in MATLAB within brackets, and the elements may be real or complex numbers, functions or expressions. For simplicity, matrices with elements consisting of numerical-real numbers or character strings are considered first in this chapter.

The *input* statement can be used to create a matrix, such as

$$X = input \text{ ('Enter the value for the matrix } X \text{ in brackets')}$$

When the *input* instruction is executed, the text 'Enter the value for the matrix X in brackets' will be displayed on the screen. The user can then enter the element values (in brackets), row by row. All those values will be assigned to the matrix X.

R.3.4 For example, use MATLAB to create the row vector X and column vector Y, defined as

$$X = [1 \quad 2 \quad 3 \quad 4] \quad \text{and} \quad Y = \begin{bmatrix} 5 \\ 6 \\ 7 \\ 8 \end{bmatrix}$$

MATLAB Solution
```
>> format compact
>> X=input ('Enter the elements of X in brackets separated by spaces')

    Enter the elements of X in brackets separated by spaces [1 2 3 4]

    X =
         1      2      3      4

>> Y= input ('Enter the elements of Y in brackets separated by
                                      semicolons ')

Enter the elements of Y in brackets separated by semicolons [5;6;7;8]

    Y =
            5
            6
            7
            8
```

R.3.5 The command $A = input('expression')$ evaluates the *expression* (in quotes) and the result is assigned to A. If the return key is pressed without entering a character, then *A* becomes an empty matrix. The empty matrix can be defined as a matrix with no elements.

R.3.6   The input statement can also be used to enter a string such as

$$B = input('Enter \ a \ string', \ 's')$$

This instruction will display *Enter a string* (whatever is in quotes) and MATLAB will wait for the user to enter a string. That string will be assigned to *B*. A string can be defined as a sequence of characters.

R.3.7   For example, let us gain some MATLAB experience by performing the following:

a. Assign to *A* the value *sqrt(pi)*

b. Assign to *My_name_is* your name (with the argument *s* present)

c. Assign to *My_name_is* the empty matrix

d. Assign to *My_name_is* your name (in quotes)

e. Assign to *My_name_is* your name (no quotes)

```
MATLAB Solution
>> A= input ('Enter the MATLAB expression')

                Enter the MATLAB expression sqrt(pi)
    A =
       1.7725

>> My _ name _ is = input ('Enter your name','s')

                        Enter your name John Smith
    My _ name _ is =
                John Smith

>> My _ name _ is = input ('Enter your name')

                        Enter your name        % press the enter key
    My _ name _ is =
                [] % observe that the empty matrix is assigned
                    to My _ name _ is

>> My _ name _ is = input ('Enter your name')

                        Enter your name John Smith
    ???  John Smith
             |
    Error: Missing operator, comma, or semicolon.

>> My _ name _ is = input ('Enter your name')

                        Enter your name 'John Smith'
    My _ name _ is =
                        John Smith
```

Note the importance of the quotes and 's', and how they are used when dealing with a string.

R.3.8   The input statement $C = input('string')$ assigns to the variable C the text *'string'*. The text *string* may contain one or more '$\backslash n$' that must be placed in quotes. The sequence '$\backslash n$' performs the following action: skips to the beginning of the next line.

R.3.9   When a matrix or vector is input, the value of the matrix is displayed on the screen unless it is suppressed by placing a semicolon (;) at the end of the instruction.
   For example,

```
>> format compact
>> A = [1   2; 3    4]                        % displays matrix A

   A =
        1        2
        3        4

>> A = [1    2; 3    4];                       % suppresses the display of A
```

R.3.10 Matrices are usually assigned a variable name. For example,

$$A = [1 \quad 2 \quad 3]$$

R.3.11 Recall that the command *who* returns a list of the variable names used and the command *whos* returns a list of the variable names used as well as their sizes (R.2.21 and R.2.22).

R.3.12 The command *length(A)* returns the number of elements of A, if A is a vector, or the largest value of either n or m, if it is an $n \times m$ matrix.

R.3.13 The MATLAB command *size(A)* returns the size of the matrix A (*number of rows by the number of column*), and the command *[row, col] = size(A)* returns the number of rows assigned to the variable *row* and the number of columns of A assigned to the variable *col*. For example,

```
>> A = [1 2;3 4;5 6];
>> size(A)

   ans =
        3   2

>>[row,col] = size(A)

   row =
        3
   col =
        2

>>length (A)

   ans =
        3
```

R.3.14 In its simplest form, a vector is a vertical or horizontal sequence of numbers separated by commas (or spaces) or semicolons. When the sequence is vertical it is called a column vector, or a row vector when it is horizontal. For example, use MATLAB to represent $A = [1 \ 2 \ 3]$ and $B = [1; \ 2; \ 3]$ as a row and column vector, respectively.

```
MATLAB Solution
>> A = [1 2 3]

   A =
        1       2       3

>> B = [1;2;3]

   B =
        1
        2
        3
```

R.3.15 The notation $A(n, m)$ is used to identify the element that is located in the intersection of the n rows and m columns of A. For example, let

$$A = \begin{bmatrix} 1 & 4 & 7 \\ 2 & 5 & 8 \\ 3 & 6 & 9 \end{bmatrix}$$

then $A(1, 2) = 4$, $A(2, 2) = 5$, and $A(2, 3) = 8$. (R.3.16 through R.3.18 use matrix A as an example.)

R.3.16 Colons when used as an argument identify a range over a row or a column depending on its location.
   For example,

$$B = A(1:2, 2:3) = \begin{bmatrix} 4 & 7 \\ 5 & 8 \end{bmatrix}$$

Note that matrix B is defined by the intersection of the rows 1 and 2 and the columns 2 and 3 of matrix A. Therefore, B is a 2×2 matrix.

R.3.17 When dealing with rows and columns one can specify the last element (row or column) by using the keyword *end*, and the second element next to the last by *end -1*, and so on.
   Therefore, an alternate way to specify B is by using the following command:

   $A(1:end-1, 2:end)$

Any element of matrix A can be specified in terms of the keyword *end*.
For example, $A(end-2, end-1) = A(1, 2) = 4$.

R.3.18 When a matrix index is replaced by a colon, the colon represents depending on its location, either *all the rows or all the columns*. For example,

$$A(:, 2:3) = \begin{bmatrix} 4 & 7 \\ 5 & 8 \\ 6 & 9 \end{bmatrix} \qquad A(1:2, :) = \begin{bmatrix} 1 & 4 & 7 \\ 2 & 5 & 8 \end{bmatrix}$$

R.3.19 Colons can also be used to generate a sequence. For example, $n = 1{:}5$ returns the vector n, consisting of the sequence of elements from 1 to 5 in ascending order, with unit increments. Then n becomes a 1×5 (row) vector indicated as follows:

$$n = [1 \quad 2 \quad 3 \quad 4 \quad 5]$$

R.3.20 When a command has two colons that separate three numerical arguments, following the format:

$$n = [initial{:}\ increment{:}\ final]$$

then the command returns the vector n, with elements that follow the sequence: start with the *initial* value all the way to the *final* value, with *successive* increments defined by *increment*, illustrated as follows:

$$n = [1{:}0.1{:}2], \text{ returns the following sequence}$$

$$n = [1.0 \quad 1.1 \quad 1.2 \quad 1.3 \quad 1.4 \quad 1.5 \quad 1.6 \quad 1.7 \quad 1.8 \quad 1.9 \quad 2.0]$$

The preceding sequence can also be generated without the brackets or by replacing the brackets with parenthesis. For example,

$$n = 1{:}0.1{:}2 \text{ is equivalent to } n = [1{:}0.1{:}2] \text{ and } n = (1{:}0.1{:}2)$$

Note that the *increment variable* when negative is used to generate a decreasing sequence, illustrated as follows:

```
>> n = 2:-0.1:1

n =
    Columns  1 through 7
    2.0000   1.9000   1.8000   1.7000   1.6000   1.5000   1.4000
    Columns  8 through 11
    1.3000   1.2000   1.1000   1.0000
```

R.3.21 A multiple line command that starts in one line is continued on the next line by placing three consecutive periods (...) called ellipsis at the end of the first line. Continuation across several lines can be accomplished by using ellipsis repeatedly at the end of each line, but no instruction should exceed 4096 characters.

R.3.22 The elements of different vectors or matrices can be concatenated to form new extended vectors or matrices. The extended matrix is defined in terms of other vectors or matrices. For example, let $u = [0 \quad 1]$ and $v = [2 \quad 3]$.
Execute and evaluate the responses of the following MATLAB commands:

a. $s = [u\ v]$

b. $ss = [u;\ v]$

c. $sss = [ss, ss; ss, ss]$

```
MATLAB Solution
>> u = [0 1];
>> v = [2 3];
>> s = [u v]

    s =
         0      1      2      3

>> ss = [u;v]

    ss =
         0      1
         2      3

>> sss = [ss,ss;ss,ss]

    sss =
         0      1      0      1
         2      3      2      3
         0      1      0      1
         2      3      2      3
```

(R.3.23 through R.3.27 use the vector s = [0 1 2 3] as example.)

R.3.23 The elements of a vector are identified by a single index. For example,

$$s(3) = 2 \quad or \quad s(end-1) = 2$$

R.3.24 Any element can be changed by redefining it with a new value.
For example,

let $s(4) = -1$ (is entered), then vector s becomes

$$s = [0 \ 1 \ 2 \ -1]$$

R.3.25 Defining an element outside its range can be used to expand the range or length of a matrix or vector. For example, by defining $s(6) = -2$, s becomes a six-element vector, illustrated as follows:

$$s = [0 \ 1 \ 2 \ -1 \ 0 \ -2]$$

R.3.26 The elements not specifically defined in a vector or matrix are assigned the default value of zero. Observe that in R.3.25, $s(5) = 0$.

R.3.27 A set of rows or columns can be deleted by using the null vector ([ ]). For the vector s defined in R.3.25, the instruction $s(2:4) = [\ ]$ would delete elements 2, 3, and 4, and s would be $s = [0 \ 0 \ -2]$. Another way to generate the same sequence is by executing the following command:

$$s = [s(1) \quad s(5) \quad s(6)]$$

R.3.28 An alternate way to generate a sequence n that is a row vector is by using the command: *linspace (initial, final, m_points)*, where *initial* and *final* correspond to the start and end of the sequence, respectively, defined by m points, equally spaced over the range *initial/final*.

For example, $n = linspace\ (1, 2, 11)$ returns the sequence

$$n = [1.0 \quad 1.1 \quad 1.2 \quad 1.3 \quad 1.4 \quad 1.5 \quad 1.6 \quad 1.7 \quad 1.8 \quad 1.9 \quad 2.0]$$

Note that the above expression was also defined as $n = [1:0.1:2]$ in R.3.20.

R.3.29 A row vector with elements following a logarithmic sequence with length L can be generated by using the MATLAB command: $U = logspace\ (X, Y, L)$, where the initial element of U is defined by 10^X, the final element is 10^Y, and L is its length given by the number of elements.

For example, let $U = logspace(.1,3,5)$, then MATLAB returns the row vector U illustrated as follows:

```
>> U = logspace(.1,3,5)

   U =
      1.0e+003 *
      0.0013      0.0067      0.0355      0.1884      1.0000
```

R.3.30 A column (or row) vector B can be appended to a matrix A if B has the same length as the columns (or rows) of A. For example, let

$$A = \begin{bmatrix} 1 & 2 & 3 \\ 5 & 6 & 7 \end{bmatrix} \quad \text{and} \quad B = \begin{bmatrix} 4 \\ 8 \end{bmatrix}$$

Execute the following commands using MATLAB and observe and evaluate the responses:

a. $C = [A\ B]$

b. $D = [A\ A]$

c. $E = [B\ B]$

d. $F = [A;\ A]$

e. $G = [B;\ B]$

f. $H = [A;\ B]$

MATLAB Solution
```
>> A = [1 2 3;5 6 7]                        % matrix A

   A =
        1      2      3
        5      6      7

>> B = [4;8]                                % column vector B

   B =
        4
        8

>> C = [A B]                                % part(a)

   C =
        1      2      3      4
        5      6      7      8
```

```
>> D = [A A]                                      % part(b)

   D =
        1      2      3      1      2      3
        5      6      7      5      6      7

>> E = [B B]                                      % part(c)

   E =
        4      4
        8      8
        9

>> F = [A;A]                                      % part(d)

   F =
        1      2      3
        5      6      7
        1      2      3
        5      6      7

>> G = [B;B]                                      % part(e)

   G =
        4
        8
        4
        8

>> H = [A;B]                                      % part(f)

??? Error using ==> vertcat
All rows in the bracketed expression must have the same
number of columns.
```

R.3.31 Array operations are operations performed on the individual element of a given matrix. They are indicated by a dot (.), followed by the operation (.*, ./, .^).

The ./, and .\ indicate two distinct array divisions called the right and left division, respectively. For the case of addition and subtraction, the dot is optional (not required).

For example, let

$$A = \begin{bmatrix} 1 & 2 \\ 3 & 4 \end{bmatrix} \quad \text{and} \quad B = \begin{bmatrix} 6 & 7 \\ 8 & 9 \end{bmatrix}$$

Then

$$C = A + B = \begin{bmatrix} 7 & 9 \\ 11 & 13 \end{bmatrix}, \quad \text{where } C(i, j) = A(i, j) + B(i, j)$$

$$D = A. * B = \begin{bmatrix} 6 & 14 \\ 24 & 36 \end{bmatrix}, \text{ where } D(i, j) = A(i, j). * B(i, j)$$

$$E = A./B = \begin{bmatrix} 1/6 & 2/7 \\ 3/8 & 4/9 \end{bmatrix} = \begin{bmatrix} 0.1667 & 0.2857 \\ 0.3750 & 0.4444 \end{bmatrix}, \text{ where } E(i, j) = A(i, j)./B(i, j)$$

$$F = A.\backslash B = \begin{bmatrix} 6 & 7/2 \\ 8/3 & 9/4 \end{bmatrix} = \begin{bmatrix} 6.0000 & 3.5000 \\ 2.6667 & 2.2500 \end{bmatrix}, \text{ where } F(i, j) = B(i, j)./A(i, j)$$

$$G = A - B = \begin{bmatrix} -5 & -5 \\ -5 & -5 \end{bmatrix}, \text{ where } G(i, j) = A(i, j) - B(i, j)$$

$$H = A.\wedge B = \begin{bmatrix} 1^6 & 2^7 \\ 3^8 & 4^9 \end{bmatrix} = \begin{bmatrix} 1 & 128 \\ 6561 & 262144 \end{bmatrix}, \text{ where } H(i, j) = A(i, j). \wedge B(i, j)$$

$$I = A.\wedge 2 = \begin{bmatrix} 1 & 2^2 \\ 3^2 & 4^2 \end{bmatrix} = \begin{bmatrix} 1 & 4 \\ 9 & 16 \end{bmatrix}, \text{ where } I(i, j) = A(i, j). \wedge 2$$

R.3.32 Addition, subtraction, multiplication, division, and exponentiation (+, −, *, /, ^) of arrays can only be performed when the arrays involved have the same size.

R.3.33 An array X can be an argument of a function, resulting in a matrix B, obtained by applying the function to each element of X. For example, let the range of X be over $2\pi \le X \le 4\pi$, consisting of eight linearly spaced points organize as a 2×4 matrix. Obtain the matrix Y consisting of the natural logarithmic value of each of the elements of X.

MATLAB Solution
```
>> X = [2*pi:pi/3:3*pi;3*pi:pi/3:4*pi]

    X =
        6.2832      7.3304      8.3776      9.4248
        9.4248     10.4720     11.5192     12.5664

>> Y = log(X)

    Y =
        1.8379      1.9920      2.1256      2.2433
        2.2433      2.3487      2.4440      2.5310
```

R.3.34 The MATLAB command $dot\_prod = dot(X, Y)$ returns the scalar dot product of the two vectors X and Y, where X and Y must have the same length. If they do not, MATLAB returns an error message. If X and Y are real column vectors, then MATLAB returns the standard inner product $X' * Y$ or $Y' * X$, as the $dot\_prod = dot(X, Y)$, where $dot\_prod = \Sigma_i X_i * Y_i$ over all i's.

MATLAB does not compute the inner product of complex vectors in the standard way. The dot product of two nonzero vectors with a common origin returns a scalar given by $dot(X, Y) = \|X\| \cdot \|Y\| \cos(XY)$, * where $\|X\| = sqrt(X * X')$, $\|Y\| = sqrt(Y * Y')$, and $cos(XY)$ is the cosine of the angle between X and Y.

The notation $\|V\|$ is commonly referred as *norm* and defines the length of V.

For example, let $X = [1\ 2\ 3]$ and $Y = [4\ 5\ 6]$, then $dot\_prod = dot(X, Y)$ returns $dot\_prod = 4 * 1 + 2 * 5 + 3 * 6 = 32$.

The dot product can be used to determine whether two vectors (X and Y) are orthogonal. Recall that two vectors are orthogonal if the angle between them is 90°. Then two nonzero vectors are orthogonal if the dot product is zero.

For example, let

$$X = \begin{bmatrix} 1 \\ 2 \\ 3 \\ 4 \end{bmatrix} \quad \text{and} \quad Y = \begin{bmatrix} 0 \\ -1 \\ -2 \\ -3 \end{bmatrix}$$

Write a program that returns the angle between X and Y in radians and degrees.

MATLAB Solution
```
>> X= [1;2;3;4];
>> Y= [ 0;-1;-2;-3];
>> COS _ X _ Y= dot(X,Y)/(norm(X)*norm(Y));
>> Angle _ X _ Yin _ rad = acos(COS _ X _ Y)

   Angle _ X _ Yin _ rad =
                        2.9216

>> Angle _ X _ Yin _ degree = Angle _ X _ Yin _ rad*180/pi

   Angle _ X _ Yin _ degree =
                        167.3956
```

R.3.35 The MATLAB function *cross_prod = cross(X, Y)* returns the cross product of the two vectors X and Y (where X and Y must have the same length).

In the physical sciences the cross product of two nonzero vectors with a common origin is given by

$X \times Y = \|X\| \cdot \|Y\| sin(XY)\ \vec{m}$, where $\|X\| = sqrt(X \cdot X')$, $\|Y\| = sqrt(Y \cdot Y')$, $sin(XY)$ is the sine of the angle between X and Y, and the $\vec{m}$ is a unit vector that indicates the resulting direction.

The direction ($\vec{m}$) is perpendicular to the intersection of vectors X and Y.

Note that the cross product can only be defined, and makes sense in a 3-D space.

R.3.36 The dot product is used to find the component of one vector in the direction of another, or the projection of one nonzero vector along another.

The cross product is used to implement the so-called right-handed system of axes. For example, in physics it can be used to represent angular momentum, or

---

* Trigonometric functions are presented and discussed in Chapter 4.

in field theory defines the direction of the force resulting to a moving charge in a magnetic field.

R.3.37 Array operations extended to the n-dimensional vectors A and B are illustrated as follows:

$$\text{Let } A = [a_1 \quad a_2 \quad a_3 \quad \ldots \quad a_n] \quad \text{and} \quad B = [b_1 \quad b_2 \quad b_3 \quad \ldots \quad b_n]$$

Then

$$C = A + B = [a_1 + b_1 \quad a_2 + b_2 \quad a_3 + b_3 \quad \ldots \quad a_n + b_n]$$

$$D = A - B = [a_1 - b_1 \quad a_2 - b_2 \quad a_3 - b_3 \quad \ldots \quad a_n - b_n]$$

$$E = A. * B = [a_1 * b_1 \quad a_2 * b_2 \quad a_3 * b_3 \quad \ldots \quad a_n * b_n]$$

$$F = A./B = [a_1/b_1 \quad a_2/b_2 \quad a_3/b_3 \quad \ldots \quad a_n/b_n]$$

$$G = A. {\wedge} B = [a_1 {\wedge} b_1 \quad a_2 {\wedge} b_2 \quad a_3 {\wedge} b_3 \quad \ldots \quad a_n {\wedge} b_n]$$

$$H = A. {\wedge} 2 = [a_1^2 \quad a_2^2 \quad a_3^3 \quad \ldots \quad a_n^2]$$

R.3.38 As an additional example, let

$$A = \begin{bmatrix} 1 & 2 & 3 \\ 4 & 5 & 6 \\ 7 & 8 & 9 \end{bmatrix}$$

Evaluate the following instructions using MATLAB:

a. $B = 2. * A$

b. $C = A. {\wedge} 2$

c. $D = 1./A$

ANALYTICAL Solution

a. $B = \begin{bmatrix} 2 & 4 & 6 \\ 8 & 10 & 12 \\ 14 & 16 & 18 \end{bmatrix}$

b. $C = \begin{bmatrix} 1 & 4 & 9 \\ 16 & 25 & 36 \\ 49 & 64 & 81 \end{bmatrix}$

c. $D = \begin{bmatrix} 1/1 & 1/2 & 1/3 \\ 1/4 & 1/5 & 1/6 \\ 1/7 & 1/8 & 1/9 \end{bmatrix}$

R.3.39 The dot (·) preceding the operation symbol (*, /, ^) tells MATLAB to perform element-by-element array operations. Operations without the dot indicate matrix operations, which are quite different from array operations, discussed later in this section.

R.3.40 Let C be the matrix product of the matrices A and B (indicated as C = A * B), then the element C(i, j) is obtained by multiplying the elements of the *i*th row of A by the corresponding elements of the *j*th column of B, and then adding them up, illustrated as follows using the dot product.

$$C(i \cdot j) = dot\ [A(i, :), B\ (:, j)],\ \text{for all } i\text{'s and } j\text{'s}$$

The product between a row (from A) and a column (from B) requires that the number of elements of the row must be equal to that of the column.

For example, let A be an $n \times m$ matrix and B an $m \times r$ matrix, then the product C = A * B returns an $n \times r$ matrix.

For example, let

$$A = \begin{bmatrix} 1 & 2 & 3 \\ 4 & 5 & 6 \end{bmatrix} \text{ and } B = \begin{bmatrix} 1 \\ 0 \\ 1 \end{bmatrix}$$

then

$$C = A * B = \begin{bmatrix} 1*1+2*0+3*1 \\ 4*1+5*0+6*1 \end{bmatrix} = \begin{bmatrix} 1+3 \\ 4+6 \end{bmatrix} = \begin{bmatrix} 4 \\ 10 \end{bmatrix}$$

Note that A is a 2×3 matrix, B is a 3×1 matrix, then C = A * B results in a 2×1 matrix. Also note that MATLAB evaluates the product A * B and returns a result, if and only if the number of rows of A equals the number of columns of B.

Note that if A * B exists, B * A may not exist and in general $A * B \neq B * A$.

R.3.41 The following example illustrates some of the concepts just presented. Let

$$A = \begin{bmatrix} 2 & -3 & 5 \\ 6 & -9 & 7 \end{bmatrix} \text{ and } B = \begin{bmatrix} 4 & 5 \\ -3 & 1 \\ 6 & -9 \end{bmatrix}$$

Execute the following matrix operations:

a. C = A * B

b. D = B * A

c. Note that C is a 2×2 matrix, whereas D is a 3×3 matrix

MATLAB Solution
```
EDU>> A =  [2 -3 5;6 -9 7]

     A =
           2       -3       5
           6       -9       7
```

```
EDU>> B = [4 5;-3 1;6 -9]

    B =
          4         5
         -3         1
          6        -9

EDU>> C = A*B

    C =
         47       -38
         93       -42

EDU>> D = B*A

    D =
         38       -57        55
          0         0        -8
        -42        63       -33
```

R.3.42 A matrix in which the number of rows is the same as the number of columns is called a square matrix. The order of a square matrix is given by the number of rows (or columns). An $n \times n$ matrix is also referred as an *n-square* matrix.

R.3.43 Two matrices A and B are said to be equal, if and only if, they are of the same size and the corresponding elements are equal, that is, $A(i, j) = B(i, j)$ for any i and j.

R.3.44 A matrix A is called diagonal, if A is a square matrix in which all the off-diagonal elements are zeros, that is, $A(i, j) \neq 0$ for $i = j$, and zero otherwise.

R.3.45 The main diagonal also referred as the diagonal of an *n-square* matrix consists of all the elements defined by the sequence: $A(1, 1), A(2, 2), A(3, 3), \ldots, A(n, n)$.

R.3.46 An *n-square* matrix with 1s along the main diagonal and zeros everywhere else is called the unit matrix, denoted by I. The matrix I plays the same role in matrix multiplication as the number 1 does in multiplication of real numbers, that is,

$$A = A * I = I * A$$

Note that I is a special diagonal matrix, that can be created by the MATLAB command *eye*.

R.3.47 An *n-square* matrix consisting of zeros everywhere is called a zero matrix, denoted by 0. The matrix 0 plays the same role in matrix multiplication as the number 0 does in real number multiplication, that is,

$$A * 0 = I * 0 = 0$$

But beware that $A * B = 0$ does not imply that A or B is equal to the zero (0) matrix.

R.3.48 The transpose of a matrix or vector A is denoted by $A.'$ and is a new matrix where the columns of A become rows of $A.'$. For example,

$$\text{Let } A = \begin{bmatrix} 1 & 2 \\ 3 & 4 \end{bmatrix}, \text{ then } A.' = \begin{bmatrix} 1 & 3 \\ 2 & 4 \end{bmatrix}$$

$$\text{and let } V = [1 \quad 2 \quad 3 \quad 4], \text{ then } V.' = \begin{bmatrix} 1 \\ 2 \\ 3 \\ 4 \end{bmatrix}$$

For the case of a square matrix, the transpose can be obtained by reflecting the elements across the main diagonal. Mathematicians usually indicate the transpose operation by an exponent T. For example, $A.' = A^T$.

Note that if A is an $n \times m$ matrix, then $A.'$ becomes an $m \times n$ matrix.

R.3.49 Let A be a complex matrix, then the MATLAB operation A' returns the complex conjugate transpose, often referred as the *Hermitian* transpose, and is denoted by the superscript H ($A^H = A'$), where the element a_{nm} of A becomes a*$_{mn}$ of A^H (* denotes complex conjugate).

For example, let

$$A = \begin{bmatrix} 1 - j & 2 + 3j \\ 4 & -5j \end{bmatrix}$$

then

$$A' = \begin{bmatrix} 1 + j & 4 \\ 2 - 3j & 5j \end{bmatrix}$$

Note that if A is a real matrix, then $A = A.'$. If $z = a + jb$, then the complex conjugate of $z = a - jb$ where a and b are real numbers (complex numbers are presented in Chapter 6).

R.3.50 The transpose operations present interesting properties some of which are indicated as follows.

Let A and B be two matrices (with the same size), then

1. $A = (A^T)^T$
2. $(kA)^T = k^T A^T$, where k is a scalar
3. $(AB)^T = B^T A^T$
4. $(AB)^H = B^H A^H$
5. $(A^{-1})^T A = (A^T)^{-1}$
6. $(A^{-1})^H A = (A^H)^{-1}$
7. $trace(A\ B) = trace(B\ A)$
8. $(A + B)^T = B^T + A^T$

R.3.51 A symmetric matrix is a square matrix that remains the same if the rows and columns are interchanged, that is, $A(i, j) = A(j, i)$ for all i's and j's.

R.3.52 Let A be a symmetric square matrix, then it follows that $A = A'$.

R.3.53 An antisymmetric or skew matrix is a square matrix that satisfies the following relation: $A(j, i) = -A(i, j)$, for all i's and j's.

R.3.54 The determinant of a square matrix A is a scalar. The following example illustrates the procedure followed in the computation of the determinant for a 2×2 matrix.

For example, if $A = \begin{bmatrix} 1 & 2 \\ 3 & 4 \end{bmatrix}$ then *determinant (A) = 1 * 4 − 3 * 2 = −2* (product of the elements in the main diagonal minus the product of the elements of the other diagonal). The MATLAB command *det(A)* returns the value of the determinant of *A*. The concept of determinant applies only for the case of square matrices.

R.3.55 The process of evaluating the determinant of third, fourth, and higher order matrices can be defined and evaluated by using symbolic variables* shown as follows:

```
>> syms a b c d e f g h k
>> A =                                          % 2ᵈ order
          a        b
          c        d
>> det (A)

     ans =
            a*d-b*c
>> B =                                          % 3th order
          a        b        c
          d        e        f
          g        h        k
>> det (B)

     ans =
            a*e*k-a*f*h-d*b*k+d*c*h+g*b*f-g*c*e
```

Observe that symbolic variables can be used to define the rule followed to evaluate a determinant.

R.3.56 Let *A* and *B* be two *n-square* matrices, then

$$det(A * B) = det(A) * det(B)$$

R.3.57 Some useful properties of the determinant of the matrix *A* are

a. *det(A′) = det(A)*.

b. If *A* has a row or column of zeros, then *det(A) = 0* (*A* is then referred as singular).

c. If *A* is a triangular matrix (defined later in this section) then *det(A) = trace(A)*.

d. det(eye) = 1.

e. If *A* is orthogonal then det(*A*) = 1.

f. The *det(A)* changes sign by exchanging two of its rows.

R.3.58 A square matrix *A* is said to have an inverse, if there exists a matrix *B* that satisfies the following relation:

$$A * B = B * A = I \ (identity)$$

Matrix *B* exists and is unique provided that *A* is not singular, that is, det(*A*) ≠ O; then *B* is called the inverse of *A* denoted as A^{-1}.

---

* Symbolic variables are presented and discussed in Chapter 7. At this point it is sufficient to know that a non-numerical element can be a qualitative or symbolic variable.

R.3.59 The MATLAB command $B = inv(A)$ or $A^\wedge -1$ returns the matrix B, that is, the inverse of the square matrix A.*

R.3.60 Let A and B be n-square invertible matrices, then the inverse of the product is the product of the inverses in reverse order indicated as follows:

$$inv(A\ B) = inv(B) * inv(A)$$

R.3.61 The command $B = rats(A)$ returns the matrix B, consisting of the elements of A converted to rational fraction approximations.

R.3.62 The following example illustrates some of the concepts and MATLAB commands presented earlier.

Execute and evaluate the responses of the following commands:

a. $B = inv(A)$

b. $C = rats(B)$

c. $D = A * B$

d. $E = B * A$

for the following matrix:

$$A = \begin{bmatrix} 1 & 3 & 8 \\ 2 & 0 & 6 \\ 1 & -5 & -3 \end{bmatrix}$$

MATLAB Solution
```
>> A = [1 3 8;2 0 6;1 -5 -3]

        A =
           1       3       8
           2       0       6
           1      -5      -3

>> B = inv(A)                                        % part(a)

        B =
          -2.1429       2.2143     -1.2857
          -0.8571       0.7857     -0.7143
           0.7143      -0.5714      0.4286

>> C = rats(B)                                       % part(b)

        C =
          -15/7        31/14       -9/7
           -6/7        11/14       -5/7
            5/7        -4/7         3/7
```

---

* $inv(A) = A^\wedge -1$ is not equal to $1/A$ and $1/A$ will cause an error.

```
>> D = A*B                              % part(c)

    D =
       1.0000              0         -0.0000
       0              1.0000         -0.0000
       0                   0          1.0000

>> E = B*A                              % part(d)

    E =
       1.0000         0.0000          0.0000
      -0.0000         1.0000               0
       0.0000              0          1.0000
```

R.3.63 Matrix A is an orthogonal matrix, if A is a square matrix that possesses an inverse that satisfies the following relation: $inv(A) = A^T$, or its equivalent relation $A^T A = A A^T = I$.

Observe that matrix A is orthogonal if its rows or columns form an orthonormal system.

R.3.64 For example, verify that matrix A, defined as follows, constitutes an orthogonal matrix for any value of x.

$$A = \begin{bmatrix} \cos(x) & -\sin(x) \\ \sin(x) & \cos(x) \end{bmatrix}$$

Let us test orthogonality over the range $0 \leq x \leq 4\pi$, using five linearly spaced points, by writing the script file: *orthonormal*.

MATLAB Solution
```
% Script file:orthogonal
disp('****************************************************')
disp('B=trans(A)*A is evaluated for x=0:pi/4:pi')
disp('****************************************************')
for x =0:pi/4:pi;
   A= [cos(x) -sin(x);sin(x) cos(x)];
   B =A'*A
end
```

The script file: orthogonal* is executed and the results are shown as follows:

```
>>  orthogonal
****************************************************
B=trans(A)*A is evaluated for x=0:pi/4:pi
****************************************************
B =
    1    0
    0    1
B =
    1    0
    0    1
```

---

* The command *for-end* returns A and B for the five different values of x, and returns B instead of repeating five times the same set of instructions. The *for-end* is presented in Chapter 8 and trigonometric functions are presented in Chapter 4.

```
B =
     1        0
     0        1
B =
     1        0
     0        1
B =
     1        0
     0        1
```

R.3.65 Let A be a Hermitian matrix, defined by $A = A^H$. (In general, a square complex matrix with the property that the transposed conjugate of A equals A.) Then all its diagonal elements must be real.

R.3.66 A matrix A is said to be a skew Hermitian matrix if it satisfies the following relation: $A^H = -A$.

R.3.67 A square matrix A, with zeros for all the elements below the main diagonal is called an upper triangular matrix, that is, $A(i, j) = 0$, for $i > j$.

R.3.68 Similarly a square matrix A, with zeros for the elements above the main diagonal is called a lower triangular matrix, that is, $A(i, j) = 0$ for $i < j$.

R.3.69 Recall that A is a diagonal matrix if all the elements that are not on the main diagonal are zeros, that is, $A(i, j) = 0$, for $i \neq j$.

   Note that a diagonal matrix is both an upper and lower triangular matrix.

R.3.70 Elementary row operations are used to systematically transform a given matrix A, into another equivalent matrix B, where B presents a structure that can more easily be solved.

   Matrix A may represent a set of simultaneous equations, where each row represents an equation. Elementary row operations consist of

   a. Interchanging any equation (or rows)

   b. Multiplying any equation by a nonzero constant (multiply all the elements of a row by a constant)

   c. Adding an equation to another equation (or add two rows)

R.3.71 The Gauss–Jordan elimination method consists of solving the linear matrix equation of the form $A * X = Y$, by systematically performing elementary row operations, where A is usually an n-square matrix, and X and Y are n by 1 column vectors. The Gauss–Jordan method consists of reducing A into an upper triangular matrix.

   MATLAB uses the *LU* decomposition, which is based on a variation of the Gaussian elimination method, presented later in this chapter. The final step of the Gauss–Jordan method leads to a unique solution if one exists, and the final structure of the matrix is referred as the reduced row echelon form, denoted by RREF.

R.3.72 The *RREF* matrix has the following properties:

   a. The first nonzero entry in each row is 1, referred to as a leading 1.

   b. Each leading 1 is the only nonzero entry in each column.

   c. Rows consisting of zeros are placed at the bottom of the matrix.

R.3.73 The *RREF* of the square matrix A is used to determine the existence and uniqueness of the solution of the often-encountered matrix equation $A * X = Y$.

R.3.74 The rank of a matrix A is equal to the number of nonzero rows of the *RREF* of the matrix A. If A is an n-square nonsingular matrix then the $rank(A) = n$.

R.3.75 The rank of a matrix *A* can be determined by executing the MATLAB instruction *rank(A)*. The rank of a matrix *A* can be viewed as the number of linearly independent rows or columns of *A*. If *A* is an *n*-square nonsingular matrix [*det(A)* ≠ 0], then the rank of *A* is *n*.

R.3.76 The MATLAB command *rank(A)* returns the number of nonzero rows of the *RREF* of the matrix *A*.

R.3.77 If *A* is a nonsquare matrix, then the command *rank(A)* is the number of nonzero rows of the reduced matrix, or the number of linearly independent rows.

R.3.78 The MATLAB instruction *rref(A)* returns the reduced form for *A*.

R.3.79 The command *rrefmovie(A)* activates a movie showing the execution of the sequence of commands leading to *rref(A)*.

R.3.80 The MATLAB command *[RR, k] = rref(A)* returns the reduced form of *A* as the matrix *RR*, as well as the rank of *A*, denoted by *k*.

R.3.81 Matrix *A* is said to be *ill* condition, if matrix *A* is close to singularity and its inverse then becomes inaccurate. The number associated with the singularity condition of a matrix is called its condition number.

Two MATLAB instructions provide an estimation of the condition of a matrix *A*, where *A* need not be square. They are

$$cond(A) \quad and \quad rcond(A)$$

The function *cond(A)* returns the condition of matrix *A* such that

a. *cond(A)* ≈ *1*, then *A* is in *well* condition and the inverse exists.

b. *cond(A)* ≈ *large*, then *A* is in *ill* condition, *det(A)* = 0, and the inverse does not exist or may be impossible to evaluate its value with precision, due to numerical computational errors.

The function *rcond(A)*, the reciprocal estimator, provides the same information as *cond(A)* but is less reliable and involves less computations.

The function *rcond(A)* returns the condition of the matrix *A* with a number between one and zero, such that

$$rcond(A) = 1 \ or \ close \ to \ 1, \text{ implies that } A \text{ is } well \text{ condition}$$

and

$$rcond(A) = 0 \ or \ close \ to \ 0, \text{ implies that } A \text{ is } ill \text{ condition}$$

The smaller the value of *rcond* the worse the condition for *A*; meaning that matrix *A* does not possess an inverse or *det(A)* = 0. Some of the matrix concepts presented earlier are illustrated in the next example.

Observe that the command *rank(A)* can also be used to estimate the condition of a matrix. The condition number for *A* is defined as the product of the norm of *A* with A^{-1}.

R.3.82 For example, let *A* = [3 0 0; 0 2 0; 0 0 1] and *B* = [1 4 7; 2 5 8; 3 6 −9].
Use MATLAB and determine:

a. *cond(A)* and *rcond(A)*

b. *det(B)*, *cond(B)*, *rcond(B)*, and *1/rcond(B)*

MATLAB Solution
```
>> A = [3 0 0; 0 2 0; 0 0 1]

   A =
       3        0        0
       0        2        0
       0        0        1

>> [cond(A)  rcond(A)]

   ans =
          3.0000      0.3333

>> B = [1 4 7;2 5 8;3 6 -9]

   B =
       1        4        7
       2        5        8
       3        6       -9

>> det(B)  % det(B) ≠ 0, then the inv(B) exist

   ans =
          54

>> [cond(B)     rcond(B)   1/rcond(B)]

   ans =
       34.2158      0.0163      61.3333
```

R.3.83 The following problem is frequently encountered in science and engineering: Given a set of n linearly independent algebraic equations in terms of n unknowns, obtain a solution to these equations for each of the unknowns.

For example, given the following set of equations: (three equations with three unknowns)

$$x + 2y + z = 0$$
$$2x - y + z = 5$$
$$4x + 2y + 5z = 6$$

Solve for the unknowns x, y, and z.

The preceding set of equations can be expressed in matrix form as

$$\begin{bmatrix} 1 & 2 & 1 \\ 2 & -1 & 1 \\ 4 & 2 & 5 \end{bmatrix} \begin{bmatrix} x \\ y \\ z \end{bmatrix} = \begin{bmatrix} 0 \\ 5 \\ 6 \end{bmatrix}$$

Let

$$\text{matrix } A = \begin{bmatrix} 1 & 2 & 1 \\ 2 & -1 & 1 \\ 4 & 2 & 5 \end{bmatrix}, \text{ vector } B = \begin{bmatrix} 0 \\ 5 \\ 6 \end{bmatrix}, \text{ and vector } V = \begin{bmatrix} x \\ y \\ z \end{bmatrix}$$

Then the matrix equation becomes $A * V = B$.

The existence and uniqueness of a solution for V can be checked if the augmented matrix $C = [A\ B]$, an $(n) \times (n + 1)$ matrix satisfies the relation

$$rank(A) = rank([A\ B]) = n$$

If in contrast $rank(A) = r < n$, then an infinite number of solutions exist for V. Assume that a unique solution for the matrix equation $A * V = B$ exists and can be obtained $(rank(A) = n)$ by applying either of the following MATLAB commands:

a. $V = inv(A) * B$

b. $V = A/B$, or

c. $Gauss = rref(C)$, where $C = [A\ \ B]$, and $V = Gauss\ (:, end)$

The solutions (a) and (b) follow standard matrix algebra operations, whereas solution (c) uses the command *rref*. Recall that this technique is based on the Gaussian elimination method (using the augmented matrix) and the Gauss–Jordan reduction procedure, where the last column of the Gauss matrix provides the solution V.

The backlash command $A\backslash B$ returns the solution vector if A is not singular.

If A is not square, the backlash operation returns a solution using least square approximations.

The following MATLAB program illustrates the steps followed in obtaining the solution for the preceding example.

MATLAB Solution
```
>> format compact
>> A = [1 2 1; 2 -1 1; 4 2 5];          % matrix A
>> B = [0; 5; 6];                        % vector B
>> Solution _ A = inv(A)*B

   Solution _ A =
                    2.0000
                   -1.0000
                    0

>> Solution _ B =A\B

   Solution _ B =
                    2
                   -1
                    0

>> C = [A B];                            % augmented matrix
>> Gauss = rref (C);                     % reduced form
>> Solution _ C = Gauss(:,4),

   Solution _ C =
                    2
                   -1
                    0
```

Note that the solutions (a), (b), and (c) return identical numerical answers.

R.3.84 Let us use the command *rrefmovie(C)* to show the steps involved in obtaining the *rref(C)* for the matrix *C* defined in R.3.83 as follows (used in the last example):

```
MATLAB Solution
>> C = [1 2 1 0;2 -1 1 5;4 2 5 6]

    C =
        1           2           1           0
        2          -1           1           5
        4           2           5           6

>> rrefmovie(C)

        Original matrix
    A =
        1           2           1           0
        2          -1           1           5
        4           2           5           6
Press any key to continue. . .
swap rows 1 and 3
    A =
        4           2           5           6
        2          -1           1           5
        1           2           1           0
Press any key to continue. . .
pivot = A(1,1)
    A =
        1          1/2         5/4         3/2
        2          -1           1           5
        1           2           1           0
Press any key to continue. . .
eliminate in column 1
    A =
        1          1/2         5/4         3/2
        2          -1           1           5
        1           2           1           0
Press any key to continue. . .
    A =
        1          1/2         5/4         3/2
        0          -2         -3/2          2
        0          3/2        -1/4        -3/2
Press any key to continue. . .
pivot = A(2,2)
    A =
        1          1/2         5/4         3/2
        0           1          3/4         -1
        0          3/2        -1/4        -3/2
Press any key to continue. . .
eliminate in column 2
    A =
        1          1/2         5/4         3/2
        0           1          3/4         -1
        0          3/2        -1/4        -3/2
```

```
Press any key to continue. . .
  A =
     1              0              7/8            2
     0              1              3/4           -1
     0              0             -11/8           0
Press any key to continue. . .
  A =
     1              0              0              2
     0              1              0             -1
     0              0              1              0
Press any key to continue. . .
```

Note that the last column of the reduction process represents the solution of the given system of equations.

R.3.85 A set of n equations are linearly dependent (where each equation represents a row of a matrix equation), if any one of the equations can be expressed as a linear combination of any subset of the n equations.

   If the set of n equations are not linearly dependent, then they are linearly independent.

R.3.86 The MATLAB command $B = max(A)$ and $C = min(A)$, where A is an arbitrary $n \times m$ matrix, returns a row vector B or C with m columns in which each element is either the maximum or minimum element of each one of the columns of A.

   For example, let

$$A = \begin{bmatrix} 1 & 3 & 4 \\ 2 & 1 & 5 \\ 3 & 2 & -1 \end{bmatrix}$$

then the command $C = min(A)$ returns $C = [1\ 1\ -1]$ and the command $B = max(A)$ returns $B = [3\ 3\ 5]$.

   For the case of a vector V, the commands $max(V)$ and $min(V)$ return the maximum and minimum value of V, respectively.

   The functions max and min can be used not only to determine the maximum or minimum value but also the location in the array where the maximum and minimum occurs, by employing the format

$$[Vmax, index] = max(V)$$

$$[Vmin, index] = min(V)$$

R.3.87 Given the vector $V = [v_1\ v_2\ \dots\ v_n]$, the command $Y = sum(V)$ returns Y as the sum of all the elements in V, that is, $Y = \sum_{i=1}^{n} v_i$.

R.3.88 For example, let

$$V = [1\quad 3\quad 0.1\quad 8\quad 5\quad 12\quad 13\quad 6\quad 3.9\quad -9\quad 1.3\quad -5.4\quad 0.2\quad 13.8]$$

Execute and observe the responses of the following commands using MATLAB:

a. $Y = sum(V)$

b. $Vmax = max(V)$

c. $[maxv, indexmax] = max(V)$

d. $Vmin = min(V)$

e. $[minv, indexmin] = min(V)$

MATLAB Solution
```
>> Y = sum(V)                    % returns the sum of all the elements
                                   in V

   Y =
       52.900

>> Vmax = max(V)                 % returns the largest element
                                   in V

   Vmax =
           13.8000

>> [maxv, indexmax] = max(V)     % returns the value and location of
                                   the largest element in V
   maxv =
           13.8000
   indexmax =
               14

>> Vmin = min(V)                 % returns the value of the smallest
                                   element in V

   Vmin =
           -9

>> [minv, indexmin] = min(V)     % returns the value and location of
                                   the smallest element in V

   minv =
           -9
   indexmin =
               10
```

R.3.89 Let A be an $n \times m$ matrix, then the command $B = sum(A)$ returns the row vector B consisting of m columns, where the elements of B consist of the sum of each column of A.

For example, let

$$A = \begin{bmatrix} 0 & 1 & 2 \\ 3 & 4 & 5 \\ 6 & 7 & 8 \end{bmatrix}$$

Use MATLAB to evaluate the sum of each column of A.

```
MATLAB Solution
>> A = [0 1 2;3 4 5;6 7 8]

   A =
        0      1      2
        3      4      5
        6      7      8

>> sum(A)
   ans =
        9      12     15
```

R.3.90 Given the vector $V = [v_1 \ v_2 \ \dots \ v_n]$, the command $Y = prod(V)$ returns Y the numerical product of all the elements in V, that is, $Y = \prod_{i=1}^{n}(v_i)$.

Vector V defined in R.3.88 is used through R.3.100 as illustrations. For example,

```
>> prod(V)

   ans =
        7.6385e+006
```

R.3.91 Given the vector $V = [v_1 \ v_2 \ \dots \ v_n]$, the command $Y = sort(V)$ returns the vector Y, consisting of the elements of V arranged in ascending order.

For example,

```
>> sort(V)

   ans =
   Columns 1 through 7
   -9.0000   -5.4000   0.1000   0.2000   1.0000   1.3000   3.0000
   Columns 8 through 14
   3.9000    5.0000    6.0000   8.0000   12.0000  13.0000  13.8000
```

R.3.92 The MATLAB command $[y, z] = sort(V)$ returns two vectors y and z, where y consists of all the elements of V sorted in ascending order and z represents the indexes of the sorted elements of y. For example,

```
>> V =[1   3   0.1   8   5   12   13   6   3.9   −9   1.3   −5.4   0.2   13.8]

   V =
      Columns 1 through 10
      1.0000    3.0000    0.1000    8.0000    5.0000    12.0000   13.0000
      6.0000    3.9000    -9.0000
      Columns 11 through 14
      1.3000    -5.4000   0.2000    13.8000

>> [y,z] = sort(V)

   y =
      Columns 1 through 10
      -9.0000   -5.4000   0.1000    0.2000    1.0000   1.3000    3.0000
       3.9000    5.0000    6.0000
      Columns 11 through 14
      8.0000    12.0000   13.0000   13.8000

   z =
        10    12    3    13    1    11    2    9    5    8    4    6    7    14
```

R.3.93 In MATLAB version 7, or higher, the sort direction can be specified (in quotes) by the following command $[y, z] = sort(V, 'descend')$ or $[y, z] = sort(V, 'ascend')$.

R.3.94 Given the matrix A, the command *sortrows(A)* returns the sorted rows of A in ascending order, according to the elements of the first column of A. For example, let

$$A = \begin{bmatrix} 1 & 2 & 3 & 4 \\ 0 & 3 & -4 & 5 \\ 4 & 7 & 8 & 9 \\ -3 & -5 & 0.7 & 5 \end{bmatrix}$$

Execute the command *sortrows(A)* and observe the response.

```
>> B = sortrows(A)

   ans =
      -3.0000       -5.0000        0.7000        5.0000
       0             3.0000       -4.0000        5.0000
       1.0000        2.0000        3.0000        4.0000
       4.0000        7.0000        8.0000        9.0000
```

R.3.95 The command $B = sort(A(:, :))$ returns the matrix B, consisting of the sorted elements of each of the column of A in ascending order. For example,

```
>> B = sort(A (:, :))

   B =
      -3.0000       -5.0000       -4.0000        4.0000
       0             2.0000        0.7000        5.0000
       1.0000        3.0000        3.0000        5.0000
       4.0000        7.0000        8.0000        9.0000
```

R.3.96 The command $B = sort (A(:, c))$ returns the column vector B, consisting of the sorted elements of column c of A, in ascending order. For example, sorting the second column of the matrix A defined in R.3.94 yields

```
>> B = sort(A(:, 2))     % returns column 2 of A, sorted in ascending
                           order

   B =
       -5
        2
        3
        7
```

Similarly, the command $D = sort (A(r, :))$ returns the row vector D, consisting of the sorted elements of row r of A, in ascending order.

R.3.97 Given the vector $V = [v_1 \ v_2 \ \dots \ v_n]$, the instruction *mean(V)* returns the average value of all elements of V, where $mean(V) = \frac{1}{n} \sum_{i=1}^{n} v_i$.

R.3.98 Similarly, the command *median(V)* returns the median value of the set of elements in *V*.

For example, let $V = [1\ 3\ 0.1\ 8\ 5\ 12\ 13\ 6\ 3.9\ -9\ 1.3\ -5.4\ 0.2\ 13.8]$, then

```
>> Ave = mean(V)              % Average Ave of all the elements in V

    Ave =
          3.7786

>> median _ V = median(V)

    median _ V =
              3.45            % median _ V of the elements in V
```

R.3.99 The command $M = median(A)$, where A is an $n \times m$ matrix, returns the row vector M with m elements, where each element is the *median* value of each of the corresponding column of A. For example, let

$$A = [1\quad 2\quad 3;4\quad -5\quad 6;7\quad 8\quad -9]$$

Execute and observe the response of the command $M = median(A)$.

```
>> A = [1 2 3;4 -5 6;7 8 -9]

    A =
          1      2      3
          4     -5      6
          7      8     -9

>> M = median(A)

    M =
          4      2      3
```

R.3.100 The variance denoted by *Var* and the standard deviation denoted by *Stander* of the elements in vector *V* are defined as

$$Var = G^2 = \sum_{x=1}^{n} \frac{(V(x) - mean(V))^2}{n-1}$$

$$Stander = G = \sum_{x=1}^{n} \left(\frac{(V(x) - mean(V))^2}{n-1} \right)^{1/2}$$

The standard deviation is an index of the scattering of the samples (data) given by the set of elements in the vector *V*.

The greater the standard deviation the greater the spread of the elements in *V* (more scattered).

The MATLAB command *std(V)* returns the standard deviation of the elements in *V*.

For example, the *std(V)* for the *V* defined in R.3.98 is illustrated as follows:

Recall that $V = [1\ 3\ 0.1\ 8\ 5\ 12\ 13\ 6\ 3.9\ -9\ 1.3\ -5.4\ 0.2\ 13.8]$, then

```
>> std(v)

    ans  =
           6.5964                               % standard deviation
```

R.3.101 The command $C = std(A)$, where *A* is a matrix, returns a row vector *C*, with the value of the standard deviation for each column of *A*. For example, let

$$A = \begin{bmatrix} 1 & 1 & 1 & 1 \\ 0 & 3 & -4 & 5 \\ 4 & 7 & 8 & 9 \\ -3 & -5 & 0.7 & 5 \end{bmatrix}$$

Execute the command *std(A)* and observe the response.

```
>> A = [1   1   1   1; 0   3   -4   5; 4   7   8   9; -3   -5   .7   5]

    A =
           1.0000      1.0000      1.0000      1.0000
                0      3.0000     -4.0000      5.0000
           4.0000      7.0000      8.0000      9.0000
          -3.0000     -5.0000      0.7000      5.0000

>> std (A)

    ans =
           2.8868      5.0000      4.9453      3.2660
```

R.3.102 Given the vector $V = [v_1\ v_2\ \dots\ v_n]$, the command *cumsum_V = cumsum(V)* returns the row vector *cumsum_V*, consisting of the cumulative sums of the elements in *V*. For example, let $V = [1\ 3\ 0.1\ 8\ 5\ 12\ 13\ 6\ 3.9\ -9\ 1.3\ -5.4\ 0.2\ 13.8]$, then

```
>> cumsum _ V = cumsum(V)

    cumsum _ V =
           Columns 1 through 7
           1.0000   4.0000   4.1000   12.1000   17.1000   29.1000
           42.1000
           Columns 8 through 14
           48.1000   52.0000   43.0000   44.3000   38.9000   30.1000
           52.9000
```

R.3.103 The command $P = cumprod(V)$ returns a row vector with the cumulative products of the elements in *V*. For example, the command $P = cumprod(V)$ is executed as follows for the vector $V = [1\ 3\ 0.1\ 8\ 5\ 12\ 13\ 6\ 3.9\ -9\ 1.3\ -5.4\ 0.2\ 13.8]$.

```
>> P = cumprod(V)

   P =
       1.0e+006   *
       Columns  1  through   7
       0.0000    0.0000    0.0000     0.0000    0.0000    0.0001    0.0019
       Columns  8   through    14
       0.0112    0.0438    -0.3942    -0.51125  2.7676    0.5535    7.6385
```

R.3.104 The norm of the vector V expressed as $\|V\|$, sometimes referred to as the length of V, is defined by applying the generalized Pythagorean theorem given by

$$\|V\| = \left(\sum_{i=1}^{n} v_i^2\right)^{1/2} = \sqrt{v_1^2 + v_2^2 + \cdots + v_n^2}$$

The MATLAB instruction *norm(V)* returns the values of

[the norm of the vector V] = $sqrt(V^T * V) = sqrt(dot(V, V))$

Recall that the concept of norm was first introduced when the dot product was defined.

R.3.105 A useful and often employed relation in a real inner product is the Cauchy–Schwarz inequality that states

$$[dot(X, Y)]^2 \leq dot(X, X) \cdot dot(Y, Y)$$

or

$$|dot(X, Y)| \leq \|X\| \|Y\|$$

The angle between the vectors X and Y can be defined as

$$cos(\text{angle between } X \text{ and } Y) = dot(X, Y)/(\|X\| \|Y\|)^*$$

Note that X and Y are orthogonal if the angle between them is 90° (or $\pi/2$ radians), implying that

$$cos(90°) = 0 = dot(X, Y)/(norm\ (X*.norm(Y)), \text{ or } dot(X, Y) = 0$$

The principle of an orthogonal set (S) is applied to vectors as well as functions, and constitutes an important concept in a variety of applications in science and engineering, and can easily be tested by using the dot product.

S is an orthogonal set if $dot(X_i,Y_j) = 0$, for any $i \neq j$, then all the vectors are mutually orthogonal and S is linearly independent.

---

* Trigonometric functions such as cosine are treated in Chapter 4.

The norm of an arbitrary vector V is often encountered in mechanical and electrical applications. For example, the root-mean-square (or rms) value of an AC signal with n components can be defined as the norm divided by $\sqrt{n}$.

R.3.106 The MATLAB instruction *abs(A)*, when A is a matrix, returns the matrix A, in which each element of A is replaced by its absolute value. For example,

a. Let

$$A = \begin{bmatrix} 1 & -3 \\ 0 & -2 \end{bmatrix}$$

then

$$abs(A) = \begin{bmatrix} 1 & 3 \\ 0 & 2 \end{bmatrix}$$

(note that A is a real matrix).

b. Let

$$B = \begin{bmatrix} 1 & 1+j \\ 3+2j & -3 \end{bmatrix}$$

then

$$abs(B) = \begin{bmatrix} 1 & 1.4242 \\ 3.6056 & 3 \end{bmatrix}$$

(note that B is a complex* matrix).

R.3.107 The MATLAB command *diag(A)* returns a column vector whose elements are the elements of the main diagonal of the square matrix A. For example, let

$$A = \begin{bmatrix} 1 & 0 & 3 \\ 4 & 3 & 8 \\ 9 & 6 & 5 \end{bmatrix}, \text{ then } diag(A) = \begin{bmatrix} 1 \\ 3 \\ 5 \end{bmatrix}$$

The MATLAB command $D = diag(A, d)$ returns the column vector D, when the diagonal is moved d positions to the right for a positive d, or d positions to the left when d is negative. For example, the command *diag(A, 1)* returns

$$diag(A, 1) = \begin{bmatrix} 0 \\ 8 \end{bmatrix}$$

---

* Observe that $abs(a + jb) = sqrt(a^2 + b^2)$. For example, the $abs(B(2, 1)) = abs(3 + 2j) = sqrt(3^2 + 2^2) = 3.6056$. See Chapter 6 for information regarding complex numbers.

and the command *diag(A, −1)* returns

$$diag(A, \ -1) = \begin{bmatrix} 4 \\ 6 \end{bmatrix}$$

R.3.108 The command $S = diag(V, d)$, where V is a row vector, returns the square matrix S, which consists of the elements of V placed along the diagonal moved d positions to the right of its main diagonal when d is positive, or moved to the left when d is negative, where all the remaining elements of the matrix S are zeros. For example, let

$$V = [1 \quad 2 \quad 3 \quad 4]$$

then

$$A = diag(v, 0)$$

returns the following 4 × 4 matrix:

$$A = \begin{bmatrix} 1 & 0 & 0 & 0 \\ 0 & 2 & 0 & 0 \\ 0 & 0 & 3 & 0 \\ 0 & 0 & 0 & 4 \end{bmatrix}$$

and the command $B = diag(v, 1)$ returns the following 5 × 5 matrix:

$$A = \begin{bmatrix} 0 & 1 & 0 & 0 & 0 \\ 0 & 0 & 2 & 0 & 0 \\ 0 & 0 & 0 & 3 & 0 \\ 0 & 0 & 0 & 0 & 4 \\ 0 & 0 & 0 & 0 & 0 \end{bmatrix}$$

Observe that the command diagonal plays a double purpose illustrated by the following examples:

a. $diag(A) = V$

b. $diag(V) = A$

c. $diag(diag(A)) = A$

R.3.109 The command $C = A(:)$ returns the matrix A as a column vector C, whose elements are arranged as: first column of A, followed by the second column of A, ... followed by the last column of A.

For example, let

$$A = \begin{bmatrix} 1 & 4 & 7 \\ 2 & 5 & 8 \\ 3 & 6 & 9 \end{bmatrix}$$

then the command $B = A(:)$ returns

$$B = \begin{bmatrix} 1 \\ 2 \\ 3 \\ 4 \\ 5 \\ 6 \\ 7 \\ 8 \\ 9 \end{bmatrix}$$

R.3.110 The command *trace(A)* returns the sum of the elements of the main diagonal of A.
For example, the command $B = trace(A)$ returns $B = 9$ for the matrix A defined in R.3.107.

R.3.111 Given the vector $V = [v_1 v_2 \dots v_n]$, the command $Y = diff(V)$ returns the row vector Y consisting of $n - 1$ elements, where each element of $Y(i) = V(i + 1) - V(i)$ for $1 \le i \le n$ arranged in order starting with $Y(1) = V(2) - V(1)$. This command can be used to approximate the numerical derivative.* For example, let

$$V = [1 \quad 3 \quad 0.1 \quad 8 \quad 5 \quad 12 \quad 13 \quad 6 \quad 3.9 \quad -9 \quad 1.3 \quad -5.4 \quad 0.2 \quad 13.8]$$

then

```
>> y = diff(V)

y =
    Columns 1 through 7
    2.0000    -2.9000    7.9000    -3.0000    7.0000    1.0000    -7.0000
    Columns 8 through 13
    -2.10000   -12.9000   10.3000   -6.7000    5.6000    13.6000
```

R.3.112 The instruction *Area = trapz(V)* returns the area under the curve defined by the elements of V (magnitude on the *y*-axis), assuming that the spacing on the *x*-axis is unity. To evaluate the area for different spacing execute the following command:
Area = trapz(x,V), where x and V must be vectors of the same length. For example, let

$$V = [1 \quad 3 \quad 0.1 \quad 8 \quad 5 \quad 12 \quad 13 \quad 6 \quad 3.9 \quad -9 \quad 1.3 \quad -5.4 \quad 0.2 \quad 13.8]$$

```
>> Area = trapz(V)                          % shown in Figure 3.1

    Area =
        45.5000
```

---

* The concept of derivative is presented and discussed in Chapter 7.

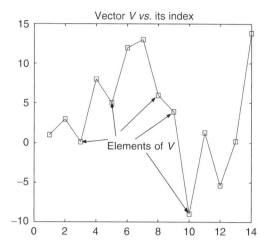

FIGURE 3.1
trapz(V) returns the area under the curve defined by the points over x = 1:14.

Suppose that the x scale is compressed by half as shown in Figure 3.2. What is the new area under the curve?

MATLAB Solution
```
>> x = 1:0.5:7.5
>> Area = trapz(x,V)

   Area =
        22.7500
```

R.3.113 The command *cumtrapz(V)* returns the cumulative area under the points defined by V in R.3.112 assuming a unity spacing.

```
>> cumtrapz(V)

   ans =
        Columns 1 through 7
        0   2.0000 3.5500   7.6000    14.1000   22.6000   35.1000
        Columns 8 through 14
        44.6000    49.5500  47.0000   43.1500   41.1000   38.5000   45.5000
```

R.3.114 Let A be a matrix, then the command *trapz(A)* returns the area of A as defined in R.3.112, for each column of A, and the instruction *cumtrapz(A)* returns the cumulative area of A, for each column of A, illustrated as follows:
For example, let

$$A = \begin{bmatrix} 1 & 2 & 3 & 4 \\ 0 & 3 & -4 & 5 \\ 4 & 7 & 8 & 9 \\ -3 & -5 & -0.7 & 5 \end{bmatrix}$$

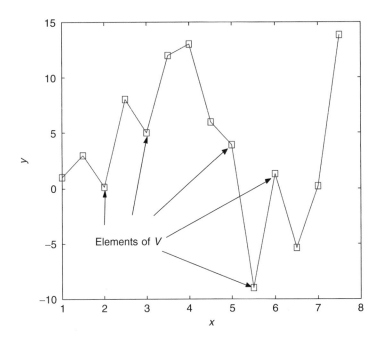

FIGURE 3.2
trapz(x, V) returns the area under the curve defined by vector *V* over *x* = 1:0.5:7.5.

then

```
>> trapz(A)

   ans =
          3.0000      8.5000      5.8500     18.5000

>> cumtrapz(A)

   ans =
              0           0           0           0
         0.5000      2.5000     -0.5000      4.5000
         2.5000      7.5000      1.5000     11.5000
         3.0000      8.5000      5.8500     18.5000
```

R.3.115 The MATLAB command $B = reshape(A, u, v)$, where A is an $n \times m$ matrix, returns the elements of the matrix A rearranged into a new $u \times v$ matrix B, if and only if $n \times m = u \times v$, where the elements of B are the columns of A, arranged in sequential order by columns.

For example, let A be a 4×4 matrix defined as

$$A = \begin{bmatrix} 1 & 5 & -1 & -4 \\ 2 & 6 & -2 & -6 \\ 3 & 7 & -3 & -7 \\ 4 & 8 & -4 & -8 \end{bmatrix}$$

Then use MATLAB to reshape *A* into

   a. A 2 by 6 matrix

   b. A 2 by 8 matrix

   c. An 8 by 2 matrix

Note that $B = reshape(A, 2, 8)$ will return a 2×8 matrix *B* and $C = reshape(A, 8, 2)$ will return an 8×2 matrix *C*, but *reshape(A, 2, 6)* will return an error message, as indicated below.

MATLAB Solution

```
>> A= [1  5  -1  -4;2  6  -2  -6;3  7  -3  -7;4  8  -4  -8]

   A =
      1      5     -1     -4
      2      6     -2     -6
      3      7     -3     -7
      4      8     -4     -8

>> B = reshape(A,2,6)

??? Error using ==> reshape
To RESHAPE the number of elements          % note that 4x4 ≠ 2x6
must not change.

>> B = reshape(A,2,8)

   B =
      1      3      5      7     -1     -3     -4     -7
      2      4      6      8     -2     -4     -6     -8

>> C = reshape(A,8,2)

   C =
      1     -1
      2     -2
      3     -3
      4     -4
      5     -4
      6     -6
      7     -7
      8     -8
```

R.3.116 The MATLAB command $B = rot90(A)$ returns matrix *B*, which consists of elements of the matrix *A* rotated 90° following the algorithm: the first column of *A* becomes the last row of matrix *B*, the second column of *A* becomes the next to the last row of *B*, and so on, and the last column of *A* becomes the first row of *B* indicated as follows (using matrix *A* defined in R.3.115).

```
>> B = rot90(A)

   B =
     -4     -6     -7     -8
     -1     -2     -3     -4
      5      6      7      8
      1      2      3      4
```

R.3.117 The MATLAB command $D = fliplr(A)$ returns matrix D, with the columns of A flipped from left to right, as indicated by the following example:

```
>> C = fliplr(A)    % where A was defined in R.3.115

   C =
        -4       -1       5       1
        -6       -2       6       2
        -7       -3       7       3
        -8       -4       8       4
```

R.3.118 The command $E = flipud(A)$ returns matrix E, which consists of the rows of A interchanged, following the algorithm: the last raw of A becomes the first row of E, the row next to the last row of A becomes the second row of E, and so on, and the first row of A becomes the last row of E, illustrated as follows:
        For example,

```
>> E = flipud(A)                          % where A was defined in R.3.115

   E =
        4        8       -4      -8
        3        7       -3      -7
        2        6       -2      -6
        1        5       -1      -4
```

R.3.119 The MATLAB command $F = find(A)$ returns the column vector F consisting of the indexes of all nonzero elements of A. For example, using A defined in R.3.115

```
>> F = find(A)

   F =
        1
        2
        3
        4
        5
        6
        7
        8
        9
       10
       11
       12
       13
       14
       15
       16
```

Since A has no zero value elements, consider now a simpler example with a new matrix A (with a zero element) defined as follows:

$$\text{Let } A = \begin{bmatrix} 0 & 1 \\ 2 & 3 \end{bmatrix}$$

then $F = find(A)$ returns

$$F = \begin{bmatrix} 2 \\ 3 \\ 4 \end{bmatrix}$$

the indexes of the nonzero elements in A.

R.3.120 The MATLAB command $G = exp(A)$ returns matrix G, whose elements are $exp = 2.71$ raised to each element of A, respectively, illustrated, for the following case:

$$\text{Let } A = \begin{bmatrix} 0 & 1 \\ 2 & 3 \end{bmatrix}$$

then

```
>> G = exp(A)

   G =
        1.00           2.7183
        7.3891         20.0855
```

R.3.121 The MATLAB command $H = expm(a)$ returns the matrix H, whose elements are evaluated using the following power series:

$$e^a = I + a + \frac{a^2}{2!} + \frac{a^3}{3!} + \cdots + \frac{a^n}{n!}$$

For example, using matrix $A = \begin{bmatrix} 0 & 1 \\ 2 & 3 \end{bmatrix}$, let's execute the following command $H = expm(A)$ and observe its response.

```
>> H = expm(A)

   H =
        5.2892          8.4033
       16.8065         30.4990
```

R.3.122 The command $J = log(A)$ returns the matrix J, where each element of J is the logarithm of the corresponding element of A.

For example,

```
>> J = log(A)                              %  for A defined in R.3.121


    J =
        -Inf              0
        0.6931           1.0986
```

R.3.123 The MATLAB command $K = logm(A)$ returns the matrix K, whose elements satisfy the matrix relation $e^K = A$, illustrated as follows for matrix A defined in R.3.121.

```
>> K= logm (A)


    K =
        -0.3255    +    2.7137i      0.4480    -    0.7619i
         0.8960    -    1.5239i      1.0186    +    0.4279i
```

R.3.124 The MATLAB command $L = sqrt(A)$ returns the matrix L consisting of the square root of each element of A. For example, let

$$A = \begin{bmatrix} 0 & 1 \\ 2 & 3 \end{bmatrix}$$

then

```
>> L = sqrt(A)


    L =
        0                 1.0000
        1.4142            1.7321
```

R.3.125 The MATLAB command $M = sqrtm(A)$ returns the matrix M, whose elements satisfy the matrix product $A = M * M$. For example, using the matrix A defined in R.3.124

```
>> M = sqrtm(A)


    M =
        0.2570    +    0.6473i      0.4577    -    0.1817i
        0.9154    -    0.3635i      1.6302    +    0.1021i

>> CHECK = M*M                             %  checks if M is correct


    CHECK =
            0.0000 - 0.0000i         1.0000
        ≠   2.0000 - 0.0000i         3.0000  -0.0000i
```

R.3.126 The terms eigenvalues and eigenvectors came up often in many branches of science and engineering when solving the following matrix equation: $Ax = \lambda x$, where A is an $n \times n$ matrix, λ is an unknown scalar, and x is an unknown column vector consisting of *n* elements. An alternate of the preceding equation can be written as $A - \lambda Ix = 0$, where the nonzero values of λ, which satisfy the equation $A - \lambda I = 0$, are called as eigenvectors and the *x*s their eigenvalues, by mathematicians. The eigenproblem represents an homogeneous system of equations $(A - \lambda I)x = 0$, where the solution is nontrivial $x \neq 0$. A square homogeneous system has a nontrivial solution, if and only if $det(A - \lambda I) \neq 0$. The procedure of computing eigenvalues and eigenvectors is long and complex when done by hand, but relatively simple with MATLAB. In fact the original purpose of MATLAB, back in the 1980s, was to build a software capable to solve the eigenproblem. Eigenvalues and eigenvectors are often encountered in different disciplines of the physical sciences and engineering such as controls, servos, linear systems, electronics, and communications.

For example, eigenvalues may represent the vibration in a mechanical system, or the frequencies of oscillations in some electrical systems.

R.3.127 The evaluation of the determinant $A - \lambda I$ yields an *n*-degree polynomial in λ (Hill and Zitarelli, 1996), called the characteristic polynomial of the matrix *A*, and its zeros or roots represent the λs. The eigenvectors of the characteristic equation are the solutions of the homogeneous system $(A - \lambda I)x = 0$.

For example, let

$$A = \begin{bmatrix} 1 & 2 & 3 \\ 2 & 3 & 4 \\ 5 & 6 & 8 \end{bmatrix}$$

Write a program that returns the characteristic polynomial* as well as its eigenvalues.

MATLAB Solution
```
>> A = [1 2 3;2 3 4; 5 6 8]

   A =
        1      2      3
        2      3      4
        5      6      8

>> poly _ A = poly(A)

   poly _ A =
        1.0000   -12.0000    -8.0000    1.0000

>> eigenvalues _ A = roots(poly _ A)

   eigenvalues _ A =
                    12.6273
                    -0.7350
                     0.1077
```

---

* Polynomials are presented in Chapter 7. At this point for example, it is sufficient to know that $Y = 3X^3 + 4X^2 + 5X + 6$ is a polynomial that can be represented using MATLAB as a row vectors using its coefficients as elements such as $Y = [3 \quad 4 \quad 5 \quad 6]$.

Observe that the function *poly(A)* returns the coefficients of the characteristic polynomial. The polynomial is then $poly\_A(\lambda) = \lambda^3 - 12\lambda^2 - 8\lambda + 1$.

The roots or zeros are the eigenvalues of the characteristic equation that can be obtained by using the function *roots*.*

R.3.128 An alternate way to evaluate the eigenvalues of the characteristic equation as well as the characteristic equation is by using the symbolic functions: *poly(sym(A)), factor* or *solve* illustrated as follows:

```
>> poly_A = poly(sym(A))

   poly_A =
              x^3-12*x^2-8*x+1


>> factor (poly_A)

          (x- 12.6273)*(x+0.7350)*(x- 0.1077)


>> solve (poly_A)

     ans
        [12.6273   ]
        [-0.7350   ]
        [ 0.1077   ]
```

R.3.129 The MATLAB function *eig(A)* returns the eigenvalues or/and eigenvectors of the matrix *A*. The *eig* function can take different forms. The most frequently used are

a. *eigval = eig(A)*

b. *[eigvec, eigval] = eig(A)*

where *eigval* is a vector consisting of the eigenvalues of *A*.

Form (b) returns a diagonal matrix with the eigenvalues on the main diagonal and *eigvec* that is a matrix consisting of all the eigenvectors of *A* as column vectors.

R.3.130 The MATLAB functions *eigvec = eigs(A)* and *[eigvec, eigval] = eigs(A)* return a few selected eigenvalues, or eigenvectors of *A*.

R.3.131 Some observations about the eigensolutions of the system matrix *A*.

a. If *A* is a real symmetric matrix ($A^T = A$), its eigenvalues and eigenvectors are real (positive or negative), and the eigenvalues corresponding to distinct eigenvalues are orthogonal.

b. If $A^T = -A$, then the eigenvalues are imaginary or zero.

c. If $A^T = A^{-1}$, all eigenvalues have unity magnitude.

d. If *A* is a real nonsymmetric matrix, its eigenvalues and eigenvectors are either real or appear in complex conjugate pairs.

e. The sum of the eigenvalues of *A* is equal to the *trace(A)*. Recall that *trace(A)* returns the sum of all the elements of the main diagonal of *A*.

---

* See Chapter 7 for additional details.

f. The product of the eigenvalues of *A* is equal to the *det(A)*.

g. Matrix *A* is singular if one of its eigenvalues is zero.

h. If matrix *A* is a triangular matrix, then its eigenvalues are its diagonal elements.

i. Eigenvalues and eigenvectors can be complex, even if the matrix *A* is real.

j. If *A* is Hermitian, then its eigenvalues are real and its eigenvectors are orthogonal.

R.3.132 For example, write a program that returns the eigenvalues and eigenvectors for the following system matrix:

$$A = \begin{bmatrix} 1 & 3 & 5 \\ 2 & 3 & 6 \\ 3 & 4 & 8 \end{bmatrix}$$

MATLAB Solution

```
>> A = [1 3 5;2 3 6;3 4 8];
>> [eigvec,eigval] = eig(A)

   eigvec =
           -0.4418      0.9305       0.5031
           -0.5327      0.0779       0.6869
           -0.7218     -0.3579      -0.5245
   eigval =
           12.7881      0            0
           0           -0.6717       0
           0            0           -0.1164
```

The results indicate that the eigenvalue: $\lambda = -0.1164$, that is, located on the third column of *eigval*, is associated with the eigenvector defined by the third column of the *eigvec* matrix. Then,

$$x = \begin{bmatrix} 0.5031 \\ 0.6869 \\ -0.5245 \end{bmatrix}$$

R.3.133 The eigenvectors associated to a system matrix *A* are not unique. Furthermore, the eigenvectors obtained by using the *eig* command have the property that they are normalized, a statement that is verified by using the matrix defined in R.3.132 as example.

MATLAB Solution

```
>> A = [1 3 5;2 3 6;3 4 8];
>> [eigvec,eigval]=eig(A);
>> norm(eigvec(:,1))
```

```
    ans =
          1.0000

>> norm(eigvec(:,2))

    ans =
          1.0000

>> norm(eigvec(:,3))

    ans =
          1
```

Observe also that any scalar that multiplies an eigenvector is still an eigenvector.

R.3.134 Let *A* be a random matrix of order 5 that can be created by the command *A = rand* (5). Create the MATLAB *script file: eig_val_vec,* that

a. Creates *A*

b. Solves the system matrix equation $Ax = \lambda x$

c. Verifies the solutions obtained

MATLAB Solution
```
% Script file:eig _ val _ vec
n = 5;
disp('*****************************************')
disp(' Matrix A is given by:')
A = rand(5)
[V,D] = eig(A);
check(1) = sum(A*V(:,1)-D(1,1)*V(:,1));
check(2) =sum(A*V(:,2)-D(2,2)*V(:,2));
check(3) = sum(A*V(:,3)-D(3,3)*V(:,3));
check(4) = sum(A*V(:,4)-D(4,4)*V(:,4));
check(5) = sum(A*V(:,5)-D(5,5)*V(:,5));
disp('*****************************************')
disp('The eigenvalues of A are :')
disp(D)
disp('*****************************************')
disp(' The eigenvectors of A are :')
disp(V)
disp('*****************************************')
disp(' Checks the solutions by using the characteristic equation')
disp(check)
disp('*****************************************')
```

The script file *eig_val_vec* is executed and the results are shown as follows:

```
>> eig _ val _ vec

*****************************************
 Matrix A is given by:
  A =
       0.1934   0.6979   0.4966   0.6602   0.7271
       0.6822   0.3784   0.8998   0.3420   0.3093
       0.3028   0.8600   0.8216   0.2897   0.8385
```

```
        0.5417  0.8537  0.6449  0.3412  0.5681
        0.1509  0.5936  0.8180  0.5341  0.3704
*********************************************

The eigenvalues of A are :
Columns 1 through 4
 2.7925          0                0                  0
    0      -0.2315 + 0.2215i       0                  0
    0            0         -0.2315 - 0.2215i         0
    0            0                0               0.2850
    0            0                0                  0
Column 5
    0
    0
    0
    0
 -0.5094

*********************************************

The eigenvectors of A are :
Columns 1 through 4
 0.4402     -0.3071 + 0.3661i    -0.3071 - 0.3661i    0.5499
 0.4262      0.6267 - 0.0339i     0.6267 + 0.0339i    0.1874
 0.4935     -0.0007 + 0.0187i    -0.0007 - 0.0187i   -0.5687
 0.4686      0.0051 - 0.2963i     0.0051 + 0.2963i    0.5415
 0.4018     -0.5377 - 0.0187  i   -0.5377 + 0.0187i   -0.2140
Column 5
 -0.2797
  0.7496
 -0.4570
 -0.3460
  0.1771
*********************************************
Checks the solutions by using the characteristic equation

 1.0e-014 *
Columns 1 through 4
 -0.4663  0.4534 - 0.1883i  0.4534 + 0.1883i  -0.1124
Column 5
 0.2193
*********************************************
```

R.3.135 The null space of matrix A, denoted in MATLAB as $Z = null(A)$, is the orthonormal vector Z, such that $A * Z = 0$. The command $Z = null(A, 'r')$, where r stands for the rational basis for the null space is obtained from the reduced row echelon form of the matrix equation $A * Z = 0$. For example, execute $Z = null(A)$ and $Z = null(A, 'r')$, for the matrix

$$A = \begin{bmatrix} 1 & 2 & 3 \\ 1 & 2 & 3 \\ 1 & 2 & 3 \end{bmatrix}$$

and observe the MATLAB responses.

```
MATLAB Solution
>> A = [1 2 3;1 2 3;1 2 3]; then
>> Z = null(A)

    Z =
            -0.1690      -0.9487
             0.8452       0.0000
            -0.5071       0.3162

>> ZZ = null (A,'r')

    ZZ =
            -2   -3
             1    0
             0    1
```

R.3.136 A string vector can be created by typing (entering) characters within quotes.
For example, create the *stringvee* consisting of *this is a string*.

```
>> stringvec = 'this is a string'

    stringvec =
                this is a string
```

stringvec is a 16-character row string, where spaces are valid characters.

The commands *length(stringvec)*, or *size(stringvec)* can be used to determine the number of elements in *stringvec*. For example,

```
>> size(stringvec)

    ans =
            1    16
```

meaning that *stringvee* is a row vector consisting of 16 characters.

R.3.137 String matrices can be created, as long as each row of the matrix contains exactly the same number of characters. For example,

a. Create the string matrix $A = \begin{bmatrix} big \\ red \\ car \end{bmatrix}$, and

b. Check the size of A

```
MATLAB Solution
>> A = ['big';   'red';  'car']

    A =
        big
        red
        car

>> size(A)

    ans =
            3    3
```

Observe that A is then a 3×3 string matrix.

R.3.138 If the rows of a string matrix do not have the same number of characters, MATLAB returns an error message. For example, let

$$B = \begin{bmatrix} blue \\ red \\ green \end{bmatrix}$$

Then when entered and executed, MATLAB returns an error message indicated as follows:

```
>> B = [ 'blue' ; 'red' ; 'green']

??? All rows in the bracketed expression must have
the same number of columns.
```

R.3.139 When the rows of a string matrix do not have the same number of characters, blanks can be inserted with the objective of creating a compatible string matrix having the same number of characters per row. For example, convert matrix *B* defined in R.3.138 into a string matrix.

```
>> B = ['blue   '; 'red   ' ; 'green' ']

B =
    blue
    red
    green
```

R.3.140 The MATLAB function *C = char(string_1, string_2, ..., string_n)* automatically inserts spaces where and when required in each *string*. For example, create the following string matrix:

$$C = \begin{bmatrix} blue \\ red \\ green \\ black \quad and \quad white \end{bmatrix}$$

```
MATLAB Solution
>> b = 'blue';
>> r = 'red';
>> g = 'green';
>> bw = 'black and white';
>> C = char(b,r,g,bw)

C =
    blue
    red
    green
    black and white
```

R.3.141 The individual elements or sequence of elements in a string matrix can be identified following the rules defined earlier in this chapter for the case of numerical
matrices. Using as example the matrix defined in R.3.140, write a set of commands
to obtain from matrix *C* the following:

  a. The third row

  b. The second character of the third row

  c. The seventh to ninth character of the fourth row

  d. The fourth character of the second row

MATLAB Solution
```
>> C(3,:)

    ans =
          green

>> C(3,2)

    ans =
          r

>> C(4,7:9)

    ans =
          and

>> C(2,4)          % returns a blank character

    ans =
```

R.3.142 Cell arrays can be created by using curly brackets indicated by {}, when the rows
or columns consist of string (defined in quotes). For example, create the following
cell array:

$$C = \begin{bmatrix} red \\ blue \\ green \\ yellow \end{bmatrix}$$

MATLAB Solution
```
>> D = {'red';'blue';'green';'yellow'}

    D =
         'red'
         'blue'
         'green'
         'yellow'
```

R.3.143 The rows of the cell array are identified by a numerical argument in brackets.
Once a row is defined then the elements of a row can be identified.
  For example, identify the second row and the fourth character of the second row
of matrix *D* defined in R.3.142.

```
MATLAB Solution
>> second = D(2)   % observe that blue is returned in quotes

    second =
            'blue'

>> sec = D{2}      % observe that blue is returned without quotes

    sec =
          blue

>> sec (4)            % observe that individual character can be iden-
                        tified using standard techniques

    ans =
          e
```

R.3.144 The command *double(stringvec)* or *abs(stringvec)* converts the characters of the *stringvec* into the ASCII code defined in Table 3.1.

For example, create the *stringvec* = *'this is a string'* and encode the result into ASCII.

```
MATLAB Solution
>> stringvec = 'this is a string'

    stringvec =
              this is a string

  >> ASCII = double(stringvec)
    ASCII =
            Columns 1 through 12
            116   104 105 115  32 105 115  32  97  32  115  116
            Columns 13 through 16
            114 105 110 103
```

Analyzing the preceding response, it can be observed that the character code *32*, in columns 5, 8, and 10 is the ASCII character for space (Table 3.1). Observe also that *stringvec* has three spaces.

R.3.145 String arrays can be manipulated like ordinary numerical arrays, defined earlier in this chapter. For example, observe that the sequence *string* can be filtered out from the string array *stringvec*, by executing the following command:

```
>> filter = stringvec(10:16)      % stringvec was defined in R.3.144
    filter =
            string
```

R.3.146 The command *char(ASCII)* converts the string ASCII from the ASCII code, back to English characters, indicated as follows

```
>> char (ASCII)                   % ASCII was defined in R.3.144
    ans =
          this is a string
```

TABLE 3.1

American Standard Code for Information Interchange

0	NUL	32	[space]	64	@	96	'
1	SOH	33	!	65	A	97	a
2	STX	34	>>	66	B	98	b
3	ETX	35	#	67	C	99	c
4	EOT	36	$	68	D	100	d
5	ENQ	37	%	69	E	101	e
6	ACK	38	&	70	F	102	f
7	BEL	39	'	71	G	103	g
8	BS	40	(	72	H	104	h
9	HT	41	)	73	I	105	i
10	LF	42	*	74	J	106	j
11	VT	43	+	75	K	107	k
12	FF	44	,	76	L	108	l
13	CR	45	–	77	M	109	m
14	SO	46	.	78	N	110	n
15	SI	47	/	79	O	111	o
16	DLE	48	0	80	P	112	p
17	DC1	49	1	81	Q	113	q
18	DC2	50	2	82	R	114	r
19	DC3	51	3	83	S	115	s
20	DC4	52	4	84	T	116	t
21	NAK	53	5	85	U	117	u
22	SYN	54	6	86	V	118	v
23	ETB	55	7	87	W	119	w
24	CAN	56	8	88	X	120	x
25	EM	57	9	89	Y	121	y
26	SUB	58	:	90	Z	122	z
27	ESC	59	;	91	[	123	{
28	FS	60	<	92	\	124		
29	GS	61	=	93	]	125	}
30	RS	62	>	94	^	126	~
31	US	63	?	95	-	127	DEL

Note: NUL, Null/Idle; SOH, start of heading; STX, start of text; ETX, end of text; EOT, end of transmission; ENQ, enquiry; ACK, acknowledge; BEL, audible or attention signal; BS, backspace; HT, horizontal tabulation; LF, line feed; VT, vertical tabulation; FF, form feed; CR, carriage return; SO, shift out; SI, shift in; DLE, data link escape; DC1, DC2, DC3, DC4, special device control codes; NAK, negative acknowledge; SYN, synchronous idle; ETB, end of transmission block; CAN, cancel; EM, end of medium; ESC, escape; FS, file separator; GS, group separator; RS, record separator; US, unit separator; and DEL, delete.

R.3.147 Some additional useful string commands are summarized in Table 3.2.

R.3.148 The MATLAB command *numb = dec2base(D, B)*, defined in Table 3.2, returns *numb*, the decimal number D converted to base B as a string.

R.3.149 The MATLAB command *numbase = base2dec(N, B)* returns *numbase*, the number N expressed in base B, converted into its decimal equivalent, where B must be an integer between 2 and 36. The following examples illustrates some conversions.

R.3.150 Use MATLAB and perform the following conversions:

a. $a = 13578_{10}$ to binary, and assign to variable (b)

b. Convert back (b) to decimal, and assign to variable (c)

c. Convert a to hexadecimal, and assign to variable (d)

TABLE 3.2

Additional String Commands

Instruction	Description
Bin2dec(x)	Converts string x from binary to decimal
Dec2bin(x)	Converts string x from decimal to binary
Dec2hex(x)	Converts string x from decimal to hexadecimal
Hex2dec(x)	Converts string x from hexadecimal to decimal
Dec2base(x, y)	Converts string x (decimal) to base y
base2dec(x, y)	Converts string x (base y) to decimal
num2str	Converts number to string
Str2num	Converts string to number
Findstr(x1, x2)	Finds one string within another
Strvcat(x1, x2, x3, …)	Vertical string concatenation of x1, x2, x3
Strcat(x1, x2, x3, …)	Horizontal string concatenation of x1, x2, x3

d. Convert d back from hexadecimal to decimal, and assign variable e

e. Convert a into the following bases: 7, 20, and 36, by assigning variables f, g, and h

f. Convert h back to decimal from base 36, by assigning variable k

MATLAB Solution

```
>> b = dec2bin(13578)     % conversion of 13578 (decimal) to binary

   b =
      11010100001010

>> c = bin2dec(a)         % conversion from binary to decimal

   c =
      13578

>> d = dec2hex(13578)     % conversion of 13578 (decimal) to hex

   d =
      350A

>> e = hex2dec(c)         % conversion from hex. to decimal

   e =
      13578

>> f = dec2base(13578,7)  % conversion from decimal to base 7

   f =
      54405

>> g = dec2base(13578,20) % conversion from decimal to base 20

   g =
      1DII
```

```
>> h = dec2base(d,36)          % conversion from decimal to base 36

h =
     AH6

>> k = base2dec(h,36)          % conversion from base 36 to decimal

k =
     13578
```

R.3.151 A square matrix A can be factored into a lower (L) and an upper (U) triangular matrices, when possible, by using the command $[L, U] = lu(A)$, where $L * U = A$.

R.3.152 For example, factor matrix A, into an upper (U) and a lower (L) triangular matrix and verify its decomposition for

$$A = \begin{bmatrix} 0 & 1 & 2 \\ 3 & 4 & 5 \\ 6 & 7 & 8 \end{bmatrix}$$

MATLAB Solution
```
>> A = [0 1 2;3 4 5;6 7 8]

A =
     0   1   2
     3   4   5
     6   7   8

>> [L,U] = lu(A)

L =
          0        1.0000        0
     0.5000        0.5000   1.0000
     1.0000        0             0
U =
     6   7   8
     0   1   2
     0   0   0

>> L*U                    % check the factorization

ans =
          0   1   2
          3   4   5
          6   7   8
```

R.3.153 The MATLAB command $[L, U, P] = lu(A)$ returns three matrices L, U, and P such that $L * U = P * A$, where L is the lower triangular matrix and U is its upper triangular matrix.

R.3.154 For example, let

$$A = \begin{bmatrix} 0 & 1 & 2 \\ 3 & 4 & 5 \\ 6 & 7 & 8 \end{bmatrix}$$

Use the command $[L, U, P] = lu(A)$ and verify that $L * U = P * A$.

MATLAB Solution
```
>> [L,U,P] = lu(A)

   L =
       1.0000             0             0
       0             1.0000             0
       0.5000        0.5000        1.0000
   U =
       6    7    8
       0    1    2
       0    0    0
   P =
       0    0    1
       1    0    0
       0    1    0

>> L*U                      % verify results

   ans =
       6    7    8
       0    1    2
       3    4    5

>> P*A                      % verify results

   ans =
       6    7    8
       0    1    2
       3    4    5
```

R.3.155 The command $B = triu(A)$ returns matrix B, whose elements above the main diagonal are the elements of A, and the elements below the main diagonal are replaced by zeros. The command $C = triu(A, d)$ returns the matrix C, whose elements below the main diagonal moved d positions to the right for positive d are replaced by zeros (for a negative d, the main diagonal is moved d positions to the left), and the remaining elements are the elements of A.

R.3.156 The command $D = tril(A)$ returns the matrix D with size of matrix A, consisting of the lower triangular part of A, and the elements above the main diagonal are replaced by zeros. The main diagonal can be shifted to the right and left, as presented for the case of $triu(A)$ further controlling the elements above and below the diagonal.

R.3.157 For example, let

$$A = \begin{bmatrix} 1 & 2 & 3 \\ 4 & 5 & 6 \\ 7 & 8 & 9 \end{bmatrix}$$

Execute and observe the responses of the following commands:

a. *B* = *triu(A, 1)*

b. *C* = *triu(A. –1)*

c. *D* = *tril(A)*

d. *E* = *tril(A, 1)*

e. *F* = *tril(A, –1)*

MATLAB Solution
```
>> A =  [1 2 3; 4 5 6; 7 8 9]

   A =
       1   2   3
       4   5   6
       7   8   9

>> B = triu(A,1)

   B =
       0   2   3
       0   0   6
       0   0   0

>> C = triu(A,-1)

   C =
       1   2   3
       4   5   6
       0   8   9
```

The execution of parts (c), (d), and (e) is left as an exercise to the reader (similar to parts (a) and (b)).

R.3.158 The command *[ort, U]* = *qr(A)* returns the orthogonal matrix *ort*, and the upper triangular matrix *U*, such that *ort * U = A*.

R.3.159 The command *[ort1, diag, ort2]* = *svd(A)* (single value decomposition) factors the matrix *A* into two orthogonal matrices: *ort1* and *ort2*, and the diagonal matrix *diag* such that

$$ort1 * diag * ort2 = A$$

R.3.160 For example, let

$$A = \begin{bmatrix} 1 & 2 & 3 \\ 4 & 5 & 6 \\ 7 & 8 & 9 \end{bmatrix}$$

Execute and observe the responses of the following commands:

a. *[ort, tri]* = *qr(A)*

b. *B* = *ort * tri* (checks the decomposition of part (a))

c. *[ort1, diag, ort2]* = *svd(A)*

d. *C* = *ort1 * diag * ort2'* (checks the decomposition of part (c))

```
MATLAB Solution
>> A = [1 2 3;4 5 6;7 8 9];
>> [ort,tri] = qr(A)

    ort =
            -0.1231   0.9045   0.4082
            -0.4924   0.3015  -0.8165
            -0.8616  -0.3015   0.4082
    tri =
            -8.1240   -9.6011   -11.0782
             0         0.9045     1.8091
             0         0         -0.0000

>> B = ort*tri                    % verify

    B =
            1.0000   2.0000   3.0000
            4.0000   5.0000   6.0000
            7.0000   8.0000   9.0000

>> [ort1,diag,ort2]=svd(A)

    ort1=
            0.2148   0.8872  -0.4082
            0.5206   0.2496   0.8165
            0.8263  -0.3879  -0.4082
    diag =
            16.8481     0         0
             0          1.0684    0
             0          0         0.0000
    ort2=
            0.4797  -0.7767   0.4082
            0.5724  -0.0757  -0.8165
            0.6651   0.6253   0.4082

>> C = ort1*diag*ort2             % verify

    C =
            1.0000   2.0000   3.0000
            4.0000   5.0000   6.0000
            7.0000   8.0000   9.0000
```

R.3.161 The command *maxi = max(X, Y)*, where *X* and *Y* are arbitrary matrices of the same size, returns the matrix *maxi*, where each element *maxi(i, j)* is the maximum of $X(i, j)$, $Y(i, j)$, for all *i*'s and *j*'s.

R.3.162 The command *mini = min(X, Y)*, where *X* and *Y* are arbitrary matrices of the same size, returns the matrix *mini*, where each element *mini (i, j)* is the minimum of $X(i, j)$, $Y(i, j)$, for all possible *i*s and *j*s.

R.3.163 The following example illustrates the action of the commands *max(X, Y)* and *min(X, Y)* for the matrices *X* and *Y* defined as follows:

Let

$$X = \begin{bmatrix} 1 & 3 & 5 & 7 \\ 2 & 4 & 6 & 8 \end{bmatrix} \text{ and } Y = \begin{bmatrix} 1 & 2 & 3 & 4 \\ 5 & 6 & 7 & 8 \end{bmatrix}$$

MATLAB Solution
```
>> X= [1 3 5 7;2 4 6 8]

   X =
       1   3   5   7
       2   4   6   8

>> Y = [1 2 3 4;5 6 7 8]

   Y =
       1   2   3   4
       5   6   7   8

>> mini = min(X,Y)

   mini =
           1   2   3   4
           2   4   6   8

>> maxi = max(X,Y)

   maxi =
           1   3   5   7
           5   6   7   8
```

R.3.164 The command $B = A(V1, :)$ returns B consisting of the rows of matrix A permuted following the indexing of vector $V1$. Similarly, the command $C = A(:, V2)$ returns the matrix C consisting of the columns of matrix A permuted following the indexing set by vector $V2$. Observe that if A is an $m \times n$ matrix, then $length(V1) = m$ and $length(V2) = n$.

R.3.165 For example, let

$$A = \begin{bmatrix} 1 & 2 & 3 \\ 4 & 5 & 6 \\ 7 & 8 & 9 \\ 10 & 11 & 12 \\ 13 & 14 & 15 \end{bmatrix}, V1 = [2 \ 1 \ 4 \ 5 \ 3], \text{ and } V2 = [3 \ 2 \ 1]$$

Execute and observe the responses of the following commands:

a. $B = A(V1, :)$
b. $C = A(:, V2)$

MATLAB Solution
```
>> A = [1 2 3;4 5 6;7 8 9;10 11 12;13 14 15]

   A =
           1       2       3
           4       5       6
           7       8       9
          10      11      12
          13      14      15
```

```
>> V1 = [2 1 4 5 3]

   V1 =
       2   1   4   5   3

>> B = A(V1,:)              % observe that the first row of B is the
                             second row of A, ...., etc.

   B =
           4       5       6
           1       2       3
          10      11      12
          13      14      15
           7       8       9

>> V2 = [3 2 1]            % observe that the first column of C is
                            the 3rd. column  of A, ..., etc

   V2 =
       3   2   1

>> C = A(:,V2)

   C =
           3       2       1
           6       5       4
           9       8       7
          12      11      10
          15      14      13
```

R.3.166 MATLAB has a number of built-in special functions that are used to create either specific or special matrices* that occur frequently in matrix manipulations.

Some of these matrices are magic, eye, ones, zeros, hilbert, pascal, vender, rand, and randn.

For a list of the special matrices available in MATLAB, use the help command followed by *specmat*.

R.3.167 The command *magic(n)*, where *n* is an integer smaller than 32, returns an $n \times n$ matrix with elements that constitute a magic square.[†] A magic matrix is a special square matrix where the sums along each diagonal, columns, or rows return the same constant.

The number of rows (or the number of columns) constitutes the order of the magic matrix. Since their discovery, magic squares have been the source of many mathematical puzzles and games. Also as a result of the research done on magic

---

* There are over 50 special matrices available in MATLAB. For a complete list type *help gallery*.

† Magic squares have been known from antiquity. The first magic square appeared on the back of a tortoise which was discovered by the Chinese Emperor Yu around 2200 BC. In ancient India people used to wear stone or metal ornaments engraved with arrays forming magic squares. According to the Jewish Kabbalah teaching, a specific magic square, called a *kameas*, was associated with each of the following planets: Saturn, Jupiter, Mars, Venus, and Mercury, in addition to the sun and moon (from a 3 × 3 for Saturn to a 9 × 9 matrix for the moon). This is probably the reason why ancient civilization believed that certain planets possess power to influence human events. In old Persia, magicians were also doctors that employed magic squares for medical purposes.

squares, a number of useful mathematical concepts have been discovered, although no special powers are attributed to them, but it cannot be denied their important role in certain religions and cultures.

MATLAB returns magic squares of at least order 3. There is no magic square of order 2. The sum obtained by adding the rows, columns, or diagonals is referred to as the constant of the square. Note that the total number of sums needed to verify if a matrix of order n is magic is $2(n + 1)$, and it is usually required that magic squares be formed from the consecutive numbers, 1 to n^2.

R.3.168 For example, let us use MATLAB to create a magic matrix of order 3, and verify using MATLAB if indeed the returned matrix is magic by performing all the eight required additions $\{2(n + 1)\}$, and also verify that the magic constant is given by $(n^3 + n)/2$.

MATLAB Solution
```
>> A = magic(3)

   A =
        8    1    6
        3    5    7
        4    9    2

>> sumcol1 = sum(A(:,1))

   sumcol1 =
           15

>> sumcol2 = sum(A(:,2))

   sumcol2 =
           15

>> sumcol3 = sum(A(:,3))

   sumcol3 =
           15

>> sumrow1 = sum(A(1,:))

   sumrow1 =
           15

>> sumrow2 = sum(A(2,:))

   sumrow2 =
           15

>> sumrow3 = sum(A(3,:))

   sumrow3 =
           15
```

```
>> sumdiag1 = trace(A)

    sumdiag1 =
                15

>> aflip = fliplr(A)

    aflip =
              6    1    8
              7    5    3
              2    9    4

>> sumdiag2 = trace(aflip)

    sumdiag2 =
                15

>> n = 3;
>> magic _ constant = (n^3+n)/2

    magic _ constant =
                    15
```

R.3.169 The magic matrix of order 5, given as follows, requires 12 additions, where each one of the additions yields 65. It is left for the reader as an exercise to verify if indeed the returned matrix is magic.

```
>> magic (5)

    ans =
           17   24    1    8   15
           23    5    7   14   16
            4    6   13   20   22
           10   12   19   21    3
           11   18   25    2    9
```

R.3.170 The command *pascal(n)*, where n is an integer, returns an $n \times n$ symmetric, positive, definite matrix with integers entries, made up from the *Pascal* triangle.*

The Pascal triangle is named after the seventeenth century mathematician, Blaise Pascal. The Pascal triangle is shown in Table 3.3.

Observe that each element in the Pascal triangle is found by adding the pair of elements from the row immediate above. Also, observe that the sum of the elements in each row is a power of two.

---

* Blaise Pascal (1623–1662) was a brilliant French mathematician and physicist. He began the study of mathematics at the age of 12 and at 13 discovered the Pascal triangle. By 16 years of age, he stated the Pascal Theorem, and at the age of 17 used the theorem to derive 400 propositions. At the age of 19 he invented a calculating machine. But by 1654, he made a radical switch from mathematics and physics (hydrostatics) into theology. For a short period (1658–1659), he returned to mathematics where he came very close to discover calculus, which was developed years later by Leibniz and Newton.

TABLE 3.3

Pascal's Triangle

Rows								Total Sum per Row
1				1	1			$2 = 2^1$
2			1	2	1			$4 = 2^2$
3		1	3	3	1			$8 = 2^3$
4	1	4	6	4	1			$16 = 2^4$
5	1	5	10	10	5	1		$32 = 2^5$
6	1	6	15	20	15	6	1	$64 = 2^6$

The Pascal triangle for $n = 5$ is indicated as follows:

```
>> Pascal(5)
```

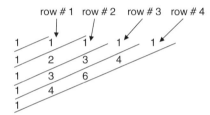

The elements of the Pascal triangle constitute the coefficients of a binomial (Tahan,) expression raised to the power n.

For example, let us assume that the binomial is $(a + b)^4$, then the coefficients are shown in row 4 of the Pascal triangle, given by *1, 4, 6, 4, 1*.

Similarly, the coefficients of $(x + y)^5$ are given by *1, 5, 10, 10, 5, 1*.

The coefficients of the binomial $(a^2 + 3b^2)^4$ are 1, 4, 6, 4, 1 and the corresponding polynomial can then be expressed as indicated below

$$(a^2 + 3b^2)^4 = (a^2)^4 + 4(a^2)^3\,(3b^2) + 6(a^2)^2\,(3b^2)^2 + 4(a^2)\,(3b^2)^3 + (3b^2)^4$$

$$(a^2 + 3b^2)^4 = a^8 + 12a^6b^2 + 54a^4b^4 + 108a^2b^6 + 81b^8$$

In general, the coefficients of the binomial expansion $(a + b)^n$ can be expanded as (Balador, 2000) follows:

$$(a + b)^n = a^n + n\,a^{n-1}b + \frac{n(n - 1)}{1*2}a^{n-2}b^2 + \frac{n(n - 1)(n - 2)a^{n-3}b^3}{1*2*3} + \cdots$$

R.3.171 Newton's formula can also be used to evaluate the coefficients of a binomial of the form $(a + b)^n$, indicated by

$$(a + b)^n = \binom{n}{0}a^nb^0 + \binom{n}{1}a^{n-1}b^1 + \binom{n}{2}a^{n-2}b^2 + \cdots + \binom{n}{n}a^0b^n$$

or

$$(a + b)^n = \sum_{k=0}^{n} \binom{n}{k} a^{n-k} b^k$$

Recall that

$$\binom{n}{k} = \binom{n}{n-k} = \frac{n!}{(n-k)!(n-(n-k))!} = \frac{n!}{(n-k)!k!}$$

R.3.172 For example expand $(a + b)^4$, using Newton's formula resulting in

$$(a + b)^4 = a^4 + 4a^3b + 6a^2b^2 + 4ab^3 + b^4$$

$$(a + b)^4 = \binom{4}{0} a^4b^0 + \binom{4}{1} a^3b^1 + \binom{4}{2} a^2b^2 + \binom{4}{3} a^1b^3 + \binom{4}{4} a^0b^4$$

Observe that the coefficients obtained using Newton's formula in the preceding example fully agree with the results obtained by using Pascal's triangle.

R.3.173 MATLAB has a number of built-in functions that return the dates and times in the form of vectors or arrays. Some of the functions are: *calendar, now,* and *datenum.*

R.3.174 The command *calendar(date)* returns a 7×7 array of the month specified by the date, where date is expressed as '*mm/dd/yyyy.*'

For example obtain the months specified by 05/05/1941 and 04/09/1946.

MATLAB Solution
```
>> calendar ('05/05/1941')

                May 1941
     S    M   Tu    W   Th    F    S
     0    0    0    0    1    2    3
     4    5    6    7    8    9   10
    11   12   13   14   15   16   17
    18   19   20   21   22   23   24
    25   26   27   28   29   30   31
     0    0    0    0    0    0    0

>> calendar ('04/09/1946')

                Apr 1946
     S    M   Tu    W   Th    F    S
     0    1    2    3    4    5    6
     7    8    9   10   11   12   13
    14   15   16   17   18   19   20
    21   22   23   24   25   26   27
    28   29   30    0    0    0    0
     0    0    0    0    0    0    0
```

R.3.175 The command *now* returns the number of days from the year zero. For example,

```
>> now

    ans =
            7.3097e+005
```

R.3.176 The command *datenum('date')* returns the number of days since year zero up to *date*, where *date* is expressed using the format: *'mm/dd/yyyy hh:mm:ss.'*
For example,

```
>> datenum('04/09/1946')        % number of days from year zero to
                                  4/9/1946

    ans =
            710861


>> datenum('04/09/1946 12:43')  % number of days from year zero
>>                              % time elapsed from zero to 4/9/1946-
                                  time 12:43

    ans =
            7.1086e+005
```

R.3.177 The command *hilb(n)*, where *n* is an integer smaller than 32, returns a special $n \times n$ matrix referred as the Hilbert matrix. The Hilbert matrix is defined as follows:

$$
\begin{bmatrix}
1 & 1/2 & 1/3 & \cdots & 1/n \\
1/2 & 1/3 & 1/4 & \cdots & 1/(n+1) \\
\cdots & \cdots & \cdots & \cdots & \cdots \\
1/n & 1/(n+1) & 1/(n+2) & \cdots & 1/2n
\end{bmatrix}
$$

Hilbert matrices are known to be ill-conditioned.

R.3.178 The MATLAB command *invhilb(n)* returns the inverse matrix of *hilb(n)*.
For example, use MATLAB and perform the following:

a. $A = hilb(5)$

b. $B = invhilb(5)$

c. Verify if B is the inverse of A

d. Evaluate $C = det(A)$, and check if A is near singularity

e. Evaluate $D = cond(A)$ and $E = cond(B)$, and observe that $D = E$

MATLAB Solution
```
>> A = hilb(5)                  % part(a)

    A =
          1.0000   0.5000   0.3333   0.2500   0.2000
          0.5000   0.3333   0.2500   0.2000   0.1667
          0.3333   0.2500   0.2000   0.1667   0.1429
          0.2500   0.2000   0.1667   0.1429   0.1250
          0.2000   0.1667   0.1429   0.1250   0.1111

>> B = invhilb(5)               % part(b)
```

```
   B =
           25        -300        1050       -1400         630
          -300        4800      -18900       26880      -12600
          1050      -18900       79380     -117600       56700
         -1400       26880     -117600      179200      -88200
           630      -12600       56700      -88200       44100

>> hilb(5)*invhilb(5)          % part(c) checks if I =A*B

   ans =
         1.0000        0           0           0           0
         0          1.0000        0           0           0
         0.0000        0        1.0000      -0.0000        0
         0             0           0        1.0000        0
         0             0           0           0        1.0000

>> C = det(A)             % part(d), observe that det(A) is small

   C =
       3.7493e-012

>> D = cond(A)                 % part(e)

   D =
       4.7661e+005

>> E = cond (invhilb(5))

   E =
       4.7661e+005
```

R.3.179 The MATLAB function *gallery('special_matrix', n)* is another way to access over 50 special matrices, where *special_matrix* defines the matrix name and *n* its order.
For example, *gallery('hilb', 5)* returns the *hilbert* matrix of order 5.
A partial list of special matrices that can be of interest to the reader are *cauchy, chebspec, house, lehmer, poisson, vander, wilk*, etc.

R.3.180 Given a vector $V = [v_1 \ v_2 \ ... \ v_n]$, the command $A = vander(V)$ returns the vandermonde matrix, defined as follows:

$$vandermonde \ matrix \ A = \begin{bmatrix} v_1^{n-1} & ... & v_1 & 1 \\ v_2^{n-1} & ... & v_2 & 1 \\ v_3^{n-1} & ... & v_3 & 1 \\ v_n^{n-1} & ... & v_n & 1 \end{bmatrix}$$

R.3.181 For example, let $V = [1 \ 2 \ 3 \ 4 \ 5]$. Use MATLAB to obtain the matrix *vander(V)*.

```
>> V = 1:1:5

   V =
         1   2   3   4   5

>> vander(V)
```

```
ans =
       1     1     1     1     1
      16     8     4     2     1
      81    27     9     3     1
     256    64    16     4     1
     625   125    25     5     1
```

R.3.182 Recall that a symmetric matrix is a square matrix that is equal to its transpose $(A = A^T)$; therefore $A(i, j) = A(j, I)$. For example,

$$A = \begin{bmatrix} 1 & 3 & 4 \\ 3 & 2 & 6 \\ 4 & 6 & 7 \end{bmatrix}$$

is a symmetric matrix.

A quick way to generate a symmetric matrix is by multiplying any square matrix by its transpose $(A * A'\ or\ A' * A)$. For example, let

$$A = \begin{bmatrix} 1 & 2 & 3 \\ 4 & 5 & 6 \\ 7 & 8 & 9 \end{bmatrix}$$

Use MATLAB to verify that $C = A * A^T$ and $D = A^T * A$ are symmetric matrices, where $C \neq D$.

MATLAB Solution
```
>> A =  [1 2 3; 4 5 6; 7 8 9]

   A =
       1    2    3
       4    5    6
       7    8    9

>> B = A'

   B =
       1    4    7
       2    5    8
       3    6    9

>> C = A*B                   % observe that C = A*A' yields a symmetric
                             matrix

   C =
      14    32    50
      32    77   122
      50   122   194

>> D = B*A                   % observe that D = B*A, returns a symmetric
                             matrix, but D ≠ C

   D =
      66    78    90
      78    93   108
      90   108   126
```

R.3.183 Recall that if $A = B * C * D$, where A, B, C, and D are matrices with compatible sizes, (and operations are possible) then $A^{-1} = B^{-1} * C^{-1} * D^{-1}$, or in terms of MATLAB $inv(A) = inv(B) * inv(C) * inv(D)$.

R.3.184 The concepts of transpose, inverse, and symmetric are used to define special matrices often encountered in science and engineering.

Some of these special matrices are defined as follows:

Skew Symmetric	*if*	$A^1 = -A$
Orthogonol	*if*	$A^1 = inv(A)$
Nilpotent	*if*	$A^n = 0$ for $n = 1, 2, \ldots$
Idempotent	*if*	$A^2 = A$

R.3.185 The MATLAB command *randn(n)* returns an $n \times n$ random matrix, with elements chosen from the normal Gaussian distribution with a mean value of zero and a variance of one.

Similarly, *randn(n, m)* returns an $n \times m$ normal random matrix.

R.3.186 The command *rand(n, n)* returns a pseudo random $n \times n$ matrix with a uniform distribution on the interval zero to one. Similarly, *rand(n, m)* returns a pseudo $n \times m$ randomly matrix.

R.3.187 For example, execute and observe the responses of the following MATLAB commands:

a. *randn(3)*

b. *rand(3)*

c. $x = rand(1, 10)$

d. $y = randn(1, 10)$

e. $u = randn(2, 3)$

```
>> randn (3)

    ans =
          -0.4326    0.2877    1.1892
          -1.6656   -1.1465   -0.0376
           0.1253    1.1909    0.3273

>> rand (3)

    ans =
           0.9501    0.4860    0.4565
           0.2311    0.8913    0.0185
           0.6068    0.7621    0.8214

>> x = rand (1,10)

    x =
          Columns 1 through 7
           0.9501    0.2311    0.6068    0.4860    0.8913    0.7621    0.4565
          Columns 8 through 10
           0.0185    0.8214    0.4447

>> y = randn (1,10)
```

```
y =
     Columns 1 through 7
     -0.4326  -1.6656   0.1253   0.2877  -1.1465   1.1909   1.1892
     Columns 8 through 10
     -0.0376   0.3273   0.1746

>> u = randn(2,3)

u =
     -0.1867  -0.5883  -0.1364
      0.7258   2.1832   0.1139
```

R.3.188 The command *randperm(n)* returns a vector with *n* elements randomly permuted.
For example, execute three times the command *randperm(10)*, and observe that
the responses are indeed random.

MATLAB Solution
```
>> A = randperm(10)

A =
     5   6   9   1   4   2  10   8   3   7

>> B = randperm(10)

B =
     4   1   5   7  10   6   9   2   8   3

>> C = randperm(10)

C =
     1   9   6   3   5   8  10   2   7   4
```

R.3.189 The command *B = circshift(A, C)* returns the matrix *B*, consisting of the elements
of *A*, circular shifted according to the vector *C*.
The concept of circular shifting is best understood by means of an example.

a. If *C = 1*, the last column of *A* becomes the first row of *B*, the first column of *A*
becomes the second row of *B*, the second column of *A* becomes the third row of
B, and so on.

b. If *C = [0 1]*, the last row of *A* becomes the first column of *B*, the first row of
A becomes the second column of *B*, the second row of *A* becomes the third
column of *B*, and so on.

R.3.190 For example, let

$$A = \begin{bmatrix} 1 & 2 & 3 \\ 4 & 5 & 6 \\ 7 & 8 & 9 \end{bmatrix}$$

Execute and observe the responses of the following circular shift commands:

a. *B = circshift(A, 1)*
b. *C = circshift(A, [0 1])*

```
>> A = [1 2 3;4 5 6;7 8 9];
```

```
>> B = circshift(A,1)

   B =
      3   6   9
      1   4   7
      2   5   8

>> C = circshift (A,  [0 1])

   C =
      7   1   4
      8   2   5
      9   3   6
```

It is left as an exercise for the reader to execute and observe the responses of the following commands $B = circshift(A, 3)$ and $circshift(A, [0\ 3])$.

R.3.191 The creation of matrices consisting exclusively of ones or zeros can be accomplished by using the following special MATLAB built-in functions:

a. *zeros(n, m)*, returns an $n \times m$ matrix consisting of zeros.

b. *ones(n, m)*, returns an $n \times m$ matrix of ones.

R.3.192 The MATLAB command *eye(n)* returns the $n \times n$ identity matrix.

R.3.193 The creation of the identity matrix, as well as matrices consisting of zeros and ones are illustrated below. For example, execute and observe the responses of the following MATLAB commands:

a. *zeros(3)*

b. *zeros(2, 3)*

c. *ones(3)*

d. *ones(2, 3)*

e. *eye(3)*

f. *eye(2, 3)*

g. *eye(4, 3)*

MATLAB Solution

```
>> zeros(3)

   ans =
        0   0   0
        0   0   0
        0   0   0

>> zeros(2,3)

   ans =
        0   0   0
        0   0   0

>> ones(3)

   ans =
        1   1   1
        1   1   1
        1   1   1
```

```
>> ones (2,3)

   ans =
           1   1   1
           1   1   1

>> eye(3)

   ans =
           1   0   0
           0   1   0
           0   0   1

>> eye(2,3)

   ans =
           1   0   0
           0   1   0

>> eye(4,3)

   ans =
           1   0   0
           0   1   0
           0   0   1
           0   0   0
```

R.3.194 As an additional example, let us create the matrix A defined as

$$A = \begin{bmatrix} 5 & 5 & 5 \\ 5 & 5 & 5 \end{bmatrix}$$

by using the commands *ones* and *zeros*.

MATLAB Solution
```
>> B = ones(2,3).*5

   B =
        5        5        5
        5        5        5

>> C = zeros(2,3)+5

   C =
        5        5        5
        5        5        5
```

R.3.195 The MATLAB command $a = isequal(X, Y)$ compares the matrices X with Y and returns $a = 1$, if $X = Y$ and $a = 0$ otherwise ($X \neq Y$).

R.3.196 For example, let $X = rand(1, 10)$ and $Y = rand(1, 10)$

a. Test if $X = Y$

b. Test if $X. * Y = Y. * X$

MATLAB Solution

```
>> X = rand(1,10)

    X =
        Columns 1 through 7
        0.9501  0.2311  0.6068  0.4860  0.8913  0.7621  0.4565
        Columns 8 through 10
        0.0185  0.8214  0.4447

>> Y = rand(1,10)

    Y =
        Columns 1 through 7
        0.6154  0.7919  0.9218  0.7382  0.1763  0.4057  0.9355
        Columns 8 through 10
        0.9169  0.4103  0.8936

>> a = isequal (X,Y)

    a =
        0

>> A = X.*Y

    A =
        Columns 1 through 7
        0.5847  0.1830  0.5594  0.3588  0.1571  0.3092  0.4270
        Columns 8 through 10
        0.0170  0.3370  0.3974

>> B = Y.*X

    B =
        Columns 1 through 7
        0.5847  0.1830  0.5594  0.3588  0.1571  0.3092  0.4270
        Columns 8 through 10
        0.0170  0.3370  0.3974

>> b = isequal(A,B)

    b =
        1
```

R.3.197 The elements of an array may include the following special characters:
NaN (Not A Number), *Inf* (infinity), or *empty*.
NaN is clearly defined by *IEEE* mathematical standards, and it states that any operation involving *NaN* results in *NaN*. *Inf* involves the division by zero (for example, *Inf = 1/0*).

R.3.198 Empty arrays are arrays with no matrix element and zero length in one or more dimensions. *NaN* and *empty* cannot be compared to another entry of *NaN*, or with another empty matrix.

R.3.199 For example, consider the matrix

$$A = \begin{bmatrix} 0 & 1 \\ 2 & NaN \end{bmatrix}$$

and perform the operations indicated as follows:

a. $B = A.^2$

b. $C = ones(2)./A$

c. $D = logm(A)$

MATLAB Solution
```
>> A = [ 0 1;2 NaN]

    A =
       0   1
       2   NaN

>> B = A.^2

    B =
       0   1
       4   NaN

>> C = ones(2)./A

    Warning: Divide by zero.
    C =
       Inf       1.0000
       0.5000    NaN

>> D = logm(A)

    Warning: Log of zero.
    D =
       NaN   NaN
       NaN   NaN
```

R.3.200 The function *isempty(A)* is used to test whether a matrix is empty. MATLAB returns then a *1* if *A* is empty, and a *0* otherwise.

R.3.201 Let us define an empty array and explore some of its characteristics by executing the following MATLAB commands and observing their responses.

a. $A = []$

b. *size(A)*

c. *length(A)*

d. $B = zeros(0, 3)$

e. *size(B)*

f. $C = 5*ones(3, 0)$

g. *size(C)*

h. *length(C)*

i. *isempty(A)*

j. *isempty(B)*

k. *isempty(C)*

l. $D = A + B$

m. $E = A * B$

n. $F = 3 * A$

o. $G = Inf * A$

MATLAB Solution

```
>> A = []              % part(a) empty array

   A =
      []

>> size(A)             % part(b)

   ans =
        0   0

>> length(A)           % part(c)

   ans =
        0

>> B = zeros(0,3)    % part(d), empty array with 3 columns

   B =
      Empty matrix: 0-by-3

>> size(B)             % part(e)

   ans =
        0   3

>> C = 5*ones(3,0)   % part(f), empty array with 3 rows

   C =
      Empty matrix: 3-by-0

>> size(C)             % part(g)

   ans =
        3   0

>> length(C)           % part(h)

   ans =
        0

>> isempty(A)           % part(i)

   ans =
        1
```

```
>> isempty(B)            % part(j)

   ans =
        1

>> isempty(C)            % part(k)

   ans =
        1

>> D = A+B               % part(l)

   ??? Error using ==> +
   Matrix dimensions must agree.

>> E = A*B               % part(m)

   E =
      Empty matrix: 0-by-3

>> F = 3*A               % part(n)

   F =
      []

>> G = Inf*A             % part(o)

   G =
      []
```

R.3.202 The MATLAB command *B = unique(A)* returns the column vector *B*, consisting of the elements of *A*, sorted in ascending order, with single value elements (duplicated elements are removed).

R.3.203 For example, let

$$A = \begin{bmatrix} 1 & 3 & 4 \\ 3 & 4 & 6 \end{bmatrix}$$

Execute the command *B = unique(A)* and observe the response.

```
>> A = [1 3 4; 3 4 6]

   A =
      1   3   4
      3   4   6

>> B = unique(A)

   B =
      1
      3
      4
      6
```

R.3.204 The command $B = repmat(A, r, c)$ returns the matrix B, with the elements of A replicated, where

A can be a scalar, vector, or matrix.

r is the number of times the rows of A will be replicated.

c the number of times the columns of A will be replicated.

R.3.205 For example, create a row vector consisting of eight elements with the same value 5, using the command *repmat*.

```
>> repmat(5,1,8)

    ans =
         5   5   5   5   5   5   5
```

Obviously *ones(1, 8)*5* would also return the sequence consisting of eight 5's.

R.3.206 The function *repmat* can be used to create a matrix A that repeats a given matrix D a number of times, illustrated as follows:

For example, let $D = [1\ 2;3\ 4]$. Create the matrix A consisting of matrix D repeated 15 times in a grid like structure consisting of 3 rows by 5 columns.

MATLAB Solution
```
>> D = [1 2;3 4]

    D =
         1     2
         3     4

>> A = repmat (D,3,5)

    A =
         1   2   1   2   1   2   1   2   1   2
         3   4   3   4   3   4   3   4   3   4
         1   2   1   2   1   2   1   2   1   2
         3   4   3   4   3   4   3   4   3   4
         1   2   1   2   1   2   1   2   1   2
         3   4   3   4   3   4   3   4   3   4
```

The matrix A can also be created by the following command $A = [D\ D\ D\ D\ D;$ $D\ D\ D\ D\ D; D\ D\ D\ D\ D]$.

R.3.207 Let $V = [1, 2, 3, 4]$, create a matrix A, consisting of the row vector V repeated three times.

MATLAB Solution
```
>> V = [1 2 3 4]

    V =
         1   2   3   4

>> A = repmat (V,3,1)

    A =
         1   2   3   4
         1   2   3   4
         1   2   3   4
```

R.3.208 Given the following two vectors $V = [v_1\ v_2\ v_3\ v_4]$ and $W = [w_1\ w_2]$, the MATLAB function $[X, Y] = meshgrid(V, W)$ returns two matrices: X and Y, with sizes *length* (W) by *length* (V), where X consists of the *row* vector V repeated *length* (W) times, and Y consists of *length* (V) columns, where each column consists of the elements of W', illustrated as follows: Let us assume that $V = [v_1\ v_2\ v_3\ v_4]$ and $W = [w_1\ w_2]$, then *meshgrid(V, W)* returns

$$X = \begin{bmatrix} v_1 & v_2 & v_3 & v_4 \\ v_1 & v_2 & v_3 & v_4 \end{bmatrix}$$

and

$$Y = \begin{bmatrix} w_1 & w_1 & w_1 & w_1 \\ w_2 & w_2 & w_2 & w_2 \end{bmatrix}$$

R.3.209 For example, let $V = [1\ 2\ 3\ 4]$ and $W = [5\ 6\ 7]$. Observe the response when the following command is executed $[X, Y] = meshgrid(V, W)$:

MATLAB Solution
```
>> V = [1 2 3 4]

    V =
        1   2   3   4

>> W = [5 6 7]

    W =
        5   6   7

>> [X,Y] = meshgrid(v,w)

    X =
        1   2   3   4
        1   2   3   4
        1   2   3   4
    Y =
        5   5   5   5
        6   6   6   6
        7   7   7   7
```

The *meshgrid* command is generally used to create 3-D plots (see Chapter 5).

R.3.210 A 3-D matrix can be created by specifying the three indexes as $A(n, m, p)$, where n and m identify the row and column, and p the page or layer number.

For example, a $3 \times 3 \times 3$ identity matrix A and a random $2 \times 5 \times 3$ matrix B are created below as follows:

```
>> A = eye(3, 3, 3);        % returns a 27-element identity matrix.
>> B = randn(2, 5, 3);      % returns a 30-element random normalize matrix.
```

R.3.211 The algebraic rules defined for 2-D arrays are equally valid for 3-D arrays. Observe that a 4-D matrix uses four *indexes* to identify each element of the matrix, whereas a 5-D matrix uses five *indexes*, and so on.

R.3.212 Matrix algebra rules for *d*-dimensional matrices (for $d > 3$) are not defined, and should not be used in a MATLAB environment. For example, let $A = ones(3, 3, 3)$ and $B = eye(3, 3, 3)$. Execute and observe the responses of the following MATLAB commands:

a. *A(1, 1, 1)*

b. *B(1, 1, 1)*

c. $C = A + B$

d. *C(1, 1, 1)*

e. $D = A. * B$

f. *D(2, 1, 1)*

g. $E = sqrt(C)$

h. *E(1, 1, 1)*

MATLAB Solution
```
>> A = ones(3,3,3);
>> B = eye(3,3,3);
>> C = B+A
>> A(1,1,1)

   ans =
         1

>> B(1,1,1)

   ans =
         1

>> C(1,1,1)

   ans =
         2

>> D = A.*B;
>> D(2,1,1)

   ans =
         0

>> E = sqrt(C);
>> E(1,1,1)

   ans =
             1.411111
```

R.3.213 The MATLAB function *num = ndim(A)* returns *num*, the number that identifies the dimension of the matrix *A*.

R.3.214 The MATLAB command $A = kron(u, v)$, often referred as the Kronecker tensor product *u v*, returns the matrix *A*, whose elements are obtained by multiplying each element of the first argument (*u*) by the matrix represented by the second argument (*v*).

R.3.215 For example, let

$$u = \begin{bmatrix} 1 & 2 \\ 3 & 4 \end{bmatrix} \quad \text{and} \quad v = \begin{bmatrix} 5 & 6 \\ 7 & 8 \end{bmatrix}$$

Execute and observe the responses of the following MATLAB instructions:

a. *A = kron(u, v)*

b. $B = \begin{bmatrix} u_{11} * v & u_{12} * v \\ u_{21} * v & u_{22} * v \end{bmatrix}$

c. Observe that *A = B*

MATLAB Solution
```
>> u = [1 2;3 4];
>> v = [5 6;7 8];
>> A = kron(u,v)

   A =
          5    6   10   12
          7    8   14   16
         15   18   20   24
         21   24   28   32

>> B = [u(1,1)*v u(1,2)*v;u(2,1)*v u(2,2)*v]

   B =
          5    6   10   12
          7    8   14   16
         15   18   20   24
         21   24   28   32
```

Observe that indeed *A* is equal to *B*.

R.3.216 A matrix is said to be sparse if a high number of its elements are zeros. These matrices are used to reduce storage resources and computational time, when operations are performed. There is no rule to decide when a matrix should be declared *sparse*. In general, a matrix is sparse if it has more zeros than nonzero elements.

Sparcity is usually decided by the reader/programmer based on experience, and not by a definition.

R.3.217 The terms sparse and full matrix refer exclusively to the way memory is allocated to the declared variables. Mathematically speaking these two matrices are equivalent.

R.3.218 The MATLAB function *B = sparse(A)* converts the full matrix *A* into a sparse matrix *B*.

R.3.219 The MATLAB function *C = full(B)* converts the sparse matrix *B* into a full matrix *C*.

R.3.220 The MATLAB function *eye(n)* returns the identity as a full matrix. The identity matrix is a good candidate to be converted to a sparse matrix, if desired, illustrated as follows:

```
>> A = eye(4)                    % full matrix

    A =
         1   0   0   0
         0   1   0   0
         0   0   1   0
         0   0   0   1

>> B = sparse(A)                 % sparse matrix

    B =
        (1,1)       1
        (2,2)       1
        (3,3)       1
        (4,4)       1
```

R.3.221 The MATLAB function *Ispar = speye(n)* returns *Ispar*, that is, the sparse identity matrix.

R.3.222 The MATLAB function *nnz(A)* returns the number of nonzero elements in a given sparse or full matrix *A*.

R.3.223 The MATLAB function *z = nonzeros(A)* returns the vector *z* consisting of the nonzero elements of the sparse matrix *A*.

R.3.224 The MATLAB function *nzmax(A)* returns the maximum number of nonzero elements in the sparse matrix *A*.

R.3.225 The MATLAB function *issparse(A)* checks the structure of the matrix *A*, and returns a 1, if *A* is sparse (true), and a 0, if A is full (false).

R.3.226 The MATLAB function *spy(A)* returns the structure of the matrix *A*, displaying symbolically the nonzero elements as dots in a 2-D array.

R.3.227 The MATLAB function *A = sparse(row, col, values, total_row, total_col)* returns the sparse matrix *A*, where the vectors *row* and *col* indicate the positions of the nonzero elements specified by the vector *values*. The arguments: *total_row* and *total_col* define the size of the sparse matrix *A*.

R.3.228 The MATLAB function *A = spalloc(row, col, nonzero)* allocates sufficient memory for the sparse matrix *A*, specified by the number of rows and columns that have nonzero elements defined by the variables *rows, col*, and *nonzero*, respectively.

R.3.229 The following example illustrates the action of the sparse functions just defined earlier.

First, create a 5 × 5 sparse matrix *A*, whose elements have the value 5 in each diagonal and zeros elsewhere. Then convert the matrix *A* to a full matrix *B*, and test the matrices *A* and *B* for sparsity, and for the number of nonzero elements in each matrix.

MATLAB Solution
```
>> clear
>> row = 1:5;
>> col = 1:5;
>> A = sparse(row,col,values,5,5)
```

```
>> values = 5*ones(1,5);
>> A = sparse(row,col,values,5,5)     % 5's on main diagonal

   A =
      (1,1)    5
      (2,2)    5
      (3,3)    5
      (4,4)    5
      (5,5)    5

>> colrev = 5:-1:1;
>> AA = sparse(row,colrev,values,5,5)    % 5's on other diagonal

   AA =
        (5,1)    5
        (4,2)    5
        (3,3)    5
        (2,4)    5
        (1,5)    5

>> A=A+AA;
>> A(3,3) = 5;                   % defines the element at the
                                 %    intersection of the two diagonal
>> A                            % non zero elements of A

   A =
      (1,1)    5
      (5,1)    5
      (2,2)    5
      (4,2)    5
      (3,3)    5
      (2,4)    5
      (4,4)    5
      (1,5)    5
      (5,5)    5

>> B = full(A)

   B =
      5   0   0   0   5
      0   5   0   5   0
      0   0   5   0   0
      0   5   0   5   0
      5   0   0   0   5

>> checksparse = [issparse(A)  issparse(B)]

   checksparse =
                1   0

>> checknnz = [nnz(A)  nnz(B)]

   checknnz =
                9   9
```

R.3.230 Illustrate the concentrations of nonzero elements and explore the corresponding densities of the matrices *A* and *B* of the preceding example using the *spy* function (Figure 3.3).

MATLAB Solution
```
>> subplot(1,2,1); % see footnote*
>> spy(A);
>> title('Matrix structure for spy(A)')
>> subplot(1,2,2)
>> spy(B);
>> title('Matrix structure for spy(B)')
```

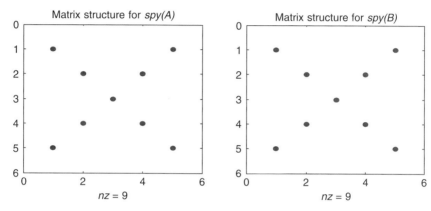

FIGURE 3.3
Plots of *spy(A)* and *spy(B)* of R.3.230.

Observe that the spy function works equally well with sparse and nonsparse (full) matrices.

R.3.231 The MATLAB function $A = sprandn(n, m, den)$ returns an $n \times m$ sparse matrix *A* with normally distributed nonzero element with density *den*, where *den* is in the range 0–1.

R.3.232 The MATLAB function $A = sprandsym(n, den)$ returns an $n \times n$ sparse symmetric matrix *A* with normally distributed nonzero elements of density *den*, where *den* is in the range 0–1.

R.3.233 The density function *den* is defined as

$$den = \frac{nonzero\ elements\ of\ A}{total\ number\ of\ elements\ of\ A}$$

R.3.234 The following example illustrates and reviews some of the concepts just presented earlier in this section. For example, create

a. A 3×5 sparse matrix *A*, with normally distributed nonzero elements with density *den = 0.30*

b. A 5×5 symmetric sparse matrix *B*, with normally distributed nonzero elements with density *den = 0.5*

---

* The *subplot* command is discussed in Chapter 5. The reader can test this program by suppressing the *subplot* command, without any loss of generality.

c. Convert the sparse matrices *A* and *B* to full matrices *AA* and *BB*, respectively, and observe the densities by obtaining a *spy* diagram

d. Verify numerically the densities *den* of each matrix, and observe that the specified densities are not equal to the calculated one but very close

MATLAB Solution

```
>> A = sprandn(3,5,0.3)          % observe that from is elements
                                    five are non-zero

    A =
        (2,1)    -1.6656
        (1,3)    -0.4326
        (3,3)     0.2877
        (3,4)    -1.1465
        (2,5)     0.1253


>> B = sprandsym(5,0.5)          % observe that from 25 elements
                                    12 are non-zero

    B =
        (3,1)    -0.1867
        (5,1)     0.8628
        (4,2)     0.3273
        (5,2)    -0.5883
        (1,3)    -0.1867
        (4,3)     1.1909
        (2,4)     0.3273
        (3,4)     1.1909
        (5,4)     1.1892
        (1,5)     0.8628
        (2,5)    -0.5883
        (4,5)     1.1892

>> AA = full(A)

    AA =
         0         0      -0.4326    0          0
        -1.6656    0       0         0          0.1253
         0         0       0.2877   -1.1465     0

>> BB = full(B)

    BB =
         0         0      -0.1867    0          0.8628
         0         0       0         0.3273    -0.5883
        -0.1867    0       0         1.1909     0
         0         0.3273  1.1909    0          1.1892
         0.8628   -0.5883  0         1.1892     0

>> check_denA = nnz(A)/15

    check_denA =
                   0.3333    % den-error=0.333-0.3=0.033

>> check_denB = nnz(B)/25

    check_denB =
                   0.4800    % den-error=0.50-0.48=0.02
```

```
>> subplot(1,2,1)
>> spy(A);
>> title('structure of matrix A')
>> subplot(1,2,2)
>> spy(B);
>> title('structure of matrix B')
```

The corresponding *spy* plots are shown in Figure 3.4.

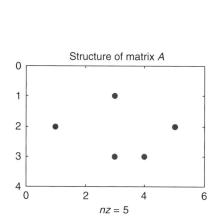
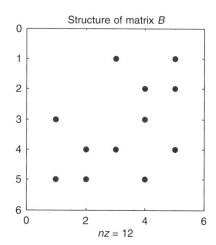

FIGURE 3.4
spy(A) and *spy(B)* of R.3.234.

R.3.235 Once a sparse matrix is created all the MATLAB functions defined for full matrices work equally well for the sparse case.

R.3.236 MATLAB operations can be performed involving mixed matrices, full, and sparse. The general rules followed by MATLAB are

a. Operations involving mixed matrices result in a full matrix.

b. Operations involving sparse matrices result in a sparse matrix.

c. MATLAB tries to preserve sparsity when possible.

R.3.237 The following example illustrates reviews and explores some of the concepts presented earlier in this section.

Execute each of the following instructions and observe the responses:

a. Create the following sparse matrices: $A = sprandn(50, 50, 0.1)$ and $B = sprandn(50, 50, 0.5)$

b. Convert the sparse matrices A and B into the full matrices: AA and BB, respectively

c. Perform the operation $ADDmix = AA + B$ and $ADDsparse = A + B$

d. Perform the operation $PRODmix = AA. * B$ and $PRODsparse = A * B$

e. Check if your software uses the *IEEE* floating point arithmetic

f. Execute the command $det(A)$ and $det(AA)$, and compare the results

g. Execute the command *whos*, and observe which of the results obtained are sparse or full as well as the storage allocated to each class of matrices

MATLAB Solution

```
>> A = sprandn(50,50,0.1);
>> B = sprandn(50,50,0.5);
>> AA = full(A);
>> BB = full(B);
>> ADDmix = AA+B;
>> ADDsparse = A+B;
>> PRODmix = AA.*B;
>> PRODsparse = A*B;
>> issparse(ADDmix)      % observe that the mix sum results in a
                            full matrix

   ans =
        0

>> issparse(ADDsparse)   % observe that the resulting matrix is
                            sparse

   ans =
        1

>> issparse(PRODmix)     % observe that the mix product results in
                            sparse

   ans =
        1

>> issparse(PRODsparse)  % observe that the sparse product result
                            in sparse

   ans =
        1

>> isieee                % checks IEEE floating point arithmetic

   ans =
        1

>> detsparse = det(A)

   detsparse =
                0.0015

>> detfull = det(AA)     % observe  that det(AA)=det(A)
                            as expected

   detfull =
                0.0015

>> whos
     Name        Size        Bytes      Class
     A           50x50       3132       sparse array
     AA          50x50       20000      double array
     ADDmix      50x50       20000      double array
     ADDsparse   50x50       14820      sparse array
```

B	50x50	11892	sparse array
BB	50x50	20000	double array
PRODmix	50x50	11892	sparse array
PRODsparse	50x50	28788	sparse array
ans	1x1	8	double array (logical)
detfull	1x1	8	double array
detsparse	1x1	8	double array

```
Grand total is 13295 elements using 130548 bytes
```

Observe that MATLAB allocates 3132 bytes to store the spare matrix *A*, whereas the equivalent full version, matrix AA employs 20,000 bytes. Also note that the sparse matrix PRODmix uses 11,892 bytes, whereas the equivalent matrix PRODsparse employs 28,788 bytes of memory. Also note that these results do not look logical.

R.3.238 The MATLAB function *B = sponces(A)* returns matrix *B* consisting of all the non-zero elements of the sparse matrix *A* replaced with ones.

R.3.239 The MATLAB function *A = spdiags(diag_ele, diag_loc, m, n)* returns the sparse matrix *A* with size *m* × *n*, with the diagonal elements identified by the location specs, *diag_ele* and *diag_loc*, (value and location) respectively.

R.3.240 Let us illustrate the action of the functions defined earlier in this section by executing and observing the responses to each of the following:

a. Create a 100 × 100 sparse matrix *A*, whose elements have the values of ones in the main diagonal and the subdiagonals shifted up and down by two units, and zeros elsewhere

b. Repeat step (a) for the other diagonal

c. Create a 100 × 100 sparse matrix *C*, whose elements have the values of one for the columns 48, 50, and 52

d. Repeat part (c) by replacing the columns by the rows

e. Obtain for each structure a *spy* diagram

f. Store the preceding commands in the script file: *spypattern*. Execute this file and display the results

MATLAB Solution
```
% Script file:spypatterns
a = ones(100,1);
b = ones(100,1);
c = ones(100,1);
subplot(2,2,1)
A = spdiags([a b c ],[-2,0,2],100,100);
spy(A);
title('Matrix structure for spy(A)')
subplot(2,2,2)
B = flipud(A);
spy(B)
title('Matrix structure for flipud(A)')
subplot(2,2,3)
row =[1:100 1:100 1:100];
col =[48*ones(1,100) 50*ones(1,100) 52*ones(1,100)];
values = ones(1,300);
C = sparse(row,col,values,100,100);
```

```
spy(C)
title('Matrix structure for spy(C)')
subplot(2,2,4)
D = rot90(C);
spy(D)
title('Matrix structure for rot90(C)')
```
The resulting plots are shown in Figure 3.5.

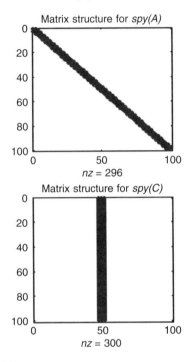
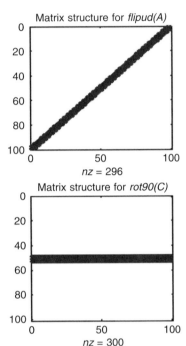

FIGURE 3.5
spy diagrams for R.3.240.

R.3.241 Symbolic commands can be used to create symbolic matrices, that is, matrices whose elements are symbolic (see Chapter 7 for additional information).

A symbolic element can be defined as an element with no assigned numerical value, that must be declared as symbolic (sym) before it is used.

R.3.242 The following example illustrates some of the commands presented earlier in this section used now on symbolic matrices. For example, create the script file: *sym_oper* that returns the following:

a. The symbolic matrix $A = [a \; b; c \; d]$, where a, b, c, and d are symbolic elements.

b. The characteristic polynomial equation for matrix A.

c. The eigenvalues of A.

MATLAB Solution
```
% Script file:sym _ oper
syms a b c d;
disp ('*****************************')
disp('The symbolic matrix is')
sym _ A = [a b;c d]
```

```
disp('The characteristic equation is :')
char _ eq = poly(sym _ A)
disp ('Its eigenvalues are:')
ieg _ values _ sym _ A = eig(sym _ A)
disp('****************************')
```

The script file: *sym_oper* is executed and the results are shown as follows:

```
>> sym _ oper

*********************************************
The  symbolic  matrix  is :
sym _ A =
       [ a, b]
       [ c, d]
The  characteristic  equation  is
char _ eq =
       x^2-x*d-a*x+a*d-b*c
Its  eigenvalues  are:
        ieg _ values _ sym _ A =
[  1/2*a+1/2*d+1/2*(a^2-2*a*d+d^2+4*b*c)^(1/2)]
[  1/2*a+1/2*d-1/2*(a^2-2*a*d+d^2+4*b*c)^(1/2)]
```

3.4   Examples

Example 3.1

Write and run a program that evaluates the following sequence:

$$\sum_{n=0}^{N-1} a^n \quad \text{for } a = 0.5 \text{ and } N = 11 \text{ and } 21.$$

Verify if the above series converges to 2 for any large value of N.

MATLAB Solution
```
>> n = 0:10;              % creates the vector n = [0 1 2 3 ... 10]
>> y1 = (0.5) . ^ n;      % creates the sequence y1 = [0.5^0 0.5^1 0.5^1 ...
                             0.5^10]
>> sum10 = sum(y1);       % computes the sum of all the elements of y1
>> m = 0:20;              % creates a vector   m = [0 1 2 3 ... 20]
>> y2 = (0.5) . ^ m;      % creates the sequence y2 = [0.5^0 0.5^1 0.5^2 ...
                             0.5^20]
>> sum20 = sum(y2);       % computes the sum of all the elements of y2
>> sum10                  % returns sum10

   sum10 =
           1.9990
```

```
>> sum20              % returns sum20

  sum20 =
        2.0000
```

Observe that the sum of the given series converges to 2 as N approaches infinity.

Example 3.2

Write a program that evaluates the *n* factorial (*n*!) for *n* = 10 and 20, where

$$n! = \prod_{i=1}^{n} i = 1 * 2 * 3 * 4 * \cdots * n$$

Matlab Solution
```
>> n=1:10;              % creates the vector n = [1 2 3 ... 10]
>> FACT10 = prod(n);    % FACT10 is the product of the elements of n
>> FACT10               % displays the product of the elements of the
                           sequence n

  FACT10 =
        3628800


>> m = 1:20;            % creates the vector m = [1 2 3 ... 20]
>> FACT20 = prod (m);   % FACT20 is the product of the sequence
                           given by m

>> FACT20


  FACT20 =
        2.4329e+018
```

Example 3.3

Write a MATLAB program that returns the sequence consisting of square of the first 50 even numbers (0 4 16 36 64 100 ... 9604), and identify and display the first, fifth, and tenth element of that sequence.

MATLAB Solution
```
>> n = 0:2:98;          % creates the vector n = [0 2 4 6 ... 98]
>> y = n.^2             % creates and displays the vector y = [0 2² 4²
                            6² ... 98²]


  y =
    Columns 1 through 6
    0          4      16        36       64       100
    Columns 7 through 12
    144      196     256       324      400       484
    Columns 13 through 18
    576      676     784       900     1024      1156
    Columns 19 through 24
    1296    1444    1600      1764     1936      2116
    Columns 25 through 30
    2304    2500    2704      2916     3136      3364
```

```
          Columns 31 through 36
          3600     3844     4096     4356     4624     4900
          Columns 37 through 42
          5184     5476     5776     6084     6400     6724
          Columns 43 through 48
          7056     7396     7744     8100     8464     8836
          Columns 49 through 50
          9216     9604
```

```
>> a1 = y(1) ;              % first element of y
>> a5 = y(5) ;              % fifth element of y
>> a10 = y(10) ;            % tenth element of y
>> format compact          % suppresses extra line-feeds when
                             display a1,a5,and a10


>> a1


   a1 =
        0


>> a5


   a5 =
        64


>> a10


   a10 =
        324
```

Example 3.4

Write a Matlab program that creates matrix A composed of four submatrices $B, C, D,$ and E arranged forming the structure indicated as follows

$$A = \begin{bmatrix} B & C \\ E & D \end{bmatrix}$$

where $B, C, D,$ and E are 5×5 matrices, with the following characteristics:

1. B is the identity matrix.
2. C consists of ones in the first two columns and the rest elements are zeros.
3. D consists of zeros in the first three columns and the remaining two columns are ones.
4. E consists of the sequence of integers 1 through 25 in ascending order, column by column, from left to right.

Write a MATLAB program that returns the following:

a. The matrix A
b. The elements: A(5, 5), A(3, 10), and A(10, 2)
c. The size of A
d. The transpose of A, where $F = A'$
e. The following elements: $F(1, 2)$, $F(4, 2)$, and $F(5, 5)$

MATLAB Solution

```
>> format compact            % suppress extra line-feeds when
                               displaying results
>> B = [eye(5)];             % creates the identify matrix B
>> C = [ones(5,2) zeros(5,3)];  % creates the matrix C
>> D = [zeros(5,3) ones(5,2)];  % creates the matrix D
>> G = [1:5;6:10;11:15;16:20;21:25];  % creates the matrix G
>> E = G';                   % creates the matrix E
>> A = [B C; E D]            % creates and displays the matrix A
```

```
A =
     1   0   0   0   0   1   1   0   0   0
     0   1   0   0   0   1   1   0   0   0
     0   0   1   0   0   1   1   0   0   0
     0   0   0   1   0   1   1   0   0   0
     0   0   0   0   1   1   1   0   0   0
     1   6  11  16  21   0   0   0   1   1
     2   7  12  17  22   0   0   0   1   1
     3   8  13  18  23   0   0   0   1   1
     4   9  14  19  24   0   0   0   1   1
     5  10  15  20  25   0   0   0   1   1
```

```
>> a5 _ 5 = A(5,5)            % returns A(5,5)


   a5 _ 5 =
           1


>> A3 _ 10 = A(3,10)         % returns A(3,10)


   A3 _ 10 =
             0


>> A10 _ 2 = A(10,2)         % returns A(10,2)


   A10 _ 2 =
            10


>> size _ A = size(A)        % returns the size of matrix A


   size _ A =
            10  10
```

```
>> F = A'                                    % displays the transpose of A

   F =
        1    0    0    0    0    1    2    3    4    5
        0    1    0    0    0    6    7    8    9   10
        0    0    1    0    0   11   12   13   14   15
        0    0    0    1    0   16   17   18   19   20
        0    0    0    0    1   21   22   23   24   25
        1    1    1    1    1    0    0    0    0    0
        1    1    1    1    1    0    0    0    0    0
        0    0    0    0    0    0    0    0    0    0
        0    0    0    0    0    1    1    1    1    1
        0    0    0    0    0    1    1    1    1    1

>> F1 _ 2 = F(1,2)                           % displays F(1,2)

   F1 _ 2 =
            0

>> F4 _ 2 = F(4,2)                           % displays F(4,2)
   F4 _ 2 =
            0

>> F5 _ 5 = F(5,5)                           % displays F(5,5)

   F5 _ 5 =
            1
```

Example 3.5

Write a MATLAB program that returns the 4×4 random matrix R, with random elements consisting of integers between 1 and 100, then determine:

a. The maximum and minimum value for each row and column of R.
b. The elements on the main diagonal of R.
c. The elements of the other diagonal of R.
d. The sum and product of the elements of the main diagonal of R.
e. The average and median of the elements of the main diagonal of R.
f. The maximum and minimum values of the elements on the main diagonal of R.
g. The maximum and minimum values of all the elements in R.
h. The determinant of R.
i. The rank of R and if possible the inverse of R (R^{-1}).
j. The 2×2 matrix that is located at the center of R.
k. The matrix $R_{SQ} = R^2$.
l. Square each element of R.
m. Reshape matrix R into a 2×8 and an 8×2 matrices.
n. The matrix e^R.
o. The matrix e raise to each element of R.
p. The matrix consisting of the square root of each element of R.
q. The matrix $R_x = \sqrt{R}$, where $R_x * R_x = R$.

MATLAB Solution

```
>> R = fix(rand(4)*100)          % R consist of random integers

   R =
        95  89  82  92
        23  76  44  73
        60  45  61  17
        48   1  79  40


>> x = max(R)                    % maximum of each column of R, part (a)

   x =
        95  89  82  92

>> y = min(R)                    % returns the minimum of each column of R

   y =
        23 1 44 17


>> RR = R'

   RR =
        95  23  60  48
        89  76  45   1
        82  44  61  79
        92  73  17  40

>> minrow = min(RR)              % returns the minimum of each row of R

   minrow =
             82  23  17   1

>> maxrow = max(RR)              % returns the maximum of each row of R

   maxrow =
             95  76  61  79

>> vectdiR = diag(R)             % returns the diagonal of A, part (b).

   vectdiR =
             95
             76
             61
             40

>> flipR = fliplR(R)             % flips the columns of R from left to
                                 % right

   flipR =
             92  82  89  95
             73  44  76  23
             17  61  45  60
             40  79   1  48
```

```
>> otherdiag = diag(flipR)          % returns the second diagonal of R,
                                       part (c)

   otherdiag =
                92
                44
                45
                48


>> SumdiR = sum(vectdiR)            % sum of the elements of the main
                                       diagonal, part (d)

   SumdiR =
             272

>> prodiR = prod(vectdiR)           % product of the elements of the main
                                       diagonal

   prodiR =
            17616800

>> avediR = mean(vectdiR)           % average of the elements of the main
                                       diagonal, part (e)

   avediR =
             68

>> meddiR = median(vectdiR)         % median of the elements of the main
                                       diagonal

   meddiR =
             68.5000

>> maxvalue = max(x)                % maximum value in R, part (g)

   maxvalue =
              95

>> minvalue = min(y)                % minimum value in R

   minvalue =
              1

>> maxdia = max(vectdiR)            % largest element in the main
                                       diagonal, part (f)

   maxdia =
             95
```

```
>> mindia = min(vectdiR)          % smallest element in the main
                                    diagonal

   mindia =
        40

>> detR = det(R)                  % determinant of R, part (h)

   detR =
        11580761

>> rank(R)                        % rank of R, part (i)

   ans =
        4

>> H = inv(R)                     % inverse of R, part (i).

   H =
        0.0224   -0.0236   -0.0045   -0.0067
       -0.0076    0.0122    0.0169   -0.0120
       -0.0201    0.0142    0.0154    0.0137
        0.0129   -0.0001   -0.0255    0.0063

>> cond(R)

   ans =
        12.5427

>> I = H*R                        % checks for the identity matrix

   I =
        1.0000   -0.0000   -0.0000   -0.0000
        0.0000    1.0000    0.0000    0.0000
       -0.0000    0         1.0000   -0.0000
       -0.0000   -0.0000   -0.0000    1.0000


>> II = R*H

   II =
        1.0000    0        -0.0000    0.0000
        0         1.0000   -0.0000    0.0000
       -0.0000    0         1.0000   -0.0000
       -0.0000    0.0000    0.0000    1.0000

>> Rcenter = R(2:3,2:3)           % returns  the 2x2 centered matrix of
                                    R, part (j)

   Rcenter =
            76   44
            45   61

>> RSQ = R*R                      % part (k)

   RSQ =
        20408   19001   23976   20311
        10077    9876   13681   11332
        11211   11522   11964   10522
        11243    7943   11959    7432
```

```
>> Rsq = R.^2                      % square each element of R,
                                     part (l)


   Rsq =
        9025    7921    6724    8464
         529    5776    1936    5329
        3600    2025    3721     289
        2304       1    6241    1600

>> reshape2x8 = reshape(R,2,8)     % returns a 2x8 matrix with the ele-
                                     ments of R. part (m)

  reshape2x8 =
            95  60  89  45  82  61  92  17
            23  48  76   1  44  79  73  40


>> reshape8x2 = reshape(R,8,2)


  reshape8x2 =
            95  82
            23  44
            60  61
            48  79
            89  92
            76  73
            45  17
             1  40


>> expR = expm(R)                  % evaluates the series I+R+R^2/
                                     2!+R^3/3!+, part (n)


   expR =
        1.0e+099 *
        5.7012  5.2376  6.6039  5.3458
        3.0171  2.7717  3.4948  2.8290
        3.0906  2.8393  3.5800  2.8980
        2.7321  2.5099  3.1646  2.5618


>> expeleR = exp(R)                % returns e raised to each element in
                                     R, part (o)


  expeleR =
            1.0e+041 *
            1.8112  0.0045  0.0000  0.0902
            0.0000  0.0000  0.0000  0.0000
            0.0000  0.0000  0.0000  0.0000
            0.0000  0.0000  0.0000  0.0000


>> sqrteleR = sqrt(R)              % square root of each element of R,
                                     part (p)
```

```
sqrteleR =
        9.7468        9.4340        9.0554        9.5917
        4.7958        8.7178        6.6332        8.5440
        7.7460        6.7082        7.8102        4.1231
        6.9282        1.0000        8.8882        6.3246

>> Rx = sqrtm(R)                    % square root of R


Rx =
    8.3899 - 0.0000i 5.2142 + 0.0000i 3.2680          4.7286 + 0.0000i
    0.6054 + 0.0000i 8.7517 + 0.0000i 1.0163          4.6934 - 0.0000i
    3.8929 - 0.0000i 1.5143 - 0.0000i 7.0883 - 0.0000i -0.6346+ 0.0000i
    1.8466          -1.1267 - 0.0000i 5.5233 - 0.0000i 6.3294 + 0.0000i

>> check = Rx*Rx                    % checks  the result


check =
95.0000-0.0000i 89.0000 + 0.0000i 82.0000 - 0.0000i 92.0000 + 0.0000i
23.0000+0.0000i 76.0000 + 0.0000i 44.0000 - 0.0000i 73.0000 + 0.0000i
60.0000-0.0000i 45.0000 - 0.0000i 61.0000 - 0.0000i 17.0000 + 0.0000i
48.0000-0.0000i  1.0000 - 0.0000i 79.0000 - 0.0000i 40.0000 + 0.0000i
```

Example 3.6

Write a program that returns an equivalence table of temperatures in terms of degree Celsius (°C), Fahrenheit (°F), and Kelvin (K), over the range 0°C (freezing point) and 100°C (boiling point), linearly spaced every 10°C.

From Chapter 2, Table 2.5, the following relations are known:

$$1°F = 1.8°C + 32$$

$$1K = 1°C + 273.15$$

MATLAB Solution

```
>> Celsius = 0:10:100;              % returns an 11 elements Celsius
                                       array
>> Farhnt = 1.8*Celsius+32          % returns the equivalent Farhnt
                                       array
>> Kelvin = Celsius+273.15;         % returns the Kelvin equivalent array
>> A= [ 'Celsius  Farhnt    Kelvin'];
>> % display the table, where the first column is ....
>> % the temperature in Celsius, the second column is in Fahrenheit,
                                               and the third ...
>>                               column is in Kelvin degrees
>> disp('*****************************')
>> disp('          Tables of temperatures             ')
>> disp('*****************************')
>> B = [Celsius'  Farhnt'  Kelvin'];
>> disp('*****************************')
>> disp(A), disp(B)
>> disp('*****************************')
```

```
**********************************
      Tables of temperatures
**********************************
  Celsius     Farhnt    Kelvin
**********************************
      0        32.0000   273.1500
  10.0000      50.0000   283.1500
  20.0000      68.0000   293.1500
  30.0000      86.0000   303.1500
  40.0000     104.0000   313.1500
  50.0000     122.0000   323.1500
  60.0000     140.0000   333.1500
  70.0000     158.0000   343.1500
  80.0000     176.0000   353.1500
  90.0000     194.0000   363.1500
 100.0000     212.0000   373.1500
**********************************
```

Example 3.7

The equivalent resistance R_{eq_s} of an electrical network consisting of n resistors connected in series is given by

$$R_{eq\_s} = \sum_{i=1}^{n} R_i$$

where the $R_{i's}$ are the series resistor (see Chapter 2* for additional details). The equivalent resistance R_{eq_p} of an electrical network consisting of n resistors connected in parallel is given by

$$R_{eq\_p} = \frac{1}{\sum_{i=1}^{n} 1/R_i}$$

where the $R_{i's}$ are the resistors connected in parallel.

Write a MATLAB program that evaluates R_{eq} for the series and parallel cases, for an electrical network consisting of four resistors with the following values $R_1 = 2.5$, $R_2 = 4$, $R_3 = 11$, and $R_4 = 6.3$ (in Ohms).

MATLAB Solution
```
>> R1 = 2.5;R2 = 4;R3 = 11; R4 =6.3;  % values of the resistors in Ohms
>> X = [R1 R2 R3 R4];                 % X is the resistor  network array
>> reqseries = sum(X)                 % equivalent resistance when
                                        connected in series

    reqseries =
             23.8000

>> Y = 1./X;                          % transform the resistances to
                                        admittances
>> Ys = sum(Y);                       % total admittance of the network
>> reqparal = 1/Ys                    % equivalent resistance connected
                                        in parallel

    reqparal =
             1.1116
```

---

* MATLAB applications for engineers.

Example 3.8

Given the vector *V* specified below by

$$V = [1\ 3\ 7\ 9\ 12\ 10\ 8\ 5\ 3\ -1\ -2\ 0\ 2\ 3\ 1\ 2]$$

Write a program using MATLAB that returns the following:

a. the row vector *V*
b. the size and length of *V*
c. *V* converted into a column vector
d. the elements of *V* sorted in ascending order
e. the addition of all the elements of *V*
f. the product of all the elements of *V*
g. the maximum and minimum value of *V* and their locations in the array
h. the sequence that consists of squaring each element of *V*
i. the mean and median of *V*
j. the variance and standard deviation of *V*
k. the cumulative sum and cumulative product of *V*
l. the norm of *V* ($\|V\|$)
m. the area under the curve defined by the elements of *V* with unit spacing
n. the cumulative area under *V*
o. a new vector consisting of the first four elements of *V* (remove elements fifth through sixteenth)

MATLAB Solution
```
>> V = [1   3   7   9   12   10   8   5   3   -1   -2   0   2   3   1   2];
>> V _ size = size (V)

   V _ size =
                 1    16

>> V _ len = length(V)

   V _ len =
                16

>> V _ col  = V'

   V _ col =
                 1
                 3
                 7
                 9
                12
                10
                 8
                 5
                 3
```

```
                -1
                -2
                 0
                 2
                 3
                 1
                 2

>> V _ sort = sort (V)

   V _ sort =
                Columns 1  through 12
                -2  -1   0   1   1   2   2  3  3  3  5  7
                Columns 13  through 16
                 8  9   10   12

>> V _ sum = sum(V)

   V _ sum =
                63

>> V _ prod = prod(V)

   V _ prod =
                0

>> [V _ max,max _ position] = max(V)

   V _ max =
                12
    max _ position =
                       5

>> [V _ min,min _ position] = min(V)

   V _ min =
                -2
   min _ position =
                     11

>> V _ square = V.^2

   V _ square =
                Columns 1 through 12
                1 9 49 81 144 100 64 25 9 1 4 0
                Columns 13 through 16
                4 9 1  4

>> V _ mean = mean(V)

   V _ mean =
                3.9375

>> V _ median = median(V)

   V _ median =
                  3
```

```
>> V _ stand _ dev = std(V)

   V _ stand _ dev =
                        4.1387

>> V _ variance = V _ stand _ dev^2

   V _ variance =
                     17.1292

>> V _ cum _ sum = cumsum(V)

   V _ cum _ sum =
                   Columns 1 through 12
                   1 4 11 20 32 42 50 55 58 57 55 55
                   Columns 13 through 16
                   57 60 61 63

>> V _ cum _ prod = cumprod(V)

   V _ cum _ prod =
                   Columns 1 through 6
                   1 3 21 189 2268 22680
                   Columns 7 through 12
                   181440 907200 2721600 −2721600 5443200 0
                   Columns 13 through 16
                   0 0 0 0

>> V _ norm = norm(V)

   V _ norm =
                22.4722

>> V _ area = trapz(V)

   V _ area =
                61.5000

>> V _ cum _ area = cumtrapz(V)

   V _ cum _ area =
                   Columns 1 through 7
                   0 2.0000 7.0000 15.0000 25.5000 36.5000 45.5000
                   Columns 8 through 14
                   52.0000 56.0000 57.0000 55.5000 54.5000 55.5000
                   58.0000
                   Columns 15 through 16
                   60.0    61.5000

>> V(5:end) = []            % removes elements 5 through 16 from V

   V =
        1   3   7   9
```

```
>> V _ new = V          % new vector with  elements 1 through 4

   V _ new =
            1   3   7   9
```

Example 3.9

Given the following matrices:

$$A = \begin{bmatrix} 1 & 3 & 0 \\ 4 & 8 & 2 \\ 3 & -1 & -1 \end{bmatrix} \quad \text{and} \quad B = \begin{bmatrix} 1 & 6 & 5.3 \\ 1.8 & 7 & 3 \\ 2 & 5 & 0 \end{bmatrix}$$

Write the MATLAB commands that return the following:

a. matrices A and B
b. $C = A + B$
c. $D = A - B$
d. multiply array A by B and assign the result to E
e. divide array A by B and assign the result to F
f. add the second column of A to the first column of B
g. the condition number for matrices A and B
h. the determinant of A and B
i. if possible, the inverse of A and B
j. matrix A is converted to column and row vectors
k. the indexes of the nonzero elements of A and B
l. the matrix G consisting of the maximum values of either A or B
m. the matrix H consisting of the minimum values of either A or B
n. the matrix I that consists of multiplying every element of A by 3
o. the composite matrices $V = [A\ B]$ and $W = \begin{bmatrix} A \\ B \end{bmatrix}$
p. the matrix J by replacing all the elements above the main diagonal of A by zeros
q. the matrix K by replacing all the elements below the main diagonal of A by zeros

MATLAB Solution
```
>> A =  [1 3 0; 4 8 2; 3 -1 -1];          % matrix A
>> B = [1 6 5.3; 1.8 7 3; 2 5 0];         % matrix B
>> C = A+B                                % add A to B

   C =
         2.0000        9.0000        5.3000
         5.8000       15.0000        5.0000
         5.0000        4.0000       -1.0000

>> D = A-B                                % subtract B from A

   D =
              0       -3.0000       -5.3000
         2.2000        1.0000       -1.0000
         1.0000       -6.0000       -1.0000
```

```
>> E = A.*B                               % array product

   E =
       1.0000      18.0000            0
       7.2000      56.0000       6.0000
       6.0000      -5.0000            0

>> F = A./B                               % array division

   Warning: Divide by zero.
   F =
       1.0000       0.5000            0
       2.2222       1.1429       0.6667
       1.5000      -0.2000         -Inf

>> A2 = A(:, 2)                           % second column of A

   A2 =
        3
        8
       -1

>> B1 = B(:, 1)                           % first column of B

   B1 =
       1.0000
       1.8000
       2.0000

>> G = A2+B1                              % adds the second column
                                          %   of A to the
                                          %   first column of B

   G =
       4.0000
       9.8000
       1.0000

>> cond(A)                                % condition number of A.

   ans =
        12.9106

>> cond(B)                                % condition number of B

   ans =
        86.5485

>> det(A)                                 % determinant of A

   ans =
        24
```

```
>> det(B)                              % determinant of B

   ans =
        -5.5000

>> inv_A = inv(A)                      % inverse of A.

   inv_A =
        -0.2500      0.1250      0.2500
         0.4167     -0.0417     -0.0833
        -1.1667      0.4167     -0.1667

>> K = inv(B)                          % inverse of B

   K =
         2.7273     -4.8182      3.4727
        -1.0909      1.9273     -1.1891
         0.9091     -1.2727      0.6909

>> CA = A(:)                           % column vector with the
                                       %   elements of A

   CA =
         1
         4
         3
         3
         8
        -1
         0
         2
        -1

>> RA = CA'                            % transforms the columns
                                       %   vector into a row vector

   RA =
         1   4   3   3   8   -1   0   2   -1

>> find(A)                             % indexes of the nonzero
                                       %   elements of A

   ans =
         1
         2
         3
         4
         5
         6
         8
         9
```

```
>>                                          % note that the 7th. element
                                            is missing
>> find(B)                                  % indexes of the nonzero
                                            elements of B

   ans =
         1
         2
         3
         4
         5
         6
         7
         8

>>                                          % note that the 9th element
                                            is missing
>> G = max(A,B)                             % maximum values of A or B

   G =
      1.0000      6.0000      5.3000
      4.0000      8.0000      3.0000
      3.0000      5.0000           0

>> H = min(A,B)                             % minimum values of A or B

   H =
      1.0000      3.0000           0
      1.8000      7.0000      2.0000
      2.0000     -1.0000     -1.0000

>> I = A.*3                                 % multiplies the elements of
                                            A by 3

   I =
      3      9      0
     12     24      6
      9     -3     -3

>> V = [A B]                                % V becomes  a 3x6 matrix

   V =
      1.0000      3.0000           0      1.0000      6.0000      5.3000
      4.0000      8.0000      2.0000      1.8000      7.0000      3.0000
      3.0000     -1.0000     -1.0000      2.0000      5.0000           0

>> W = [A; B]                               % W becomes a 6x3 matrix

   W =
      1.0000      3.0000           0
      4.0000      8.0000      2.0000
      3.0000     -1.0000     -1.0000
      1.0000      6.0000      5.3000
      1.8000      7.0000      3.0000
      2.0000      5.0000           0
```

```
>> J = tril(A)                          % elements above the main
                                          diagonal of A become zeros


   j =
        1        0        0
        4        8        0
        3       -1       -1

>> K = triu(A)                          % elements below the diagonal
                                          of A become zeros


     ans =
          1        3        0
          0        8        2
          0        0       -1
```

Example 3.10

Write a set of MATLAB commands that return the following:

a. The month of October 2000
b. The magic 3×3 matrix
c. The first six rows of the Pascal triangle
d. The 5×5 random matrix, using the commands: *randn* and *rand*
e. A 5×5 matrix with all elements equal to -2
f. The 5×5 identity matrix (*eye*)
g. The 5×5 Hilbert matrix and label it H
h. The condition numbers for matrix H and the inverse of H
i. The eigenvalues and eigenvectors for the matrix H
j. The present year using the command *now*

MATLAB Solution
```
>> calendar(2000,10)                    % returns 10/2000


   Oct 2000
   S     M    Tu     W    Th     F     S
   1     2     3     4     5     6     7
   8     9    10    11    12    13    14
  15    16    17    18    19    20    21
  22    23    24    25    26    27    28
  29    30    31     0     0     0     0
   0     0     0     0     0     0     0

>> magic(3)                             % the magic 3x3 matrix


     ans =
          8        1        6
          3        5        7
          4        9        2
```

```
>> pascal(7)                                % the Pascal 7x7 matrix


   ans =
         1     1     1     1      1     1      1
         1     2     3     4      5     6      7
         1     3     6    10     15    21     28
         1     4    10    20     35    56     84
         1     5    15    35     70   126    210
         1     6    21    56    126   252    462
         1     7    28    84    210   462    924


>> randn(5)                                % the 5x5 random Gaussian
                                             matrix


   ans =
        -0.4326      1.1909     -0.1867      0.1139      0.2944
        -1.6656      1.1892      0.7258      1.0668     -1.3362
         0.1253     -0.0376     -0.5883      0.0593      0.7143
         0.2877      0.3273      2.1832     -0.0956      1.6236
        -1.1465      0.1746     -0.1364     -0.8323     -0.6918


>> rand(5)                                 % the 5x5 random matrix


   ans =
         0.9501      0.7621      0.6154      0.4057      0.0579
         0.2311      0.4565      0.7919      0.9355      0.3529
         0.6068      0.0185      0.9218      0.9169      0.8132
         0.4860      0.8214      0.7382      0.4103      0.0099
         0.8913      0.4447      0.1763      0.8936      0.1389


>> eye(5)                                  %  the 5x5 identity matrix


   ans =
         1     0     0     0     0
         0     1     0     0     0
         0     0     1     0     0
         0     0     0     1     0
         0     0     0     0     1


>> M = ones(5).*-2                         % the 5x5 vector with elements
                                             of -2


   M =
        -2    -2    -2    -2    -2
        -2    -2    -2    -2    -2
        -2    -2    -2    -2    -2
        -2    -2    -2    -2    -2
        -2    -2    -2    -2    -2


>> H = hilb(5)                             % the 5x5 Hilbert matrix
```

```
H =
    1.0000      0.5000      0.3333      0.2500      0.2000
    0.5000      0.3333      0.2500      0.2000      0.1667
    0.3333      0.2500      0.2000      0.1667      0.1429
    0.2500      0.2000      0.1667      0.1429      0.1250
    0.2000      0.1667      0.1429      0.1250      0.1111
```

>> cond(H)   % condition number for H

```
    ans =
        4.7661e+005
```

>> inv _ H = inv(H)   % inverse of H

```
    inv _ H =
            1.0e+005 *
        0.0002    -0.0030     0.0105    -0.0140     0.0063
       -0.0030     0.0480    -0.1890     0.2688    -0.1260
        0.0105    -0.1890     0.7938    -1.1760     0.5670
       -0.0140     0.2688    -1.1760     1.7920    -0.8820
        0.0063    -0.1260     0.5670    -0.8820     0.4410
```

>> cond(inv _ H)   % conditional number for
                                           inv _ H

```
    ans =
        4.7661e+005
```

>> [vec, valu] = eig(H)   % eigenvector and eigenvalues
                                           of H

```
    vec =
        0.0062      0.0472      0.2142     -0.6019      0.7679
       -0.1167     -0.4327     -0.7241      0.2759      0.4458
        0.5062      0.6674     -0.1205      0.4249      0.3216
       -0.7672      0.2330      0.3096      0.4439      0.2534
        0.3762     -0.5576      0.5652      0.4290      0.2098
    valu =
        0.0000           0           0           0           0
             0      0.0003           0           0           0
             0           0      0.0114      0.2085           0
             0           0           0           0      1.5671
```

>> x = now   % number of days since year
                                           zero

```
    x =
        7.3297e+005
```

>> present _ year = x./365.2604   % 1 year = 365 days +6.25hrs

```
    present _ year =
                2.0067e+003             % estimates the present year
```

Example 3.11

Given the following MATLAB equations:

$$X = (2.^5.^2)$$
$$Y = (2.^5).^2$$
$$Z = 2.^(5.^2)$$

Write a set of MATLAB commands that return the following:

a. X, Y, and Z as string vectors
b. the length of each string
c. the string matrix as $A = \begin{bmatrix} X \\ Y \\ Z \end{bmatrix}$
d. the size of matrix A
e. evaluate X, Y, and Z
f. the ASCII code for X, Y, and Z
g. determine if string X contains the string $V = $ '2.^5'
h. concatenate the string X, Y, and Z into a row vector
i. concatenate the strings X, Y, and Z into a column vector

MATLAB Solution
```
>> format compact
>> X = '(2.^5.^2)'                          % strings X, Y and Z are
                                              created

    X =
       (2.^5.^2)
>> Y = '(2.^5).^2'

    Y =
       (2.^5).^2

>> Z = '2.^(5.^2)'

    Z =
       2.^(5.^2)

>> lenX = length(X), lenY = length(Y), lenZ = length(Z)

    lenX =
            9
    lenY =
            9
    lenZ =
            9

>> A = [X;Y;Z]                              % creates the sting matrix A

    A =
       (2.^5.^2)
       (2.^5).^2
       2.^(5.^2)

>> [row, col] = size (A)

    row =
            3
    col =
            9
```

```
>> eval(X)                          % evaluates each expression:
                                      X, Y, and Z

   ans =
        1024

>> eval (Y)

   ans =
        1024

>> eval (Z)

   ans =
        33554432

>> double(X)                        % converts  to ASCII.

   ans =
        40    50    46    94    53    46    94    50    41

>> double (Y)

   ans =
        40    50    46    94    53    41    46    94    50

>> double (Z)

   ans =
        50    46    94    40    53    46    94    50    41

>> V = '2.^5'                       % string V

   V =
        2.^5

>> findstr(X,V)                     % finds if V is contained in X

   ans =
        2

>> strcat(X,Y,Z)                    % concatenates strings X, Y, Z
                                      as a row

   ans =
        (2.^5.^2)(2.^5).^22.^(5.^2)

>> strvcat(X;Y;Z)                   % concatenates strings X, Y, Z
                                      as a column

   ans =
        (2.^5.^2)
        (2.^5).^2
        2.^(5.^2)
```

Example 3.12

Create the MATLAB script file *British_flag* that returns the British flag by first creating a sparse matrix and then displaying its structure using the *spy* function.

MATLAB Solution
```
% Script file: British _ flag
a = ones(100,1);
A = spdiags([a a a a a ],[-6,-2,0,2,6],100,100);
figure(1);
B = flipud(A);
Row = [1:100 1:100 1:100 1:100 1:100];
Col = [ 46*ones(1,100)  49*ones(1,100)  50*ones(1,100)    51*ones(1,100)
54*ones(1,100)];
values = ones(1,500);
C = sparse(row,col,values,100,100);
D = rot90(C);
E = A+B+C+D;
spy(E);
title('British flag')
```
The resulting *spy* plot is shown in Figure 3.6.

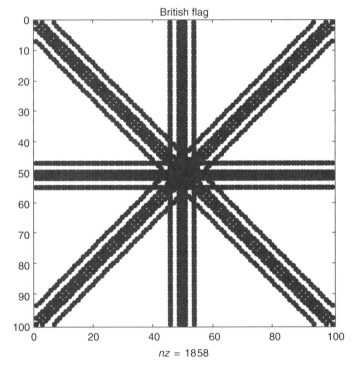

FIGURE 3.6
spy diagram of the British flag of Example 3.12.

Example 3.13

Create the MATLAB script file *big_V* that returns the plot of the letter *V* by creating an appropriate 50 × 100 sparse matrix and by using the *spy* function.

MATLAB Solution
```
% Script file:big_V
a = ones(50,1);
A = spdiags([a a a a a ],[-4,-2,0,2,4],50,50);
figure(1);
B = flipud(A);
C = [A B];
spy(C);
title('Big V')
```

The resulting *spy* plot is shown in Figure 3.7.

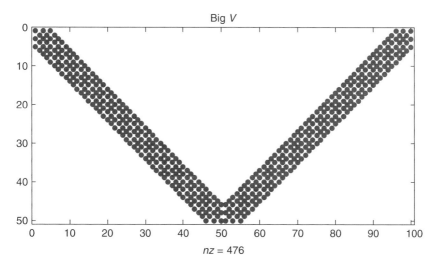

FIGURE 3.7
spy diagram of Example 3.13.

3.5 Further Analysis

Q.3.1 Load and run the program of Example 3.1.

Q.3.2 Run the program of Example 3.1 without the semicolons (;).

Q.3.3 Run Example 3.1 for $N = 50$ and 100.

Is the series divergent? Comment.

Q.3.4 Run Example 3.1 for the case of $a = 0.6$ and 0.75. Are the series convergent? If yes, what do they converge to?

Q.3.5 What variable is used to test for convergence?

Q.3.6 Load and run the program of Example 3.2.

Q.3.7 Load and run the program of Example 3.2 without the semicolons (;).

Q.3.8 Determine the sizes and lengths of the vectors n and m.

Q.3.9 Change the program of Example 3.2 to compute $(n - 1)!/n!$ and $n!/(n - 1)!$

Q.3.10 Load and run the program of Examples 3.3.

Q.3.11 List the variables used and their sizes.

Q.3.12 Determine by hand the first, fifth, and tenth elements of the vector y and compare your results with the values of the variables a1, a5, and a10.

Q.3.13 What is the difference between the statements $n. \wedge 2$ and $n \wedge 2$?

Q.3.14 Modify the program of Example 3.3 for the case of odd numbers and obtain the square of the first, fifth, and tenth elements. Run the modified program and display the sequence y.

Q.3.15 Determine by hand computation of the first, fifth, and tenth elements of the sequence generated in Q.3.14.

Q.3.16 Load and run the program of Example 3.4 and obtain matrix A.

Q.3.17 What is the size and total number of elements of A?

Q.3.18 Determine by hand elements $A(5, 5)$, $A(3, 10)$, and $A(10, 2)$, and compare your results with the values of variables of a5_5, a3_10, and a10_2.

Q.3.19 Using MATLAB list all the variables used in Example 3.4 and their sizes.

Q.3.20 Rerun the program of Example 3.4, instruction by instruction, and verify the comments placed next to each instruction.

Q.3.21 Determine by hand the transpose of $A(F = A')$ and compute $F(1, 2)$, $F(4, 2)$, and $F(5, 5)$, and compare them with F1_2, F4_2, and F5_5.

Q.3.22 What should be the rank of A if A has an inverse?

Q.3.23 What is the meaning and purpose of the instruction cond(A)?

Q.3.24 Describe the composition of matrix A in Example 3.4. How many submatrices constitute matrix A? Can you decompose matrix A into six submatrices? Define and construct matrix A in terms of those submatrices.

Q.3.25 Load and run the program of Example 3.5.

Q.3.26 Check by hand, if the elements of the vectors x and y represent the maximum and minimum value, for each column of R, respectively.

Q.3.27 Evaluate by hand the vector that consists of the elements of the main diagonal of R and compare your result with the values of vectdiR.

Q.3.28 Evaluate sumdiR without using the sum instruction.

Q.3.29 Evaluate by hand the average value of the elements of the main diagonal of R, and compare your result with avediR.

Q.3.30 Are the results obtained for the determinant, rank, and condition of R compatible? Discuss.

Q.3.31 Does the center, 2×2 matrix, and Rcenter possess an inverse?

Q.3.32 Without performing any computations, what is the rank of Rcenter?

Q.3.33 Besides reshaping matrix R with an 8×2 and a 2×8 matrices, indicate other possibilities of reshaping R.

Q.3.34 What is the difference between the instructions $expm(R)$ and $exp(R)$?

Q.3.35 What is the difference between the instructions $sqrt(R)$ and $sqrtm(R)$?

Q.3.36 Load and run the program of Example 3.6.

Q.3.37 What is the size of vector Celsius?

Q.3.38 What is the total number of variables used in the program? Can you run Example 3.6 with fewer variables?

Q.3.39 Verify each one of the comments placed next to the instructions.

Q.3.40 Load and run the program of Example 3.7.

Q.3.41 Rewrite the program of Example 3.7, for the case of incorporating two additional resistors $R5 = 1$ and $R6 = 9$ (in Ohms), for the series and parallel cases.

Q.3.42 Compare the result obtained in Example 3.7 with the result obtained by executing Q.3.41. Can you derive any general conclusion?

Q.3.43 Load and run the program of Example 3.8.

Q.3.44 Compute by hand:

    a. The size of V

    b. The smallest and largest element of V

    c. The sum, product, mean, and median of all the elements of V

    d. Norm of V

Q.3.45 What specific instructions in Example 3.8 provide the answers for Q.3.44?

Q.3.46 List the variable names used in Example 3.8.

Q.3.47 Reshape V as a square matrix different from the one shown in Example 3.8.

Q.3.48 Reshape V as a 2×8 and an 8×2 matrices.

Q.3.49 Partition vector V into two vectors V_1 and V_2, where V_1 consists of elements 1 through 8 and V_2 consists of elements 9 through 16 and modify and rerun Example 3.8.

Q.3.50 Load and run the program of Example 3.9.

Q.3.51 What are the sizes of matrices A and B?

Q.3.52 What are the condition numbers for A and B?

Q.3.53 What do the condition numbers indicate?

Q.3.54 According to the condition numbers of Q.3.52, which matrix is better conditioned?

Q.3.55 Is matrix A a symmetric matrix? If not, by what other matrix should A be multiplied by to obtain a symmetric matrix?

Q.3.56 Write a program that interchanges the main diagonals of matrix A with matrix B.

Q.3.57 Load and run the program of Example 3.10.

Q.3.58 Define the Magic and Pascal matrices.

Q.3.59 Write a program that verifies that indeed the 3×3 matrix obtained by using the MATLAB function *magic(3)* is *magic*.

Q.3.60 Define what is meant by eigenvector and eigenvalue.

Q.3.61 Define the Hilbert matrix and generate *hilb(n)*, for an $m = 10$.

Q.3.62 What is the size of the matrix generated by the command $H = hilb(10)$?

Q.3.63 Write a set of instructions that exchange the last and the first row of the matrix H.

Q.3.64 Repeat Q.3.63 for the case of the columns of H.

Q.3.65 Load and run the program of Example 3.11.

Q.3.66 Define what is meant by a string matrix?

Q.3.67 When can the instruction *eval* be used on a string vector?

Q.3.68 Why is X equal to Y, but not equal to Z? Explain.

Q.3.69 What is meant by a code word and an ASCII-coded character.

Q.3.70 Describe the characters of the ASCII code.

Q.3.71 Encode the word MATLAB using ASCII by hand.

Q.3.72 Write a program that returns the encoded string MATLAB in ASCII.

Q.3.73 Load and run the script file *British_flag* of Example 3.12.

Q.3.74 Modify the program of Example 3.12, so that only the upper portion of the cross is displayed.

Q.3.75 Repeat Q.3.74 for the lower portion.

Q.3.76 Modify the program of Example 3.12 to display only the half right side.

Q.3.77 Load and run the script file *big _V* of Example 3.13.

Q.3.78 Modify and run the program of Example 3.13 that will return the plot of the letter W.

3.6 Application Problems

P.3.1   Let us

a. Construct a column vector V_1 with the following elements: $-1, 0, -2$, and 3

b. Construct a column vector V_2 with the following elements: $3, -1, -7$, and 9

c. Construct a matrix A, whose columns are V_1 and V_2

d. Construct a matrix B, whose columns are V_1, V_1, V_2, V_2, and $V_1 + V_2$

P.3.2   Let $x = 1{:}10$. Evaluate the following commands:

a. x-x

b. x.^x

c. x. * x

d. x * x′

e. x′ * x

f. x.\x

g. x./x

h. x = 'x'

P.3.3   Evaluate by hand and then check by using MATLAB the output y generated by the following sequence:

$$x = [\text{-pi:pi}/2{:}2 * \text{pi}]$$

$$y = \text{x.^2-pi}$$

P.3.4   Analyse the command *sum* $= A + B$, and discuss the necessary and sufficient conditions for the *sum* to exist.

P.3.5   Discuss the various conditions for the existence of the following commands:

$$P_1 = A * B$$

$$P_2 = A. * B$$

$$P_2 = A.{^\wedge}B$$

$$P_3 = A/B$$

$$P_4 = A./B$$

P.3.6   Evaluate and give a descriptive comment next to each instruction of the following program:

$$X = [\text{-}10\text{:}1\text{:}10]$$

$$A = \text{eye(length(x))}$$

$$B = \text{fix(rand(size(A))}$$

$$C = \text{zeos(size(A))} + B$$

$$D = \text{tril(ceil(rand(4)))}$$

$$E = \text{tril(fix(5 } * \text{ randn(4)))}$$

$$F = \text{diag(d). } * \text{ diag(E)}$$

P.3.7   Create the following matrices using MATLAB:

$$A = \begin{bmatrix} 1 & 2 & 3 \\ 4 & 5 & 6 \\ 7 & 8 & 9 \end{bmatrix} \quad \text{and} \quad B = \begin{bmatrix} 1 & 2 \\ 3 & 4 \\ 5 & 6 \end{bmatrix}$$

Determine using MATLAB commands
1. Size of A and B
2. Rank of A
3. Determinant of A
4. Transpose of A
5. Inverse of A
6. $C = A * B$
7. Maximum and minimum values of the elements in C
8. Append B to A to return a 3×5 matrix
9. Create a 3×2 array D, consisting of all the elements in the first two columns of A

P.3.8   Given matrices A and B are

$$A = \begin{bmatrix} 1 & 2 \\ 3 & 4 \end{bmatrix} \quad \text{and} \quad B = \begin{bmatrix} 5 & 6 \\ 7 & 7 \end{bmatrix}$$

a. Evaluate the following using MATLAB:
   1. $C = A * B$
   2. $D = B * A$
   3. $E = A * 3$
   4. $F = B. * A$

b. Create a 2 × 3 array consisting of all the elements of *A* and the first column of *B*

c. Create the vectors *V* and *W* consisting of

$$V = [A \quad B]$$

and

$$W = \begin{bmatrix} A \\ B \end{bmatrix}$$

P.3.9   Using matrices *A* and *B* from P.3.8 evaluate

$$C = A \wedge 3$$

$$D = A. \wedge 3$$

Observe and discuss the differences.

P.3.10   Execute the following MATLAB command: $x = 10 * rand(1, 20)$ and record vector *x*. Write a set of MATLAB commands that return

a. The maximum value of *x*

b. The minimum value of *x*

c. The sum of all the elements of *x*

d. The product of all the elements of *x*

e. The average value of the elements of *x*

f. The median value of the elements of *x*

g. The size of *x*

h. Identify and display the value of the 7th element of *x*

i. Rearrange the values of *x* in ascending and descending order

P.3.11   Enter the MATLAB command $x = rand(1, 100)$ that returns a row vector consisting of 100 random elements. Write a MATLAB program that adds all the elements of *x* with even indexes indicated by the following equation:

$$sum\_even = \sum_{k=1}^{50} x(k * 2)$$

P.3.12   Repeat P.3.11 for the case of the elements with odd indexes indicated by:

$$sum\_odd = \sum_{k=1}^{50} x(k * 2 + 1)$$

P.3.13   Write a MATLAB program that returns $sum\_series = \sum_{n=1}^{N} ka^n$, for $a = 0.5$ and $N = 10$ and 20. Verify if the preceding series converges to $2 * k$ for any integer *k*.

P.3.14 Evaluate and record the response of each of the following MATLAB commands:

a. X = 1:1:25

b. Y = rand(5)

c. V = randn(5)

d. U = sqrt(X)

e. R = [1:7, 3:3:27, 5:.5:15]

f. S = [-pi:pi/2:pi]

g. T = [Y ones(size(Y)); zeros(size(Y)), eye(size(Y))]

P.3.15 Write a MATLAB program or a script file that returns the following sequence $y = \sum_{n=1}^{N}(1/n)$ for $N = 10$ and 50.

P.3.16 Write and run a program that approximates e^x, by using the following equation:

$$e^x = \sum_{n=0}^{k} \frac{x^n}{n!}$$

for $k = 10$, 15, and 30, and compare the approximations with the MATLAB built-in function $exp(1)$.

P.3.17 Given

$$A = [1 \quad 2 \quad 3], B = [-5 \quad 0 \quad -1], \text{ and } C = \begin{bmatrix} 1 & 4 & 7 \\ 2 & 5 & 8 \\ 3 & 6 & 9 \end{bmatrix}$$

Use MATLAB to create A, B, and C, then execute and explain the action of the following commands:

a. *max (A)*

b. *min (B)*

c. *dot (A, B)*

d. *D = C ∗ B*

e. *min (A, B)*

f. *cumprod (B)*

g. *max (A, B)*

h. *sort((−1). ∗ B + A)*

i. *sort(C)*

j. *mean (B)*

k. *median (A)*

l. *cumsum (B)*

P.3.18 The following equation defines a geometric series given by

$$B_n = \sum_{i=0}^{n} az^i = a(1 + z + z^2 + \cdots + z^{n-1})$$

Write a MATLAB program that verifies that $B_n = a(1 - z^n)/1 - z$, for a finite n ($n \neq \infty$) and $B_n = a/(1 - z)$, for $n = \infty$, converges for $-1 < z < 1$, and diverges otherwise.

Verify the convergence for various values of z such as -0.5, 0.2, and 0.7, and divergence for -1.3, 2.0, and 3.0.

P.3.19 The following equation defines an arithmetic series, given by

$$B_n = \sum_{i=1}^{n}(a + (i - 1)d)$$

Write a MATLAB program that verifies that the above arithmetic infinite series always diverges for any a and d (test your program for $a = 1$ and $d = 2$, and rerun it for any arbitrary a or d).

P.3.20 A power series is defined by the following equation:

$$B_n = \sum_{k=1}^{n}\frac{1}{k^x} = 1 + \frac{1}{2^x} + \frac{1}{3^x} + \frac{1}{4^x} + \cdots + \frac{1}{n^x}$$

Write a MATLAB program that verifies if the series converges or diverges for

a. The infinite series for $x = 2$

b. The infinite series for $x = 5$ and 1

P.3.21 Use MATLAB and evaluate the following series: $y = \sum_{n=1}^{N}(4n - 3)/(3n - 4)$, for $N = 50, 100$, and 200, and verify if the series converges to 1.33.

P.3.22 Evaluate the following series: $y = \sum_{n=1}^{N}(-3)^{n-1}/4^n$, for $N = 50, 100$, and 200, and verify if the series converges to 0.1429.

P.3.23 Evaluate the following series: $y = \sum_{n=1}^{N}(-5)^n/n!$, for $N = 50, 100$, and 200, and show that the series converges to zero.

P.3.24 Recall that the determinant of a square matrix is a scalar. For a 2×2 matrix A, the determinant of A is defined as follows in terms of its elements. Let

$$A = \begin{bmatrix} a & c \\ b & d \end{bmatrix}, \text{ then } \det(A) = a * b - b * c$$

For a 3×3 matrix B, the determinant of B is defined as follows in terms of its elements:

$$B = \begin{bmatrix} a & d & g \\ b & e & h \\ c & f & i \end{bmatrix}, \text{ then } \det(B) = a * \begin{bmatrix} e & h \\ f & i \end{bmatrix} - b * \begin{bmatrix} d & g \\ f & i \end{bmatrix} + c * \begin{bmatrix} d & g \\ e & h \end{bmatrix}$$

$$\det(B) = a * (e * i - f * h) - b * (d * i - f * g) + c * (d * h - e * g)$$

$$\text{Given } B = \begin{bmatrix} 3 & -1 & -2 \\ 2 & 3 & -4 \\ -4 & 7 & -6 \end{bmatrix}$$

a. Determine by hand the determinant of B.

b. Using MATLAB create the matrix B and obtain the *det(B)*. Compare your result with that of part (a).

c. Determine by hand the transpose of B.

d. Use MATLAB to evaluate the transpose of B. Compare your result with that of part (c).

e. Create a random 3 × 3 matrix C, and using MATLAB verify that *det(B*C) = det(B) * det(C)*.

P.3.25 The inverse of a matrix A is a matrix B if and only if A * B = I, where I is the identity matrix. Recall that the identity matrix is defined as a square matrix with ones along the left to right diagonal and zeros elsewhere.

A 2 × 2 matrix A is defined as follows in terms of its elements, as well as its inverse B.

$$A = \begin{bmatrix} a & b \\ c & d \end{bmatrix}$$

then the inverse of A is

$$B = \frac{1}{D}\begin{bmatrix} d & -b \\ -c & a \end{bmatrix}$$

where $D = \det(A) = a * d - c * b$.

Given

$$A = \begin{bmatrix} 2 & 3 \\ 4 & 7 \end{bmatrix}$$

a. Evaluate by hand B, the inverse of A

b. Verify by hand that A * B = I

c. Verify using MATLAB parts (a and b)

P.3.26 Verify using MATLAB that A + B = B + A, but A * B ≠ B * A, for the following matrices:

$$A = \begin{bmatrix} 3 & 1 & 1 \\ 2 & -1 & 1 \\ 5 & 2 & 3 \end{bmatrix} \quad \text{and} \quad B = \begin{bmatrix} 4 & 0 & -1 \\ 0 & 0 & 2 \\ 2 & 1 & 4 \end{bmatrix}$$

P.3.27 Use MATLAB to express the following expressions as single matrices:

a. $4 * \begin{bmatrix} 2 & 1 \\ 1 & 1 \end{bmatrix} + 3 * \begin{bmatrix} -1 & 1 \\ 2 & 4 \end{bmatrix} =$

b. $\begin{bmatrix} 1 & 2 & 1 & 0 \\ 0 & 3 & 1 & 0 \end{bmatrix} + 3 * \begin{bmatrix} -1 & 1 & 0 & 1 \\ 1 & 2 & 1 & 0 \end{bmatrix} =$

c. $-3 * \begin{bmatrix} 1 & 1 \\ 3 & 2 \\ 4 & 6 \end{bmatrix} - \dfrac{1}{2} * \begin{bmatrix} -4 & 6 \\ -3 & 8 \\ 4 & 9 \end{bmatrix} =$

P.3.28 Given are the matrices:

$$A = \begin{bmatrix} 2 & 1 \\ 1 & 1 \end{bmatrix}, B = \begin{bmatrix} 1 & 2 & -1 \\ 1 & 1 & 1 \end{bmatrix}, C = \begin{bmatrix} 6 & 1 & 2 & 1 \\ 4 & 0 & 1 & -1 \\ 2 & 1 & 3 & 0 \end{bmatrix}, \text{ and } D = \begin{bmatrix} 4 & 2 & 1 \\ 1 & 2 & 0 \\ 2 & -3 & -1 \\ 6 & 1 & 1 \end{bmatrix}$$

Show that

$$A * B = \begin{bmatrix} 3 & 5 & -1 \\ 5 & 8 & -1 \end{bmatrix} \text{ and } C * D = \begin{bmatrix} 35 & 8 & 5 \\ 12 & 4 & 2 \\ 15 & -4 & -1 \end{bmatrix}$$

a. By hand calculation
b. Using MATLAB

P.3.29 Verify using MATLAB that $A * B \neq B * A$, for the following matrices:

a. $A = \begin{bmatrix} 2 & 1 & 5 \\ 1 & 2 & 3 \end{bmatrix}$ and $B = \begin{bmatrix} 1 & 2 \\ 3 & 1 \\ -1 & 4 \end{bmatrix}$

b. $A = \begin{bmatrix} 1 & 1 \\ 0 & 1 \end{bmatrix}$ and $B = \begin{bmatrix} 1 & -3 \\ 0 & 1 \end{bmatrix}$

P.3.30 Verify using MATLAB that $A^m * A^n = A^{m+n}$ and $(A^m)^n = A^{m*n}$, for $A = \begin{bmatrix} 1 & 2 \\ -1 & 1 \end{bmatrix}$, $m = 3$, $n = 2$, and $m = n = 2$.

P.3.31 Let the column vectors be

$$X = \begin{bmatrix} 1 \\ 2 \\ 3 \\ 4 \end{bmatrix} \text{ and } Y = \begin{bmatrix} 0 \\ -1 \\ -2 \\ 3 \end{bmatrix}$$

Use MATLAB to evaluate the angle between X and Y.

P.3.32 Given the vectors

$$A = \begin{bmatrix} 1 \\ 3 \\ 5 \\ 7 \end{bmatrix} \text{ and } B = \begin{bmatrix} 2 \\ 4 \\ 6 \\ 8 \end{bmatrix}$$

Evaluate the following using MATLAB:

a. Are the vectors A and B mutually orthogonal?

b. The norms of A and B

c. The angle between A and B

d. The dot and cross product of A and B

e. Verify Cauchy–Schwartz inequality

P.3.33   Let A be an arbitrary matrix. Define conditions for which $A * A^T$ exists.

P.3.34   Let

$$A = \begin{bmatrix} 1 & 2 & 3 \\ -3 & -5 & 9 \end{bmatrix}$$

use MATLAB and obtain

a. $B = A * A^T$

b. $C = A^T * A$

P.3.35   Verify that $(A * B)^T = B^T * A^T$ (recall that T stands for transpose), using the following matrices:

$$A = \begin{bmatrix} 1 & 4 \\ 3 & 7 \end{bmatrix} \quad \text{and} \quad B = \begin{bmatrix} 2 & 6 \\ 8 & 9 \end{bmatrix}$$

P.3.36   Let A be on abitrary matrix. Define conditions for which $A * A^{-1}$ exist.

P.3.37   Given

$$A = \begin{bmatrix} 1 & -1 \\ 1 & -1 \end{bmatrix}$$

verify that

$$A * A = A^2 = \begin{bmatrix} 0 & 0 \\ 0 & 0 \end{bmatrix}$$

Discuss the results obtained.

P.3.38   Given the diagonal matrix

$$A = \begin{bmatrix} 1 & 0 & 0 \\ 0 & 2 & 0 \\ 0 & 0 & 3 \end{bmatrix}$$

show that

$$A^n = \begin{bmatrix} 1^n & 0 & 0 \\ 0 & 2^n & 0 \\ 0 & 0 & 3^n \end{bmatrix} \text{ for any } n.$$

Use MATLAB to verify the above statement for $n = 3$ and 10.

P.3.39 Let A and B be two diagonal matrices. Verify that $A * B = B * A$, using the following matrices:

$$A = \begin{bmatrix} 1 & 0 & 0 & 0 \\ 0 & 2 & 0 & 0 \\ 0 & 0 & 3 & 0 \\ 0 & 0 & 0 & 4 \end{bmatrix} \text{ and } B = \begin{bmatrix} 5 & 0 & 0 & 0 \\ 0 & 6 & 0 & 0 \\ 0 & 0 & 7 & 0 \\ 0 & 0 & 0 & 8 \end{bmatrix}$$

P.3.40 Let

$$A = \begin{bmatrix} 1 & 0 & 0 \\ 0 & 2 & 0 \\ 0 & 0 & 3 \end{bmatrix}$$

Verify using MATLAB that $A^2 \neq 0$, but $A^n = 0$, for any integer $n > 2$.

P.3.41 Let

$$A = \begin{bmatrix} 1/2 & 1/2 \\ 1/2 & 1/2 \end{bmatrix}$$

Verify if $A^n = A$ for any integer n.

P.3.42 Let

$$A = \begin{bmatrix} -1 & 0 \\ 0 & 1 \end{bmatrix}$$

Verify if

$$A^n = I = \begin{bmatrix} 1 & 0 \\ 0 & -1 \end{bmatrix} \text{ for any integer } n \geq 2$$

P.3.43 Let A be a diagonal matrix given by

$$A = \begin{bmatrix} a_{11} & & & & \\ & a_{22} & & 0 & \\ & & a_{33} & & \\ & 0 & & \ddots & \\ & & & & a_{nn} \end{bmatrix}, \text{ then } inv(A) = \begin{bmatrix} 1/a_{11} & & \cdots & & \\ & 1/a_{22} & & 0 & \\ & & 1/a_{33} & & \\ \cdots & 0 & & & \cdots \\ & & & & a_{nn} \end{bmatrix}$$

Use MATLAB to verify the preceding statement for

$$A = \begin{bmatrix} 1 & 0 & 0 & 0 \\ 0 & 2 & 0 & 0 \\ 0 & 0 & 3 & 0 \\ 0 & 0 & 0 & 4 \end{bmatrix}$$

P.3.44 For

$$A = \begin{bmatrix} 1 & 1 & 1 \\ 1 & -1 & 1 \\ -1 & 1 & 1 \end{bmatrix} \quad \text{and} \quad B = \begin{bmatrix} 1/2 & 0 & 1/2 \\ 1/2 & 1/2 & 0 \\ 0 & 1/2 & 1/2 \end{bmatrix}$$

Verify using MATLAB that if $A * B = B * A$, then A is the inverse of B.

P.3.45 Show that $(A^T)^{-1} = (A^{-1})^T$ for matrix A of P.3.42 (recall that T indicates transpose and -1 indicates inverse).

P.3.46 Verify using MATLAB that $(A * B)^{-1} = (B^{-1} * A^{-1})$, for the matrices:

$$A = \begin{bmatrix} 1 & 2 \\ 3 & 4 \end{bmatrix} \quad \text{and} \quad B = \begin{bmatrix} 1 & 6 \\ 5 & 7 \end{bmatrix}$$

P.3.47 Let the following system of equations be

$$x - 0.5y + 0.5z = 4$$

$$0.33x + 0.67y + z = 3$$

$$x - 0.33z = 1$$

a. Convert the set of linear equations into a matrix equation
b. Create and process the matrix equation using MATLAB
c. Solve numerically for the unknowns x, y, and z
d. Solve symbolically for the unknowns x, y, and z

P.3.48 Let

$$A = \begin{bmatrix} 5 & -8 & -1 \\ 4 & -7 & -4 \\ 0 & 0 & 4 \end{bmatrix}$$

a. Define the characteristic equation
b. Determine the characteristic equation
c. Show that the characteristic polynomial is given by:

$$\lambda^3 - 2\lambda^2 - 11\lambda + 12$$

d. Verify that the eigenvalues of A are $\lambda = 4, -3,$ and 1

e. Verify that the eigenvectors of A are

$$\begin{bmatrix} 1 \\ 0 \\ 1 \end{bmatrix}, \begin{bmatrix} 1 \\ 1 \\ 0 \end{bmatrix}, \text{ and } \begin{bmatrix} 2 \\ 1 \\ 0 \end{bmatrix}$$

P.3.49 Create an upper random triangular matrix of order 3, and determine its eigenvalues and eigenvectors.

P.3.50 Repeat problem P.3.49, for a third order lower random triangular matrix.

P.3.51 Create random diagonal matrices of order 3, 4, and 5, and evaluate in each case its eigenvalues and eigenvectors. Discuss the results.

P.3.52 Create a random vector consisting of 100 elements with a uniform random distribution between 0 and 10, and determine the average, the median, the variance, and the standard deviation.

P.3.53 A third-order magic square matrix can be formed using the integers 1 through 9. Construct seven other third-order square matrices from these integers.

P.3.54 Show that the constants for a fourth-order magic matrix constructed with the integers 1 through 16 is 34.

P.3.55 Verify using MATLAB that

$$\binom{n}{0} = \binom{n}{n} = 1, \text{ and } \binom{n}{1} = \binom{n}{n-1} = n$$

P.3.56 Determine the binomial coefficients for $n = 5, 6,$ and 7 using the Pascal's triangle and verify using Newton's formula.

P.3.57 Determine the binomial coefficients using Pascal's triangle and Newton's formula for the following expressions:

a. $(a^2 + 3\sqrt{b})^3$

b. $(2\sqrt{a} + x)^4$

c. $(1 + 3a)^5$

P.3.58 (a) Write a program with a minimum number of instructions that would generate a 10×10 matrix where all the columns consist of the identical sequence 1, 2, 3, ..., 10. (b) Repeat part (a) by substituting the word columns with rows.

P.3.59 Let us

a. Create a 3-D array X whose three layers (A, B, C) are given by the following MATLAB commands:

$$A = [1:3; 4:6; 7:9]$$

$$B = [2:2:6; 3:2:7; 4:2:8]$$

$$C = [-1:-1:-3; -2:-2:-6; -3:-3:-9]$$

   b. Display the elements of the array X

   c. Display the elements of all the diagonals

   d. Display the elements located at the second row

   e. Display the elements located at the second column

P.3.60 Create a sparse 200 × 200 matrix *A*, with about 10% of its elements consisting of random numbers. Next set all the elements of the main diagonal to 3, all the elements of the other diagonal to –1, and the element located at the intersection of the two diagonals to 10, while all the remaining elements are zero

   a. Evaluate the array density.

   b. Convert the sparse matrix *A* into a full matrix *B*.

   c. Compare the memory requirements to store *A* and *B*. Which is more efficient?

   d. Create the vector *C* = *randn(1, 200)*, and solve the system of equations *A* * *x* = *C* and *B* * *x* = *C*, and compare the two results. Which process is more efficient? And define the term efficiency.

P.3.61 Write a program that draws the U.S. flag by creating first a sparse matrix and displaying its structure using the MATLAB function spy (for simplicity draw only one star).

4

Trigonometric, Exponential, Logarithmic, and Special Functions

It is not in the nature of things for any one man to make a sudden violent discovery; science goes step by step, and every man depends on the work of his predecessor.

Scientists do not depend on the ideas of a single man, but on the combined wisdom of thousands of men.

Ernst Rutherford

4.1 Introduction

MATLAB® can be used as a scientific calculator in the sense that it offers, besides the means of evaluating the common arithmetic operations $(+, -, *, /, ^)$, it can be use to evaluate logarithmic, exponential, and trigonometric functions.

Most functions are executed just by calling them by using the proper syntax. Each function usually performs an operation that would otherwise take several programming instructions.

Because trigonometric, exponential, and logarithmic functions are often used in engineering and the sciences, it is convenient to define them as MATLAB functions and call them when needed, without having to write a program for each of them separately each time they are called. These MATLAB functions use easy to remember notations because their mnemonics closely resemble the function and considerably reduce the labor involved in writing a program.

Some of the MATLAB functions can perform complicated tasks such as rem(x, y), which returns the remainder after dividing x by y, where x and y could be polynomials.

These built-in functions are usually identified by three or four lower case letters, often referred to by the mnemonic that defines their action.

For example, if the cosine of x is desired to be computed, the following MATLAB instruction can be used:

$$b = cos(x)$$

where the variable b is assigned the value of the cosine, whose angle is specified by x given in radians. MATLAB works only in radians where

$$360 \; degrees = 2\pi \; radians = 1 \; clockwise \; revolution$$

$$1 \; degree = \{1/360\} \; revolution$$

$$1 \; radian = \{180/\pi\} \; degrees = 57.3 \; degrees$$

$$1 \; degree = \{\pi/180\} \; radians = 0.0175 \; radians$$

The argument of the MATLAB function is always placed in parenthesis and is preceded by the function's name. A function can be an argument of another function as long as the syntax and the function/parenthesis convention is maintained. For example,

$$b = cos(cos(x))$$

One of the family of functions presented in this chapter is the trigonometric functions. Trigonometry is a very old discipline, which dates back to the time of the old Greek astronomers such as Menelaus of Alexandria, as early as AD 100. Around the second century BC, Hipparchus and Ptolemy were credited as the founders of this branch of mathematics called trigonometry.

Trigonometry basically deals with angles and sides of triangles. These concepts were originally developed to serve astronomers, but over time it has evolved to serve in a variety of other applications, such as in navigation, surveying, and military and civilian constructions. In more modern times, trigonometry is used in a variety of additional applications involving the modeling of brain waves, sound waves, wave propagation and antennas, ocean tides, oscillations and vibrations, and many other phenomenas.

MATLAB uses the MacLaurin series representation to evaluate trigonometric expressions.

Because numbers, relations, and functions, in particular trigonometric, logarithmic, and exponential, are important in science and engineering, a brief presentation and chronological evolution of the major developments and contributions are summarized.

Although older civilizations used numbers and mathematical relations before the Greeks, the Greeks get the credit for tabulating, recording, and leaving historic proofs of much of the early discoveries and applications in science, philosophy, ethics, music, drama, logic, and mathematics.

The first human identified as having made a significant contribution to philosophy, logic, and mathematics was Thales of Miletus (634–548 BC). He is credited with establishing one of the first centers of learning at Miletus. At that time, Miletus was a Greek city on the west coast of Asia Minor, with strong commercial and cultural connections with the ancient civilizations of Egypt and Babylon (Newman, 1956).

Thales taught mathematics, philosophy, and logic, and is given credit for many early discoveries. Thales was considered one of the seven wise men of ancient Greece, and one of his major contribution was the use of the deductive method. This method of reasoning became the hallmark of Greek thought and philosophy, and centuries later of western logic.

One of Thales' students was Pythagoras (580–500 BC) from the island of Samos.

He became the engine of Greek philosophical thinking. Pythagoras philosophy was based on logical reasoning, relations, and numbers. In fact, for Pythagoras numbers were atoms of the universe and the prime cause of almost any event. For Pythagoras numbers could be used to explain, control, and influence all events, and were the building blocks of reality. In fact, Pythagoras philosophy evolved into a religious–philosophical order where discoveries were kept secret.

Pythagoras greatly influenced the Greek society as well as Plato, the father of idealism.

Probably, one of the greatest inventions ever made by man was the concept of numbers and the numbering system. But different civilizations over time developed different numbering systems.

In fact, the great ancient civilizations were great because they had their own unique numbering system, such as the Egyptians, Babylonians, Romans, and Mayans. Numbers and equations are the first mathematical achievement of mankind, and the early records were found in old Babylon in the third millennium BC and in ancient Egypt around 1800 BC. In old Babylon, mathematics was often used to settle legal questions involving the sharing of wealth and inheritance. According to the inheritance laws of old Babylon, the firstborn always receives the largest share, the second a little less, and so on, following a strict proportional sequence.

The modern numbering system is based on the Hindu–Arabic (Stein, 1964) system, believed to be developed in India and brought to Europe by the Arab traders. Equations and algebraic symbolic language evolved over centuries incorporating concepts, such as the equal sign (=), a relatively modern concept first proposed by Robert Recorde (1510–1558), a royal court physician, in his publication *The Whetstone of Witte* (1557).

One of the first known numbers since antiquity is the constant called π.

The history of π is in some respect the history of the evolution of man, in the great ancient civilizations that span over the past 4000 years. The modern definition of π is a simple ratio of a circumference of a circle to its diameter.

But π is much more complex than what its simple definition indicates. π was the object of study for thousands of years by the best human minds. Pi (π) happens to be an irrational number, that is, it cannot be expressed as an integer or a fraction. The earliest written record about π was found in Egypt around 1700 BC, and suggested that π was the ratio of $(16/9)^2 = 256/81$ or $3.16049....$

(Proofs are found in the *Rhind* or *Ahmes Papyrus* from Thebes, now in British and Brooklyn museums)

Archimedes from ancient Syracuse is credited with the following estimation for π (Petr, 1971).

$$3.140845 < \pi < 3.142857, \text{ around 220 BC}$$

Around AD 125, in ancient China, Chang Hong estimated π as $\pi = (10)^{1/2} = 3.162$.

Around AD 265, Wan Fan estimated π as $\pi = 142/45 = 3.1555$ and around 480 BC Ch'ung-Chih and his son Tsu Keng-Chih expressed π as $\pi = 355/113 = 3.1415929$, which is almost the correct value ($3.1415926 < \pi < 3.1415927$); an accuracy that was not attained in Europe until the sixteenth century.

The French mathematician Francois Vieta (AD 1592) was the first to express π as an infinite series. In 1655, the English mathematician John Wallis came out with a simpler series version, and later, William Bouncker, an Irish mathematician, came out with a more compact one.

Gottfried Leibniz (1646–1716) showed that

$$\pi = 4\left[1 - \frac{1}{3} + \frac{1}{5} - \frac{1}{7} + \frac{1}{9} - \frac{1}{9} - \frac{1}{11}...\right]$$

In 1873, William Shank, an English mathematician, evaluated π to 707 decimal places.

In 1948, John W. French Jr. (United States) and D. F. Fergunson (England) evaluated π to 808 decimal places, and in 1950, the value of π was calculated by an electronic computer to 2000 places.

The value of π is of critical importance in trigonometric relations and applications.

This chapter deals with trigonometry, exponentiation, logarithm, and special MATLAB functions. One of the special MATLAB functions is *prime(n)*, which returns the list of prime numbers up to the number n, a problem that has occupied mathematicians for thousands of years.

Why are prime numbers important?

Because all the natural numbers can be generated by multiplying the prime numbers, and in this way the sequence of numbers can be controlled.

This question was first studied by Euclid of ancient Greece, but only in the past 200 years serious research has been done in this area by some of the best mathematicians of all times, such as Riemann, Euler, Legendre, Gauss, and many others, leading to other discoveries such as the Zeta function by Euler (1737), Riemann functions, Riemann hypothesis, Mobius functions, and many others.

What is amazing is that all these functions are interrelated.

A brief summary of major events and contributions are listed as follows in chronological order starting with Pythagoras.

- Pythagorean theorem* (Greece, 540 BC) (Newman, 1956).
- Euclid (300 BC) in *Elements*, volume 13, used the law of cosines.
- $\sin(a + b)$ was effectively used (around 300 BC).
- Ptolemy used the law of sines (around 150 BC).
- Menelaus of Alexandria (AD 100) was one of the first individuals who used extensible trigonometric functions.
- The sine function was introduced in India (around AD 300).
- Nasir ed-din (AD 1250), a Persian astronomer, published the first book containing a systematic treatment of trigonometric functions.
- Regiomontanus (1436–1475) made trigonometry a part of mathematics.
- Copernicus (1473–1543) improved Regiomontanus' work.
- Rhaeticus (1514–1576) was the first to define the six trigonometric functions as they are presently known.
- John Napier (1550–1617) and Jobst Burgi (1552–1632) invented the logarithms.
- Thomas Finck (1583) defined the trigonometric functions with the present name.
- Roger Cote's formula: $\cos\theta = 1/2\ (e^{j\theta} + e^{-j\theta})$.
- John Napier (1614), from Scotland, introduced the base $e = 2.71\ldots$.
- Henry Briggs (1615), from Oxford in England, introduced the log base 10.
- James Stirling (1730) first introduced the MacLaurin series.
- Leonard Euler (1707–1783)[†] established the present notations and the famous relation $e^{j\theta} = \cos(\theta) + j\sin(\theta)$ (1743).
- Henry Briggs published the first table of common logarithms.
- Gauss (1792), at the age of 15, estimated the function prime(n) as $n/ln(n)$, and also estimated prime(n) as n approaches infinity.
- Andrien Marie Legendre (1798) estimated the function prime(n) as $n/[ln(n) - 1.08366]$ (Clawson, 1996).

---

[*] Recall that Pythagoras was a pupil of Thales (640–550 BC). Little is known about Thales or Pythagoras; neither left any known writing. Without Thales there would not have been a Pythagoras, and without Pythagoras there would not have been a Plato (427–347 BC), and without Plato the ancient Western civilization would have been different, deprived of many wonderful ideas.

[†] Leonhard Euler is one of the greatest mathematicians who ever lived. He received his bachelor's degree at 15 and his master's at 16 from the University of Baqsel. At 18 he published his first mathematics paper, and at 25 a two-volume text on mechanics. He became partially blind and later in his life he lost his sight completely. Yet he published 400 papers, enough maths to fill 90 volumes. He was a dedicated husband and a loving father to his 13 children.

- Friedrich Bernhard Riemann (1826–1866) discovered the zeta functions and estimated prime(n).
- August Ferdinand Mobius* (1832) developed the Mobius function, the reciprocal of the zeta functions.
- M. Deleglise (1992) estimated that prime(n) = 2,625,557,157,654,233, for $n = 10^{16}$.

4.2 Objectives

After completing this chapter the reader should be able to

- Understand the concept of degree and radian as units of angle measurements
- Convert from degrees to radians and vice versa
- Know the ratio definition of the basic trigonometric functions for the right triangle (cos, sin, tan, cot, csc, sec)
- Know the values of key trigonometric angles (cos(0), cos($\pi/4$), sin(π), ...)
- Draw the plots of the basic trigonometric functions
- Understand the law of sines
- Understand the law of cosines
- Use a series approximations to evaluate e = 2.71, ..., sin(x), cos(x), etc.
- Understand the concept of reciprocal
- Understand the trigonometric relations for angles located in any of the quadrants of the Cartesian plane
- Know that the trigonometric functions are periodic
- Know the range, domain, period, amplitude, and frequency of the basic six trigonometric functions
- Know the basic trigonometric identities that relate the sum, difference, doubling, and half of angles
- Know that trigonometric functions can be expressed in terms of exponentials
- Know that an exponential can be approximated in terms of a Mac Larin's series or a binomial
- Understand exponential and logarithmic functions ($\log_{10}(x)$, log(x), exp(x))
- Understand the rounding off of MATLAB functions (fix(x), floor(x), ceil(x), etc.)
- Define and use the hyperbolic functions (sinh(x), cosh(x), tanh(x), etc.)
- Define and use the inverse trigonometric functions (acos(x), asin(x), etc.)
- Use the inverse hyperbolic functions (acosh(x), asinh(x), atanh(x), etc.)
- Use the special-purpose MATLAB arithmetic functions (prime(n), factor, rem(x, y), gcd(x))
- Understand the close relations between exponential, trigonometric, and logarithmic functions
- Use the power of MATLAB to solve classes of exponential, trigonometric and logarithmic problems

---

* Student of Gauss and was considered by Gauss as his most talented student.

4.3 Background

R.4.1   A MATLAB function is generally assigned to a variable name located on the left of an equality, and the function itself is located at the right of the equality, with the argument in parenthesis, such as

$$y = sin(x)$$

where y is the variable name, *sin* the function, and x the argument.

R.4.2   Functions are expressed using lower case letters. Recall that MATLAB is case sensitive (*casesen on/off*).

R.4.3   The units frequently used to express angles are in degrees, minutes, and seconds. One complete circular revolution is defined as 360° 0′ 0″, where

$$1° = 60' \text{ (the symbol } ' \text{ stands for minutes)}$$

$$1' = 60'' \text{ (the symbol } '' \text{ stands for seconds)}$$

R.4.4   An alternate unit of measuring or expressing angles is the radian, where

$$1 \text{ radian} = (180/\pi) \text{ degrees} = 57.296° = 57° 17' 45''$$

$$1 \text{ degree} = (\pi/180 \text{ radians}) = 0.01745 \text{ radians}$$

MATLAB accepts only the radian as argument.

R.4.5   To convert x-radians to degrees, it is necessary to multiply x by 57.296 or divide by 0.01745.

R.4.6   To convert y-degrees to radians, multiply the number y by 0.01745 or divide by 57.296.

R.4.7   Some useful equivalences between degrees and radians are summarized in Table 4.1.

R.4.8   The six basic trigonometric functions that collectively apply to right triangles are sine, cosine, tangent, cotangent, secant, and cosecant. The terminology, syntax, and function definition as applied to the right triangle of Figure 4.1 are indicated below:

Function Name		MATLAB Notation		Value
sine of A	=	$sin(A)$	=	a/c
cosine of A	=	$cos(A)$	=	b/c
tangent of A	=	$tan(A)$	=	a/b
secant of A	=	$sec(A)$	=	c/b
cosecant of A	=	$csc(A)$	=	c/a
cotangent of A	=	$cot(A)$	=	b/a

TABLE 4.1

Degree–Radian Conversion

Degrees	0	30	45	60	90	180	270	360
Radians	0	$\pi/6$	$\pi/4$	$\pi/3$	$\pi/2$	π	$3\pi/2$	2π

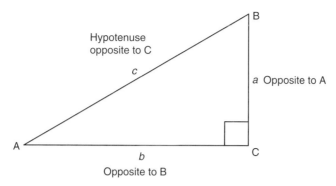

FIGURE 4.1
Right triangle of R.4.8.

R.4.9 Two acute angles are referred to as complementary if their sum is 90°. In the triangle shown in Figure 4.1, $\angle A$ (angle A) and $\angle B$ (angle B) are complementary angles, because $\angle A + \angle B = 90°$.

The relations shown in the following indicate the pairing of the trigonometric functions in the case of complementary angles

$$\sin(A) = \frac{a}{c} = \cos(B)$$

$$\cos(A) = \frac{b}{c} = \sin(B)$$

$$\tan(A) = \frac{a}{b} = \cot(B)$$

$$\sec(A) = \frac{c}{b} = \csc(B)$$

$$\csc(A) = \frac{c}{a} = \sec(B)$$

$$\cot(A) = \frac{b}{a} = \tan(B)$$

R.4.10 The paired functions, such as $\sin(A) = \cos(B)$ are referred as cofunctions. Any function of an acute angle x is equal to the corresponding cofunction of its complement of the angle $(90° - x)$.

R.4.11 The trigonometric values of various angles are summarized in Table 4.2.

The trigonometric values of any angle can be computed by using the appropriate MATLAB syntax with arguments in radians. For example, use MATLAB to determine the values of $\sin(\pi)$, $\sin(\pi/2)$, $\sin\pi$, and $\sin\pi/2$.

MATLAB Solution
```
>> sin.(pi)

    ans =
         1.2246e-016
```

TABLE 4.2

Trigonometric Values of Some Angles

Angle Degree/Radian	sin	cos	tan	sec	csc	cot
$0° = 0\ rad$	0	1	0	1	∞	∞
$30° = \frac{\pi}{6}\ rad$	$1/2$	$\sqrt{3}/2$	$\sqrt{3}/3$	$2\sqrt{3}/3$	2	$\sqrt{3}$
$45° = \frac{\pi}{4}\ rad$	$\sqrt{2}/2$	$\sqrt{2}/2$	1	$\sqrt{2}$	$\sqrt{2}$	1
$60° = \frac{\pi}{3}\ rad$	$\sqrt{3}/2$	$1/2$	$\sqrt{3}$	2	$2\sqrt{3}/3$	$\sqrt{3}/3$
$90° = \frac{\pi}{2}\ rad$	1	0	∞	∞	1	0
$120° = \frac{2\pi}{3}\ rad$	$\sqrt{3}/2$	$-1/2$	$-\sqrt{3}$	-2	$2\sqrt{3}/3$	$-\sqrt{3}/3$
$135° = 3\frac{\pi}{4}\ rad$	$\sqrt{2}/2$	$-\sqrt{2}/2$	-1	$-\sqrt{2}$	$\sqrt{2}$	-1
$150° = \frac{5\pi}{6}\ rad$	$1/2$	$-\sqrt{3}/2$	$-\sqrt{3}/3$	$-2\sqrt{3}/3$	2	$-\sqrt{3}$
$180° = \pi\ rad$	0	-1	0	-1	∞	∞
$270° = \frac{3\pi}{2}\ rad$	-1	0	∞	∞	-1	0
$360° = 2\pi\ rad$	0	1	0	1	∞	∞

```
>> sin.(pi/2)

  ans =
        1
>> sin pi/2

  ans =
        -0.8900    -0.9705     0.1236    -0.2624
>> sin pi

  ans =
        -0.8900    -0.9705
```

Note that some MATLAB responses are unexpected, when the argument is not in paranthesis.

R.4.12 Trigonometric functions such as sine, cosine, and tangent are evaluated using the MacLaurin series representation as follows:

$$sin(x) = x - \frac{x^3}{3!} + \frac{x^5}{5!} - \frac{x^7}{7!} + \frac{x^9}{9!} \cdots$$

$$cos(x) = 1 - \frac{x^2}{2!} + \frac{x^4}{4!} - \frac{x^6}{6!} + \frac{x^8}{8!} \cdots$$

$$tan(x) = \frac{sin(x)}{cos(x)} = x - \frac{x^3}{3!} + \frac{x^5}{5!} - \frac{x^7}{7!} + \frac{x^9}{9!} \cdots \bigg/ 1 - \frac{x^2}{2!} + \frac{x^4}{4!} - \frac{x^6}{6!} + \frac{x^8}{8!} \cdots$$

R.4.13 From the definitions of the six basic trigonometric functions, the following relationships can be observed:

a. *sin(A)* and *csc(A)* are reciprocal functions {*sin(A)* = *1/csc(A)*}.

b. *cos(A)* and *sec(A)* are reciprocal functions {*cos(A)* = *1/sec(A)*}.

c. *tan(A)* and *cot(A)* are reciprocal functions {*tan(A)* = *1/cot(A)*}.

R.4.14 If one trigonometric function is known, then the remaining trigonometric functions can be evaluated.

R.4.15 For example, let $sin(A) = 3/5$, as shown in Figure 4.1, then $a = 3$, $c = 5$, and $b = \sqrt{c^2 - a^2} = 4$.

Then,

$$cos(A) = 4/5$$
$$tan(A) = 3/4$$
$$sec(A) = 5/4$$
$$csc(A) = 5/3$$
$$cot(A) = 4/3$$

R.4.16 The Cartesian plane can be defined in terms of quadrants, where the plane consists of a complete rotation of 360°, and each quarter is called a quadrant.

An angle is said to be in a standard or quadrant position if its vertex is at the origin of the rectangular coordinate system, and one side coincides with the positive x-axis, and the other side forms an angle α, with a range of revolutions from 0° to 360°.

The four quadrants are defined as follows:

- The first quadrant is when the angle α has a range defined by $0° \leq \alpha \leq 90°$.
- The second quadrant is when α has a range defined by $90° < \alpha \leq 180°$.
- The third quadrant is when α has a range defined by $180° < \alpha \leq 270°$.
- The fourth quadrant is when α has a range defined by $270° < \alpha < 360°$.

R.4.17 All the trigonometric functions are periodic, and any two angles that differ in $360° = 2\pi$ radians have the same trigonometric value (the period is 2π). For example,

$$sin(\alpha + 2\pi) = sin(\alpha)$$
$$cos(\alpha + 2\pi) = cos(\alpha)$$
$$tan(\alpha + 2\pi) = tan(\alpha)$$

R.4.18 As an additional example, evaluate the following trigonometric functions: $sin(390)$, $cot(3645°)$, $sin(730°)$, $tan(3903° \; 20')$, and $-tan(56° \; 40')$.

ANALYTICAL Solution

$$sin(390) = sin(360° + 30°) = sin(30°) = 0.5$$
$$cot(3645°) = cot(360° * 10 + 45°) = cot(45°) = 1$$
$$sin(730°) = sin(730° - 2 * 360°) = sin(10°) = 0.1736$$
$$tan(3903° \; 20') = tan(3903° \; 20' - 10 * 360°) = tan(303° \; 20') = -tan(56° \; 40') = -1.5224$$

R.4.19 Any angle on the second, third, and fourth quadrant can be reduced to an acute positive angle on the first quadrant.

R.4.20 The process of reducing an angle α from the second quadrant to the first quadrant consists of finding its complement $180° - \alpha = \beta$. The trigonometric functions of the resulting acute angle β, sign and magnitude are illustrated as follows:

$$sin(\alpha) = sin(180° - \alpha) = sin(\beta)$$

$$cos(\alpha) = -cos(180° - \alpha) = -cos(\beta)$$

$$tan(\alpha) = -tan(180° - \alpha) = -tan(\beta)$$

$$cot(\alpha) = -cot(180° - \alpha) = -cot(\beta)$$

$$sec(\alpha) = -sec(180° - \alpha) = -sec(\beta)$$

$$csc(\alpha) = csc(180° - \alpha) = csc(\beta)$$

R.4.21 For example, reduce to the first quadrant, the following second quadrant trigono-metric functions: $sin(135°)$, $sin(150°)$, $cos(135°)$, $cos(120°)$, $tan(135°)$, $cot(150°)$, $sec(135°)$, and $csc(135°)$.

ANALYTICAL Solution

$$sin(135°) = sin(180° - 135°) = sin(45°) = \frac{\sqrt{2}}{2}$$

$$sin(150°) = sin(180° - 150°) = sin(30°) = 0.5$$

$$cos(135°) = -cos(180° - 135°) = -cos(45°) = -\frac{\sqrt{2}}{2}$$

$$cos(120°) = -cos(180° - 120°) = -cos(60°) = -0.5$$

$$tan(135°) = -tan(180° - 135°) = -tan(45°) = -1$$

$$cot(150°) = -cot(180° - 150°) = -cot(30°) = -\sqrt{3}$$

$$sec(135°) = -sec(180° - 135°) = -sec(45°) = -\sqrt{2}$$

$$csc(135°) = csc(180° - 135°) = csc(45°) = \sqrt{2}$$

R.4.22 The process of reducing an angle α from the third quadrant to the first quadrant consists of finding its complement $180° - \alpha = \beta$. The trigonometric functions of the resulting acute angle β, sign and magnitude are illustrated as follows:

$$sin(\alpha) = -sin(\alpha - 180°) = -sin(\beta)$$

$$cos(\alpha) = -cos(\alpha - 180°) = -cos(\beta)$$

$$tan(\alpha) = tan(\alpha - 180°) = tan(\beta)$$

$$cot(\alpha) = cot(\alpha - 180°) = cot(\beta)$$

$$sec(\alpha) = -sec(\alpha - 180°) = -sec(\beta)$$

$$csc(\alpha) = -csc(\alpha - 180°) = -csc(\beta)$$

R.4.23 For example, reduce to the first quadrant and evaluate the following third quadrant trigonometric functions: $sin(210°)$, $sin(240°)$, $cos(210°)$, $tan(225°)$, and $sec(210°)$.

ANALYTICAL Solution

$$sin(210°) = -sin(210° - 180°) = -sin(30°) = -0.5$$

$$sin(240°) = -sin(240° - 180°) = -sin(60°) = -\frac{\sqrt{3}}{2}$$

$$cos(210°) = -cos(210° - 180°) = -cos(30°) = -\frac{\sqrt{3}}{2}$$

$$tan(225°) = tan(225° - 180°) = tan(45°) = 1$$

$$sec(210°) = -sec(210° - 180°) = -sec(30°) = -\frac{2\sqrt{3}}{3}$$

R.4.24 The process of reducing an angle α from the fourth quadrant to the first quadrant consists of finding the angle $\beta = 360° - \alpha$. The trigonometric functions of the resulting acute angle β consisting of sign and magnitude are illustrated as follows:

$$sin(\alpha) = -sin(360° - \alpha) = -sin(\beta)$$

$$cos(\alpha) = cos(360° - \alpha) = cos(\beta)$$

$$tan(\alpha) = -tan(360° - \alpha) = -tan(\beta)$$

$$cot(\alpha) = -cot(360° - \alpha) = -cot(\beta)$$

$$sec(\alpha) = sec(360° - \alpha) = sec(\beta)$$

$$csc(\alpha) = -csc(360° - \alpha) = -csc(\beta)$$

R.4.25 For example, reduce to the first quadrant and evaluate the following fourth quadrant trigonometric functions: $sin(315°)$, $sin(330°)$, $cos(300°)$, $sec(315°)$, $tan(300°)$, and $csc(315°)$.

ANALYTICAL Solution

$$sin(315°) = -sin(360° - 315°) = -sin(45°) = -\frac{\sqrt{2}}{2}$$

$$sin(330°) = -sin(360° - 330°) = -sin(30°) = -0.5$$

$$cos(300°) = cos(360° - 300°) = cos(60°) = 0.5$$

$$sec(315°) = sec(360° - 315°) = sec(45°) = \sqrt{2}$$

$$tan(300°) = -tan(360° - 300°) = -tan(60°) = -\sqrt{3}$$

$$csc(315°) = -csc(360° - 315°) = -csc(45°) = -\sqrt{2}$$

R.4.26 A triangle is a structure consisting of three sides and three angles. If one side and two angles, or two sides and one angle are known, then the other three unknown quantities can be evaluated.

R.4.27 Recall that the Pythagorean theorem states that in a right triangle, as illustrated in Figure 4.1,

$$c^2 = a^2 + b^2$$

where angle(A) + angle(B) = 90°.

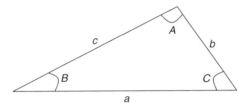

FIGURE 4.2
Oblique triangle of R.4.30.

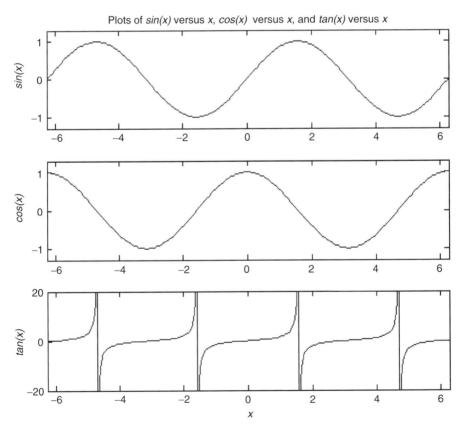

FIGURE 4.3
Plots of *sine*(x), *cosine*(x), and *tangent*(x).

R.4.28 A general triangle, also called oblique, as illustrated in Figure 4.2, is a triangle that contains no right angle.

R.4.29 Solving a triangle means finding the values of all the sides and angles. The *law of sines* and the *law of cosines* are stated below and are generally used to solve an oblique triangle (Linderburg, 1982).

R.4.30 The law of sines states that in a general oblique triangle, referred to Figure 4.2, the following relations hold:

$$\frac{\sin(A)}{a} = \frac{\sin(B)}{b} = \frac{\sin(C)}{c}$$

R.4.31 The law of cosines states that in a general oblique triangle, referred to Figure 4.2, if two sides and the angle formed by them are known, then the third side can be evaluated by the following relations:

$$a^2 = b^2 + c^2 - 2\,b\,c\,\cos(A)$$

$$b^2 = c^2 + a^2 - 2\,a\,c\,\cos(B)$$

$$c^2 = a^2 + b^2 - 2\,a\,b\,\cos(C)$$

R.4.32 Graphs of the standard trigonometric functions are shown in Figures 4.3 and 4.4. Graphs of the reciprocal functions $\sin(x) = 1/\csc(x)$ and $\cos(x) = 1/\sec(x)$ are shown in

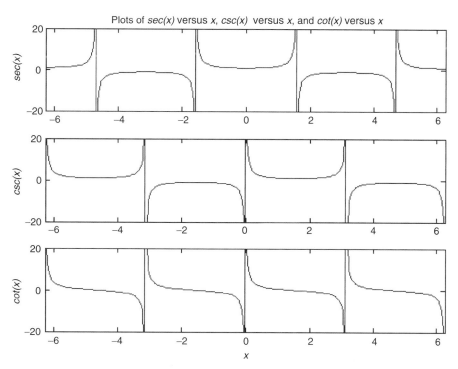

FIGURE 4.4
Plots of *secant*(x), *cosecant*(x), and *cotangent*(x).

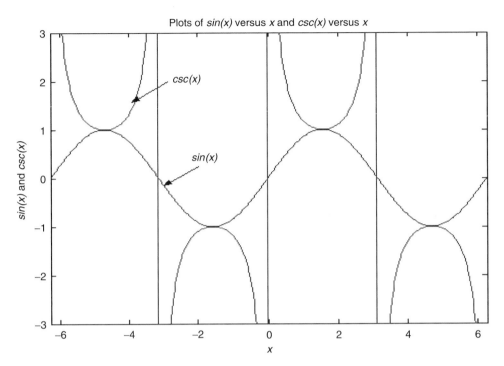

FIGURE 4.5
Plots of *sine*(*x*) and *cosecant*(*x*).

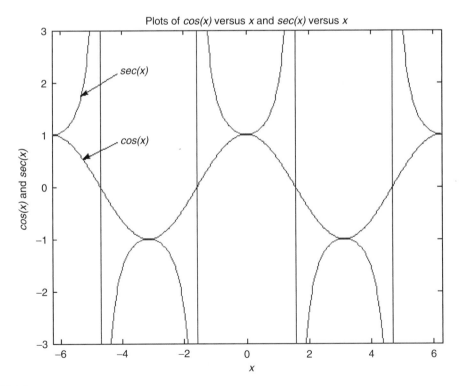

FIGURE 4.6
Plots of *cosine*(*x*) and *secont*(*x*).

Figures 4.5 and 4.6. The reader should observe the periodic nature of the trigonometric functions. Recall that a function $f(t)$ is said to be periodic, with period T, if

$$f(t) = f(t \pm nT) \quad \text{for any integer} \quad n = 1, 2, 3, \dots$$

R.4.33 From the graphs of the trigonometric functions shown in Figures 4.3 and 4.4, the period T, frequency f, and amplitude A can be evaluated. Table 4.3 summarizes the characteristics of the standard six trigonometric functions.

R.4.34 Because the trigonometric functions are periodic, the inverse trigonometric functions may not be unique, but if the domain is restricted to a particular interval, then each trigonometric function will have a unique inverse.

The mathematical inverse of $sin(x)$ is denoted in two ways, $arcsin(x)$ or $sin^{-1}(x)$, and the same notation is used for the other trigonometric functions ($acos(x)$ or $cos^{-1}(x)$, etc.).

R.4.35 The six inverse trigonometric MATLAB functions are

$$asin(x), acos(x), atan(x), acot(x), asec(x), \text{ and } acsc(x)$$

R.4.36 Table 4.4 indicates the inverse trigonometric functions, its direct functions, its domain and range.

R.4.37 Some useful and often used trigonometric identities are as follows:

a. $sin^2(x) + cos^2(x) = 1$

b. $1 + tan^2(x) = sec^2(x)$

c. $1 + cot^2(x) = csc^2(x)$

d. $cos(x) + sin(x) \, tan(x) = sec(x)$

e. $sin(2x) = 2sin(x) \cdot cos(x)$

f. $cos(2x) = cos^2(x) - sin^2(x)$

g. $sin(x) = 2sin(x/2) \cdot cos(x/2)$

h. $cos^2(x) = 1/(1 + tan^2(x))$

i. $sin(x) \cdot csc(x) = 1$

j. $cos(x) \cdot sec(x) = 1$

k. $tan(x) \cdot cos(x) = 1$

TABLE 4.3

Period, Amplitude, and Frequency of the Six Standard Trigonometric Functions

Function	Period T	Amplitude	Frequency $f = 1/T$
$y = A \, sin(\omega * t)$	$2\pi/\omega$	A	$\omega/2 * \pi$
$y = A \, cos(\omega * t)$	$2\omega/\omega$	A	$\omega/2\omega$
$y = A \, tan(\omega * t)$	π/ω	—	ω/π
$y = A \, cot(\omega * t)$	π/ω	—	ω/π
$y = A \, sec(\omega * t)$	$2\pi/\omega$	—	$\omega/2 * \pi$
$y = A \, csc(\omega * t)$	$2\pi/\omega$	—	$\omega/2 * \pi$

TABLE 4.4

Inverse and Direct Standard Trigonometric Functions

Inverse Function	Function	Domain	Range
$x = asin(y)$	$y = sin(x)$	$-1 < y < +1$	$-\pi \le x \le +\pi$
$x = acos(y)$	$y = cos(x)$	$-1 < y < +1$	$0 \le x \le +\pi$
$x = atan(y)$	$y = tan(x)$	—	$-\pi/2 \le x \le +\pi/2$
$x = acot(y)$	$y = cot(x)$	—	$0 \le x \le +\pi$
$x = asec(y)$	$y = sec(x)$	$\|y\| \ge 1$	$-\pi \le x \le 0, x \ne -\pi/2$
$x = acsc(y)$	$y = csc(x)$	$\|y\| \ge 1$	$-\pi/2 \le x \le \pi/2, x \ne 0$

R.4.38 Some useful trigonometric identities relating sums and differences of angles are as follows (Kay, 1994):

a. $sin(x + y) = sin(x) \cdot cos(y) + cos(x) \cdot sin(y)$

b. $cos(x + y) = cos(x) \cdot cos(y) - sin(x) \cdot sin(y)$

c. $tan(x + y) = (tan(x) + tan(y))/(1 - tan(x) \cdot tan(y))$

d. $sin(x - y) = sin(x) \cdot cos(y) - cos(x) \cdot sin(y)$

e. $cos(x - y) = cos(x) \cdot cos(y) + sin(x) \cdot sin(y)$

f. $tan(x - y) = [tang(x) - tan(y)]/[(1 + tan(x) \cdot tan(y)]$

g. $tan(x + y) = [tang(x) + tan(y)]/[(1 - tan(x) \cdot tan(y)]$

R.4.39 Some useful trigonometric identities that relate the doubling of angles are as follows:

a. $sin(2x) = 2 sin(x) \cdot cos(x)$

b. $cos(2x) = cos^2(x) - sin^2(x)$

c. $tan(2x) = 2 tan(x)/(1 - tan^2(x))$

R.4.40 Some useful trigonometric identities that relate half angles are as follows:

a. $sin(x/2) = \pm \sqrt{(1 - cos(x))/2}$

b. $cos(x/2) = \pm \sqrt{(1 + cos(x))/2}$

c. $tan(x/2) = \pm \sqrt{(1 - cos(x))/(1 + cos(x))} = -1 \pm \sqrt{(1 + tan^2(x))/(tan(x))}$

R.4.41 Some useful trigonometric identities relating sums and differences of angles are as follows:

a. $sin(x) + sin(y) = 2 sin((x + y)/2) \cdot cos((x - y)/2)$

b. $sin(x) - sin(y) = 2 cos((x + y)/2) \cdot sin((x - y)/2)$

c. $cos(x) + cos(y) = 2 cos((x + y)/2) \cdot cos((x - y)/2)$

d. $cos(x) - cos(y) = -2 sin((x + y)/2) \cdot sin((x - y)/2)$

R.4.42 Some useful trigonometric products are as follows:

a. $sin(x) \cdot cos(y) = 1/2 (sin(x + y) + sin(x - y))$

b. $cos(x) \cdot sin(y) = 1/2 (sin(x + y) - sin(x - y))$

c. $cos(x) \cdot cos(y) = 1/2 (cos(x + y) + cos(x - y))$

d. $sin(x) \cdot sin(y) = -1/2 (cos(x + y) - cos(x - y))$

R.4.43 Some useful inverse trigonometric identities are as follows:

a. $arcsin(x) + arccos(x) = \pi/2$

b. $arctan(x) + arccot(x) = \pi/2$

c. $arcsec(x) = arccos(1/x)$

d. $arccsc(x) = arcsin(1/x)$

e. $arcsec(x) + arccsc(x) = \pi/2$

f. $arccot(x) = arctan(1/x)$

g. $arcsin(-x) = -arcsin(x)$

h. $arccos(-x) = \pi - arccos(x)$

i. $arccsc(-x) = arcsec(x)$

j. $arctan(-x) = -arctan(x)$

k. $arccot(-x) = \pi - arccot(x)$

l. $arcsec(-x) = \pi - arcsec(x)$

R.4.44 Trigonometric functions can be expressed in terms of exponentials of the irrational number $e = 2.71828182$. Recall that e^x is expressed in MATLAB as $exp(x)$.

R.4.45 The number e can be defined as the limit of the following relation:

$$e = \left(1 + \frac{1}{n}\right)^n, \text{ for a large } n \text{ (approaching infinity)}$$

TABLE 4.5

Approximations for $e = (1 + 1/n)^n$ for Different n's

n	1	2	3	4	5	6	10	1,000	10,000
e	2	2.25	2.37	2.47	2.488	2.55	2.5937	2.7169	2.71814

Table 4.5 illustrates how e approaches 2.7182812 as n increases.

R.4.46 The value of e can also be computed by using the MacLaurin series as follows:

$$exp(x) = 1 + x + \frac{x^2}{2!} + \frac{x^3}{3!} + \frac{x^4}{4!} + \cdots + \frac{x^n}{n!}$$

Recall that $n! = n * (n-1) * (n-2) * \cdots 3 * 2 * 1$ is called the n-factorial. Therefore, if $x = 1$,

$$exp(1) = e = 1 + \frac{1}{1} + \frac{1}{2!} + \frac{1}{3!} + \frac{1}{4!} + \cdots + \frac{1}{n!}$$

$$e = 1 + 1 + \frac{1}{2} + \frac{1}{6} + \frac{1}{24} + \frac{1}{120} + \frac{1}{720} + \cdots$$

The value of e converges faster to 2.7182812 when the MacLaurin's series is used instead of $(1 + 1/n)^n$.

R.4.47 Hyperbolic functions are exponential functions of the form e^x and e^{-x}.

The hyperbolic functions present properties that are similar to the trigonometric functions, but are simpler and more straightforward. The combination of e^x and e^{-x} appears regularly in certain types of engineering and science problems, and to preserve simplicity the six hyperbolic functions are defined as follows:

$$sinh(x) = \frac{(e^x - e^{-x})}{2}$$

$$cosh(x) = \frac{(e^x + e^{-x})}{2}$$

$$tanh(x) = \frac{e^x - e^{-x}}{e^x + e^{-x}} = \frac{sinh(x)}{cosh(x)}$$

$$coth(x) = \frac{e^x + e^{-x}}{e^x - e^{-x}} = \frac{cosh(x)}{sinh(x)}$$

$$sech(x) = \frac{2}{e^x + e^{-x}} = \frac{1}{cosh(x)}$$

$$csch(x) = \frac{2}{e^x - e^{-x}} = \frac{1}{sinh(x)}$$

R.4.48 Graphs of the hyperbolic functions are shown in Figures 4.7 and 4.8.

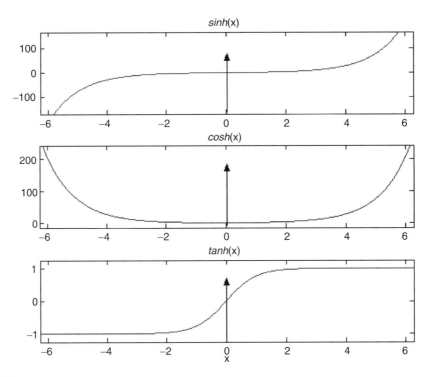

FIGURE 4.7
Hyperbolic plots of *sine*(x), *cosine*(x), and *tangent*(x) over the range −1 ≤ x ≤ 1.

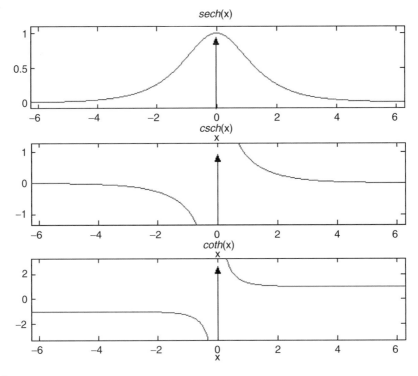

FIGURE 4.8
Hyperbolic plots of *secant*(x), *cosecant*(x), and *cotangent*(x) over the range −6.26 < x < 6.28.

R.4.49 Observe the similarity between the standard trigonometric with the hyperbolic functions. For example, the *cosh(x)* (pronounce "*kosh*") is an even function, whereas *sinh(x)* is an odd function.

$$(cosh(x) = cosh(-x) \text{ and } sinh(x) = -sinh(-x))$$

The exact same relation holds for the standard functions such as: $sin(x) = -sin(-x)$ and $cos(x) = cos(-x)$.

R.4.50 Some useful hyperbolic identities are as follows:

    a. $cosh^2(x) - sinh^2(x) = 1$

    b. $1 - tanh^2(x) = sech^2(x)$

    c. $e^x = cosh(x) + sinh(x)$

    d. $e^{-x} = cosh(x) - sinh(x)$

    e. $cosh(x + y) = cosh(x)\, cosh(y) + sinh(x)\, sinh(y)$

    f. $sinh(x + y) = sinh(x)\, cosh(y) + cosh(x)\, sinh(y)$

R.4.51 The inverse hyperbolic function $sinh^{-1}(x)$ or $arcsinh(x)$ denoted in MATLAB as $asinh(x)$ is defined by

$$asinh(x) = ln\left(x + \sqrt{(x^2 + 1)}\right) \quad \text{for } -\infty \leq x \leq +\infty$$

R.4.52 The other inverse hyperbolic functions are defined as follows:

$$acosh(x) = ln\left(x + \sqrt{(x^2 - 1)}\right) \quad \text{for } x \geq 1$$

$$atanh(x) = ln\left(\frac{1 + x}{1 - x}\right)^{1/2} \quad \text{for } |x| \leq 1$$

$$acoth(x) = ln\left(\frac{x + 1}{x - 1}\right)^{1/2} \quad \text{for } x > 1 \quad \text{or} \quad x < -1$$

$$asech(x) = ln\left(\frac{1 + \sqrt{(1 - x^2)}}{x}\right) \quad \text{for } 0 < x \leq 1$$

$$acsch(x) = ln\left(\frac{1 + \sqrt{(1 + x^2)}}{x}\right) \quad \text{for } x \neq 0$$

R.4.53 The graphs of the six inverse hyperbolic functions are plotted using MATLAB, and are shown in Figure 4.9, over the range $-2 \leq x \leq 2$. Observe that $acoth(x) = atanh(x)$.

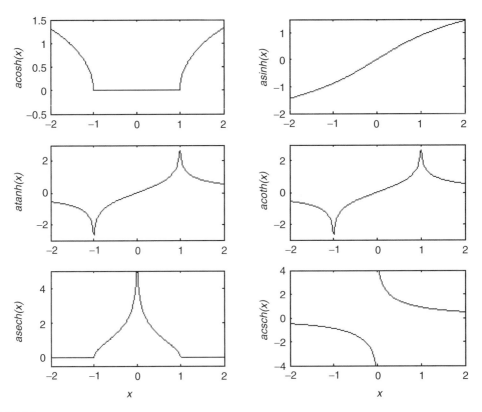

FIGURE 4.9

Plots of the six inversed hyperbolic functions over the range $-2 \leq x \leq 2$.

R.4.54 For example, the MATLAB script file *inver_hyper* below returns in a table like format the six inverse hyperbolic functions, shown in Figure 4.9, over the range $-2 \leq x \leq 2$, with linear spacing of $\Delta x = 0.5$:

MATLAB Solution

```
% Script file: inver _ hyper
x = -2:0.5:2;
aco = acosh(x);
asi = asinh(x);
ata = atanh(x);
acot = acoth(x);
ase = asech(x);
acsc = acsch(x);
disp('*************************************************************')
disp('   x( rad )    asinh(x)    acosh(x)    atanh(x)    acoth(x)   ')
disp('*************************************************************')
results = [x' asi' aco' ata' acot']   % the columns correspond to x, asinh(x),
                                      % acosh(x), atanh(x), acoth(x) ,
disp('*************************************************************')
disp('  asec(x)      acsch(x)  )'
disp('*************************************************************')
res = [ase' acsc']              % asech(x), acsch(x)
```

The script file *inver_hyper* is executed and the results are as follows:

```
*************************************************************
   x( rad )   asinh(x)    acosh(x)     atanh(x)         acoth(x)
*************************************************************
 results =
 Columns 1 through 5
 -2.0000     -1.4436    1.3170 - 3.1416i  -0.5493            -1.5708i -0.5493
 -1.5000     -1.1948    0.9624 - 3.1416i  -0.8047            -1.5708i -0.8047
 -1.0000     -0.8814       0 - 3.1416i    NaN            +   NaN   -Inf
 -0.5000     -0.4812       0 - 2.0944i    -0.5493            -0.5493  1.5708i
  0            0           0 - 1.5708i     0                 0       1.5708i
  0.5000      0.4812       0 - 1.0472i     0.5493             0.5493  1.5708i
  1.0000      0.8814       0              NaN +   NaNi    NaN +     NaN
  1.5000      1.1948    0.9624            0.8047 - 1.5708i 0.8047
  2.0000      1.4436    1.3170            0.5493 - 1.5708i 0.5493
        ***************************
             asec(x)      acsch(x)
        ***************************
           res=
           Columns 1    through 2
           0 - 2.0944i      -0.4812
           0 - 2.3005i      -0.6251
           0 - 3.1416i      -0.8814
           1.3170 - 3.1416i -1.4436
           Inf              Inf
           1.3170           1.4436
           0                0.8814
           0 - 0.8411i      0.6251
           0 - 1.0472i      0.4812
```

R.4.55 The trigonometric *sin(x)* and *cos(x)* are related to the hyperbolic *sinh(x)* and *cosh(x)* by the following relations:

$$cosh(jx) = cos(x)$$

$$sin(x) = -jsinh(-jx)$$

where $\sqrt{-1} = j$.

R.4.56 Some useful mathematical relations involving exponentials are as follows:

a. *exp(0) = 1*

b. *exp(a) * exp(b) = exp(a + b)*

c. *exp(a)/exp(b) = exp(a − b)*

d. *(exp(a))^b = exp(a * b)*

R.4.57 Some useful mathematical relations, as well as the corresponding MATLAB commands are summarized in Table 4.6.

R.4.58 The inverse of an exponential function is a logarithmic function. For example, let $x = log_a(y)$, then $a^x = y$, where *a* is referred as the base of the logarithmic function.

R.4.59 The equations $y = a^x$ and $log_a(y) = x$ express exactly the same relation between *x* and *y*.

R.4.60 Table 4.7 states some important relations between dependent variable *y* and *log(x)*, where *x* is the independent variable for $y = log(x)$.

TABLE 4.6

Mathematic/MATLAB Translation

Mathematical Relations	MATLAB Instruction
$x^0 = 1$	$x \wedge 0$
$x^{-a} = 1/(x^a)$	$x \wedge (-a)$
$\sqrt[n]{x} = x^{1/n}$	$x \wedge (l/n)$
$\sqrt{x} = x^{1/2}$	$sqrt(x)$
$\sqrt[n]{\sqrt[m]{x}} = \sqrt[n \cdot m]{x}$	$x \wedge (1/(n * m))$
$x^a \cdot x^b = x^{a+b}$	$x \wedge (a + b)$
$x^a/x^b = x^{a-b}$	$x \wedge (a - b)$
$(x^a)^b = x^{ab}$	$x \wedge (a * b)$
$x^{a/b} = \sqrt[b]{x^a}$	$x \wedge (a/b)$
$(x/y)^a = x^a/y^a$	$(x/y) \wedge a$

TABLE 4.7

y and x Relation for $y = log(x)$

y	$log(x)$
Does not exist	$x < 0$
Negative	$0 < x < 1$
Positive	$x > 1$

R.4.61 Observe that in general

a. If $y = log(x)$, then the larger the positive value of x, the larger the value of y.

b. The function f and its inverse f^{-1} are symmetrical functions with respect to $y = x$. For example, 2^x and $log_2(x)$, or $cos(x)$ and $acos(x)$ are symmetrical functions with respect to $y = x$.

R.4.62 The following numerical examples provide some insight into the nature of logarithms:

$log_2(8) = 3$ since $2^3 = 8$

$log_{(0.5.)}(0.125) = 3$ since $(0.5)^3 = 0.125$

$log_{(10)}(100) = 2$ since $10^2 = 100$

$log_{(-2)}(10)$ does not exist

$log_5(1/25) = -2$ since $5^{-2} = 1/25$

$log_e(23) = 3.135$ since $e^{3.135} = 23$

$log_{10}(1) = 0$ since $10^0 = 1$

$log_e(1) = 0$ since $e^0 = 1$

$log_{10}(0.000001) = -6$ since $10^{-6} = 0.000001$

$log_e(-23)$ does not exist

$log_{10}(0.1) = -1$ since $10^{-1} = 0.1$

$log_{10}(\sqrt[5]{81}) = 0.38$ since $10^{0.38} = \sqrt[5]{81}$

R.4.63 Some useful logarithmic properties are as follows (for $a > 0$, $a \neq 1$, and $x > 0$):

$log_a(x) = b$ (then $a^b = x$)

$log_a(x)^{-1} = -log_a(x)$

$log_a(a) = 1$

$log(1) = 0$

TABLE 4.8

MATLAB Rounding Functions

MATLAB Function	Description
$y = round(x)$	Rounds x to the nearest integer; $y = round(3.7) = 4$
$y = ceil(x)$	Rounds x to the nearest greatest positive integer; $y = ceil(3.7) = 4$
$y = fix(x)$	Rounds x toward the nearest lowest integer; $y = fix(3.7) = 3$
$y = floor(x)$	Rounds x toward $-\infty$; $y = floor(3.7) = 3$
$y = sign(x)$	Returns $y = 1$, if $x > 0$
	$y = 0$, if $x = 0$
	$y = -1$, if $x < 0$

$log_a(a^x) = x$

$log_a(x)^n = nlog_a(x)$

$log_a^n \sqrt{x} = 1/n(log_a(x))$

$log_a(x \cdot y) = log_a(x) + log_a(y)$

$log_a(x/y) = log_a(x) - log_a(y)$

$ln(x) = log_{10}(x)/log_{10}(e) = 2.3026 \cdot log_{10}(x)$

$log(x) = .4343 ln(x)$

$log_a(x) = log(x)/log(a) = ln(x)/ln(a)$

$log_b(x) = (1/log_a(b)) \, log_a(x)$ for $b > 0$ and $b \neq 0$

Recall that $ln(x) = log_e(x)$.

R.4.64 Most of the practical logarithmic functions use *10*, *e*, or *2* as their base.

The logarithm of base *10* is usually expressed as *log(x)* and is called decimal or logarithm of Brigg. MATLAB uses the format *log10(x)* to indicate that the base is *10*.

The logarithm of base *e* is called natural or Naperian.* MATLAB uses the format *log(x)* to indicate that the base is *e*. The logarithm of base *2* is called binary and the MATLAB syntax is *log2(x)*.

R.4.65 An important property of the logarithmic function is that the plot of $y = log_a(x)$ for any (base) *a* passes through the point *(1, 0)*.

R.4.66 Some special MATLAB functions are the rounding functions, which examine the argument of the independent variable (*x*) and return an approximation value for the dependent variable *y*.

The most common MATLAB rounding functions are presented in Table 4.8 with a brief description and a short example.

R.4.67 The least common multiple and greatest common divisor given the integers *x* and *y* can be obtained by using the command *gcd(x, y)* and *lcm(x, y)*.

For example, *gcd(20, 25)* will return *5*, and *lcm(20, 25)* will return *100*.

R.4.68 The function *rem(y, x)* returns the remainder *r* after dividing *y* by *x* as indicated by

$$\frac{y}{x} = c + \left(\frac{r}{y}\right)$$

For example, *rem(20, 3)* will return a *2*, while *rem(20, 0)* returns *NaN*.

---

* The abbreviation *ln* is from Latin *logarithmic naturalis*.

R.4.69 The MATLAB command *factor(n)*, returns a vector containing the prime factors of *n*.
For example, use MATLAB and factor the numbers *121* and *120*.

```
>>factor(121)

  ans =
      11   11

>>factor(120)

  ans =
       2      2      2      3      5
```

R.4.70 The command *primes(n)* returns a vector consisting of all the prime numbers
between zero and *n*. Recall that a prime number is one that can only be divided
by itself or by 1, with zero remainder. For example, use MATLAB to obtain the
sequence of prime numbers that are less than 100.

```
>> first _ 100 _ prime = primes(100)

   first _ 100 _ prime=
       Columns 1 through 12
       2  3  5  7  11  13  17  19  23  29  31  37
       Columns 13 through 24
       41  43  47  53  59  61  67  71  73  79  83  89
       Column 25
       97
```

R.4.71 The command *isprime(n)* checks if *n* is a prime number, in which case MATLAB
returns a 1 (one), otherwise MATLAB returns a 0 (zero).
For example, use MATLAB and check if the numbers 13 and 14 are prime
numbers.

```
>>isprime(13)

  ans =
          1

>>isprime(14)

  ans =
        0
```

R.4.72 The command *[N, D] = rat(n)*, where *n* is a number, returns the rational approximation for *n*, consisting of two integers such that *N/D* is a close approximation for *n*.
When the instruction *rat(n)* is executed, MATLAB returns the rational approximation for *n* consisting of sums.
For example, use MATLAB to obtain the rational approximation of *e* and π.

```
MATLAB Solution
>> rat(pi)

   ans =
        3 + 1/(7 + 1/(16))

>> [N,D] = rat(pi)

   N =
       355
```

```
    D =
        113
>>  [N,D] = rat(exp(1))

    N =
        1457

    D =
        536

>> check = 1457/536

    check =
        2.7183         % observe that the rational approximation is very close
                         to e
```

4.4 Examples

Example 4.1

Create the script file *rad_deg* that returns a table of the angle x expressed in degrees and radians as well as the corresponding values of $sin(x°)$ and $cos(x°)$, over the range $0° \le x \le 360°$, in linear increments of $\Delta x = 0.314$ radians.

MATLAB Solution
```
% Script file: rad _ deg
x = 0:0.314:2*pi;                    % creates the row vector x
y1 = sin(x);                         % creates the row vector  sin(x)
y2 = cos(x);                         % creates the row vector  cos(x)
z = x*180/pi;                        % converts radians to degrees
disp('**************************')
disp('x( deg)x (rad)   sin(x)   cos(x)')
disp('**************************')
[z' x' y1' y2]                       % displays in table format
disp('**************************')
>>  rad _ deg                        % in the command window
**************************************
  x( deg)    x (rad)    sin(x)    cos(x)
**************************************
     0          0         0       1.0000
   17.9909     0.3140    0.3089    0.9511
   35.9817     0.6280    0.5875    0.8092
   53.9726     0.9420    0.8087    0.5882
   71.9635     1.2560    0.9509    0.3096
   89.9544     1.5700    1.0000    0.0008
  107.9452     1.8840    0.9514   -0.3081
  125.9361     2.1980    0.8097   -0.5869
  143.9270     2.5120    0.5888   -0.8083
  161.9179     2.8260    0.3104   -0.9506
  179.9087     3.1400    0.0016   -1.0000
  197.8996     3.4540   -0.3074   -0.9516
  215.8905     3.7680   -0.5862   -0.8101
```

```
233.8814      4.0820     -0.8078    -0.5895
251.8722      4.3960     -0.9504    -0.3111
269.8631      4.7100     -1.0000    -0.0024
287.8540      5.0240     -0.9518     0.3066
305.8449      5.3380     -0.8106     0.5856
323.8357      5.6520     -0.5901     0.8073
341.8266      5.9660     -0.3119     0.9501
359.8175      6.2800     -0.0032     1.0000
   *************************************
```

Example 4.2

Create the script file *exp_appr* that returns in a table like format the evaluated value of $e = 2.718182$, using successive approximations from up to 10 terms, employing the following equations:

a. $e^x = 1 + (x/1) + (x^2/2!) + (x^3/3!)\cdots(x^9/9!) + (x^{10}/10!)\cdots$ for $x = 1$

b. $e = (1 + 1/n)^n$ for $n = 0, 1, 2, 3, \ldots, 9, 10$

MATLAB Solution
```
% Script file: exp _ appr
% part(a)
n =1:1:10;                    % creates the series 1 2 3 4 ........10.
den = cumprod(n);             % creates the series 1 2 6 24..........
series = cumsum(1./den);      % creates the sequence 1 1+1/2 1+1/2+1/6
                                +1/24+......
exposeri = 1+series ;         % creates the first 10 approximations of
                                the series
%  part(b)
den1 = 1./n;                  % creates the sequence 1 1/2 1/3 1/4.....
exp = (1+den1).^n;            % creates a series with the first 10
                                approximations
disp('***********R E S U L T S ***************')
disp('***************************************')
disp('n(# of terms)          e=1+1+1^2/2!...        e=(1+1/n)^n')
disp('***************************************')
[n' exposeri' exp']
disp('***************************************')
```

The script file *exp_appr* is executed and the results are shown in the following:

```
>> exp _ appr

*************R E S U L T S ***********************
*****************************************************
n(# of terms)      e=1+1+1^2/2!...      e=(1+1/n)^n
*****************************************************
ans =
       1.0000           2.0000           2.0000
       2.0000           2.5000           2.2500
       3.0000           2.6667           2.3704
       4.0000           2.7083           2.4414
       5.0000           2.7167           2.4883
```

6.0000	2.7181	2.5216
7.0000	2.7183	2.5465
8.0000	2.7183	2.5658
9.0000	2.7183	2.5812
10.0000	2.7183	2.5937

****

The preceding data is illustrated graphically in Figure 4.10.

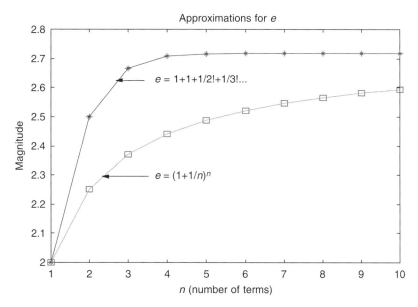

FIGURE 4.10
Plots of the two approximations of *e* of Example 4.2.

Example 4.3

Given the function $x(t) = 4e^{2t} - 5e^{-3t}$, create the script file x_of_t that returns in a table like format $x_1(t) = 4e^{2t}$, $x_2(t) = -5e^{-3t}$, and $x(t)$ over the range $0 \leq t \leq 3$, in linear increments of $\Delta t = 0.1$.

MATLAB Solution

```
% Script file: x _ of _ t
t = 0:0.1:3;         % creates an array t consisting of 31 elements
x1 = 4.*exp(2.*t);   % creates an array x1 consisting of 31 elements
x2 = 5.*exp(-3.*t);  % returns an array x2 consisting of 31 elements
x = x1-x2;           % returns  x(t) as an array of 31 elements
                         (x=x₁ - x₂)
% the values of x1 = 4exp(2t),  x2=5exp(-3t), and x1-x2
% are displayed in table like  format  over the range of t
disp('************************************************')
disp( ' **x1(t) = 4exp(-2t)** x2=5exp(-3t)**x(t) = x1(t)-x2(t)**' )
disp('************************************************')
[x1'    x2'   x']
disp('************************************************')
```

The script file *x_of_t* is executed and the results are shown in the following:

```
>> x _ of _ t
*************************************************************
**x1(t) = 4exp(-2t)** x2(t)=5exp(-3t)** x(t)= x1(t)-x2(t)**
*************************************************************
ans =
          1.0e+003  *
          0.0040     0.0050    -0.0010
          0.0049     0.0037     0.0012

          .............    .............    .............
          0.0109     0.0011     0.0098
          0.0133     0.0008     0.0125
          0.0162     0.0006     0.0156

          .............    .............    .............
          0.0539     0.0001     0.0538
          0.0658     0.0001     0.0657
          0.0803     0.0001     0.0803

          .............    .............    .............
          0.5937     0.0000     0.5936
          0.7251     0.0000     0.7251

          .............    .............    .............
          1.3212     0.0000     1.3212
          1.6137     0.0000     1.6137
*************************************************************
```

Example 4.4

Create the script file *trig_values* that returns in a table like format the first four columns of Table 4.2, that is *x* in degrees and radians as well as *sin(x)* and *cos(x)*.

MATLAB Solution
```
%Script file :trig _ values
a = [30 15 15 30];b=[90 90];
increm = [0 a a b];
angle _ deg = cumsum(increm)
convert = ones(1,11)*pi/180;
angle _ rad = angle _ deg.*convert;
sinx = sin(angle _ rad);
cosx = cos(angle _ rad);
disp('***********************************************')
disp(' **degrees ** radians ** sin(x) ** cos(x)**')
disp('***********************************************')
disp([angle _ deg' angle _ rad' sinx' cosx'   ])
disp('***********************************************')
```

The script file *trig_values* is executed and the results are shown in the following:

```
>> trig _ values
***********************************************
**degrees  ** radians  ** sin(x)  **cos(x)**
***********************************************
     0            0            0       1.0000
   30.0000      0.5236       0.5000    0.8660
   45.0000      0.7854       0.7071    0.7071
   60.0000      1.0472       0.8660    0.5000
   90.0000      1.5708       1.0000    0.0000
```

```
120.0000        2.0944        0.8660       -0.5000
135.0000        2.3562        0.7071       -0.7071
150.0000        2.6180        0.5000       -0.8660
180.0000        3.1416        0.0000       -1.0000
270.0000        4.7124       -1.0000       -0.0000
360.0000        6.2832       -0.0000        1.0000
*****************************************************
```

Example 4.5

Given the function $y(t) = 2e^{-2t} \sin(t)$, create the script file *sin_exp* that returns in a table like format $y(t)$ versus t, over the range $0 \le t \le 4\pi$, with linear increment of $\Delta t = 0.2\pi$.

MATLAB Solution
```
% Script file : sin _ exp
t = 0:.2*pi:4*pi;              % creates a 21 point- array of t
y = 2.*exp(-2*t).*sin(t);      % creates a 21 point- array of y(t)
% display in table format  t    and y(t)
disp('**************************')
disp('** t *** y(t) ************')
disp('**************************')
[t' y']
disp('**************************')
```

The script file *sin_exp* is executed and the resulting table is given below:

```
>> sin _ exp
****************************
 ** t *** y(t) ************
****************************
   ans =
          0           0
     0.6283      0.3346
     1.2566      0.1541
     1.8850      0.0439
     2.5133      0.0077
     3.1416      0.0000
     3.7699     -0.0006
     4.3982     -0.0003
     5.0265     -0.0001
     5.6549     -0.0000
     6.2832     -0.0000
     6.9115      0.0000
     7.5398      0.0000
     8.1681      0.0000
     8.7965      0.0000
     9.4248      0.0000
    10.0531     -0.0000
    10.6814     -0.0000
    11.3097     -0.0000
    11.9381     -0.0000
    12.5664     -0.0000
****************************
```

Example 4.6

Given the function $f(t) = cos(w_1 t) cos(w_2 t)$, create the script file *am_wave* that returns in a table like format the function $f(t)$ as an array consisting of 21 points over the range $0 \le t \le 2\pi$, with linear increment of $\Delta t = 0.1\pi$, for $w_1 = 2$ and $w_2 = 6$. The function $f(t)$ is referred to as an amplitude-modulated wave.

MATLAB Solution
```
%Script file: am _ wave
t = 0:.1*pi:2*pi;
w1 = 2;w2 = 6;
y1 = cos(w1.*t); y2 = cos(w2.*t);
ft = y1.*y2;
disp('***************************')
disp('        t            f(t)    ')
disp('***************************')
        [t' ft']            % return t and f(t) as column vectors
disp('**************************')
```

The script file *am_wave* is executed and the resulting table is given below:

```
>> am _ wave

***************************
      t            f(t)

***************************

ans =
      0            1.0000
 0.3142           -0.2500
 0.6283           -0.2500
 0.9425           -0.2500
 1.2566           -0.2500
 1.5708            1.0000
 1.8850           -0.2500
 2.1991           -0.2500
 2.5133           -0.2500
 2.8274           -0.2500
 3.1416            1.0000
 3.4558           -0.2500
 3.7699           -0.2500
 4.0841           -0.2500
 4.3982           -0.2500
 4.7124            1.0000
 5.0265           -0.2500
 5.3407           -0.2500
 5.6549           -0.2500
 5.9690           -0.2500
 6.2832            1.0000
***********************
```

Example 4.7

Create the script file *xy_circle* that returns 20, *xy* cartesian coordinate points in a table like format for

$$x = \cos(\beta) \text{ versus } y = \sin(\beta) \quad \text{over } 0 \le \beta \le 2\pi$$

Also, verify that the above points define a circle, with a unity radius *r*, where

$$r = \sqrt{\cos^2(\beta) + \sin^2(\beta)} = \sqrt{1} = 1$$

MATLAB Solution

```
% Script file: xy _ circle
beta = linspace(0,2*pi,20);
x = cos(beta);
y = sin(beta);
r = x.^2+y.^2;
disp('***************************************')
disp('   cos(beta)    sin(beta)      radius  ')
disp('***********************************')
[x' y' r']                   % displays:  cos(beta)
                              sin(beta)   radius by columns
disp('***********************************')
```

The script file *xy_circle* is executed and the resulting table is indicated below:

```
>> xy _ circle

***************************************
cos(beta)        sin(beta)         radius
***************************************
 1.0000              0             1.0000
 0.9458           0.3247           1.0000
 0.7891           0.6142           1.0000
 0.5469           0.8372           1.0000
 0.2455           0.9694           1.0000
-0.0826           0.9966           1.0000
-0.4017           0.9158           1.0000
-0.6773           0.7357           1.0000
-0.8795           0.4759           1.0000
-0.9864           0.1646           1.0000
-0.9864          -0.1646           1.0000
-0.8795          -0.4759           1.0000
-0.6773          -0.7357           1.0000
-0.4017          -0.9158           1.0000
-0.0826          -0.9966           1.0000
 0.2455          -0.9694           1.0000
 0.5469          -0.8372           1.0000
 0.7891          -0.6142           1.0000
 0.9458          -0.3247           1.0000
 1.0000          -0.0000           1.0000
***************************************
```

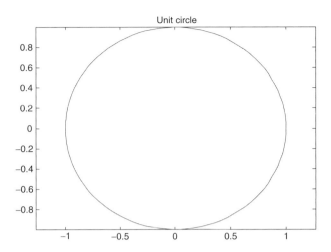

FIGURE 4.11
Plot of $x = cos(\beta)$ versus $y = sin(\beta)$, over $0 \leq \beta \leq 2\pi$ for Example 4.7.

The points obtained from the foregoing results are plotted for $x = cos(\beta)$ versus $y = sin(\beta)$, over the range $0 \leq \beta \leq 2\pi$ and the resulting graph is the unit circle shown in Figure 4.11.

Example 4.8

Create the script file *cosh_sinh* that returns 30 Cartesian coordinate points linearly spaced in a table like format, where $x = cosh(t)$ and $y = sinh(t)$, over the range $-2\pi \leq t \leq 2\pi$, and verify that $cosh^2(t) - sinh^2(t) = 1$ for any t.

MATLAB Solution
```
% Script file: cosh _ sinh
t = linspace(-2*pi, 2*pi,30);
x = cosh(t);            % x represents  cosh (t)
y = sinh(t);            % y represents sinh(t)
f = x.^2-y.^2;          % f = cosh²(x)  - sinh²(x))
disp('*****************************************')
disp('  cosh(x)    sinh(x)    [cosh(x)]²-[sinh(x)]²')
disp('*****************************************')
[x' y' f ']
disp('*****************************************')
```

The script file *cosh_sinh* is executed and the resulting table is shown below:

```
>> cosh _ sinh

***************************************************************
    cosh(x)              sinh(x)                 [cosh(x)]²-[sinh(x)]²
***************************************************************
   ans =
    267.7468             -267.7449                1.0000
    173.5947             -173.5918                1.0000

    ........               ............             ........
    12.9136              -12.8748                 1.0000
     8.3899               -8.3301                 1.0000
```

......	....	....
1.2188	-0.6967	1.0000
1.0236	-0.2184	1.0000
1.0236	0.2184	1.0000
1.2188	0.6967	1.0000
1.6465	1.3080	1.0000
2.3881	2.1687	1.0000
3.5853	3.4430	1.0000
....	....	....
173.5947	173.5918	1.0000
267.7468	267.7449	1.0000

```
*********************************************************
```

The points obtained in the foregoing results are plotted for $x = cosh(t)$ versus $y = sinh(t)$, over the range $-2\pi \leq t \leq 2\pi$. The resulting graph consists of two straight lines that intersect at (0, 0).

Note that these lines are mutually orthogonal, as illustrated in Figure 4.12.

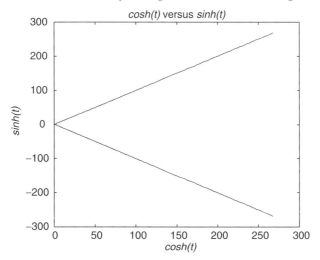

FIGURE 4.12
Plots of $x = cosh(t)$ versus $y = sinh(t)$, for $-2\pi \leq t \leq 2\pi$ of Example 4.8.

Example 4.9

Verify that if $y_1 = sin^{-1}(x)$ and $y_2 = sin(x)$ over the range $-\pi/2 \leq x \leq \pi/2$ by creating the script file *asin_sin*. Then y_1 may be complex,[*] y_2 is always real, and $y_1 \geq y_2$ for $x > 0$ and $y_2 \geq y_1$ for $x < 0$. Observe that a good approximation of y_1 can be obtained by reflecting y_2 about the line $y = x$ over the range $-1 \leq x \leq 1$.

For the sake of simplicity, consider only the real part value (the part with no i) when dealing with y_1.

MATLAB Solution
```
% Script file: asin _ sin
x = -pi/2:0.25:pi/2;
y1 = asin(x);
y2 = sin(x);
```

---

[*] Complex means that the term $i = sqrt(-1)$ is present. See Chapter 6 for additional information.

```
disp('*****************************************************')
disp('    **   x ( rad)   **   asin(x)   ******   sin(x)   ***')
disp('*****************************************************')
[x' y1'  y2' ]
disp('**************************************')
```

The script file *asin_sin* is executed and the results are shown in the following:

```
>> asin _ sin
***********************************************
**  x (rad)   **  asin(x)  *****         sin(x)  ***
***********************************************
   -1.5708     -1.5708 -  1.0232i        -1.0000
   -1.3208     -1.5708 -  0.7810i        -0.9689
   -1.0708     -1.5708 -  0.3741i        -0.8776
   -0.8208     -0.9628                   -0.7317
   -0.5708     -0.6075                   -0.5403
   -0.3208     -0.3266                   -0.3153
   -0.0708     -0.0709                   -0.0707
    0.1792      0.1802                    0.1782
    0.4292      0.4436                    0.4161
    0.6792      0.7467                    0.6282
    0.9292      1.1923                    0.8011
    1.1792      1.5708 -  0.5901i         0.9243
    1.4292      1.5708 -  0.8962i         0.9900
   ***********************************************
```

The plots of $y = x$ versus x, $y_1 = asin(x)$ versus x, and $y_2 = sin(x)$ versus x over the range $-\pi/2 \le x \le \pi/2$ are shown in Figure 4.13.

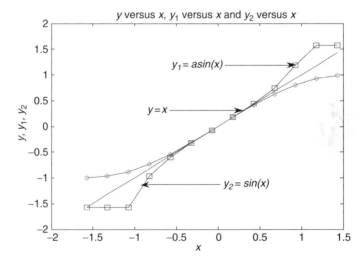

FIGURE 4.13
Plots of Example 4.9.

Example 4.10

Verify that if $y_1 = cos^{-1}(x)$ and $y_2 = cos(x)$ over the range $-\pi \le x \le \pi$ by creating the script file *acos_cos*. Then y_1 may be complex, y_2 is always real, and $y_1 \ge y_2$ for $x \le 1$.

For the sake of simplicity, consider only the real part when dealing with y_1.

MATLAB Solution

```
% Script file: acos_cos
x = -pi:0.25:pi;
y1 = asin(x);
y2 = sin(x);
disp('*************************************')
disp('** x (rad) ** acos(x)****** cos(x) ***')
disp('*************************************')
[x' y1' y2' ]
disp('*************************************')
```

The script file *acos_cos* is executed and the results are shown in the following:

```
>> acos_cos

*****************************************
** x (rad) ** acos(x) ****** cos(x) ***
*****************************************
 ans =
 -3.1416     -1.5708 - 1.8115i -0.0000
 -2.8916     -1.5708 - 1.7236i -0.2474

 ..........   ..............   ........
 -1.6416     -1.5708 - 1.0796i -0.9975
 -1.3916     -1.5708 - 0.8584i -0.9840
 -1.1416     -1.5708 - 0.5261i -0.9093
 -0.8916     -1.1009           -0.7781

 ........    ..............    ............
 0.8584      1.0322            0.7568
 1.1084      1.5708 - 0.4615i  0.8950

 ..........  ..........        ............
 2.6084      1.5708 - 1.6129i  0.5083
 2.8584      1.5708 - 1.7113i  0.2794
 3.1084      1.5708 - 1.8003i  0.0332
*****************************************
```

The plots of $y = x$ versus x, $y_1 = acos(x)$ versus x and $y_2 = cos(x)$ versus x over the range $-\pi \le x \le \pi$ are shown in Figure 4.14.

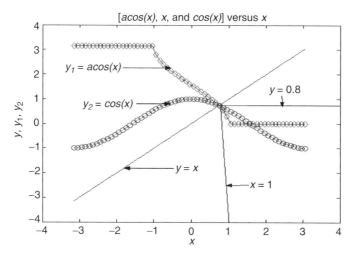

FIGURE 4.14
Plots of Example 4.10.

Example 4.11

Let $y_1 = 2^x$ and $y_2 = log_2(x)$. Use MATLAB to create the script file *log_exp* that returns a table with $y_1 = 2^x$ and $y_2 = log_2(x)$ over the range $0 \leq x \leq 3$, and verify that $y_1 \geq y_2$ for any x.

MATLAB Solution
```
% Script file: log _ exp
x = 0:0.25:3;
y = x;
y1 = 2.^x;
y2 = log2(x);
disp('**************************************')
disp('**  x (rad)   **  2^(x)******log2(x) ***')
disp('**************************************')
[x' y1' y2' ]
disp('**************************************')
```

The script file *log_exp* is executed and the results are shown in the following:

```
>> log _ exp


Warning: Log of zero.
> In C:\MATLABR11\work\A.m at line 6
**************************************************
** x (rad)      ** 2^(x)******      log2(x) ***
**************************************************
ans =
       0             1.0000            -Inf
  0.2500             1.1892         -2.0000
  0.5000             1.4142         -1.0000
  0.7500             1.6818         -0.4150
  1.0000             2.0000               0
  1.2500             2.3784          0.3219
  1.5000             2.8284          0.5850
  1.7500             3.3636          0.8074
  2.0000             4.0000          1.0000
  2.2500             4.7568          1.1699
  2.5000             5.6569          1.3219
  2.7500             6.7272          1.4594
  3.0000             8.0000          1.5850
**************************************************
```

The plots of $y = x$ versus x, $y_1 = 2^x$ versus x, and $y_2 = log_2(x)$ versus x over the range $-\pi/2 \leq x \leq \pi/2$ are shown in Figure 4.15.

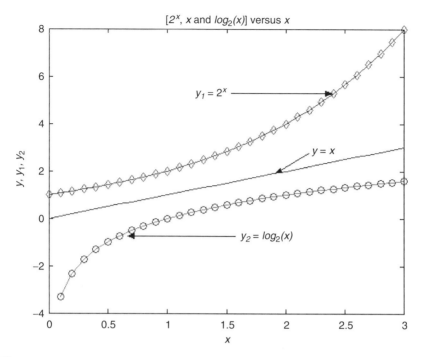

FIGURE 4.15
Plots of Example 4.11.

Example 4.12

Verify using MATLAB that $sin(x) \cdot cos(y) = 1/2[sin(x + y) + sin(x - y)]$ over the ranges $0 \le x \le \pi$ and $-\pi \le y \le \pi$, consisting of linear increments of $\Delta x = 0.3$ and $\Delta y = 0.6$, respectively by creating the script file *sin_cos*.

MATLAB Solution
```
% Script file: sin _ cos
x = 0:0.3:pi;
y = -pi:0.6:pi;
y1 = sin(x).*cos(y);
a = x-y;
b = x + y;
y2 = (1/2)*(sin(a)+sin(b));
disp('*********************************************************')
disp('*x (rad)**y (rad)***sin(x)cos(y)**(1/2)(sin(x-y)+sin(x+y))')  ****
disp('*********************************************************')
[x' y' y1' y2' ]
disp('*********************************************************')
```

The script file *sin_cos* is executed and the results are shown in the following:

```
>> sin _ cos
```

```
*************************************************************
* x (rad)   **y (rad)   *** sin(x)cos(y)**  (1/2)(sin(x-y)+sin(x+y))  ****
*************************************************************
      0        -3.1416            0                      0
  0.3000       -2.5416        -0.2439                -0.2439
  0.6000       -1.9416        -0.2046                -0.2046
  0.9000       -1.3416         0.1780                 0.1780
  1.2000       -0.7416         0.6873                 0.6873
  1.5000       -0.1416         0.9875                 0.9875
  1.8000        0.4584         0.8733                 0.8733
  2.1000        1.0584         0.4232                 0.4232
  2.4000        1.6584        -0.0591                -0.0591
  2.7000        2.2584        -0.2713                -0.2713
  3.0000        2.8584        -0.1355                -0.1355
*************************************************************
```

Observe that the last two columns of the foregoing table are identical, representing the term to the left and right of the equation $sin(x) \cdot cos(y) = 1/2[sin(x + y) + sin(x - y)]$.

Example 4.13

Verify using MATLAB that $log_2(A. * B) = log_2(A) + log_2(B)$ over the ranges $3 \leq A \leq 19$ and $2 \leq B \leq 10$, with linear increments of $\Delta A = 2$ and $\Delta B = 1$, by creating the script file *prod_sum*.

MATLAB Solution
```
%Script file: prod _ sum
A=3:2:19;
B=2:1:10;
prod = log2(A.*B);
sum = log2(A)+log2(B);
disp ('*************************************************************')
disp ('**** A ***** B ***prod=log2(A.*B)  *** sum=log2(A)+log2(B)')
disp ('*************************************************************')
[A' B' prod' sum' ]
disp('*************************************************************')
```

The script file *prod_sum* is executed and the resulting table is shown below:

```
>> prod _ sum
```

```
*****************************************************************************
**** A        ***** B      ***prod=log2(A.*B)      ***sum=log2(A)+log2(B)
*****************************************************************************
ans =
   3.0000      2.0000          2.5850                  2.5850
   5.0000      3.0000          3.9069                  3.9069
   7.0000      4.0000          4.8074                  4.8074
   9.0000      5.0000          5.4919                  5.4919
  11.0000      6.0000          6.0444                  6.0444
  13.0000      7.0000          6.5078                  6.5078
  15.0000      8.0000          6.9069                  6.9069
  17.0000      9.0000          7.2574                  7.2574
  19.0000     10.0000          7.5699                  7.5699
*****************************************************************************
```

Observe that the last two columns of the foregoing table are identical and represent the terms to the left and right of the equation $log_2(A * B) = log_2(A) + log_2(B)$.

Example 4.14

Verify using MATLAB that $log_2(A./B) = log_2(A) - log_2(B)$ over the ranges $3 \le A \le 19$ and $2 \le B \le 10$. with linear increments of $\Delta A = 2$ and $\Delta B = 1$ by creating the script file *div_sub*.

MATLAB Solution
```
% Script file: div_sub
A=3:2:19;
B =2:1:10;
div = log2(A./B);
sub = log2(A)-log2(B);
disp('***********************************************************')
disp('**** A ***** B ***log2(A./B) *** log2(A)-log2(B)')
disp('***********************************************************')
[A' B' div' sub' ]
disp('***********************************************************')
```

The script file *div_sub* is executed and the resulting table is indicated below:

```
>> div_sub
*****************************************************************
**** A           ***** B      *** log2(A./B)  *** log2(A)-log2(B)
*****************************************************************
ans =
        3.0000        2.0000        0.5850        0.5850
        5.0000        3.0000        0.7370        0.7370
        7.0000        4.0000        0.8074        0.8074
        9.0000        5.0000        0.8480        0.8480
       11.0000        6.0000        0.8745        0.8745
       13.0000        7.0000        0.8931        0.8931
       15.0000        8.0000        0.9069        0.9069
       17.0000        9.0000        0.9175        0.9175
       19.0000       10.0000        0.9260        0.9260
*****************************************************************
```

Observe that the last two columns of the foregoing table are identical and represent the term to the left and right of the equality $log_2(A./B) = log_2(A) - log_2(B)$.

4.5 Further Analysis

Q.4.1 Load and run the script file *rad_deg* of Example 4.1.

Q.4.2 Determine the number of elements of x.

Q.4.3 Do you agree with the comment statements?

Q.4.4 Draw by hand a sketch of $cos(x)$ versus x using the results obtained.

Q.4.5 What is the period T, frequency $f = 1/T$, and angular frequency $\omega = 2\pi f$ of Q.4.4 (make sure that units are included)?

Q.4.6 Sketch by hand $cos(x - \pi/4)$ versus x.

Q.4.7   Sketch by hand $cos(x + \pi/4)$ versus x.

Q.4.8   Modify Example 4.1 to obtain the coordinate points for $cos(x - \pi/4)$ versus x, and check with Q.4.6.

Q.4.9   Modify Example 4.1 to obtain the xy coordinate point for $y = cos(x + \pi/4)$ versus x, and check with Q.4.7.

Q.4.10 Load and run the program of Example 4.2.

Q.4.11 What approximation converges faster to the value of e?

Q.4.12 Estimate the minimum number of terms that returns a good result.

Q.4.13 A good result is when the sum of the terms used in the series approximation is less than 0.001. Calculate the minimum number of terms that returns such as error (smaller than 0.001).

Q.4.14 Modify the program of Example 4.2 to output a table of |error| versus [# of approximations].

Q.4.15 Load and run the program of Example 4.3.

Q.4.16 Rerun Example 4.3 without the semicolons (;).

Q.4.17 Do you agree with the comments at the end of each instruction? If not modify.

Q.4.18 Sketch $x(t)$ versus t and estimate the maximum and minimum over the range $0 \le t \le 3$.

Q.4.19 Obtain the maximum and minimum of $x(t)$ using MATLAB commands over the same range $0 \le t \le 3$. Does your answer agree with Q.4.18?

Q.4.20 Modify the program of Example 4.3 to explore the effects of the variations in the exponentials by

   a. Changing the exponent 2 to 4 and rerunning Example 4.3.

   b. Changing the exponent 2 to 1 and rerunning Example 4.3.
      Sketch $x(t)$ versus t for each one of the foregoing cases and comment/discuss your results.

Q.4.21 Load and run the script file *trig_values* of Example 4.4.

Q.4.22 Comment on the reason and purpose of the arrays a and b.

Q.4.23 Comment on the reason and purpose of the array *increm*.

Q.4.24 Can you generate the sequence *angle_deg* in a different and more efficient way? If you can explain.

Q.4.25 Complete the table of Example 4.4 by generating the columns for $tan(x)$, $cot(x)$, $sec(x)$, and $csc(x)$.

Q.4.26 Load and run the program of Example 4.5.

Q.4.27 Sketch by hand $y(t)$ versus t.

Q.4.28 Modify and run the program of Example 4.5 for the case when all the coefficients are doubled.

Q.4.29 Sketch by hand the new function as given by Q4.28.

Q.4.30 Compare the sketch of Q.4.29 with the data obtained in Q.4.27.

Q.4.31 Load and run the program of Example 4.6.

Q.4.32 Define each instruction in the form of comments (%).

Q.4.33 Draw by hand a sketch of $f(t)$ versus t.

Q.4.34 Determine the maximum and minimum values from Q.4.33.

Q.4.35 Modify the program of Example 4.6 to return $f(t)max$ and $tmax$ as well as $f(t)min$ and $tmin$.

Q.4.36 Compare the results of Q.4.35 with Q.4.34.

Q.4.37 Load and run the program of Example 4.7.

Q.4.38 From the output obtained, pick one point from each quadrant and plot it on Figure 4.11.

Q.4.39 Modify the program of Example 4.7 to produce a circle with a radius of 2.

Q.4.40 Modify Example 4.7 to produce only the upper and right half circle of Figure 4.11.

Q.4.41 Load and run the program of Example 4.8.

Q.4.42 From the output data, what range of t creates a positive slope, and what range of t create a negative slope?

Q.4.43 Indicate and discuss why $cosh^2(t) - sinh^2(t) = 1$.

Q.4.44 Load and run the program of Example 4.9.

Q.4.45 Define each of the output columns in terms of the programming variables.

Q.4.46 Do you agree that the *asin* column is showing complex values?

Q.4.47 Verify if the real angles assigned for *asin* are correct.

Q.4.48 Verify if the complex numbers assigned for *asin* are correct.

Q.4.49 Choose three points for x over the range $-1 \leq x \leq 1$ and verify the plot of Figure 4.13.

Q.4.50 Describe the symmetry of y_1 and y_2 with respect to $y = x$.

Q.4.51 Load and run the program of Example 4.10.

Q.4.52 Define each of the variables used in the program.

Q.4.53 Define each of the output columns in terms of the programming variables.

Q.4.54 Define the region in which $y_2 > y_1$.

Q.4.55 Describe the symmetry of y_1 and y_2 with respect to $y = x$.

Q.4.56 Do you observe any symmetry with respect to other lines? Discuss.

Q.4.57 Load and run the program of Example 4.11.

Q.4.58 Analyze and discuss the meaning of the warning message.

Q.4.59 Choose three points for x over the range $0.5 \leq x \leq 2.5$ and plot on Figure 4.15 y_1 and y_2.

Q.4.60 Describe the symmetry of y_1 and y_2 with respect to $y = x$.

Q.4.61 Rerun the program of Example 4.11 for $y_1 = e^x$ and $y_2 = ln(x)$.

Q.4.62 Sketch by hand y_1 versus x and y_2 versus x.

Q.4.63 Are the new plots similar to Figure 4.15? Did the symmetry change?

Q.4.64 Rerun the program of Example 4.11 for $y_1 = 10^x$ and $y_2 = log(x)$.

Q.4.65 Discuss and compare the results obtained.

Q.4.66 Load and run the program of Example 4.12.

Q.4.67 Rerun the program of Example 4.12 for a random sequences x and y.

Q.4.68 Verify the following identities:

$sin(x) + sin(y) = 2 sin((x + y)/2) \cdot cos((x - y)/2)$

$cos(x) + cos(y) = 2 cos((x + y)/2) \cdot cos((x - y)/2)$, for the sequences x and y over the ranges $-n \leq x \leq \pi$ and $-n \leq y \leq \pi$.

Q.4.69 Load and run the program of Example 4.13.

Q.4.70 Rerun the program of Example 4.13 for the random sequences A and B with the same number of elements and verify its results.

Q.4.71 Load and run the program of Example 4.14.

Q.4.72 Rerun the program of Example 4.14 for the random sequences A and B with the same number of elements and verify its results.

Q.4.73 What should be the conditions imposed on the sequences A, and B to avoid error messages?

4.6   Application Problems

P.4.1   Convert the following angles to radians:

a. $42° \; 15' =$

b. $72° \; 30' \; 30'' =$

c. $82° \; 6' \; 15'' =$

d. $35° \; 45' \; 45'' =$

P.4.2   Convert the following angles to degrees:

a. 5.2 radians =

b. 1.35 radians =

c. 6.8 radians =

d. 3134.33 radians =

P.4.3   Determine the values of the following trigonometric functions:

a. $tan(730°) =$

b. $sec((5/6)\pi) =$

c. $cos(\pi - 1) =$

d. $sin(820°) =$

P.4.4   Use MATLAB to verify the correctness of the following relations:

a. $sin(420°) \cdot cos(390°) + cos(300°) \cdot sin(-330°) = 1$

b. $(sin(120°)/sin(210°))tan(150°) + (1/sin^2(210°)) = (1 + sin^2(60°))$

P.4.5   Let $sin(60°) = \sqrt{3}/2$, $sin(45°) = \sqrt{2}/2$, and $sin(30°) = 1/2$.

Using the preceding relations, determine the six trigonometric functions for the following angles: 60°, 45°, and 30°.

P.4.6   Verify using MATLAB whether the following relations are valid:

a. $sin(60°) = 2 \, tan(30°)/(1 + tan^2(30°))$

b. $sin(30° + 60°) = sin(30°) + sin(60°)$

c. $(1 - tan^2(45°))/(2tan(45°)) = (cos(90°))/(2sin(45°)cos(45°))$

P.4.7   Determine the values of x that will satisfy the following equations:

a. $sin(x) = sin(30°) + sin(45°)$

b. $tan(x) = cos(45°) + sin(30°)$

c. $cos(x) = sin(45°) + sin(80°)$

P.4.8   Given the equation

$$2\sin(x) - 1 = 0 \quad \text{for } 0° \le x \le 360°$$

Determine the values of x that satisfy the preceding relation.

P.4.9   Repeat problem P.4.7 for the following equations:

a. $5\cos(x) + \sqrt{7} = 0$ over the range $-\pi/2 \le x \le 3\pi/2$

b. $\sin(x) = \cos(x)$ over the range $0 < x < 2\pi$

c. $3\sin(2x) = 1$

d. $4^x = 5$

e. $3\log(x) - 2\log(x) = 5$

f. $\log(7x - 9)^2 + \log(3x - 4)^2 = 2$

P.4.10   Evaluate the value of x for the following equations:

a. $\log_3(243) = x$

b. $\log_x(343) = 3$

c. $\log_{16}(x) = 1/4$

d. $\log_2(16) + \log(130) + 15 * \log_{10}(120) = x$

P.4.11   Evaluate each of the following expressions by hand and verify your result using MATLAB:

a. $\sin(45°) =$

b. $3\cos(30°) =$

c. $\sin(1) + 3\cos(.5) =$

d. $\sin(1) + 3\cos(0.5) =$

e. $\sqrt{\sin^2(1) + 3^2\cos^2(.5)} =$

f. $\sec(1) =$

g. $round(1.7) =$

h. $round(-2.8) =$

i. $fix(7.4) =$

j. $fix(-7.4) =$

k. $floor(3.6) =$

l. $floor(-3.6) =$

m. $sign(-1.3) =$

n. $sign(1.3) =$

o. $rem(17, 3) =$

p. $rem(19, 6) =$

q. $abs(3.4) =$

r. $abs(-3.4) =$

s. $abs(sign(-1.3)) =$

t. $ceil(19.2) =$

u. $ceil(\cos(1)) =$

v. $gcd(13, 260) =$

w. *lcm(13, 260) =*

x. *rem(171, 320) =*

y. *factor(1351) =*

z. *isprime(1351) =*

a1. *primes(1351) =*

b2. rat(pi-exp(1)) =

P.4.12 Write a MATLAB program that verifies if $sin(x) = cos(x - \pi/2)$ over the range $0 \le x < 2\pi$.

P.4.13 Write a MATLAB program that returns the cartesian coordinates for the following elliptic equations:

a. $x^2 + 4y^2 = 4$

b. $4x^2 + y^2 = 4$

c. $(x^2/4) + (y^2/4) = 4$

d. $(x^2/4) + y^2 = 4$

over the range being $-4 \le x \le +4$.

P.4.14 Use MATLAB to compute the Cartesian coordinates for the following (circle) equations:

a. $x^2 + y^2 = 9$

b. $(x + 2)^2 + (y - 2)^2 = 4$

i. Once enough points are obtained, sketch each circle by hand.

ii. For each circle, determine the Cartesian coordinate at its center, as well as its radius.

Recall that the standard form of the equation of a circle is given by $(x - a)^2 + (y - b)^2 = r^2$, where r is its radius and the center is located at the cartesian coordinate given by $<a, b>$.

P.4.15 Use MATLAB to compute the Cartesian coordinates of the following hyperbolas:

a. $x^2 - 4y^2 = 4$

b. $4x^2 - y^2 = 4$

c. $y^2 - 4x^2 = 4$

d. $4y^2 - x^2 = 4$

Once enough points are obtained about each function, draw each one of the hyperbolas by hand.

P.4.16 Use MATLAB to create a table of the cartesian coordinate points required for sketching the following functions:

a. $y_1 = x$

b. $y_2 = \sqrt{x}$

c. $y_3 = cos(x)$

d. $y_4 = x - \sqrt{x}$

e. $y_5 = \sqrt{x} + cos(x)$

over the range $0 \le x \le 5$, with linear increment of $\Delta x = 0.1$.

Determine in each case the maximum and minimum values of each of the preceding functions.

P.4.17 Write a MATLAB program that verifies the following trigonometric identities:

a. $sec(x) = \sqrt{(1 + tan^2(x))}$

b. $cos(x) = sin(2x)/2sin(x)$

c. $cos(x) = \sqrt{cos(2x) + sin^2(x)}$

d. $cos(x/2) = sin(x)/2sin(x/2)$

over the range $0 < x \leq 2\pi$.

P.4.18 Using MATLAB, evaluate the function $sin(x)$ by using the first five MacLaurin terms for the following arguments of $x = 15°, 30°, 45°, 60°$, and $90°$.

Compare your result with the value of $sin(x)$, for $x = 15°, 30°, 45°, 60°$, and $90°$. Recall that $sin(x) = x - (x^3/3!) + (x^5/5!) - (x^7/7!) + (x^9/9!)$, for x given in radians.

P.4.19 Using MATLAB, evaluate the value of $cos(x)$ using the first five MacLaurin terms estimate in each case its error for the following arguments of $x = 15°, 30°, 45°, 60°$, and $90°$.

P.4.20 Using MATLAB, evaluate the following functions:

a. $y = e^x$

b. $z = e^{-x} + 3e^{-2x}$

where $x = 1, 2, 3$, and 4.

i. By using the first five terms of its series expansion

ii. By direct evaluations

P.4.21 Using MATLAB, verify the following relations:

a. $sinh(x) = (e^x - e^{-x})/2$

b. $cosh(x) = (e^x + e^{-x})/2$

c. $tanh(x) = (e^x - e^{-x})/(e^x + e^{-x})$

over the range $0 < x < 2\pi$

P.4.22 Using MATLAB, verify the following identities:

a. $acosh(x) = ln[x + (x^2 - 1)^{1/2}]$ for any $5 > x \geq 1$

b. $atanh(x) = ln((1 + x)/(1 - x))^{1/2}$ for any $0 < x < 1$

c. $asech(x) = ln(1 + (1 + x^2)^{1/2})/x$ for any $0 \leq x \leq 1$

P.4.23 Let $y_1(x) = tan^{-1}(x)$ and $y_2(x) = tan(x)$ over the range $-\pi/2 \leq x \leq \pi/2$.

Determine over what ranges the following holds: $y_1 \geq y_2, y_2 \geq y_1$, and $y_2 = y_1$. Discuss if the graph of y_1 can be obtained by reflecting y_2 about the line $y = x$.

P.4.24 Using MATLAB, evaluate and verify the following equalities:

a. $log_{10}(20) = log_{10}(4) + log_{10}(5)$

b. $log_{10}(4) = log_{10}(20) - log_{10}(5)$

c. $log(2) = 2.3log_{10}(2)$

d. $log_3(6) = log_{10}(6)/log_{10}(3) = log(6)/log(3)$

e. $\sqrt[5]{\sqrt[2]{2}} = \sqrt[10]{2}$

f. $(3)^5/(3)^{-2} = 3^7$

g. $5^{3/2} = \sqrt[2]{5^3}$

h. $(2^3)^4 = 2^{12}$

P.4.25 Use MATLAB to verify that $log_2(A)^b = b.log_2(A)$ for $A = 1:1:10$ and $b = 3$.

P.4.26 Use MATLAB to generate a list of all the prime numbers between 100 and 200.

P.4.27 Use MATLAB to factor 1030 and verify the result obtained.

P.4.28 Using MATLAB, verify the following identities:

a. $cos^2(x) + cos^2(x)\, tan^2(x) = 1$

b. $(tan(x) + tan(y))/(cot(x) + cot(y)) = tan(x)tan(y)$, for $y = 0.5$, 1, and 1.5

c. $sec^4(x - 1)/tan^2(x) = tan^2(x) + 2$

d. $(sin(3x)/sin(x)) - (cos(3x)/cos(x)) = 2$

e. $(1 + tan(x))/(sin(x) + cos(x)) = sec(x)$ for $0 < x < 2\pi$

5

Printing and Plotting

The picture you create in your head often turns into the reality you hold in your hand.

Allan Hanson

5.1 Introduction

Graphic capabilities are quite important in engineering, social science, natural science, education, behavioral science, health, economy, weather, production (growth and decay), politics, biology, accounting, and business, just to mention a few and diverse disciplines.

Graphs are an important way to communicate and visualize trends and patterns that are otherwise difficult to identify, and gain valuable insight into a given relation or problem in this way. Information when given in the form of tables can be easily graphed and be used to make educated predictions and decisions.

A graph, like an equation, is the language that best helps recognize the relationship, which exists between the variables involved in a situation. Graphs, like languages, have specific rules, some of which date back to ancient civilizations.

It is believed that the coordinate system was first used in ancient times for urban planning, surveying, and astronomy in the old Egyptian and Babylonian civilizations.

For thousands of years, the rectangular coordinate system was used, not exactly the way we know it today. It was not until the 1600s, when it was rediscovered by the mathematicians of the time that geometric problems could be solved by using algebraic equations and vice versa.

Rene Descartes (1596–1650) and Pierre Fermat (1601–1665) are credited with being the first mathematicians in taking such an approach. This approach evolved over time into the Cartesian coordinate system, as it is known today.

The 2-D Cartesian system is the most frequently used system by engineers and scientists. MATLAB® offers its users simple and easy-to-use graphic commands to obtain 2-D and 3-D plots in the Cartesian coordinate system as well as in other systems. The goal of the xy Cartesian* rectangular plots (2-D) is the construction of a plot of the form $y = f(x)$. In this plot, y is plotted versus x (denoted by y versus x), where x and y are frequently referred to as the independent and dependent variables, respectively.

In general, a plot can be constructed once a relation in the form of a table exists between x and y. This relation is frequently expressed by vectors (or matrices) in MATLAB.

The command *plot (x, y)* is the most popular plotting command used in MATLAB, and MATLAB returns the plot of the points defined by $<x_i, y_i>$ for all *i*s connected by straight

---

* The rectangular coordinate system is named after the French philosopher and mathematician Rene Descartes (1596–1650). Rene Descartes first introduced the coordinate geometry (*xy* plane) in 1637 with the publication of the book, *A Discourse on the Method of Rightly Conducting the Reason and Seeking Truth in the Sciences.*

segmented lines. Clearly, this implies that the two vectors x and y must have the same dimensions.

MATLAB assumes that the points involved in the construction of the plot $y = f(x)$ will be connected by a solid blue or black line, unless defined otherwise.

5.2   Objectives

After completing this chapter, the reader should be able to

- Display messages (strings) and variables
- Convert variables to strings and vice versa
- Display sentences consisting of text and variables
- Format the display field
- Know the different options in the plot command
- Set the domain and range for $y = f(x)$
- Represent discrete points on a plane
- Define and construct functions (algebraic, trigonometric, exponential, etc.), over a given range and domain
- State the equations of a straight and a curved line
- State the requirements for plotting any arbitrary function
- Generate equidistant set of points on the x- and y-axis
- Determine when a set of ordered points is a solution of an equation or set of equations
- Set the x and y scales
- Create a linear, logarithmic, or semilogarithmic coordinate plot
- Understand the reasons for using linear, semilog, and log scales applied to either variables x and y or both
- Label the x- and y-axis
- Create a plot title
- Create 2-D plots
- Include comments on a plot
- Create a legend box on a plot
- Include text when appropriate in a graph
- Generate multiple 2-D plots on a single graph
- Use colors, markers, and line styles to identify different plots
- Understand the concept and meaning of a histogram plot
- Create *histogram* plots
- Create a *pie* diagram
- Represent a function at discrete points
- Represent a function by means of continuous, bar, and stairlike approximations

- Represent a function in a 3-D coordinate system
- Create 3-D plots
- Evaluate areas and surfaces
- View a 3-D figure or body from different reference points
- Rotate and view a 3-D figure or body
- Solve a variety of printing and plotting (2-D and 3-D) problems using the power and the many features of MATLAB

5.3 Background

R.5.1 The command *disp('text')* is used to display the string vector *text* that is enclosed in quotes, inside parenthesis. For example, display the string: *This is a string text.*

MATLAB Solution
```
>> format compact
>> text = 'This is a string text ' ;
>> disp (text)
        This is a string text
```

R.5.2 The numerical value of a *variable* can be displayed by using the command *disp(variablename).*

For example, display the values of $x = 1, \pi,$ and e.

MATLAB Solution
```
>> x =1;
>> disp(x)
        1

>> disp(pi)
        3.1416

>> disp(exp(1))
        2.7183
```

R.5.3 The command *disp(variablename)* can be used to display the numerical value of a variable, scalar, arithmetic expression, vector, or a matrix. For example, use MATLAB and display the variables: *A, B,* and *C* defined as follows:

a. $A = 313/57$

b. $B = [1\ 3\ 5\ 9\ 11]$

c. $C = \begin{bmatrix} 1 & 4 & 7 \\ 2 & 5 & 8 \\ 3 & 6 & 9 \end{bmatrix}$

Observe the responses of the *disp* command and how the semicolon at the end of a statement affects the display.

MATLAB Solution

```
>> disp(313/57)              % observe that the line ends with no
                                semicolon (;)

        5.4912

>> disp(313/57);             % observe that the line ends with a
                                semicolon

        5.4912

>> B = linspace(1,11,6)      % observe that the line ends with no
                                semicolon


   B =
        1   3   5   7   9   11

>> disp(B);                  % observe that the line ends with a
                                semicolon

        1   3   5   7   9   11

>> C = [1 4 7;2 5 8;3 6 9]   % observe that the line ends with no
                                semicolon


   C =
        1   4   7
        2   5   8
        3   6   9

>> disp(C);                  % observe that the line ends with a
                                semicolon (;)

        1   4   7
        2   5   8
        3   6   9
```

R.5.4 Strings texts and variables can be displayed in a textlike format using the command *disp(['text'])*, as long as the argument of *disp*, the *text* is a string.

R.5.5 A numerical variable can be converted to a character string using the command *num2str(x)* or *int2str(x)*, where *x* is either a number or an integer.

  For example, evaluate the values of *e* and π, and then display each one of the integrated messages.

  a. *The value of e is ????, and pi = ?????*

  b. *The value of e as an integer is ?????*

MATLAB Solution

```
>> format compact
>> x = exp(1)                % number

   x =
        2.7183

>> y = num2str (x)           % string

   y =
        2.7183
```

```
>> z = pi                          % number

    z =
        3.1416

>> v = num2str(z)                  % string

    v =
        3.1416

>> disp (['The value of e is ',y,', and pi = ',z'])

         The value of e is 2.7183, and pi = 3.1416.

>> w = int2str(x)
    w =
        3

>> disp (['The value of e as integer is ',w,])  % note that the integer
                                                   value of 2.718 ... is 3

         The value of e as an integer is 3
```

R.5.6 An alternate way to display an integrated text consisting of strings and numerical values represented by variables is by using the commands: *fprintf('text ','control', variables)*, or *sprintf('text', ' control', variables)*, where *text* is the string that will be displayed and *control* defines the format of the output *variables* as defined by Table 5.1, where *a* and *b* are integers that represent the field width (include the ".") and the number of decimal characters of the variables.

R.5.7 For example, use the commands *fprintf* and *sprintf* to display the following integrated messages:

a. *The value of e is ???????*, to six decimal places using an eight character field.

b. *Today is ?????*, using the standard MATLAB command.

MATLAB Solution
```
>> fprintf ('The value of e = %8.6f \n', exp (1))

         The value of e = 2.718282

>> fprintf ('Today is %s \n', date)

         Today is 01-Aug-2006

>> sprintf ('The value of e is %8.6f \n', exp (1))  % observe that
                                                       the response includes ans

        ans =
             The value of e is 2.718282

>> sprintf ('Today is %s \n', date)

        ans =
             Today is 01-Aug-2006
```

TABLE 5.1

Format to Display a Variable

Format	Description
%a.bd	Display as integer, decimal
%e or %E...	Display as exponential
%f	Display as floating point
%a.bg or %G...	Display the shortest version of %f or %e
%a.bx or %X...	Display in hexadecimal
%c	Display single character
%s	Display a string of characters
\n	Line feed, so that text starts in a new line

R.5.8 The control field of the *fprintf* or *sprintf* may include a sign (+ or −), the number of characters, a decimal point, and an exponential factor (following the scientific notation format).

For example, display the message, *The value of e is ??????*, include the + sign with two decimal places using six field characters expressed in exponential format.

```
>> x = exp(1)


   x =
       2.7183


>> fprintf ('The value of e is %+6.2E \n', x)

        The value of e is +2.72E+000
```

R.5.9 The special sequence \n, \r, \t, \b, \f can be used to produce *linefeed, carriage return, tab, backspace,* and *feed character,* respectively.

R.5.10 If the variable of *fprintf* or *sprintf* is the complex* number $z(z = a + jb)$, then only the real part of z will be displayed (a).

R.5.11 For example, let $z = -1 + 2i$ be a complex number. Then, display the message *The real value of z is ??????*

MATLAB Solution
```
>> z = -1+2i;
>> fprintf ('The real value of z is %2f \n',z)

        The real value of z is -1.000000

>> sprintf ('The real value of z is %2f \n',z)


     ans =
         The real value of z is -1.000000
```

Note that the imaginary part of z_1 given by 2 is ignored.

---

* Complex numbers are discussed in Chapter 6. At this point, it is sufficient to know that a complex number consists of two parts: a real and an imaginary part. The imaginary part is distinguished by the character $i = j = \sqrt{-1}$.

R.5.12 The most common 2-D command used for plotting is the command *plot(x, y)*. Recall that a number is graphed on a single line. A plane is used to graph a pair of numbers. Two perpendicular lines called axis are used to locate a point in a plane. The horizontal axis is referred to as the *x*-axis, and the vertical axis is the *y*-axis. In its simplest version, *x* and *y* can be considered vectors whose elements are ordered pairs ($<x_i, y_i>$) in the Cartesian coordinate system in honor of the French philosopher Rene Descartes (1596–1650). The axes divide the plane into four regions called "quadrants."

In the first quadrant, both coordinates are positive ($<a, b>$, $a > 0$, and $b > 0$).

In the second quadrant, the first coordinate is negative, whereas the second coordinate is positive ($a < 0, b > 0$).

In the third quadrant, both coordinates are negative ($a < 0, b < 0$).

In the fourth quadrant, the first coordinate is positive, whereas the second coordinate is negative ($a > 0, b < 0$).

The $<x_i, y_i>$ ordered pair may represent a relation, a rule, an equation, or a point on a chart. An ordered pair defines the first element as the independent variable *x*, whereas the second element represents the dependent variable *y*.

R.5.13 Recall that a function can be defined as a set of ordered pairs in which the first element of each pair is unique.

R.5.14 The set of all the values of *x* of $y = f(x)$ is called the *domain of f(x)*, whereas the set of all the values of *y* is called its *range*. To graph *y* versus *x* means to make a drawing that represents its solution.

R.5.15 In order to use the command *plot(x, y)*, the variables *x* and *y* must be stored as two separate arrays of ordered numbers in sequential order. These two sequences define a set of points on the Cartesian (*xy*) plane. For example, point p_1 is defined by $<x(1), y(1)>$, point p_2 is defined by $<x(2), y(2)>$, ..., point p_n by $<x(n), y(n)>$. The resulting graph is constructed by connecting the consecutive points with straight-line segments.

Observe that

a. The plot command can be executed if and only if the arrays *x* and *y* have the same length (number of elements).

b. The resulting plot may not be smooth, unless a sufficiently large number of points are employed.

R.5.16 A straight line can easily be plotted using the *plot* command by defining two points over a domain. Recall that two points define a line (Euclidian geometry).

R.5.17 For example, use MATLAB and create the plot of the following line defined by the equation $f(x) = y = 2x - 1$ over the range $-3 \le x \le 2$ by

a. Using the *plot* command with the argument consisting of two points on the Cartesian plane

b. Using the *plot* command with arguments *x* and *f(x)*

ANALYTICAL Solution

The chosen two points that define the line $y = 2x - 1$ over the given domain are

a. *point #1, x* $= -3, y = -7$
b. *point #2, x* $= 2, y = 3$

MATLAB Solution
```
>> x = [-3 2];
>> y = [-7 3];
>> plot(x,y)           % solution #1
>> plot(x,2.*x-1)      % solution #2
```

The two solutions are shown in Figure 5.1.

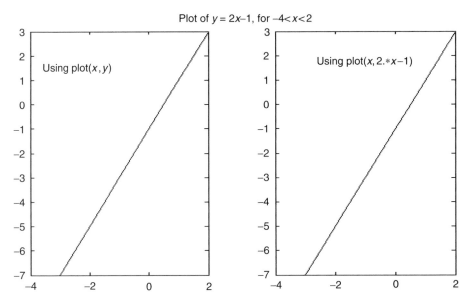

FIGURE 5.1
Linear plots of R.5.17.

R.5.18 Let us now consider a polynomial.* Use MATLAB and obtain the plots of $y = f(x) = x^3 + 4x^2 - x - 4$ over the domain $-5 \leq x \leq 2$, using 3, 4, 5, 6, 7, and 101 points (or 100 segmented approximations).

MATLAB Solution
```
>> x3 = linspace(-5,2,3);   % x1 is defined by 3 points and 2 segments
>> y3 = x3.^3+4.*x3.^2-x3-4;
>> plot (x3,y3)
>> x4 = linspace(-5,2,4);   % x2 is defined by 4 points and 3 segments
>> y4 = x4.^3+4.*x4.^2-x4-4;
>> plot (x4,y4)
>> x5 = linspace(-5,2,5);   % x3 is defined by 5 points and 4 segments
>> y5 = x5.^3+4.*x5.^2-x5-4;
>> plot (x5,y5)
>> x6 = linspace(-5,2,6);   % x4 is defined by 6 points and 5 segments
>> y6 = x6.^3+4.*x6.^2-x6-4;
>> plot (x6,y6)
>> x7 = linspace(-5,2,7);   % x7 is defined by 7 points and 6 segments
>> y7 = x7.^3+4.*x7.^2-x7-4;
```

---

* Polynomials are discussed in Chapter 7. At this point, it is sufficient to know that a polynomial is a sum of N product terms of the independent variable x that can be defined by $f(x) = \sum_{n=0}^{N} a_n x^n$.

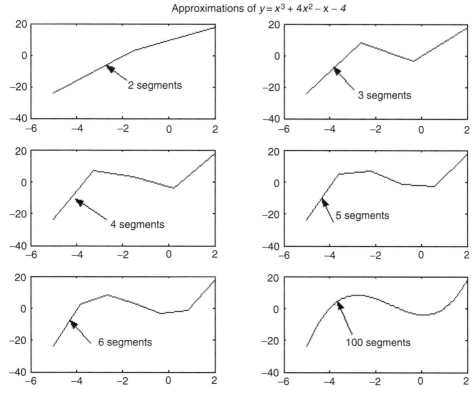

Approximations of $y = x^3 + 4x^2 - x - 4$

FIGURE 5.2
Polynomial approximations of R.5.18.

```
>> plot (x7,y7)
>> x100 = linspace(-5,2,101);    % x100 is defined by 101 points and
                                   100 segments
>> y100 = x100.^3+4.*x100.^2-x100-4;
>> plot (x100,y100)
```

The corresponding plots are shown in Figure 5.2 for each one of the six approx-imations.

R.5.19 Let us now plot the trigonometric function $f(x) = y = 1.5 \, cos(2x)$, over the domain $0 \le x \le 2\pi$, using 5, 10, 15, and 20 points, and the corresponding segmented approximations.

MATLAB Solution
```
>> x5 = linspace(0,2*pi,5);     % x5 is defined by 5 points and 4
                                  segments
>> y5 = 1.5*cos(2.*x5);
>> plot(x5,y5)

>> x10 = linspace(0,2*pi,10);   % x10 is defined by 10 points and 9
                                  segments
>> y10 = 1.5*cos(2.*x10);
>> plot(x10,y10)
```

```
>> x15 = linspace(0,2*pi,15);      % x15 is defined by 15 points and 14
                                     segments
>> y15 = 1.5*cos(2.*x15);
>> plot(x15,y15)
>> x20 = linspace(0,2*pi,20);      % x20 is defined by 20 points and 19
                                     segments
>> y20 = 1.5*cos(2.*x20);
>> plot(x20,y20)
```

The corresponding plots are shown in Figure 5.3 for each one of the four approximations.

R.5.20 Recall that a linear equation is of the form $ax + by = c$, where $a \neq 0$ and $b \neq 0$. Its graph is a straight line and as already mentioned, only two points or ordered pairs are required.

R.5.21 If an equation is not linear, then the variables x and y are raised to at least the second power, and the shape of the graph is same sort of curve. Many ordered pairs or points are usually required to decently approximate a curved line.

R.5.22 The limits for the x- and y-axis, called the domain and range, respectively, are controlled by MATLAB by the commands *xlim([xmin xmax])* and *ylim([ymin ymax])*.

R.5.23 Recall that the x-axis is the horizontal axis on the Cartesian plane and is referred as the abscissa.

R.5.24 Recall also that the y-axis is the vertical axis on the Cartesian plane and is referred to as the ordinate.

R.5.25 The x_i and y_i are referred to as the coordinates of the point P_i, where $i = 1, 2, 3, ..., n$.

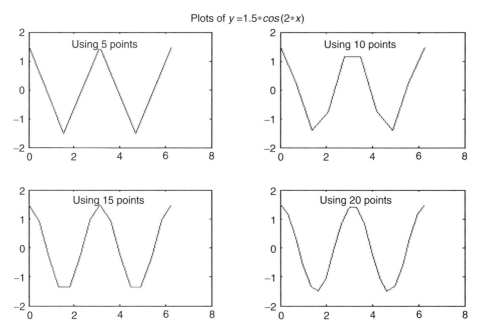

FIGURE 5.3
Linear approximations of the cosine wave of R.5.19.

R.5.26 The MATLAB command *plot (x, y)* is used to plot $y = f(x)$ when the scales of both axes are linear or equally spaced. If it is desired to plot two or more different functions such as $y_1 = f_1(x)$, $y_2 = f_2(x)$, ..., $yn = f_n(x)$ on the same graph, the command *plot (x, y_1, x, y_2,... x, y_n)* can then be used, if and only if the lengths of the arguments used are compatible. That is $length(y_1) = length(y_2) = \cdots = length(y_n) = length(x)$. Observe that the domain is common for all the functions.

R.5.27 A logarithmic scale can be created if and when it is desired that the scale spacing be equal between successive powers of 10. For example, the scale spacing between 10^{-1}–10^0 and 10^0–10^1 are equal. Observe that from $10^{-1} = 0.1$ to $10^0 = 1$, the separation is 0.9 and from 10^0 to 10^1, the separation is 9, but the spacing or the scale distances are still equal.

R.5.28 Recall that negative numbers or zero cannot be (reached) plotted on a logarithmic scale.

R.5.29 Recall that the main reason for using a logarithmic scale is to represent data that spreads over a wide range.

R.5.30 When the function $y = f(x)$ is linear with respect to x, a linear scale is usually appropriate to best illustrate the relation between x and y.

R.5.31 If the function y is of the form of $f(x) = kx^n$ and if both scales are defined as logarithmic, the plot of $f(x)$ versus x is linear.

R.5.32 If the function y is of the form of $f(x) = k(a)^{bx}$ and if the x-axis is linear, but the y-axis is logarithmic, the plot of $f(x)$ versus x is linear.

R.5.33 Depending on the nature and domain of the function to be plotted, choosing the appropriate scale can better illustrate the relation between the independent variable x and the dependent variable y.

R.5.34 The MATLAB command *semilogx(x, y)* returns the plot of y versus x, where the x-axis scale is logarithmic, but the y-axis scale is linear.

R.5.35 The MATLAB command *semilogy(x, y)* returns the plot of y versus x, where the x-axis scale is linear, but the y-axis scale is logarithmic.

R.5.36 The MATLAB command *loglog(x, y)* returns the plot of y versus x when both the axes scales are logarithmic.

R.5.37 Any of the plot commands such as *plot, semilogx, semilogy,* or *loglog* automatically activates or opens the figure window.

R.5.38 Any of the plot commands return the following:

a. The x- and y-axis

b. The plot of the points defined by the ordered pairs or points $P_i = <x_i, y_i>$, for $i = 1, 2, 3, ..., n$

c. The consecutive points (P_i with P_{i+1}) are connected with solid straight segmented lines

R.5.39 If a figure already exists, the *plot* command automatically clears the existing figure and creates a new one. The command *clf* clears the active figure window.

R.5.40 Recall that the *plot* command with multiple arguments can be used to create multiple plots on the same graph. For example, the command *plot (x, y1, x, y2)* returns the graph with the plot of $y1$ versus $x1$ and the plot of $y2$ versus $x2$, where $y1 = f(x)$ and $y2 = f(x)$.

R.5.41 Another way of creating overlay plots is by using the MATLAB command *hold on* after a *plot* command has been executed. The *hold on* command freezes the figure window and allows the placing of additional plots. The command *hold off* clears the *hold on* command.

R.5.42 For example, use MATLAB to obtain overlay plots of the following functions:

$$[y1 = 2.5\ cos(x)] \text{ versus } x \text{ and } [y2 = 3.5\ sin(x)] \text{ versus } x$$

over the domain $0 \le x \le 2\pi$ using 50 linearly spaced points.

MATLAB Solution
```
>> x = linspace(0,2*pi,50);   % x defines 50 points linearly spaced
                              %   over 0 ≤ x≤ 2π
>> y1 = 2.5*cos(x);           % y1 are the 50 values for x
>> y2 = 3.5*sin(x);           % y2 are the 50 values for x
>> plot(x,y1)                 % plot [2.5cos(x)] vs. x
>> hold on                    % holds the graph
>> plot(x,y2)                 % plot [3.5sin(x)] vs. x
```

The corresponding plots of *y1* versus *x* and *y2* versus *x* are shown in Figure 5.4.

R.5.43 The plot of a line can be added to any plot by using the MATLAB command *line(x, y, 'linestyle')*, where the *line style* is an option presented later in this section.

R.5.44 For example, let us assume that it is now desired to add the following to the two previous plots: *[2.5sin(x)]* versus *x* and *[3.5cos(x)]* versus *x* of Figure 5.4, the plot of the polynomial *[f(x) = 0.005x³ + 0.015x² + 0.01x − 1]* versus *x*. The following two instructions are added to the program of R.5.42.

```
>> fx = 0.005*x.^3+0.015*x.^2+0.01*x-1;
>> line (x, fx)
```

The resulting plots are shown in Figure 5.5.

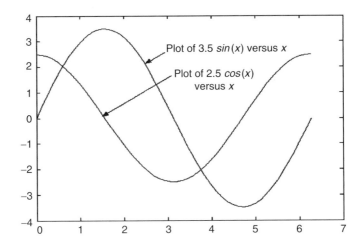

FIGURE 5.4
Plots of *sin(x)* and *cos(x).*

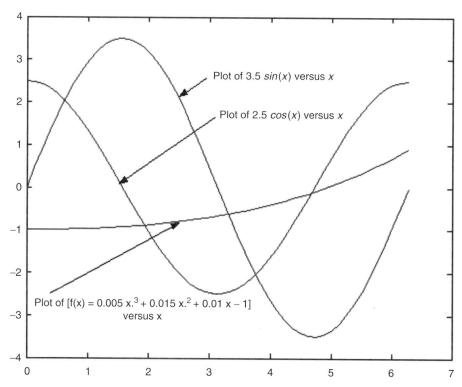

FIGURE 5.5
(See color insert following page 342.) Plots of R.5.44.

R.5.45 The command *plotyy(x, f₁(x), x, f₂(x))* returns the plots of two different functions: $f_1(x)$ versus x, and $f_2(x)$ versus x over the same (x) domain, but different (y) ranges (different vertical scales).

R.5.46 For example, let $y_1(x) = 10\ sin(x)$ and $y_2(x) = 2\ cos(x) + noise(x)$, where *noise(x)* is a random function over the range $0 \leq x \leq 3\pi$ using 100 linearly spaced points. Create a program that returns the following plots:

a. $[y_1 = 10\ sin(x)]$ versus x and $[y_2 = 2\ cos(x) + noise(x)]$ versus. x using the same y scale

b. $[y_1 = 10\ sin(x)]$ versus x and $[y_2 = 2\ cos(x) + noise(x)]$ versus x using different y scales

MATLAB Solution
```
>> x = linspace(0,3*pi,100);
>> y1 = 10*sin(x);
>> y2 = 2*cos(x) + rand(1,100);
>> plot(x,y1,x,y2)                % plots using same y-scale
```

The resulting plots are shown in Figure 5.6.

Observe that by using two different y-scales, the relation between y_1 and y_2 is better visualized. Note that the scales are represented vertically at the two ends of the graph.

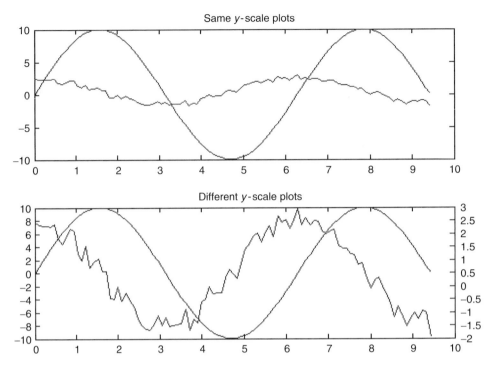

FIGURE 5.6
(See color insert following page 342.) Plots of R.5.46 using one and two scales.

R.5.47 Multiple plots can be obtained by using the command *plot(Y)*, where *Y* is an *m*
× *n* matrix. MATLAB returns *n* plots, one for each of the *n* columns of *Y* versus
its index. MATLAB returns the plots color coded with the first plot represented by
blue, the second by *green*, the third by *red*, etc.

R.5.48 Overlaid plots can also be obtained with the command *plot(Y, x)*, where *Y* is an
m × *n* matrix where *x* can be either a row vector of length *n* or a column vector of
length *m*.
　　If *x* is a row vector of length *(n)* then the *plot(Y, x)* returns *m* plots, one for each of
the rows of *Y* versus *x*, and if *x* is a vector of length *(m)* then the command *plot(Y, x)*
returns *n* plots, one for each of the columns of *Y* versus *x*.

R.5.49 The command *plot(X, Y)*, where *X* and *Y* are two matrices with the same dimensions *mxn*. MATLAB returns a set of *n* plots, each one representing the *[columns of
Y]* versus *[columns of X]*.

R.5.50 For example,

   a. Let *A* = *[1 2 3 4; 5 6 7 8; 9 10 11 12]*. Write a program that returns the plots of each
      column of A versus *its index* (see Figure 5.7).

   b. Let *B* = *[1 2 3]*; obtain the plots of the columns of *A* versus *the columns of B*
      (see Figure 5.8).

   c. Now let *B* = *[1 2 3 4]*; obtain the plots of the rows of *A* versus *the rows of B*
      (Figure 5.9).

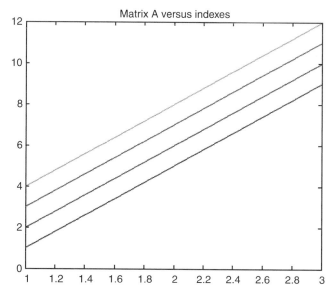

FIGURE 5.7
(See color insert following page 342.) Plot of matrix A of R.5.50(a).

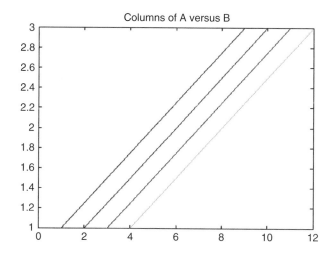

FIGURE 5.8
(See color insert following page 342.) Plot of columns of matrix A versus B of R.5.50(b).

MATLAB Solution
```
>> % part (a)
>> A= [1 2 3 4;5 6 7 8;9 10 11 12]

   A =
       1    2    3    4
       5    6    7    8
       9   10   11   12
```

```
>> plot (A)
>> title ('Matrix A vs. Indexes')

>> % part (b)
>> B = [1 2 3]

   B =
        1   2   3

>> plot (A,B)

>>   % part (c)
>> B(4) = 4

   B =
        1   2   3   4

>> plot (A,B)
```

R.5.51 The *plot* command has a number of options that are used to define its line style, markers, and colors of the plot by specifying a third argument in quotes labeled *options*. The syntax is *plot(x,y,'options')*.

R.5.52 The plot *options* consists of a set of one to three characters, labeled *a*, *b*, and *c* entered in sequential order in quotes, where *a* defines the color, *b* the marker, and *c* the line style.

The plotting options are shown in Table 5.2.

R.5.53 If no color or line style is specified, the default is a solid blue or black line.

R.5.54 Markers are used to indicate points or discrete entries. If no marker type is selected, no markers are drawn.

R.5.55 The command *scatter(X, Y, S)* returns a plot of the Cartesian points represented by circles where the location of the points are represented by the vectors X and Y. The size (area) of each marker is determined by the values assigned to the vector S.

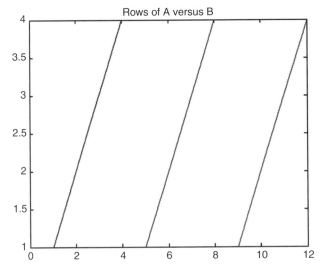

FIGURE 5.9
(See color insert following page 342.) Plot of row of matrix A versus B of R.5.50(c).

TABLE 5.2

Plotting Options

Option *a* (Color)	Option *b* (Marker)		Option *c* (Line Style)
b, Blue	.	Point/dot	- Solid line
g, Green	o	Circle	: Dotted line
r, Red	xx	Mark/cross	-. Dash-dot line
c, Cyan	+	Plus	-- Dashed line
m, Magenta	*	Star	
y, Yellow	s	Square	
k, Black	d	Diamond	
w, White	v	Triangle (down)	
	^	Triangle (up)	
	<	Triangle (left)	
	>	Triangle (right)	
	p	Pentagram	
	h	Hexagram	

The additional option *filled* assigned to *S* returns shaded markers all drawn with the same size. The default version of this command is given by *scatter(X, Y)*.

R.5.56 For example, obtain a scatter plot for the following random Cartesian coordinate points defined by commands $X = randn(1, 100). * randn(1, 100)$ and $Y = rand(100, 1)$ using the following options:

a. The *default*

b. The *filled* option

MATLAB Solution

```
>> X = randn(1,100).*randn(1,100); X=X';
>> Y = rand(100,1);
>> scatter(X,Y); box on          % default plot
>> scatter(X,Y,'filled'); box on  % plot with filled/shaded circles
```

The resulting plots are shown in Figure 5.10.

R.5.57 The command *subplot(m, n, p)* returns *m* times *n* independent subwindows, where *m* and *n* indicate that the active figure window is divided into *m* times *n* independent matrixlike subwindows and *p* is an integer over the range $1 \leq p \leq n \cdot m$. The integer *p* represents the active or current subwindow. The windows are labeled from left to right, starting from the top row.

For example, *subplot(2, 2, 3)* indicates that the figure window is divided into four subwindows (two rows by two columns) and the current plot subwindow is the third (second row by first column).

R.5.58 For example, write a program that divides the window into four subwindows and plots in each one the following functions:

$$y_1 = 27 \, e^{-2x} \quad \text{and} \quad y_2 = 15 * 0.3^x$$

over the range $1 \leq x \leq 5$.

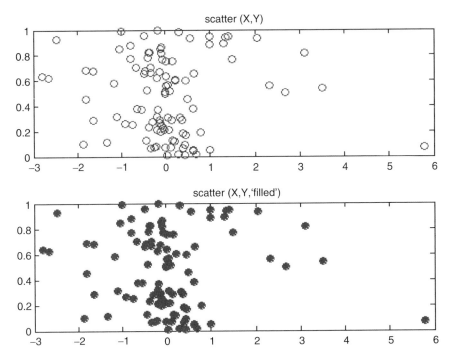

FIGURE 5.10
Scatter plots of R.5.56.

Represent the plot of y_1 by discrete points indicated by the marker "*," and y_2 by a continuous (default) solid line. Construct each plot using the following scales:

a. *Linear (plot 1)*

b. *Linear-logarithmic (plot 2)*

c. *Logarithmic-linear (plot 3)*

d. *Logarithmic-logarithmic (plot 4)*

MATLAB Solution
```
>> X = linspace(1,5,17);
>> y1 = 27*exp(-2*X);
>> y2 = 15*(0.3).^X;

>> % plot 1
>> subplot(2,2,1)
>> plot (X,y1,'*',X,y2)

>> % plot 2
>> subplot(2,2,2)
>> semilogy(X,y1,'*',X,y2)

>> % plot 3
>> subplot(2,2,3)
>> semilogx(X,y1,'*',X,y2)

>> % plot 4
>> subplot(2,2,4)
>> loglog(X,y1,'*',X,y2)
```

The resulting plots are shown in Figure 5.11.

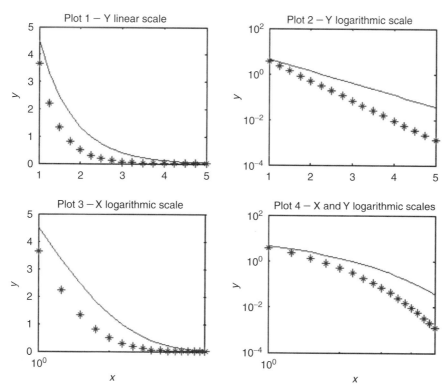

FIGURE 5.11
Plots with different scales and markers of R.5.58.

R.5.59 Let us consider a second example. Write a program that returns the following plots:

a. *[cos(x)]* versus *x* and *[sin(x)]* versus *x*, representing the discrete points with the markers "*" and "*d*," respectively

b. *[cos(x)]* versus *x* and *[sin(x)]* versus *x*, showing the points defined by the markers in part a, and by connecting the markers with a solid line
Use 20 points linearly spaced over the range $0 \leq x \leq 2\pi$.

MATLAB Solution
```
>> format compact
>> X = linspace(0,2*pi,20);
>> Y1 = cos(X);
>> Y2 = sin(X);
>> subplot (2,1,1)              % part (a)
>> plot (X,Y1,'*',X,Y2,'d')
>> subplot (2,1,2)              % part (b)
>> plot (X,Y1,'*',X,Y1,X,Y2,'d',X,Y2)
```

The resulting plots are shown in Figure 5.12.

R.5.60 Table 5.3 provides additional examples of the usage of the plot function with various options (color, markers, and line styles).

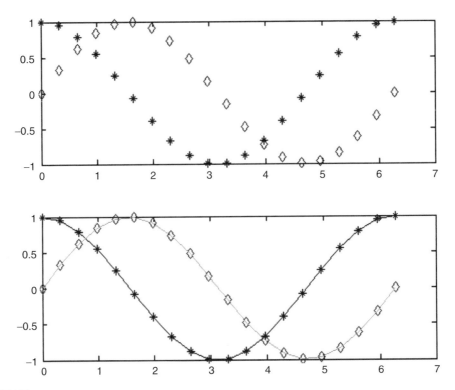

FIGURE 5.12
(See color insert following page 342.) Plots with different markers of R.5.59.

TABLE 5.3

Plotting Examples

Command	Description
plot (x, y)	Set of points $\langle x_i, y_i \rangle$ are connected with a solid blue line
plot (x, y, 'k')	Set of points $\langle x_i, y_i \rangle$ are connected with a solid black line
plot (x, y, '*')	Set of points $\langle x_i, y_i \rangle$ are indicated by stars (*)
plot (x, y, '--')	Set of points $\langle x_i, y_i \rangle$ are connected with a dashed line
plot (x, y, 'r*:')	Set of points $\langle x_i, y_i \rangle$ are connected with a dotted (:) red line and indicated with stars (*)

R.5.61 The command *colordef* defines the overall color composition of the figure window. The default is *colordef white* or *colordef* none. If a different color composition and background is required, then use the command *colordef color* and specify the *color* according to option *a* provided in Table 5.2.

R.5.62 The command *grid on* draws a set of grid lines superimposed on the active figure plot, whereas *grid off* removes the grid lines. Grid lines can be very useful when estimating or viewing a plot.

R.5.63 The command *box on* places the plot inside a box. The command *box off* removes the box.

R.5.64 It is common to label the x and y axes of a plot with a descriptive text defining the variables used and its corresponding units such as current in amperes, power in watts, frequency in hertz, or distance in meters.

The commands *xlabel('text1')* and *ylabel('text2')*, where *text1* and *text2* are string vectors (in quotes) are used to label, define, and describe the x- and y-axis and the associated variables.

R.5.65 The command *title('text3')* places the string vector *text3* at the top of the current plot as the figure title.

R.5.66 The command *text(x_a, y_a, 'text4')* places the string vector *text4* starting at the Cartesian coordinate location $<x_a, y_a>$ on the current figure window.

R.5.67 A good graph consists of properly labeled axis that includes units, a descriptive title, and any relevant information about the plot.

R.5.68 The command *gtext('text5')* opens the current figure window, places a cursor making a cross hair, and pauses. The center of the cross hair can be positioned anywhere on the active figure where the string vector *text5* can be placed by pressing the mouse button, or by pressing any key.

R.5.69 The *string* texts used in *title*, axis *labels, and gtext* may include about 100 special symbols including Greek characters. Any Greek character (lower or capital) can be called by using the back slash (\) character followed by its English version.

Table 5.4 illustrates some of the most frequently used characters and the corresponding syntax.

R.5.70 Text superscripts can be created by using the character ^, whereas text subscripts are created by the character _. Additional text control involving *font style, font size, marker size,* and *orientation* are summarized in Table 5.5.

TABLE 5.4

Special Characters Syntax

Character	Command
α	\alpha
η	\eta
ω	\omega
Ω	\Omega
μ	\mu
π	\pi
Π	\Pi
$\geq$	\geq

TABLE 5.5

Text Control Commands

Text Description	Command
Bold face	\bf
Italic	\it
Slant	\sl
Oblic	\or
Normal rom	\rm
Marker size	"markersize" (n)
Font size	"fontsize" (n)
Orientation	"rotation" (0)

R.5.71 The following example illustrates the commands associated with placing a text anywhere on a plot, a title, axis labels, etc. Let $y_1(x) = 5\,cos(2x)$ and $y_2(x) = 3\,sin(x)$ be two functions defined over the domain $0 \leq x \leq 2\pi$ using 40 linearly spaced points.

Create a program that returns the plots of $y_1(x)$ versus x and $y_2(x)$ versus x with the specs indicated as follows:

a. Use the marker "*" to indicate the points of $y_1(x)$ and connect the points with a solid line

b. Use the marker "+" to indicate the points of $y_2(x)$ and connect the points with a solid line

c. Label properly the x and y axes
d. Place the following text as title: 5 cos(2X) and 3 sin(X) versus X
e. Place the text 5 cos 2x at the (Cartesian coordinate) location <0.5, 4>
f. Place the text 3 sin(X) using the command *gtext* at the (Cartesian coordinate) location <2, 3>

MATLAB Solution

```
>> X = linspace(0,2*pi,40);   % creates X with 40 elements linearly
                                 spaced
>> Y1 = 5*cos(2.*X);          % evaluates Y1 for the 40 elements of X
>> Y2 = 3*sin(X);             % evaluates Y2 for each of the 40
                                 elements of X
>> plot (X,Y1,'*',X,Y1,X,Y2,'+',X,Y2)
                               % creates the plots of [Y1 and Y2] vs. X
>> xlabel ('X')               % creates label X
>> ylabel('Y')                % creates label Y
>> title ('5cos(2X) and 3sin(X) VS X')
                               % creates the title
>> text (0.5,4,'5cos2x')      % places text at x = 0.5, y = 4
>> gtext ('3sinx')            % places text <2,3> by the click of the
                                 mouse
>> grid on                    % adds a grid
```

The resulting plots are shown in Figure 5.13.

R.5.72 The command *legend('text_1', 'text_2', ... 'text_n')* is used to identify multiple plots on the same graph by creating a box in the upper-right corner of the graph that returns the message *text_1* on the first line, identifying the line style used for the

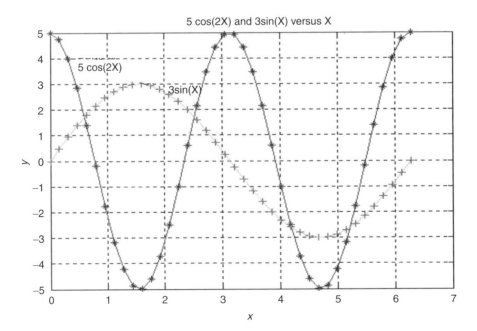

FIGURE 5.13
(See color insert following page 342.) Plots with markers and text of R.5.71.

first plot; *text_2* is placed on the second line and defines the line type used for the second plot, ..., and the *text_n* is placed on the *n*th line that identifies the line type used in the *n*th plot.

The legend box can then be moved to any location on the active figure window by clicking and holding the left mouse button near the edge of the box and dragging the box to a new location.

R.5.73 The following example illustrates the use of the commands: *legend, box, grid, xlabel, ylabel, text,* and *title* for the following plots:

$$y_1(x) = sin(x) \quad \text{and} \quad y_2(x) = sin^2(x)/x$$

over the domain $-2\pi \le x \le 2\pi$ in linear increments (spacing) of 0.4. The dotted line and star markers are used to plot [$y_1(x)$] versus *x*, and the dashed line and square markers for the plot $y_2(x)$ versus *x*.

MATLAB Solution
```
>> x = -2*pi:0.4:2*pi;
>> y1 = sin(x);
>> y2 = y1.^2./x;
>> plot (x,y1,':*',x,y2,'s--')
>> xlabel('x'), ylabel('y'),        % creates labels for x and y
>> title ('Example using legend, box, grid, labels (x & y),and title')
>> grid on; box on;                 % creates grid & box
>> legend ('y1(x)','y2(x)')         % creates the legend box
>> text(-4,0.7,'sin(x)')            % places the text sin(x) at <-4,0.7>
>> text(5,0.2,'sin(x)/x')           % places the text sin(x)/x at <5,0.2>
```

The resulting plots are shown in Figure 5.14.

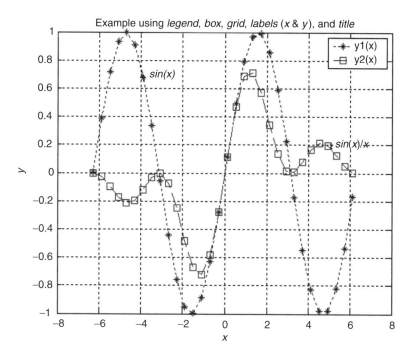

FIGURE 5.14
(See color insert following page 342.) Plots with markers, text, and legend of R.5.73.

R.5.74 The script file, *plot_enhancements* illustrates the text enhancement features in the plotting of a sine wave and a cosine wave. For example, create the script file, *plot_enhancements* that returns the following:

a. Plots of [*sin(αt)*] versus *αt* and [*cos(αt)*] versus *αt* over the range $0 \leq \alpha t \leq 2\pi/3$ using 20 points equally spaced

b. The line width of the *sin(αt)* plot to 6

c. The line width of the *cos(αt)* plot to 3 and indicate the data points with an hexagram of size 20

d. A title text using italic format with a size of 18

e. A legend box identifying each plot using bold text

f. The *x*-axis label using normal roman font at a 45° angle

g. The *y*-axis label using bold-face characters

h. A text message further identifying the sine-wave plot using bold font text with a size 13 at an angle of 45°

i. A text message further identifying the cosine-wave plot using bold font text with size 15 at an angle of 60°

MATLAB Solution
```
% Script file: plot _ enhancements
x = linspace(0,3*pi/2,20);
y1 = sin(x);y2=cos(x);
plot (x,y1,'linewidth',6);hold on;
plot (x,y2,'linewidth',3,'marker','hexagram','markersize',20)
title ('\it [sin(\alphat) & cos(\alphat)] vs \alphat','fontsize',18)
legend ('\bf sin(\alphat)','\bf cos(\alphat)')
xlabel ('\rm\alphat', 'rotation',45)
ylabel ('\bf magnitude')
text (3.2,0.0,'\bf sin(\alphat)','fontsize',13,'rotation',30)
text (1.7,0.0,'\bf cos(\alphat)','fontsize',15,'rotation',60)
box on; grid on
```

The resulting plots and enhancement features are shown in Figure 5.15.

R.5.75 Once a plot is created, additional enhancement options consisting of *text*, *arrows*, and *zoom* are available within the figure window domain.

R.5.76 The command *legend off* removes the legend box.

R.5.77 The command *refresh* redraws the current figure.

R.5.78 The command *axis on* turns on the axis; the command *axis off* turns off the axis.

R.5.79 MATLAB automatically scales the axes to accommodate the given data. The command *axis* ([x_{min} x_{max} y_{min} y_{max}]) is used to control the ranges of the *x*- and the *y*-axis, where $x_{min} \leq x \leq x_{max}$ and $y_{min} \leq y \leq y_{max}$. The *axis* command overrides the scale settings.

The *axis* command (with no argument) takes the present scale settings and uses them in subsequent plots. A second *axis* command is required to return the scale setting to the automatic mode (default).

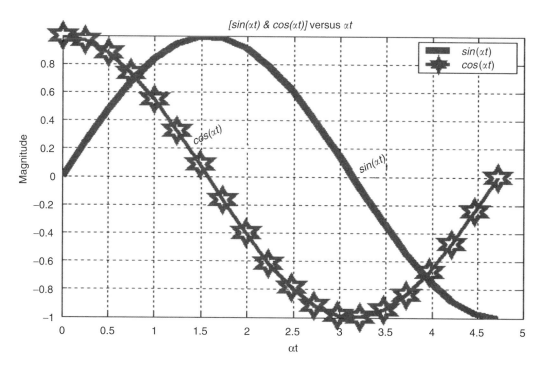

FIGURE 5.15
Plots with various enhancements of R.5.74.

R.5.80 The syntax $\pm inf$ can be used to specify the axis limits. For example, *axis* ($[-inf\ x_{max}$ $y_{min} + inf]$) specifies only the upper limit for x and the lower limit for y.

R.5.81 The command *axis equal* or *axis square* sets the same scaling factor for the x- and y-axis.

R.5.82 The command *axis normal* or *axis auto* returns the axis to the default scaling condition in which the best range and domain of the current plot are automatically set by MATLAB.

R.5.83 The command *axes on* returns a set of axes.

R.5.84 The command $[x, y] = ginput(n)$ is used to read the Cartesian coordinates of n points from the figure window by properly positioning the cursor making a cross hair over the target and by either clicking the mouse or by entering a character. The clicked point returns the coordinate value of x and y

  If no argument such as $[x, y] = ginput$ is used, then the number of target points to be read is unlimited. The *ginput* command is terminated by pressing the return key.

R.5.85 The command *zoom in* allows the expansion of a section of a 2-D plot for additional details. The *zoom* command is implemented in the figure window by clicking the left mouse button, and the plot is expanded by a factor of 2 or by clicking the right mouse button, and the plot in the figure window is compressed to half. The *zoom off* command deactivates the zoom mode.

R.5.86 The commands *legend, ginput,* and *zoom* are interdependent MATLAB functions, and restricts their use. Only one command can be used at any given time.

R.5.87 The command *fplot('f(x)', [x_{min} x_{max} y_{min} y_{max}])* returns the plot of *f(x)* versus *x* within the ranges $x_{min} \le x \le x_{max}$ and $y_{min} \le y \le y_{max}$, where the number of points are automatically chosen by MATLAB to display a *good* approximation. The term *good* depends on the particular application (see Example 5.8).

R.5.88 For example, use fplot to plot [*f(x)* = 4 *cos(x) cos(10x)*] versus *x* over the domain $0 \le x \le 2\pi$ and range $-5 \le y \le 5$.

MATLAB Solution
```
>> fplot ('4*cos(x)*cos(10*x)', [0 2*pi -5 5])
>> title ('f(x)=4*cos(x)*cos(10*x)')
>> xlabel('x')
>> ylabel('y')
```

The resulting plot is shown in Figure 5.16.

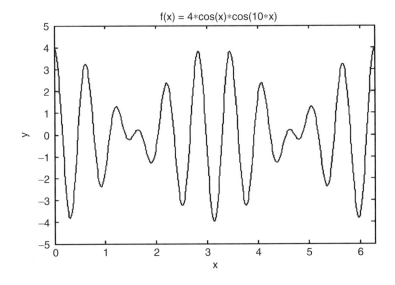

FIGURE 5.16
fplot of R.5.88.

R.5.89 The command *fplot('f1(x), f2(x), ..., fn(x).', [x_{min} x_{max} y_{min} y_{max}])* returns the multiple plots of *f1(x)* versus *x, f2(x)* versus *x, ..., fn(x)* versus *x*, within the limits, $x_{min} \le x \le x_{max}$ and $y_{min} \le y \le y_{max}$, where the number of points are automatically chosen by MATLAB.

R.5.90 For example, use *fplot* to obtain the multiple plots of the following functions: *f1(x)* = *tan(x), f2(x)* = *sec(x)* and *f3(x)* = *cot(x)* over the domain $-\pi \le x \le \pi$ and range $-3 \le y \le 3$.

MATLAB Solution
```
>> fplot ('[tan(x),sec(x),cot(x)]',[-pi pi -3 3])
>> xlabel('x');ylabel('y');
>> title('[tan(x),sec(x),cot(x)] vs x, for-pi<x<pi')
```

The resulting plots are shown in Figure 5.17.

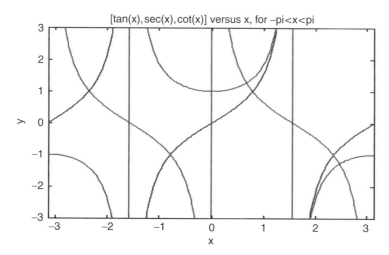

FIGURE 5.17
(See color insert following page 342.) Multiple plots using *fplot* of R.5.90.

R.5.91 The command *[x, y]* = *fplot('f(x)'*, *[x_{min} x_{max} y_{min} y_{max}]*), returns a table (with no plot) of the ordered pairs (points) $<x_i, y_i>$ over the ranges $x_{min} \leq x \leq x_{max}$ and $y_{min} \leq y \leq y_{max}$. The number of points are automatically chosen by MATLAB.

R.5.92 For example, use MATLAB and obtain a table of *cos(x)* versus *x* over the range $0.7 \leq x \leq 1$.

MATLAB Solution
```
>> [X,Y] = fplot('cos(X)',[0.7 1 -1 1]);
>> disp(' X    Y'); disp('***************');   [X Y]
```

```
        X           Y
   *****************
   ans =
      0.7000      0.7648
      0.7006      0.7645
      0.7018      0.7637
      0.7042      0.7621
      0.7090      0.7590
      0.7186      0.7527
      0.7378      0.7400
      0.7762      0.7136
      0.8530      0.6577
      0.9265      0.6006
      1.0000      0.5403
```

R.5.93 The command *ezplot('f(x)')* returns the plot of *f(x)* versus *x* over the default MATLAB interval $-2\pi \leq x \leq 2\pi$. If the function behavior of *f(x)* is over a smaller interval, then MATLAB automatically returns the plot of *f(x)* versus *x* over the smaller interval. The range of *x* can be changed by including specs in brackets. The syntax is *ezplot('f(x)'*, *[x_{min} x_{max}])*.

The command *ezplot* is similar to the command *fplot*. The only difference is that MATLAB assigns less number of points to *ezplot* in the plotting and table process.

R.5.94 For example, plot using *ezplot* the following functions: $f_1(x) = sin(x)$, $f_2(x) = sin(2x)$, and $f_3(x) = tan(x)$, without defining the range x.

MATLAB Solution
```
>> subplot(3,1,1);ezplot('sin(x)')
>> subplot(3,1,2);ezplot('sin(2*x)');
>> subplot(3,1,3);ezplot('y=tan(x)');
```

The resulting plots are shown in Figure 5.18. Observe (from Figure 5.18) that MATLAB automatically assumed a domain over $-2\pi \le x \le 2\pi$.

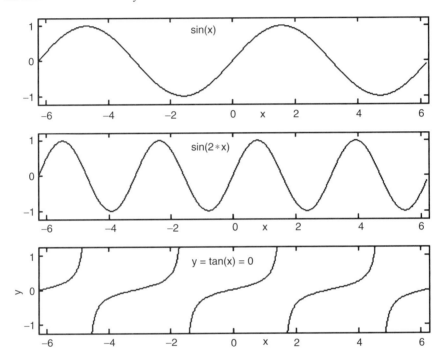

FIGURE 5.18
Plots using *ezplot* of R.5.94.

R.5.95 Let us further explore the plotting domain of *ezplot*. For example, create the plot of $f(x) = \sqrt{x}e^{-x^2}$ using the *ezplot* function without specifying its domain.

MATLAB Solution
```
>> ezplot('sqrt(x)*exp(-x^2)')
```

The resulting plot is shown in Figure 5.19.
Observe that MATLAB automatically defines the (domain) interval of interest as $0 \le x \le 2.5$.

R.5.96 Now, plot the same function $f(x) = \sqrt{x}e^{-x^2}$ using *ezplot* over $0 \le x \le 4$.

MATLAB Solution
```
>> ezplot('sqrt(x)*exp(-x^2)',[0 4])
```

The resulting plot is shown in Figure 5.20. Observe that for $x \ge 2.5$, $f(x)$ becomes a constant (zero) and the interval of interest is indeed $0 \le x \le 2.5$. Note that this interval becomes the MATLAB default interval.

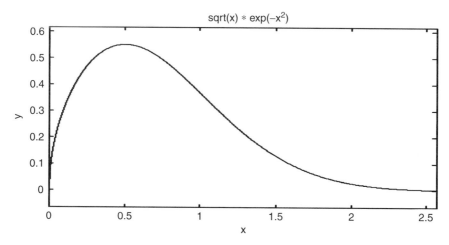

FIGURE 5.19
Plot of R.5.95 (no range).

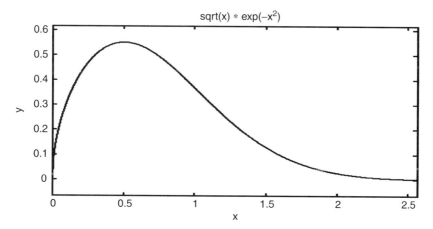

FIGURE 5.20
Plot of R.5.96 (given a range).

R.5.97 Observe that *ezplot* is an ideal tool that can be used in a quick and easy way to plot a set of simultaneous equations and be able to estimate or evaluate a system solution. The *ezplot* tool is especially useful when an algebraic approach is long and time-consuming, or yielding a complicated solution, or a solution that does not apply to a particular interval. This graphic approach is illustrated in the examples provided later in this section.

R.5.98 For example, using *ezplot*, let us estimate the Cartesian coordinates for the function $y^3 + x^2 - 3y + 2 = 0$ over the range $-6 \leq x \leq 6$ at its maximum.

MATLAB Solution
```
>> ezplot('y^3+x^2-3*y+2=0')
>> axis ([-6 6 -4 -1])
>> ginput
```

The resulting plot is shown in Figure 5.21.

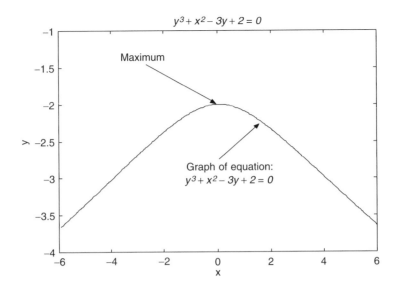

FIGURE 5.21
Plot of R.5.98.

The *ginput* placed over the target point (maximum) returns the following coordinates:

$$-0.0276 \qquad -1.9825$$

From the graph, the maximum can be estimated and occurs at the following coordinates:

$$<x = 0, y = -2>$$

R.5.99 As an additional example, let us create the script file, *sol_exp_lin* that is used to solve the following set of equations graphically: $y = x^2$ and $x + 2y - 3 = 0$ over the domain $-3 \leq x \leq 3$ (see Figure 5.22).

MATLAB Solution
```
% Script file: sol _ exp _ lin
ezplot('x^2');
axis([-3 3 -0.5 10]);hold on;
ezplot('x+2*y-3=0');
xlabel('x-axis');ylabel('y-axis');
title('Simultaneous equations: y =x^2 & x+2*y-3 = 0')
disp('************************************************************')
disp('The solution for the system of equations is the intersection
of the two lines.')
disp(' The solution can be estimated by positioning the cursor (+)')
disp('at the solution points on the plot and clicking the mouse')
disp('followed by the enter key. The xy (solutions) coordinates are: ')
disp('************************************************************')
grid on;
[x,y] = ginput
```

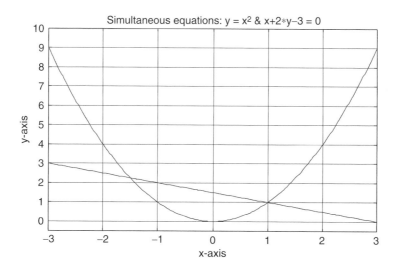

FIGURE 5.22
Plot of R.5.99.

Back in the command window, the script file, *sol_exp_lin* is executed and the results are shown as follows:

```
>> sol _ exp _ lin
```
****

The solution for the system of equations is the intersection of the two lines. The solution can be estimated by positioning the cursor (+) at the solution points on the plot and clicking the mouse followed by the enter key. The *xy* (solutions) coordinates are

****
```
    x =
      -1.5069
       0.9816
    y =
       2.1711
       0.9737
```

R.5.100 Since the graphic approach is simple, quick, and convenient in estimating the solution of a system of equations, let us use this method to estimate the solution of the following set of nonlinear equations: $2y^3 + 3x^2 - 4x + 2 = 0$ and $3y^2 - x^2 = 9$ over the domain $-3 \le x \le 3$.

MATLAB Solution
```
>> ezplot('3*y^2+7*x^2-9=0')
>> hold on
>> ezplot('2*y^3+3*x^3-6*y^2+2=0')
>> axis([-3 3 -3 3])
>> grid on
```

The resulting plot and the solutions are indicated in Figure 5.23.

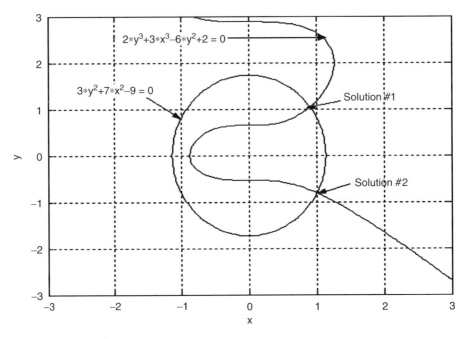

FIGURE 5.23
Plot of R.5.100.

R.5.101 The command *ezpolar* is similar to the command *ezplot*, with the exception that the *ezpolar* uses the polar coordinates system instead of the Cartesian (rectangular) coordinate system.

R.5.102 For example, using the *ezpolar* command, obtain the plot of the function $f(x) = cos(x)$ for the following cases:

a. With no domain specs

b. Over the interval $0 \le x \le 2\pi$

c. Over the interval $0 \le x \le \pi/2$

MATLAB Solution
```
% Script file:polar _ cos
subplot (1,3,1);
ezpolar('cos(x)');title('no specs')
subplot (1,3,2);ezpolar('cos(x)',[0 2*pi]);title('with specs:0-2\pi')
subplot (1,3,3);ezpolar('cos(x)',[0 pi/2]);title('with specs:0-\pi/2')
```

The three resulting plots are shown in Figure 5.24.

Observe that the *ezpolar* with no domain specs assumes the default range of $0 \le x \le 2\pi$.

R.5.103 The command *polar(beta, r)* is the numerical version of *ezpolar*, and MATLAB returns a polar plot of *beta* versus *r*. *Line color* and *line style* can be included in the function's argument that follows the same syntax as defined for the *plot* function.

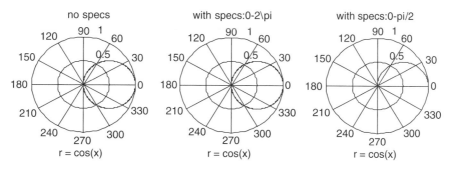

FIGURE 5.24
Plot of R.5.102.

R.5.104 The Cartesian plot [R $cos(t)$] versus [R $sin(t)$] returns a circle centered at the origin with radius R over $0 \leq t \leq 2\pi$. Since this relation is often used, let us show that this relation indeed defines a circle. Since $x = R\ cos(t)$ and $y = R\ sin(t)$ over $0 \leq t \leq 2\pi$, then

$$x^2 + y^2 = R^2\ cos^2(t) + R^2\ sin^2(x)$$
$$x^2 + y^2 = R^2[cos^2(t) + sin^2(x)]$$
$$x^2 + y^2 = R^2$$

This last equation clearly represents the equation of a circle.

R.5.105 Let us use the power of MATLAB to verify the preceding statement by executing the following script file, *circl*.

MATLAB Solution
```
% Script file:   circl
ezplot('cos(x)', 'sin(x)');
axis([-2 2 -2 2])
title('cos(x) vs sin(x)')
xlabel('cos(x)');ylabel('sin(x)');
```

The corresponding plot is shown in Figure 5.25. The circle is centered at the origin and can be moved to any location by adding constant terms. For example, observe that [$x = 4 + cos(t)$] versus [$y = 5 + sin(t)$] over $0 \leq t \leq 2\pi$ returns a circle centered at <4, 5> with unit radius as verified by the script file, *disp_circ* and its plot in Figure 5.26.

MATLAB Solution
```
%Script file: disp _ circ
ezplot(' 4+ cos(x)',' 5+ sin(x)');
axis([0 8 0 8])
title('[4+ cos(x)] vs [5+ sin(x)]')
xlabel('x-axis'); ylabel('y-axis');
```

R.5.106 Variations of the circle equation given by [$x = R\ cos(t)$] versus [$y = R\ sin(t)$] create a family of specialized curves some of which are defined as follows:

a. *Cycloid*, defined by $x = Rt - sin(t)$ and $y = R - cos(t)$

b. *Lemniscate*, defined by $x = cos(t)\sqrt{2cos(2t)}$ and $y = sin(t)\sqrt{2cos(2t)}$

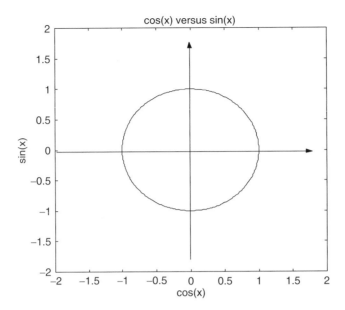

FIGURE 5.25
Plot of R.5.105.

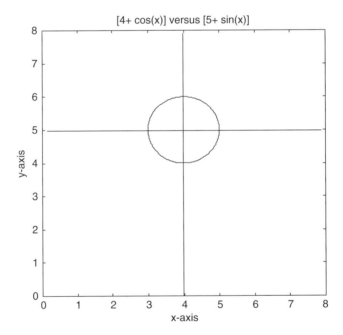

FIGURE 5.26
Plots of a shift circle.

    c. *Archimedean spiral,* defined by $x = t \cos(t)$ and $y = t \sin(t)$

    d. *Logarithmic,* defined by $x = e^{at} \cos(t)$ and $y = e^{at} \sin(t)$, where a is a constant

    e. *Cardioid,* defined by $x = 2 \cos(t) - \cos(2t)$ and $y = 2 \sin(t) - \sin(2t)$

    f. *Astroid,* defined by $x = R \cos^3(t)$ and $y = R \sin^3(t)$

    g. *Epicycloid,* defined by $x = (R + 1) \cos(t) - a \cos(t(R + 1))$ and $y = (R + 1) \sin(t) - a \sin(t(R + 1))$, where R and a are constants

h. *Hypocycloid*, defined by $x = (R - 1) \cos(t) + a \cos(t(R - 1))$ and $y = (R - 1) \sin(t) - a \sin(t(R - 1))$, where R and a are constants

R.5.107 Let us use the power of MATLAB to explore some of the equations just defined. For example, creating the script file, *astroid* returns the plots of (Novelli, 2004b)

a. $\cos^n(x)$ versus $\sin^n(x)$ for $n = 4, 3, 2, 1, 0.5, 0.25$ over $0 \le x \le \pi/2$

b. $\cos^3(x)$ versus $\sin^3(x)$ (a plot referred as *astroid*) over $0 \le x \le 2\pi$

MATLAB Solution

```
% Script file: astroid

figure (1)
x = 0:.1:pi/2;
x1 = cos(x);y1= sin(x);
x4 = x1.^4;y4 = y1.^4;x3 = x1.^3;y3 = y1.^3;x2 = x1.^2;y2 =y 1.^2;
x05 = x1.^0.5;y05 = y1.^0.5;
x25 = x1.^0.25;y25 = y1.^0.25;
plot(x4,y4,x3,y3,x2,y2,x1,y1,x05,y05,x25,y25)
title('[cos(x)]^n vs [sin(x)]^n')
xlabel('cos(x)')
ylabel('sin(x)')
legend('n=4','n=3','n=2','n=1','n=0.5','n=0.25')

figure(2)
xx = 0:0.1:2*pi;
xx3 = cos(xx).^3;yy3 = sin(xx).^3;
plot(xx3,yy3)
title('[cos(x)]^3  vs [sin(x)]^3')
xlabel('cos(x)')
ylabel('sin(x)')
```

The script file, *astroid* is executed and the results are shown in Figures 5.27 and 5.28.

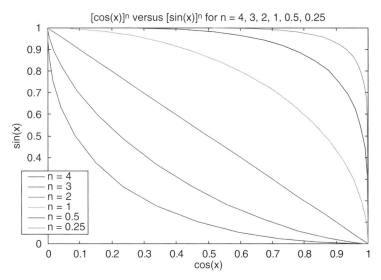

FIGURE 5.27
(See color insert following page 342.) Plots of R.5.107(a).

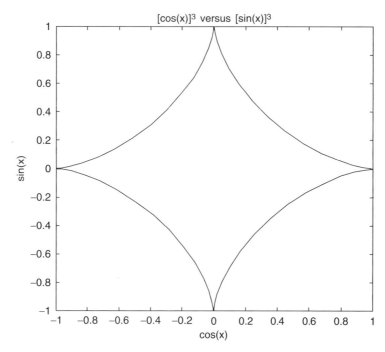

FIGURE 5.28
Plots of R.5.107(b).

R.5.108　The command *stem(x, y)* returns a discrete plot of each ordered pair $<x_i, y_i>$, for $i = 1, 2, 3, ..., n$ indicated by the marker (o) connected to a vertical line with magnitude y_i at the points $<x_i, 0>$ for all *is*.

R.5.109　The command *stairs(x, y)* returns the plot y versus x as a stair plot, where each step has a width $x_{i+1} - x_i$ and height y_i, for $i = 1, 2, 3, ..., n$.

R.5.110　The command *bar(x, y)* returns a vertical bar plot with uniform widths, where the Amplitude of the bars are given by the heights y_i, centered at each x_i for all *is*. The command *barh(x, y)* returns a horizontal plot where the x and y axes are exchanged.

R.5.111　For example, let us create the script file, *diff_plots* that returns four plots of the function $y = -x \sin(x)$ over the range $0 \le x \le 6\pi$, using 18 linearly spaced points, where each plot is implemented using one of the following commands:

a. *stem*

b. *stairs*

c. *bar*

d. *barh*

MATLAB Solution
```
%Script file: diff _ plots
axis on
X = linspace(0,6*pi,18);
Y= -X.*sin(X);
subplot (2,2,1)
stem(X,Y), title ('Plot using stem'), ylabel ('Y')
subplot(2,2,2)
stairs(X,Y), title ('Plot using stairs'), ylabel ('Y')
```

```
subplot(2,2,3)
bar (X,Y),
axis([0 20 -10 20])
xlabel ('X'), ylabel('Y'),title ('Plot using bar')
subplot (2,2,4)
barh(X,Y), xlabel('X'), ylabel('Y'), title ('Plot using barh')
axis([-20 20 0 20])
```

The script file, *diff_plots* is executed and the results are shown in Figure 5.29.

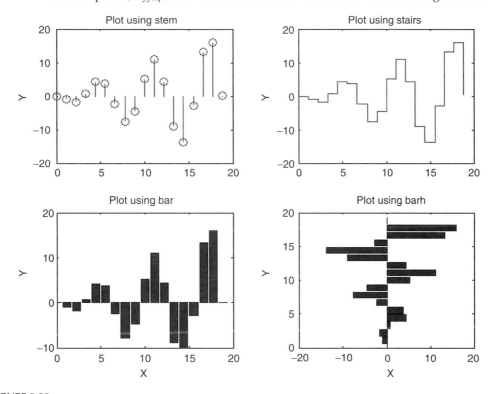

FIGURE 5.29
Plots of R.5.111.

R.5.112 The command *compass(x, y)* returns a polar plot where each point or ordered pair (x_i, y_i) is drawn as an arrow from the origin to the $<x_i, y_i>$ point. The x and y arguments can be replaced by z, where z is a complex number given by $z = x + iy$, in which case, the instruction can be expressed as *compass(z)*. If z is complex, the *compass* plot converts the rectangular form of z into polar.*

R.5.113 The command *feather(a, b)* or *feather(z)* (when z is a complex number of the form $z = a + ib$) returns the plots of the argument as arrows vectors equally spaced, on the x-axis as reference. Observe that the command *feather* is used to display a vector consisting of magnitude and direction along a path. Color and line style specs can be added that are identical to the ones defined for the *plot* command.

R.5.114 The command *polar(∅, r, 'options')* returns a polar plot of the angle ∅ versus the magnitude r where the background is a plane that consists of a grid indicating

---

* See Chapter 6 for additional information about complex numbers.

the angles every 30° apart and concentric circles that represent the magnitude. The *options* are identical as the ones defined for the *plot* command.

R.5.115 For example, write a MATLAB program that returns an array of arrows or line segments with unit magnitude over the range $0 \leq \varnothing \leq 2\pi$, with linear incremental spacing of $\Delta\varnothing = \pi/4$. Let's learn by doing and observe and analyze the respective plots returned by using each of the following commands:

a. *compass*

b. *feather*

c. *polar*

MATLAB Solution
```
>> ang = 0:pi/4:2*pi;
>> X = cos(ang);
>> Y = sin(ang);
>> R = X.^2+Y.^2;
>> subplot(2,2,1)
>> compass(X,Y),
>> title('Plot Using Compass')
>> subplot(2,2,2)
>> polar(ang,R),title ('Plot Using Polar')
>> subplot(3,1,3), feather (X,Y),
>> title ('Plot Using Feather')
```

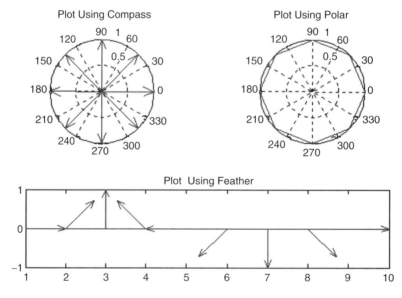

FIGURE 5.30
Plots of R.5.115.

The resulting plots are shown in Figure 5.30. Observe the following from Figure 5.30:

a. The *compass* command returns eight unit vectors, equally spaced every $\Delta\varnothing = \pi/4$ *rad*, with unit magnitude on a polar plane.

b. The *polar* command returns eight connected segments, equally spaced every $\Delta\emptyset = \pi/4$ *rad*, forming a circle with unit radius.

c. The *feather* command returns nine unit vectors with unit spacing and with angle increments of $\Delta\emptyset = \pi/4$ *rad* plotted on a Cartesian plane.

R.5.116 The command *quiver (X, Y, Z, V)* returns the plot of the velocity vectors as arrows with components $<z, v>$ at the points $<x, y>$. The matrices X, Y, Z, and V must all have the same size and contain corresponding position and velocity components (X and Y can also be vectors to specify a uniform grid). The arrows are automatically scaled to fit within the grid. The line style, markers, and colors for the velocity vectors V can be specified adding a field in *quiver*. This command can be used to illustrate the action of a physical variables such as lines of force induced by an electric or magnetic field. The option *quiver(..., 'filled')* returns a shaded (filled) plot.

R.5.117 For example create the script file, *quiver_fn* that returns the plot of the lines of force (using *quiver*) resulting from implementing $z = sqrt(-x^2 - y^2)$, given the function *meshgrid(-2:.2:2,-1:.15:1)* that specifies the XY grid.*

MATLAB Solution
```
>> % Script file: quiver _ fn
>> [x,y] = meshgrid(-2:.2:2,-1:.15:1);
>> z = sqrt(-x.^2 - y.^2);
>> contour (x,y,z), hold on
>> quiver(x,y,z), hold off
>> xlabel('x-axis'),ylabel('y-axis')
>> title('z = sqrt(-x^2 - y^2)')
```

The resulting plot is shown in Figure 5.31.

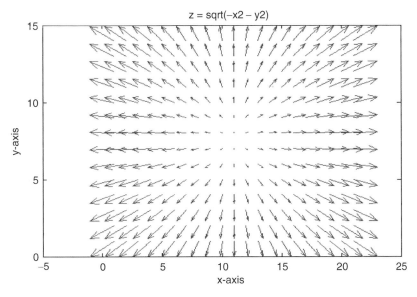

FIGURE 5.31
Plots of R.5.117.

---

* The command *meshgrid* and *countour* are defined in R.5.151 and R.5.163, respectively.

R.5.118 The command *hist(x)* returns a histogram plot of the distribution of the values of the vector x grouped in 10 bins, equally spaced over the range $x_{min} \leq x \leq x_{max}$.

R.5.119 For example, let x be a collection of 100 random, normally distributed numbers. Create the script file, *hist_gram* that returns its histogram plot.

MATLAB Solution
```
% Script file : hist _ gram
X = randn(100,1)*10;    % X consists of 100 normally random numbers
hist(X), title('Histogram plot of X, using hist(X)')
xlabel('X')
ylabel('Amplitude of X')
axis([-25 25 0 25])
```

The script file, *hist_gram* is executed and the result is shown in Figure 5.32.

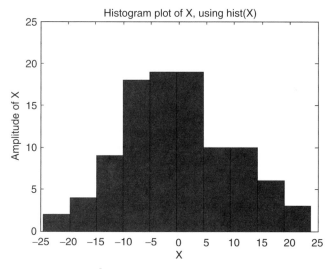

FIGURE 5.32
Histogram plot of R.5.119.

R.5.120 The command *hist(x, N)* is similar to the command *hist(x)*, but it groups the values of x in N bins.

R.5.121 Rerun the script file, *hist_gram* for the case of (N) 20 bins.

MATLAB Solution
```
% Script file: hist _ gran _ 20
X= randn(100,1)*10;    % X consist of 100 normally random distributed
                         numbers
hist(X,20)             % histogram plot with 20 bins
title('Histogram plot of X, using hist(X,20)')
xlabel('X')
ylabel('Amplitude of X')
axis([-30 30 0 15])
```

The script file *hist_gram_20* is executed and the result is shown in Figure 5.33.

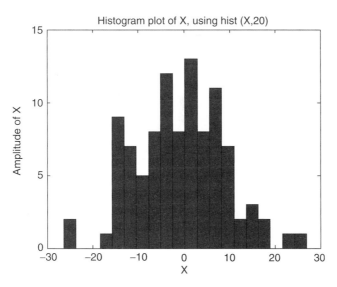

FIGURE 5.33
Histogram plot of R.5.121.

R.5.122 The command *[Nbin, Xave] = hist(x, N)* returns two row arrays.

  a. Array *Nbin*, consisting of the number of elements in each of the *N* bins

  b. Array *Xave*, consisting of the average values per bin

R.5.123 For example, the command *[Nbin, Xave] = hist(x, N)* is executed as follows for the data used in the earlier example.

MATLAB Solution
```
>> [Nbin,Xave] = hist(X,20)

    Nbin =
    Columns 1 through 13
    2   0   0   1   9   7   5   8   12   8   13   8   11
    Columns 14 through 20
    7   2   3   2   0   1   1
    Xave =
    Columns 1 through 8
    -25.0185 -22.3559 -19.6932 -17.0306 -14.3680 -11.7053 -9.0427 -6.3801
    Columns 9 through 16
    -3.7175 -1.0548 1.6078  4.2704  6.9331  9.5957 12.2583 14.9209
    Columns 17 through 20
    17.5836 20.2462 22.9088 25.5715
```

R.5.124 The command *pie(x)* returns a *pie* graph, where each slice of the pie is proportional (areawise) to the number of elements in *x* (similar to the histogram plot, where a *pie* is similar to the *bin*).

R.5.125 For example, the *pie* command and its graphical representation using the data of the earlier example (X = randn(100,1) * 10) is illustrated in Figure 5.34.

MATLAB Solution
```
>> pie(Nbin)
```

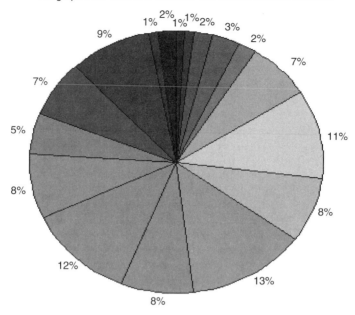

FIGURE 5.34
(See color insert following page 342.) Pie plot of R.5.125.

TABLE 5.6

Grade Distribution of a Class of 25 Students

Number of Students	3	6	10	4	2
Performance/Grade	A	B	C	D	F

R.5.126 Let us consider another example. Write a program that returns the pie plot of the academic performance of a class of 25 students with the grading distribution indicated in Table 5.6. Also identify each pie by a legend box.

The resulting pie plot is shown in Figure 5.35.

MATLAB Solution
```
>> dist = [3 6 10 4 2];          % grade distribution
>> pie(dist)
>> title('Class Performance'), legend ('A', 'B', 'C', 'D', 'F')
```

R.5.127 The command *pie(x, detach)* returns a plot similar to the command *pie(x)*, where *detach* is a logical argument in the form of a binary array (consisting of 0's and 1's) with *length(x)*. The portion of the pie that is detached corresponds to the elements of *x* represented by the ones (1s) of the *detach* argument.

R.5.128 The following example employs the command *pie(x, detach)*, where the detached pies correspond to the academic performance below *C* (*D* and *F*) using the data of the earlier example (Figure 5.36).

MATLAB Solution
```
>> dist = [3 6 10 4 2];
>> detach = [0 0 0 1 1];         % detach pies; academic performance
                                   of D and F
```

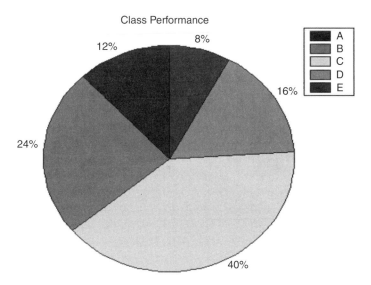

FIGURE 5.35
(See color insert following page 342.) Pie plot of R.5.126.

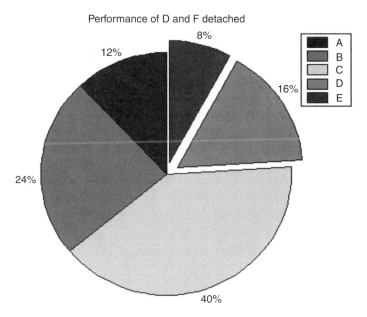

FIGURE 5.36
(See color insert following page 342.) Detached pie plot of R.5.128.

```
>> pie(dist,detach)
>> legend ('A', 'B', 'C', 'D', 'F');
>> title('Performance of D and F Detached')
```

R.5.129 The command *fill(x, y, 'a')* returns a colored 2-D plot under the points defined by the vectors *x* and *y*, with the color specified by the option *a* defined by Table 5.2.

R.5.130 For example, write a program that returns a black circle of radius 3 centered at the origin.

MATLAB Solution
```
>> Beta = linspace(0,2*pi,100);
>> X=3*cos(Beta);
>> Y=3*sin(Beta);
>> fill(X,Y,'k'); axis ('square');
>> xlabel('x-axis'); ylabel('y-axis')
>> title('Black circle')
```

The resulting plot is shown in Figure 5.37.

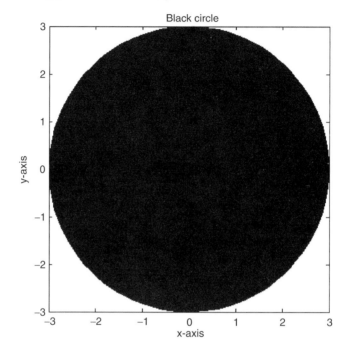

FIGURE 5.37
Plot of R.5.130.

R.5.131 The command *area(x, y)* returns a shaded plot of the area under $y = f(x)$.

R.5.132 The following example returns the shaded *area* under the positive half period of $y(x) = sin(x)$.

MATLAB Solution
```
>> x = linspace(0,pi,20);
>> y = sin(x);                    % positive half period of sin(x)
>> area(x,y), xlabel('x-axis'), ylabel('y-axis')
>> title ('Shaded  plot using area (x,y)')
>> axis([-0.5 4.0 0 1.1])
```

The resulting plot is shown in Figure 5.38.

R.5.133 To obtain a hard copy of a plot present in the active figure window, activate the figure window and then select *Print* from the *File* menu by clicking the left button of the mouse, which sends the current plot to the printer.

R.5.134 Multiple figure windows can be created by using the command *figure(n)* repeatedly and a new figure window is created for each integer *n*, where $n = 1, 2, 3, ...$ The current figure window is the active and visible figure.

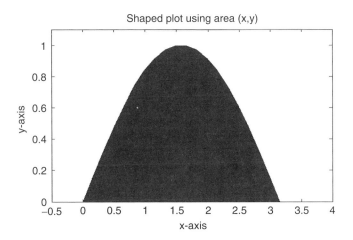

FIGURE 5.38
Plot of R.5.132.

R.5.135 The command *close* closes the current figure window; *close(n)* closes the *n* figure window; and the command *close all* closes all the figure windows.

R.5.136 The command *shg* that stands for *show graph window* selects the current figure as the active and visible figure.

R.5.137 The command *clf* clears the current (active) window.

R.5.138 The command *ribbon(X, Y, width)* returns a 2-D line as ribbons in a 3-D plane, where *X* is plotted versus *Y*. The columns of *Y* are plotted as separated ribbons in 3-D. The command *ribbon(Y)* uses the default value of *X = 1* with *width = 0.75*.

R.5.139 For example, let *A* = eye(4) and *B* = magic(4). Execute the command *ribbon(A, B)* and observe the four 3-D ribbons drawn.

MATLAB Solution
```
>> A= eye(4);
>> B = magic(4);
>> ribbon(A,B)
>> title('ribbon plot');
>> xlabel('x'); ylabel('y') ; zlabel('z');
```

The corresponding plot is shown in Figure 5.39.

R.5.140 The 2-D command *plot(x, y, 'options')* can be expanded to a 3-D space by using the following syntax *plot3(x, y, z, 'options')*, where *x, y,* and *z* are the arrays of the same length that define the points on the 3-D coordinate space (length, height, and width). The *options* are identical to the ones defined for the 2-D case (Table 5.2).

R.5.141 The commands *grid, axis, label (x, y and z),* and *title* defined for 2-D works equally well for 3-D plots (*plot3*).

R.5.142 For example, create the script file, *helix* and verify that the set of equations $x = R\cos(t)$, $y = R\sin(t)$, and $z = kt$, where *R* and *k* are real positive numbers, returns the plot of an helix. Test the script for the following cases:

a. A constant radius of $R = 1$

b. A radius-dependent function given by $R = t^2$ over the range $0 \le t \le 10\pi$, with $k = 9$

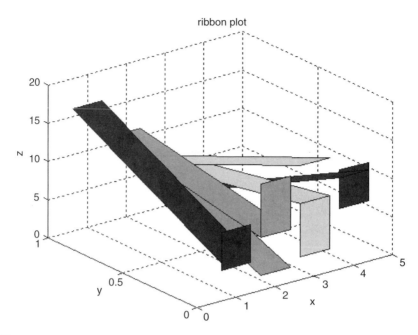

FIGURE 5.39
(See color insert following page 342.) Ribbon plots of R.5.139.

```
MATLAB Solution
% Script file: helix
% R=1, k=9
t = 0:0.1:10*pi;
x = cos(t); y = sin(t);z = 9*t;

figure(1)
plot3(x,y,z); title('Helix with constant radius R=1')
xlabel('x-axis');ylabel('y-axis');zlabel('z-axis')
grid on;

figure(2)
% R = t^2,k = 9
x1 = t.^2.*cos(t); y1 = t.^2.*sin(t);
plot3(x1,y1,z); title('Helix with variable radius R=t^2')
xlabel ('x-axis'); ylabel('y-axis');zlabel('z-axis')
grid on;
```

The script file, *helix* is executed and the resulting plots are shown in Figures 5.40 and 5.41.

R.5.143 The general equation $(x^2/a^2) + (y^2/b^2) + (z^2/c^2) = 1$ represents the ellipsoidal (Novelli, 2004b) family. Variations of the preceding equation gives rise to a number of specialized functions, some of which are defined below:

a. $(x^2/a^2) + (y^2/b^2) - z = 0$ represents the equation of an elliptic parabolic surface.

b. $(x^2/a^2) - (y^2/b^2) - z = 0$ represents the equation of an hyperbolic parabolic surface.

c. $(x^2/a^2) + (y^2/b^2) - z^2 = m > 0$ represents the equation known as one face hyperbolic function.

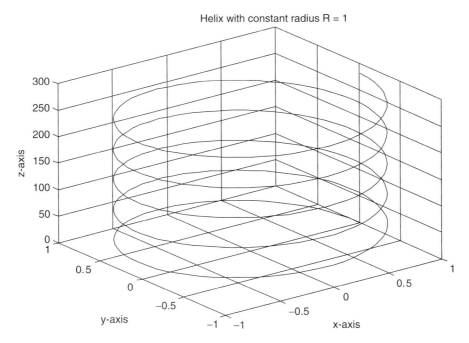

FIGURE 5.40
Plots of R.5.142(a).

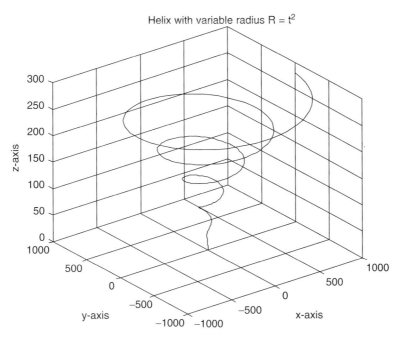

FIGURE 5.41
Plots of R.5.142(b).

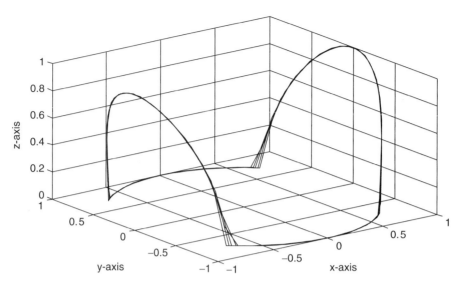

FIGURE 5.42
Plot of R.5.144.

d. $(x^2/a^2) + (y^2/b^2) - z^2 = m < 0$ represents the equation known as a two face hyperbolic function.

e. $(x^2/a^2) + (y^2/b^2) - z^2 = 0$ represents a conic surface.

R.5.144 For example, create the script file, *plot_hyper* that returns the plots of the following 3-D set of equations: $x = cos(t)$, $y = sin(t)$, and $z = \sqrt{x^2 + y^2}$ over the range $0 \leq t \leq 10\pi$.

```
% Script file: plot _ hiper
t = 0:0.1:10*pi;
x = cos(t);
y = sin(t);
z = sqrt(x.^2-y.^2);
plot3 (x,y,z);
xlabel('x-axis'); ylabel('y-axis');
zlabel('z-axis');grid on;
```

The script file, *plot_hyper* is executed and the resulting plot is shown in Figure 5.42.

R.5.145 The following example illustrates the plotting of a 3×3 identity matrix concatenated three times forming the matrix A (3×9). Write a program that returns the plot of the case: [*rows of A*] versus [*index of A*].

MATLAB Solution
```
>> A = [eye(3),eye(3),eye(3)]          % creates the identity matrix
                                       concatenated 3 times

A =
    1   0   0   1   0   0   1   0   0
    0   1   0   0   1   0   0   1   0
    0   0   1   0   0   1   0   0   1
```

```
>> plot3(A(1,:),A(2,:),A(3,:))
>> title('3D Plot')
>> xlabel('X'),ylabel('Y'),zlabel('Z')
>> grid on
```

The resulting graph is shown in Figure 5.43.

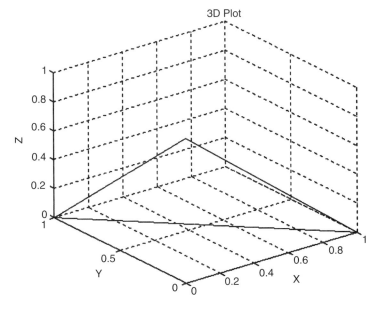

FIGURE 5.43
3-D plot of R.5.145.

R.5.146 The next example illustrates the 3-D plot of the trajectory of a particle defined by
the following set of equations:

$$x = t$$

$$y = t * cos(t)$$

$$z = e^{0.2t}$$

over the range $0 \le t \le 4\pi$.

MATLAB Solution
```
>> t = 0:0.3:4*pi;
>> x = t;
>> y = t.*cos(t);
>> z = exp(0.2.*t);
>> plot3(x,y,z)
>> grid on, xlabel('t'), ylabel('tcos(t)'), zlabel('exp (0.2t)')
>> title ('Plot of a Curve in 3D')
>> axis ('normal')
```

The resulting 3-D plot is shown in Figure 5.44.

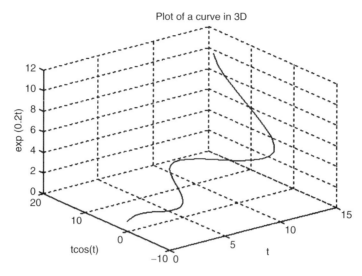

FIGURE 5.44
3-D plot of R.5.146.

R.5.147 The preceding example is now plotted using the command *stem3*.

```
MATLAB Solution
>> t = 0:0.3:4*pi;
>> X = t;
>> Y = t.*cos(t);
>> Z = exp(0.2.*t);
>> stem3(X,Y,Z,'Filled'), title('3D plot using stem3'),
>> xlabel('t'), ylabel('t cos(t)'), zlabel('exp(0.2*t)')
```

The resulting plot is shown in Figure 5.45.

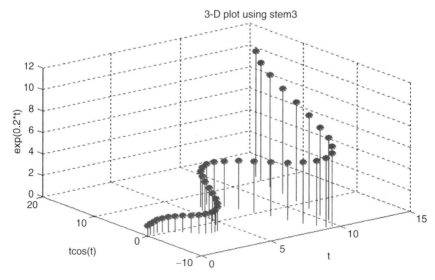

FIGURE 5.45
stem3 plot of R.5.147.

R.5.148 The 2-D commands such as *fill, pie, stem, bar,* and *barh* can be expanded to 3-D by just adding a *3* to its syntax. The corresponding 3-D commands are *fill3, pie3, stem3, bar3,* and *barh3,* respectively.

R.5.149 The following examples serve to review and illustrate some 2-D and 3-D plot commands. Let $y = sin(x)$ over $0 \leq x \leq \pi$.

Create the script file, *plots_2D_3D* that returns the plots using the following commands:

a. *plot(x,y)*

b. *stairs(x,y)*

c. *fill(x,y) and fill3(x,y,z,'k')*

d. *stem(x,y)*

e. *stem3(x,y)*

f. *bar(x,y)*

g. *bar3(x,y)*

h. *barh(x,y)*

Observe and analyze the plotting commands and the corresponding returning plots.

MATLAB Solution
```
% Script file: plots _ 2D _ 3D
x = linspace(0,pi,25);y=sin(x);

figure(1)
subplot (3,2,1);plot(x,y);
xlabel('x');ylabel('y');title('plot(x,y)')
axis([0 3.5 0 1.1])
subplot (3,2,2);stairs(x,y);
xlabel('x');ylabel('y');title('stairs(x,y) plot')
axis([0 3.5 0 1.1])
subplot (3,2,3)
fill(x,y,'k');title('fill(x,y) plot')
xlabel('x');ylabel('y');zlabel('z')
axis([0 3.5 0 1.1])
subplot (3,2,4)
z =3+[1:2:50]; fill3 (x,y,z,'k');
xlabel('x');ylabel('y');
title('fill3(x,y,z) plot')
subplot (3,2,5);stem(x,y);
xlabel('x');ylabel('y');
title('stem(x,y) plot')
axis([0 3.5 0 1.1])
subplot (3,2,6);stem3(x,y),
xlabel('x');ylabel('y');zlabel('z');
title('stem3(x,y) plot')

figure(2)
subplot (1,3,1); bar(x,y);
xlabel('x');ylabel('y');
title('bar(x,y) plot')
axis([0 3.2 0 1])
subplot (1,3,2); bar3(x,y);
```

```
xlabel('x');ylabel('y');title('bar3(x,y) plot')
subplot (1,3,3); barh(x,y);
xlabel('x');ylabel('y');
title('barh(x,y) plot')
```

The script file, *plots_2D_3D* is executed and the resulting plots are shown in Figures 5.46 and 5.47.

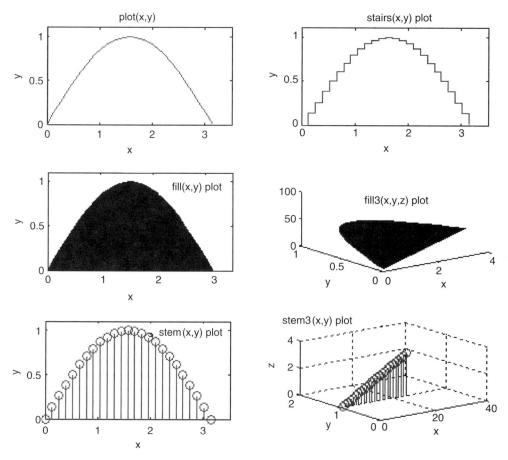

FIGURE 5.46
Plot of R.5.149(a, b, c, d and e).

R.5.150 Let us rerun the example of the pie plot that returns the performance of the class of 25 students using the academic data in Table 5.6 by employing the 3-D command *pie3*.

MATLAB Solution
```
>> % Data: A, B, C, D, F
>> dist = [3 6 10 4 2];                    % data
>> pie3(dist),title ('Class Performance')
```

The resulting plot is shown in Figure 5.48.

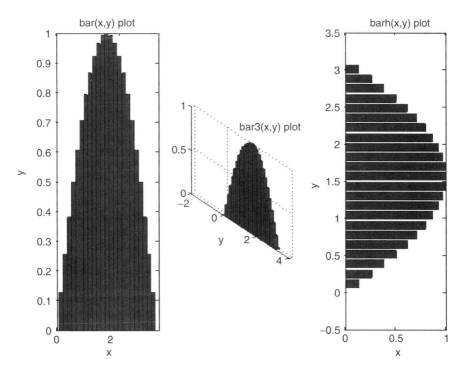

FIGURE 5.47
Plot of R.5.149(f, g and h).

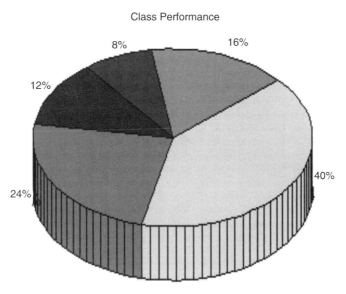

FIGURE 5.48
(See color insert following page 342.) *pie3* plot of R.5.150.

R.5.151 The command $[X, Y] = meshgrid(x, y)$ returns two rectangular matrices X and Y given the vectors x and y that define the 2-D grid points on the Cartesian plane. This command is the 2-D version of the *linspace* command. From the matrices X and Y, the Z component can be defined as a function of x and y, returning a 3-D spaced system.

R.5.152 Once the X and Y matrices are generated as a result of using *meshgrid(x, y)*, then the command *mesh(X, Y, Z)*, where $Z(x, y)$ returns a 3-D mesh structure. The command *mesh* can also be used directly with the arguments x, y, and z as *mesh(x, y, z)*.

R.5.153 A surface is generated when the function $y = f(x)$ is rotated about the *x*-axis, over the interval of interest, for example, between a and b.

R.5.154 The command *surf(X, Y, Z)* is similar to the *mesh* command, and returns a 3-D parametric surface plot where the surface color is a function of the surface height.

R.5.155 The command $[X, Y, Z] = cylinder(y, n)$ returns the matrices X, Y, and Z, which represent the coordinates of the points on the surface of revolution, where the axis of revolution is the vertical axis. The scale is from 0 to 1 over the *z*-axis. The argument n represents the number of points in each circle of revolution evenly spaced.

R.5.156 For example, write a MATLAB program that returns the 3-D plot of a cone centered at the origin with unit length using 25 equally spaced points on the x and y directions.

MATLAB Solution
```
>> x = linspace(0,2,50);
>> y = x;
>>[x,y,z] = cylinder(y,25);
>> surf(x,y,z)
>> title('Cylinder')
>> xlabel('x'); ylabel('y');zlabel('z')
```

The resulting plot is shown in Figure 5.49.

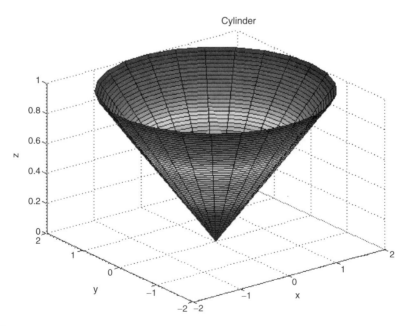

FIGURE 5.49
(See color insert following page 342.) Plot of R.5.156.

R.5.157 Repeat the above example by changing y, the surface of revolution, to $y = x^2$. Observe how the resulting shape of the cone changes from a linear to a quadratic surface. The resulting plot is shown in Figure 5.50.

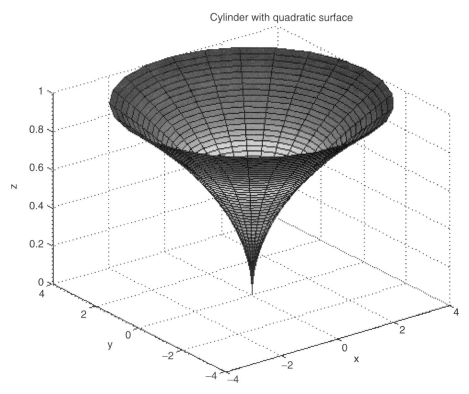

Cylinder with quadratic surface

FIGURE 5.50
(See color insert following page 342.) Plot of R.5.157.

R.5.158 The command *[X, Y, Z]= spherer(n)* returns the three n by n matrices X, Y, and Z, which represent the coordinate points of a sphere. The matrices X, Y, and Z, when used as arguments of the *surf* or *mesh* command, return the plot of a sphere of radius one, centered at the origin. The default value for n is 20.

R.5.159 For example, the following program returns the plots of the default sphere when used with the commands *surf* and *mesh*. Observe and analyze the commands and their corresponding plots.

MATLAB Solution
```
>> subplot(1,2,1)
>> [x,y,z] = sphere;
>> surf(x,y,z)
>> axis equal
>> subplot(1,2,2)
>> mesh(x,y,z)
>> axis equal
```

The resulting plots are shown in Figure 5.51.

Sphere using surf

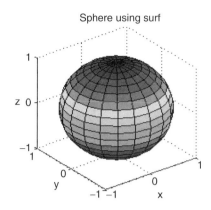

Sphere using mesh

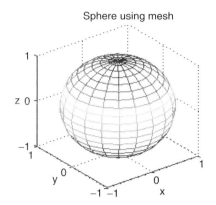

FIGURE 5.51
(See color insert following page 342.) Plot of R.5.159.

R.5.160 The command $[X, Y, Z] = ellipsoid(Xc, Yc, Zc, Xr, Yr, Zr, n)$ (distorted sphere) returns the three *matrices X, Y,* and *Z* that represent the 3-D coordinate system of the resulting body. When the *ellipsoid's* output are used as arguments of the command $surf(X, Y, Z)$, MATLAB returns the plot of an ellipsoid centered at the Cartesian coordinates *Xc, Yc,* and *Zc,* and radii *Xr, Yr,* and *Zr.* The default version for *n* is 20. When no output variables are used, MATLAB returns only the graph of the ellipsoid's surface.

R.5.161 The command *waterfall(x, y, z)* is similar to the *mesh* command, but the mesh lines are drawn in the *x* direction only.

R.5.162 The *shading* function is used to modify the color image created with the *surf, mesh,* or *fill* commands. The shading option can be accompanied with the arguments *flat, interp,* or (the default) *faceted.*

R.5.163 The command *contour(x, y, z)* returns *10* equally spaced horizontal traces (lines) according to the heights (equal heights) of the figure. A fourth argument *n* may be used to control the number of traces, such as *contour(x, y, f(x, y), n).* The different heights, line style, and color can also be controlled. The *contour* command returns plots that can be used in 2-D or 3-D with the commands *contour* and *contour3.*

R.5.164 The command $view(\emptyset_{xy}, \emptyset_z)$ returns the view of the figure image rotated by an angle $\emptyset_{xy}$ on the *xy* plane in a counterclockwise direction (azimuth), with the angle $\emptyset_{xz}$ rotated in a counterclockwise direction about the *xz* plane (elevation). Both angles $\emptyset_{xy}$ and $\emptyset_{xz}$ are specified in degrees.

R.5.165 A number of examples using the *view* command are presented and defined below:

1. *view(0, 90)* is the *xy* projection, same as *view(2).*

2. *view(0, 0)* is the *xz* projection.

3. *view(90, 0)* is the *yz* projection.

4. *view(−37.5, 30)* shows the 3-D view (default), same as *view(3).*

R.5.166 The script file, *surf_view_cont* illustrate some of the commands just presented by returning the following plots:

a. The *surf* and the *shaded surf* (*interp*) plots

b. The plots with the following view's arguments: *(0, 0), (90, 0), (−127.5, 0),* and *(−82.5, 0)*

c. The *contour* and *contour3* plots of the function defined by the following equation:

$$z = f(x, y) = \sqrt{x^2 + y^2}\,\frac{\sin(2y)}{y}$$

over the ranges $-13 \le x \le +13$, and $-13 \le y \le +13$.

MATLAB Solution

```
% Script file:surf_view_cont
x = linspace(-13,13,100);
y = x;
[X,Y] = meshgrid(x,y);
Z = sqrt(X.^2+Y.^2).*sin(2.*Y)./Y;

figure(1)                                    % surf plot
surf(X,Y,Z)
shading interp;
title('surf plot');
box on; xlabel('X'),ylabel('Y'), zlabel('Z')

figure(2)                                    % shaded surf plot
surf (X,Y,Z)
shading faceted
title ('shaded surf plot');
box on
xlabel('X'),ylabel('Y'), zlabel('Z')

figure(3)                                    % view plots
subplot(2,2,1)
surf(X,Y,Z)
shading faceted
view(0,0);
xlabel('X'),ylabel('Y'), zlabel('Z')
subplot(2,2,2)
surf(X,Y,Z)
shading faceted
view(90,0)
xlabel('X'),ylabel('Y'), zlabel('Z')
subplot(2,2,3)
surf(X,Y,Z)
shading faceted
view(-37.5-90,0)
xlabel('X'),ylabel('Y'), zlabel('Z')
subplot(2,2,4)
surf(X,Y,Z)
shading faceted
view(-37.5-45,0)
xlabel('X'),ylabel('Y'), zlabel('Z')

figure(4)                                    % contour plot
contour(X,Y,Z,20);axis square;
```

```
box on; shading interp;grid off;
title('contour plot')
xlabel('X'),ylabel('Y'), zlabel('Z')

figure(5)                                               % contour3 plot
contour3(X,Y,Z,20);axis on;grid off
axis square;box on;
shading interp;
title('contour3 plot');
xlabel('X'),ylabel('Y'), zlabel('Z')
```

The script file, *surf_view_cont* is executed and the resulting plots are shown in Figures 5.52 through 5.56.

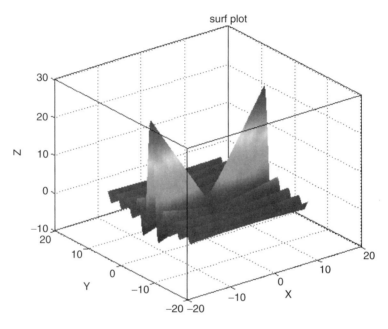

FIGURE 5.52
(See color insert following page 342.) Surf plot of R.5.159(a).

R.5.167 The command *rotate3d* permits interactive changes to view the body displayed, and by clicking the (left) mouse, the figure can then be dragged to any desired destination. The changes are shown at the left corner of the figure in terms of its *azimuth* and *elevation*. This command is closely related to the *view* command.

R.5.168 The commands *meshc(x, y, z)* and *surfc(x, y, z)* return the *mesh* and *surface* plots, but the second command in addition draws a contour plot under the surface.

R.5.169 The command *meshz(x, y, z)* returns the mesh plot and adds vertical lines to the plot, under the surface.

R.5.170 The command *Data = smooth3(data)* returns the filter *Data*, given the unfiltered *data*. The input (unfiltered) *data* can further be controlled by using a *'gaussian'* or the *'box'* (default) filter. The syntax format is *Data = smooth3(data, 'filter')*.

R.5.171 MATLAB provides the users with a number of predefined functions that are basically used for test and demo purposes. Examples of these functions are *peaks* and *humps*.

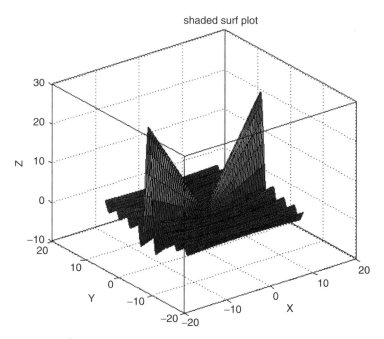

FIGURE 5.53
(See color insert following page 342.) Shaded surf plot of R.5.159(a).

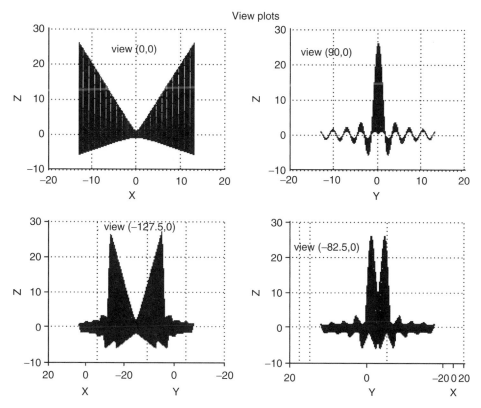

FIGURE 5.54
View plots of R.5.159(b).

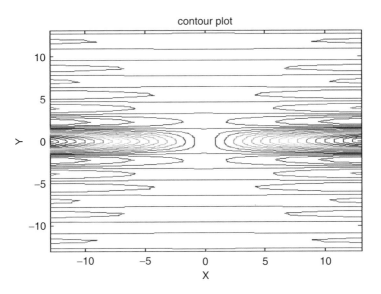

FIGURE 5.55
contour plot of R.5.159(c).

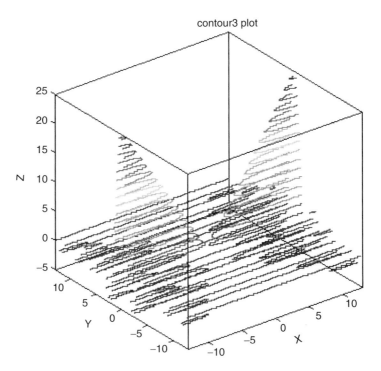

FIGURE 5.56
(See color insert following page 342.) *contour3* plot of R.5.159(c).

R.5.172 The function *peaks* returns an $n \times n$ matrix with elements taken from translating and scaling the coefficients of the Gaussian distribution function. The various syntax forms of the function *peaks* are:

$$z = peaks$$
$$z = peaks(n)$$
$$[x, y, z] = peaks$$
$$[x, y, z] = peaks(n)$$

When no argument is given, MATLAB returns a *49-by-49* matrix. When n is specified, MATLAB returns an $n \times n$ matrix, or if desired, the x, y, and z Cartesian 3-D coordinates.

R.5.173 For example, the program that returns the 2-D plot of the MATLAB function *peaks* using a *150 × 150* matrix is illustrated as follows:

MATLAB Solution
```
>> k = peaks(150);
>> plot (k)
>> title('2D plot using peaks')
>> xlabel('x-axis');ylabel('y-axis')
```

The 2-D plot of peaks is shown in Figure 5.57.

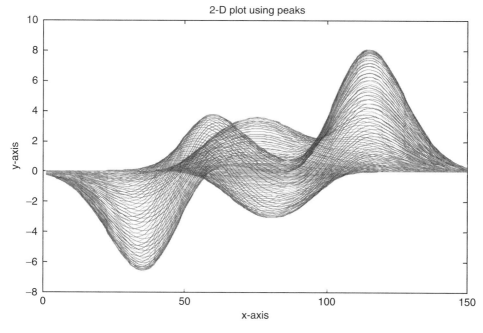

FIGURE 5.57
2-D plot of *peaks* R.5.173.

R.5.174 The function *humps* returns a smooth function with maxima at $x = 0.3$ and $x = 0.9$. The syntax and format of the function *humps* are

  a. $y = humps(x)$
  b. $y = humps$ (assuming $x = 0:0.05:1$)
  c. $[x, y] = humps(x)$

R.5.175 For example, write a program that returns the 2-D plot of the MATLAB function *humps* using the default case.

MATLAB Solution
```
>> [X,Y] = humps;
>> plot(X,Y)
>> plot(X,Y)
>> title('plot of humps')
>> xlabel('x-axis')
>> ylabel('y-axis')
```

The resulting plot is shown in Figure 5.58.

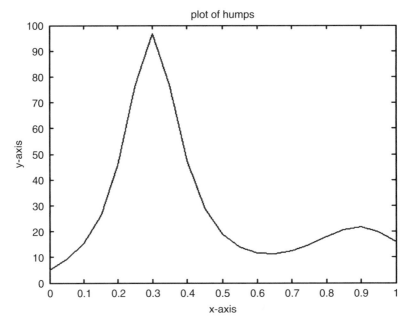

FIGURE 5.58
2-D plot of *humps* of R.5.175.

R.5.176 Let us now create a 3-D plot of the MATLAB function *peaks* using a *150 × 150* matrix for each of the variables *x*, *y*, and *z*. The plot is displayed and the size of each of the variables are checked (Figure 5.59).

MATLAB Solution
```
>> [x,y,z] = peaks(150);
>> plot3(x,y,z)
>> title('3D plot using peaks')
>> xlabel('x-axis');ylabel('y-axis');zlabel('z-axis');
>> size (x)

   ans =
          150    150

>> size (y)

   ans =
          150    150
```

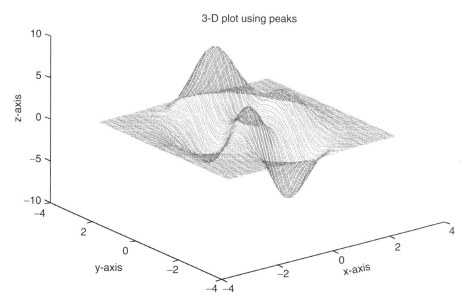

FIGURE 5.59
(See color insert following page 342.) 3-D plot of *peaks* of R.5.176.

```
>> size (z)

    ans =
            150     150
```

R.5.177 The command *trimesh (tri, X, Y, Z, c)* returns the plot of a mesh structure consisting of triangles, where *tri = delaunay(X, Y)* returns a planar structure consisting of planar triangles (2-D) that in conjunction with the Z component create a 3-D grid. The parameter *c* represents the color. The default for *c* is the color for Z = *c*; this color is proportional to the surface height.

R.5.178 The command *trisurf (tri, X, Y, Z, c)* returns the plot of the triangular surface plot where the arguments *tri, X, Y,* and *Z* are the same as in the function *trimesh*.

R.5.179 For example, write a program that returns the plot of the function *peaks* by using the commands

a. *trimesh* and *delaunay*

b. *trisurf*

MATLAB Solution
```
>> x = linspace(-13,13,15);
>> y = x;
>> [X,Y] = meshgrid(x,y);
>> Z = peaks(15);

>> figure(1)
>> tri = delaunay(X,Y);
>> trimesh(tri,X,Y,Z)
>> title('trimesh plot');
>> box on; xlabel('x-axis'),ylabel('y- axis'), zlabel('z-axis')

>> figure(2)
>> trisurf(tri,X,Y,Z)
```

```
>> title('trisurf plot');
>> box on; xlabel('x-axis'),ylabel('y- axis'), zlabel('z-axis')

>> figure(3)
>> Y= sqrt((-X.^2 - Y.^2 - Z.^2);
>> slice(X,Y,Z,v,[-1.2 .8 2],2,[-2 -.2])
```

The resulting plots are shown in Figures 5.60 and 5.61.

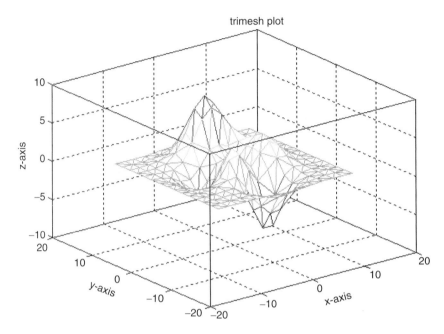

FIGURE 5.60
(See color insert following page 342.) Plot of R.5.179(a).

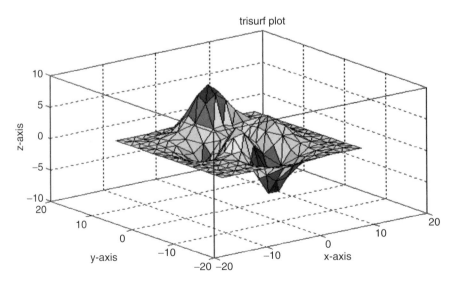

FIGURE 5.61
(See color insert following page 342.) Plot of R.5.179(b).

R.5.180 The 3-D *ez* functions are equivalent to their numerical counterpart. A list of MATLAB *ez* functions follows:

a. *ezcontour*

b. *ezcontourf*

c. *ezcontour3*

d. *ezmesh*

e. *ezmeshc*

f. *ezplot3*

g. *ezsurf*

h. *ezsurfc*

R.5.181 The command *scatter3 (x, y, z)* returns circles having the same size at the locations specified by the vectors *x, y,* and *z*. The command *scatter3 (x, y, z, 'filled')* returns shaded circles at the locations specified by the vectors *x, y,* and *z*. The color and size of the circles can be controlled by *options*.

R.5.182 Create the script file, *scatter_3D* that returns the plot of 100 randomly chosen points indicated by shaded (*filled*) circles.

MATLAB Solution

```
% Script file: scatter _ 3D
x = rand(1,100);
y = rand(1,100).*3;
z = rand(1,100).*2;

figure(1)
scatter3(x,y,z);
xlabel('x-axis')
ylabel('y-axis')
zlabel('z-axis')
title('scatter 3(x,y,z)')

figure(2)
scatter3(x,y,z,'filled');
xlabel('x-axis')
ylabel('y-axis')
zlabel('z-axis')
title('scatter 3(x,y,z,filled)')
```

The script file, *scatter_3D* is executed and the resulting plots are shown in Figures 5.62 and 5.63.

R.5.183 For additional information concerning 2-D and 3-D commands, use the *help graph2D* or *help graph3D*, and MATLAB will return a list of frequently used 2- or 3-D graph commands.

Partial lists of 2-D and 3-D commands follow, which can serve as a brief summary and review of the commands presented in this chapter.

```
>> help graph2D

Two dimensional graphs.
Elementary X-Y graphs.
plot          - Linear plot.
loglog        - Log-log scale plot.
```

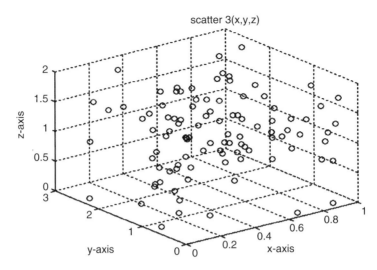

FIGURE 5.62
scatter plot of R.5.182.

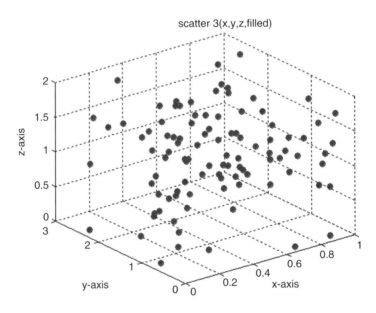

FIGURE 5.63
scatter plot of R.5.182 (*filled*).

```
semilogx       - Semi-log scale plot.
semilogy       - Semi-log scale plot.
polar          - Polar coordinate plot.
plotyy         - Graphs with y tick labels on the left and right.

Axis control.
axis           - Control axis scaling and appearance.
zoom           - Zoom in and out on a 2-D plot.
grid           - Grid lines.
```

```
box           - Axis box.
hold          - Hold current graph.
axes          - Create axes in arbitrary positions.
subplot       - Create axes in tiled positions.


Graph annotation.
legend        - Graph legend.
title         - Graph title.
xlabel        - X-axis label.
ylabel        - Y-axis label.
texlabel      - Produces TeX format from a character string
text          - Text annotation.
gtext         - Place text with mouse.


>> help graph3d


Three-dimensional graphs.
Elementary 3D Plots:

plot3         Plot lines and points in 3D space.
mesh          3D mesh surface.
surf          3D colored surface.
fill3         Filled 3D polygons.


   Color Control:

colormap      Color look-up table.
caxis         Pseudocolor axis scaling.
shading       Color shading mode.
hidden        Mesh hidden line removal mode.
brighten      Brighten or darken color map.


   Lighting:

surfl         3D shaded surface with lighting
lighting      Lighting mode.
material      Material reflectance mode.
specular      Specular reflectance.
diffuse       Diffuse reflectance.
surfnorm      Surface normals.


   Axis Control:

axis          Control axis scaling and appearance.
zoom          Zoom in and out on a 2D plot.
grid          Grid Lines.
box           Axis box.
hold          Hold current graph.
axes          Creates axes in arbitrary positions.
subplot       Creates axes in tiled positions.
```

5.4   Examples

Example 5.1

Write a program that returns the following plots:

1. $sin(x)$ versus x
2. $cos(x)$ versus x
3. $[sin(x)+cos(x)]$ versus x
4. $[sin(x)-cos(x)]$ versus x

over the range $0 \leq x \leq 2\pi$ using the following specs:

1. Twenty points to create each plot
2. Label the x- and y-axis
3. Choose color, markers, and line style for each curve
4. Create the plots in a box and without one; with and without a grid
5. *plot (x, y)* to create both the plots: $sin(x)$ and $cos(x)$
6. Limit the plotting range over $1.5 \leq y \leq -2$
7. Remove the axis
8. Identify each curve by a text string
9. Create a *legend* and then remove the *legend*
10. Plot each curve in an individual subwindow by using the *subplot* and the *stem* commands
11. Use the *stairs* command to plot $sin(x)$ versus x and $cos(x)$ versus x on separate plots

MATLAB Solution

```
>> X=linspace(0,2*pi,20);    % creates a 20 element vector X from 0 to
                               2pi

>> Y1= sin(X);               % creates a 20 element vector Y1 = sin(X)
                               for 0 < X < 2pi

>> Y2 =cos(X);               % creates a 20 element vector of Y2 =
                               cos(X) for 0 < X < 2pi

>> Y3 =Y1+Y2;
>> Y4 =Y1-Y2;
>> plot (X,Y1,'K*:')         % plots sin(X) vs X with a dotted (:) black
                               (k) line
>>                           % indicating the points with a star (*)
                               marker
>> hold on
>> plot(X,Y2,'r--')          % plots cos(X) vs X with a dashed (--) red
                               (r) line

>> hold on
>> plot(X,Y3,'b')            % plots [sin(X) + cos(X)] vs X with a solid
                               line (blue)

>> hold on
>> plot(X,Y4,'g')            % plots [sin(X) - cos(X)] vs X with a solid
                               green (g) line

>> hold off
>> box off                   %   suppresses the figure box
```

```
>> title ('Trigonometric Functions with "Box Off"')    % see Figure 5.64
>> box on                                               % turns on the
                                                          Figure Box
>> title ('Trigonometric Functions with "Box On"')     % see Figure 5.65
>> grid on                                              % turns the grid
                                                          on
>> xlabel('Independent Variable X')                     % labels the x
                                                          axis
>> ylabel('Dependent Variable Y')                       % labels the y
                                                          axis
>> title ('Trigonometric Functions with "Grid On"')    % see Figure 5.66
>> grid off
>> title ('Trigonometric Functions with "Grid off"')    % see Figure 5.67
>> axis ([0 2*pi -2 1.5])        % sets the axis 0≤ x ≤ -2pi, and  -2≤ y
                                   ≤ 1.5
>> axis off                      % removes axis
>> axis on                       % creates axis
>> legend ('sin(X)','cos(X)','sin(X)+cos(X)','sin(X)-cos(X)')
>> title('Trigonometric Functions with Fixed axis and Legend')    % see
                                                          Figure 5.68.
>> gtext('sinX') % identifies each curve with a text string
>> gtext('cos(X)')
>> gtext('sin(X)+cos(X)')
>> gtext('sin(X)-cos(X)')             % see Figure 5.69 where the curves are
                                        identified by texts
>> % plot the four functions using steam, in separate subplot
>> axis on;
>> axis([0 2*pi -1.5 2]);
>> subplot(2,2,1)
>> stem(X,Y1)
>> title('Sin(X) VS X')
>> subplot(2,2,2)
>> stem(X,Y2)
>> title('Cos(X) VS X')
>> subplot(2,2,3)
>> stem(X,Y3)
>> title('Sin(X)+Cos(X) VS X')
>> subplot(2,2,4)
>> stem(X,Y4)
>> title('Sin(X)-Cos(X) VS X')
>> title('Sin(X)+Cos(X) VS X')
>> see Figure 5.70.
>> subplot(2,1,1)
>> stairs(X,Y1)
>> title('Sin(X) VS X')
>> subplot(2,1,2)
>> stairs(X,Y2)
>> title('Cos(X) VS X')
>> % plots of :Sin(X) vs X, and Cos(X) vs X using a stair case
                                        approximation
>> % see Figure 5.71
```

Trigonometric Functions with "Box Off"

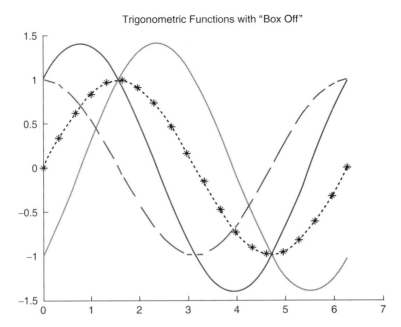

FIGURE 5.64
(See color insert following page 342.) Trigonometric plots of Example 5.1 (*"Box Off"*).

Trigonometric Functions with "Box On"

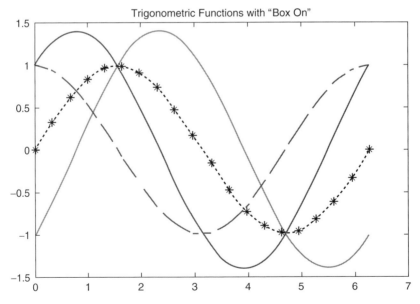

FIGURE 5.65
(See color insert following page 342.) Trigonometric plots of Example 5.1 (*"Box On"*).

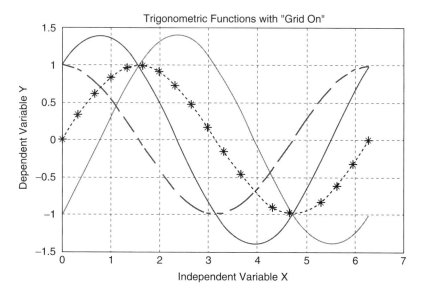

FIGURE 5.66
(See color insert following page 342.) Trigonometric plots of Example 5.1 (*"grid on"*).

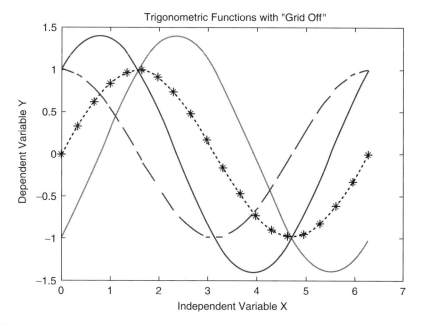

FIGURE 5.67
(See color insert following page 342.) Trigonometric plots of Example 5.1 (*"grid off"*).

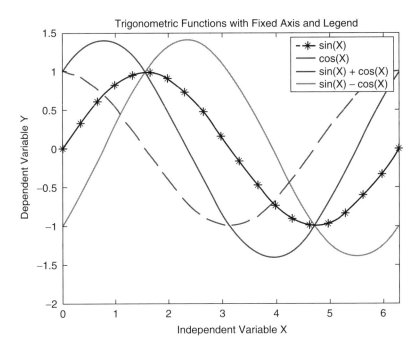

FIGURE 5.68
(See color insert following page 342.) Trigonometric plots of Example 5.1 (*axis* and *legend*).

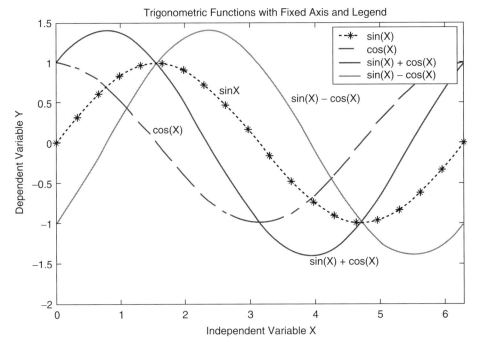

FIGURE 5.69
(See color insert following page 342.) Trigonometric plots of Example 5.1 (*axis* and *legend*).

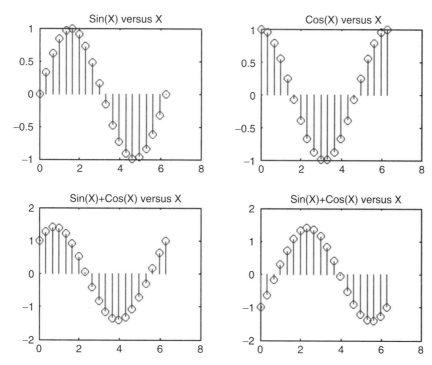

FIGURE 5.70
Stem plots of Example 5.1.

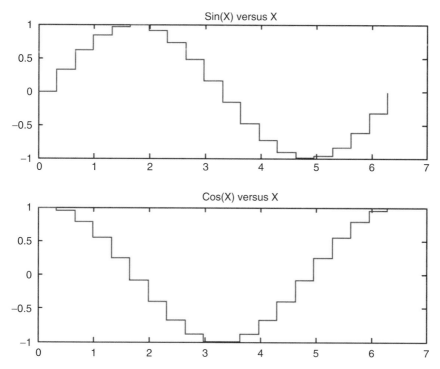

FIGURE 5.71
Stairs plots of Example 5.1.

Example 5.2

The objective of Examples 5.2 and 5.3 is to explore using the power of MATLAB the behavior of often encountered family of functions and provide on insight of their coefficients.

Write a program that returns the plots of the families of curves given in Table 5.7 (a through g) over the domain $-2\pi \le x \le 2\pi$ by using 500 linearly spaced points.

TABLE 5.7

Families of Functions for Example 5.2

a	b	c	d	e	f	g
$Y_1 = 1 + sin(x)$	$Y_4 = sin(x)$	$Y_7 = sin(x)$	$Y_{10} = 5\, sin(x) + x$	$Y_{11} = sin(x)$	$Y_{15} = sin(x)$	$cos(3x)$
$Y_2 = 2 + sin(x)$	$Y_5 = 2\, sin(x)$	$Y_8 = 3 + sin(2x)$		$Y_{12} = sin(x - \pi/2)$	$Y_{16} = -sin(x)$	versus
$Y_3 = 3 + sin(x)$	$Y_6 = 3\, sin(x)$	$Y_9 = 6 + sin(3x)$		$Y_{13} = sin(x - \pi)$	$Y_{17} = 2sin(x)$	$sin(4x)$
				$Y_{14} = sin(x - 3\pi/2)$	$Y_{18} = -2sin(x)$	

MATLAB Solution

```
>> % plot for parts: a, b, c, and d, are shown in Figure 5.72
>> format compact
>> X = linspace(-2*pi,2*pi,500);   % creates an X vector with 500 element
>> Z = sin(X);                      % creates a 500 element Z vector
>> subplot(2,2,1)                   % divides the window into 2x2
                                      sub-window
>> Y1 = Z+1;                        % activates sub-window 1,1
>> Y2 = 2+Z;
>> Y3 = 3+Z;
>> plot (X,Y1,X,Y2,X,Y3)           % plots Y1 vs X, Y2 vs X, and Y3 vs X
>> title('1+sin(X), 2+sin(X), 3+sin(X)')
>> axis([-2*pi 2*pi -1 6])
>> ylabel('Y1, Y2, Y3')
>> subplot (2,2,2)                  % activates sub-window 1,2
>> Y4 = Z;
>> Y5 = 2*Z;
>> Y6 = 3*Z;
>> plot (X,Y4,X,Y5,X,Y6)
>> title ('sin(X), 2sin(X), 3sin(X)')
>> axis([-2*pi 2*pi -4 4])
>> ylabel('Y4, Y5, Y6')
>> subplot(2,2,3)                   % activates sub-window 2,1
>> Y7 = Z;
>> Y8 = 3+sin(2*X);
>> Y9 = 6+sin(3*X);
>> plot (X,Y7,X,Y8,X,Y9)
>> title('sin(X), 3+sin(X), 6+sin(X)')
>> axis([-2*pi 2*pi -2 8])
>> ylabel('Y7, Y8, Y9')
>> subplot(2,2,4)                   % activates sub-window 2,2
>> Y10 = 5*Z+X;
>> plot (X,Y10)
>> title('5sin(X)+X')
>> axis([-2*pi 2*pi -8 8])
>> ylabel('Y10')
```

>> The resulting plots for parts a, b, c, and d are shown in Figure 5.72.

```
>> % plot for parts (e)
>> clf
>> subplot(2,1,1)
>> Y12 = sin(X-pi/2);
>> Y13 = sin(X-pi);
>> Y14 = sin(X-3*pi/2);
>> plot(X,Z,X,Y12,X,Y13,X,Y14)
>> ylabel('Y11, Y12, Y13, Y14')
>> title('Sin(X), Sin(X-pi/2), Sin(X-pi), Sin(X-3pi/2)')
>> subplot(2,1,2)
>> % plots for part (f)
>> Y16 = -Z;
>> Y17 = 2.*Z;
>> Y18 = -Y17;
>> plot(X,Z,X,Y16,X,Y17,X,Y18)
>> title('Sin(X), -Sin(X), 2Sin(X), -2Sin(X)')
>> xlabel( 'X' ),
>> ylabel( 'Y15, Y16, Y17, Y18' )
```

>> The plots for parts e and f are shown in Figure 5.73.

```
>> % plots for part (g)
>> plot (cos(3*X), sin(4*X));
>> axis ([-2 2  -1.5 1.5])
>> title ('Cos(3X) Vs Sin(4X)')
>> xlabel ('X'), ylabel ('Magnitude Sin(4X)')
```

The plot for part g is shown in Figure 5.74.

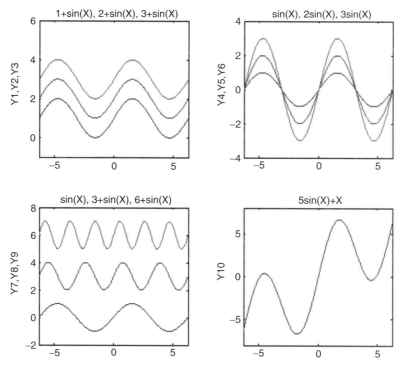

FIGURE 5.72
Plots of Example 5.2(a, b, c, and d).

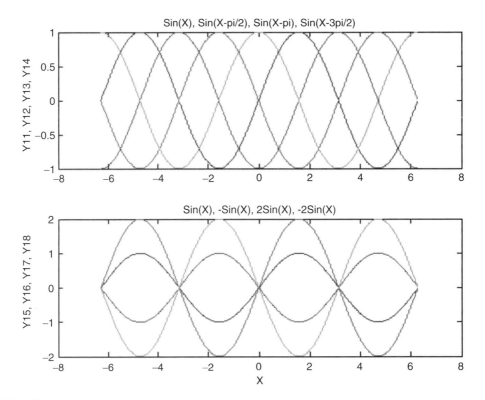

FIGURE 5.73
Plots of Example 5.2(e and f).

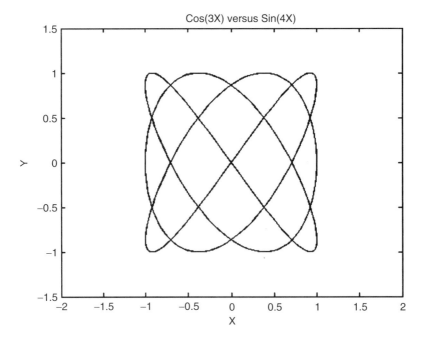

FIGURE 5.74
Plot of Example 5.2(g).

Example 5.3

Write a program that returns the plots of the following families of curves:

a. $Y_{1a} = x^a$, for $a = 1, 2, 3, 4, 5$, and 6 over the range $-2 \leq x \leq 2$ in separate subplots

b. $Y_{2b} = b * x^2$, for $b = 5, 2, 1, -5$, and -2
   $Y_{3c} = c * x^2$, for $c = -5, -2, -1, -0.5$, and -0.1
   $Y_{4d} = x^2 + d$, for $d = 2, 1, 0, -1$, and -2
   $Y_{5e} = -x^2 + e$, for $e = 2, 1, 0, -1$, and -2 over the range $-2 \leq x \leq 2$ in separate subplots

c. $Y_{6f} = f * x^2$ for, $f = 5, 2, 1, 0.5, 0.1$ over the range $-2 \leq x \leq 2$ on the same subplot

d. $Y_{7g} = e^{(g*x)}$, for $g = -5, -2, 1, -0.5$, and -0.1 over the range $0 \leq x \leq 5$ on the same subplot

MATLAB Solution

```
>> % part (a)
>> % plots of Y11, Y12, Y13, Y14, Y15, and Y16 are shown in Figure 5.75
>> X = linspace(-2,2,36);
>> Y= X;
>> subplot(2,3,1);
>> plot (X,Y)
>> title ('Y11=x');xlabel('x');
>> ylabel('Amplitude of Y11');
>> subplot(2,3,2);
>> Y= X.^2;
>> plot (X,Y)
>> title ('Y12=x^2');xlabel('x');
>> ylabel('Amplitude of Y12');
>> subplot(2,3,3);
>> Y= X.^3;
>> plot (X,Y)
>> title ('Y13=x^3');xlabel('x');
>> ylabel('Amplitude of Y13');
>> subplot (2,3,4);
>> Y= X.^4;
>> plot (X,Y)
>> title('Y14=x^4');xlabel('x');
>> ylabel('Amplitude of Y14');
>> subplot(2,3,5);
>> Y=X.^5;
>> plot (X,Y);xlabel('x');
>> ylabel('Amplitude of Y15');
>> title ('Y15=x^5')
>> ubplot(2,3,6);
>> Y= X.^6;
>> plot (X,Y)
>> title ('Y16=x^6');xlabel('x');
>> ylabel('Amplitude of Y16');
>> The plots for part a are shown in Figure 5.75.
>> % part(b), plots of Y2b, Y3c, Y4d, and Y5e
>> X = linspace(-2,2,36);
>> subplot (2,2,1);
>> Y21 = 5*X.^2;
>> Y22 = 2*X.^2;
>> Y23 = X.^2;
>> Y24 = -5*X.^2;
```

```
>> Y25 = -2*X.^2;
>> plot (X,Y21,X,Y22,X,Y23,X,Y24,X,Y25)
>> title ('Y2b = b*x^2, for b = 5,2,1,-5,-2');
>> ylabel('Amplitude of Y2b');xlabel('x');
>> subplot (2,2,2);
>> Y31 = -5*X.^2;
>> Y32 = -2*X.^2;
>> Y33 = -X.^2;
>> Y34 = -.5*X.^2;
>> Y35 = -.1*X.^2;
>> plot (X,Y31,X,Y32,X,Y33,X,Y34,X,Y35);xlabel('x');
>> title ('Y3c = c*x^2, for c = -5,-2,-1,-.5,-.1');
>> ylabel('Amplitude of Y3c');
>> subplot(2,2,3);
>> Y41 = X.^2+2;
>> Y42 = X.^2+1;
>> Y43 = X.^2;
>> Y44 = X.^2-1;
>> Y45 = X.^2-2;
>> plot (X,Y41,X,Y42,X,Y43,X,Y44,X,Y45);
>> xlabel('x');
>> title ('Y4d = x^2+d, for d = 2,1,0,-1,-2');
>> ylabel('Amplitude of Y4d');
>> subplot (2,2,4);
>> Y51 = -X.^2+2;
>> Y52 = -X.^2+1;
>> Y53 = -X.^2;
>> Y54= -X.^2-1;
>> Y55 = -X.^2-2;
>> plot (X,Y51,X,Y52,X,Y53,X,Y54,X,Y55);xlabel('x');
>> title ('Y5e = -x^2+e, for e = 2,1,0,-1,-2');
>> ylabel('Amplitude of Y5e');

>> The resulting plots are shown in Figure 5.76.

>> % part (c and d) , plots of Y6f and Y7g
>> clf
>> X = linspace(-2,2,36);
>> subplot(2,1,1);
>> Y61=5*X.^2;
>> Y62 = 2*X.^2;
>> Y63 = X.^2;
>> Y64 =.5*X.^2;
>> Y65 =.1*X.^2;
>> plot(X,Y61,X,Y62,X,Y63,X,Y64,X,Y65);
>> ylabel('Amplitude of Y6f');   xlabel('x')
>> title('Y6f = f*x^2, for f = 5,2,1,0.5,0.1')
>> subplot(2,1,2);
>> X= linspace(0,5,50);
>> Y71 = exp(-5.*X);
>> Y72 = exp(-2.*X);
>> Y73 = exp(-X);
>> Y74 = exp(-.5*X);
>> Y75 = exp(-.1*X);
>> plot(X,Y71,X,Y72,X,Y73,X,Y74,X,Y75)
```

```
>> title('Y7k = en(-gx), for g = 5,2,1,0.5,0.1');
>> ylabel('Amplitude of Y7k');  xlabel('x')
```

The plots for parts c and d are shown in Figure 5.77.

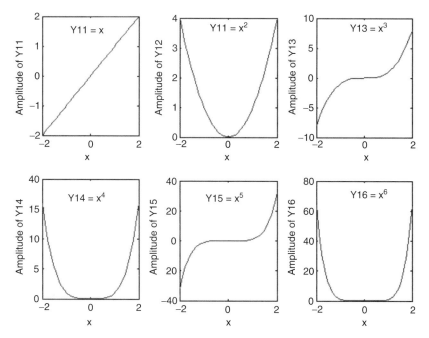

FIGURE 5.75
Plot of Example 5.3(a).

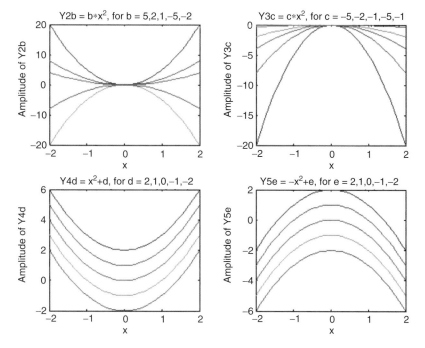

FIGURE 5.76
Plots of Example 5.3(b).

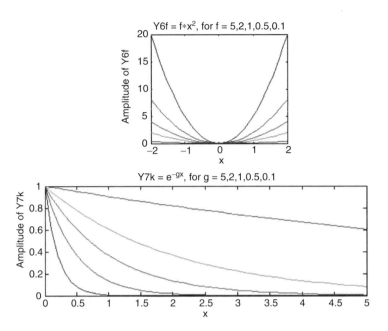

FIGURE 5.77
Plots of Example 5.3(c and d).

Example 5.4

Create the script file, *sin_x_over_x* that returns the plot of [f(x) = sin(x)/x] versus x using the following commands:

1. *fplot*
2. *ezplot*
3. *plot*
4. *ezpolar*

in four different subwindows over the domain $-5\pi \le x \le 5\pi$.

Analyze the commands used and the resulting plots in each case.

MATLAB Solution
```
% Script file: sin _ x _ over _ x
FX = 'sin(X)/X';
subplot(2,2,1);
fplot(FX, [-5*pi, 5*pi]);xlabel('x axis'); ylabel('y-axis');
title('[sin(x)/x] vs x, using fplot')
subplot(2,2,2)
ezplot(FX, [-5*pi, 5*pi])
title('[sin(x)/x] vs x, using ezplot')
subplot(2,2,3)
X = linspace(-5*pi,5*pi,100);
Y = sin(X)./X;
plot(X,Y);xlabel('x axis'); ylabel('y-axis');
title('[sin(x)/x] vs x, using plot')
subplot(2,2,4)
ezpolar(FX, [-5*pi, 5*pi])
title('[sin(x)/x] vs x, using ezpolar')
```

The script file, *sin_x_over_x* is executed and the resulting plots are shown in Figure 5.78.

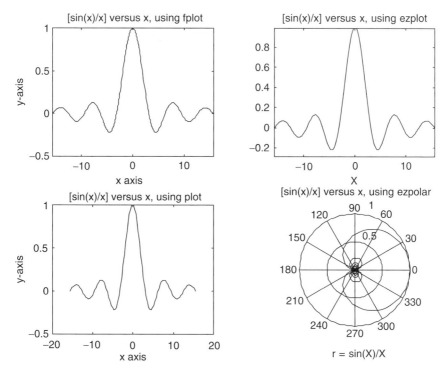

FIGURE 5.78
Plots of Example 5.4.

Example 5.5

Create the script file, *three_circ* that returns the plot consisting of three concentric circles with radius of 1, 2, and 3, where the innermost circle is shaded (filled).

MATLAB Solution
```
% Script file: three _ circ
t = 0:0.01:2*pi;
X = cos(t);
Y = sin(t);
Area (X,Y)
hold on
X=2*cos(t);
Y= 2*sin(t);
plot (X,Y)
hold on
X=3*cos(t);Y=3*sin(t);
plot (X,Y)
axis([-3.5 3.5 -3.5 3.5]);
xlabel('X'), ylabel('Y'),title('3 concentric circles')
```

The script file, *three_circ* is executed, and the resulting plot is shown in Figure 5.79.

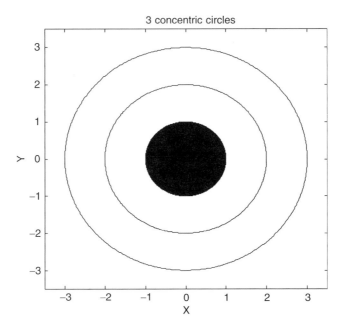

FIGURE 5.79
Plots of Example 5.5.

Example 5.6

Write a program that returns the plots of each of the following equations:

1. $r1 = 10\sin(\beta)$
2. $r2 = -10\sin(\beta)$
3. $r3 = 10\cos(\beta)$
4. $r4 = -10\cos(\beta)$

in separate subplots using the polar coordinate system over the range $0 \le \beta \le 2\pi$, and verify that, in effect each function returns a circle with a radius of 5 on each axis segment.

MATLAB Solution
```
>> Beta = linspace(0,2*pi,50);
>> R1=10*sin(Beta);
>> subplot(2,2,1)
>> polar(Beta,R1)
>> title('10sin(Beta)')
>> subplot(2,2,2)
>> R2 = -R1;
>> polar(Beta,R2)
>> title('-10sin(Beta)')
>> subplot(2,2,3)
>> R3 =10*cos(Beta);
>> polar (Beta,R3)
>> title('10cos(Beta)')
>> subplot(2,2,4)
>> R4 = -R3;
>> polar(Beta,R4)
>> title('-10cos(Beta)')
```

The resulting plots are shown in Figure 5.80.

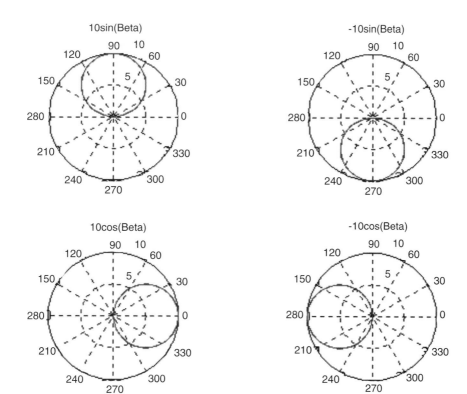

FIGURE 5.80
Plots of Example 5.6.

Example 5.7

Modify the program of Example 5.6 so that the four (circles) equations are plotted on the same polar graph.

MATLAB Solution
```
>> clf
>> Beta = linspace(0,2*pi,50);
>> R1=10*sin(Beta);
>> R2 =-R1;
>> R3 =10*cos(Beta);
>> R4 =-R3;
>> polar (Beta,R1);
>> hold on
>> polar (Beta,R2);
>> hold on
>> polar (Beta,R3);
>> hold on
>> polar (Beta,R4);
>> title ('[10sin(Beta), -10sin(Beta), 10cos(Beta), -10cos(Beta)] vs Rs,
   onone polar graph')
```

The resulting plot is shown in Figure 5.81.

[10sin(Beta), -10sin(Beta), 10cos(Beta), -10cos(Beta)] versus Rs on one polar graph

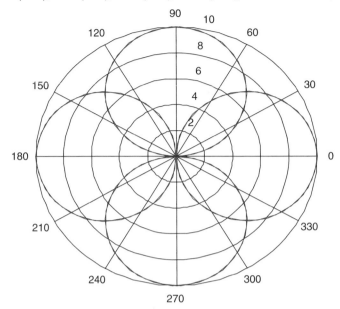

FIGURE 5.81
Plot of Example 5.7.

Example 5.8

Create the script file, *ez_xsinx* with the objective to solve the following non linear equation graphically: *x sin(x)* = *sin(1/x)*; on two intervals: inside and outside the domain $-0.15 \leq x \leq 0.15$ using ezplot, when and if possible.

ANALYTICAL Solution

The functions *x sin(x)* and *sin(1/x)* are plotted separately inside and outside $-0.15 \leq x \leq 0.15$, and the intersection of the two plots represent the points where *x sin(x)* = *sin(1/x)*, and constitute the solutions.

MATLAB Solution
```
% Script file: ez _ xsinx

figure(1)
ezplot('sin(1/x)')
hold on
ezplot('x*sin(x)')
disp('*********************************************')
disp('The solutions outside -0.15≤ x ≤0.15 are :')
ginput
disp('*********************************************')

figure(2)
ezplot('sin(1/x)')
hold on
```

```
ezplot('x*sin(x)')
axis([-0.5 0.5 -2 2])
```

The script file, *ez_xsinx* is executed and the results are shown in Figures 5.82 and 5.83.

```
**********************************************
The solutions outside -0.15≤ x ≤0.15 are :
    -6.2253    -0.1880
    -3.2429    -0.3309
     0.9555     0.8119
     3.0403     0.2881
     6.2253     0.1215
**********************************************
```

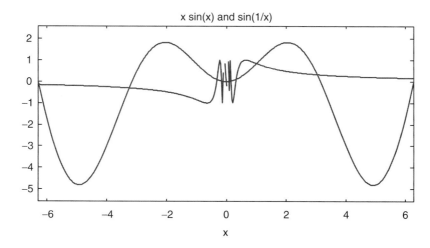

FIGURE 5.82
ezplot of Example 5.8.

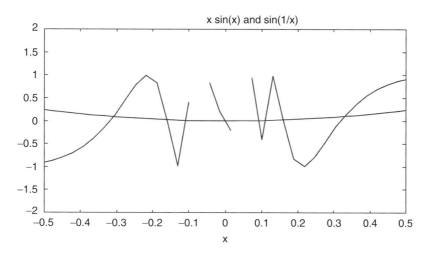

FIGURE 5.83
ezplot of Example 5.8 over −0.5 ≤ x ≤ 0.5.

Observe that the returned plot inside $-0.5 \le x \le 0.53$ is not well defined (continuous) for all values of x, since the MATLAB command *ezplot* does not assign a sufficient number of points. To obtain the solutions of $x \sin(x) = \sin(1/x)$ over the range $-0.15 \le x \le 0.15$, the numerical script file *num_xsinx* is created, indicated as follows:

MATLAB Solution
```
% Script file: num _ xsinx
x = -0.15:0.001:0.15;
y1 = x.*sin(x);
y2 = x.*sin(1./x);
plot (x,y1,x,y2,'o',x,y2)
axis equal;
legend ('y1','y2')
xlabel('x'), ylabel('y1 & y2'),title('y1=xsin(x)and y2=xsin(1/x)')
disp('***************************************************************')
disp(' Five solutions over the range -0.15≤ x ≤0.15 are shown below ')
disp('***************************************************************')
[x,y] = ginput(5)
disp('***************************************************************')
```

The script file, *num_xsinx* is executed and the solutions in the range $-0.15 \le x \le 0.15$ are many, as seen in Figure 5.84. To illustrate the process, only five numerical solutions are shown.

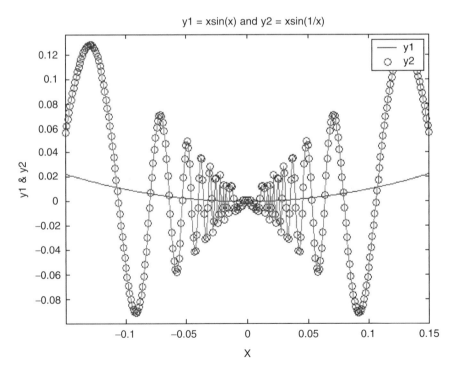

FIGURE 5.84
(See color insert following page 342.) Numerical plot of Example 5.8 over $-0.15 \le x \le 0.15$.

```
>> num _ xsinx
Warning: Divide by zero.
>> In C:\MATLAB6p1\work\A.m at line 4
****************************************************************
Five solutions over the range -0.15 ≤ x ≤ 0.15 are shown below
****************************************************************
    x =
        -0.1069
        -0.0792
        -0.0633
        -0.0522
        -0.0460
    y =
         0.0113
         0.0071
         0.0043
         0.0037
         0.0023
****************************************************************
```

Example 5.9

Create the script file, *sin_cos* that returns the following plots:

1. $x_1 = 5\,cos(2\beta)$ versus $y_1 = 5\,sin(\beta)$ over $-2\pi \le \beta \le 2\pi$
2. $x_2 = sin(2\beta + \pi/3)$ versus $y_2 = sin(\beta)$ over $-2\pi \le \beta \le 2\pi$

    1. On different subplots using the same scale
    2. On the same plot, but using different scales with 100 linearly spaced points

MATLAB Solution

```
% Script file: sin _ cos
Beta = linspace(-2*pi,2*pi,100);     % creates a 100 element Beta array

figure(1)
subplot(2,1,1)
X1 = 5*cos(2*Beta);
Y1 = 5*sin(Beta);
plot(X1,Y1)                          % returns the plot of X1 vs Y1
ylabel('Y1')
title('X1=5*cos(2*Beta) vs Y1 = 5*sin(Beta)');
subplot(2,1,2)
X2 = sin(2*Beta+pi/3);
Y2 =sin(Beta);
plot(X2,Y2)                          % returns the plot of X2 vs Y2
title('X2 = sin(2*Beta + pi /3) vs Y2 = sin(Beta)');
ylabel('Y2'), xlabel('Beta')

figure(2)
plotyy(X1,Y1,X2,Y2)                  % returns the plots of Y1 vs X &
                                     %   Y2 vs X
                                     % on different scales
xlabel ('Beta'), ylabel ('Y1,Y2')
title('X1 vs Y1 and X2 vs Y2');
```

The script file, *sin_cos* is executed, and the resulting plots are shown in Figures 5.85 and 5.86.

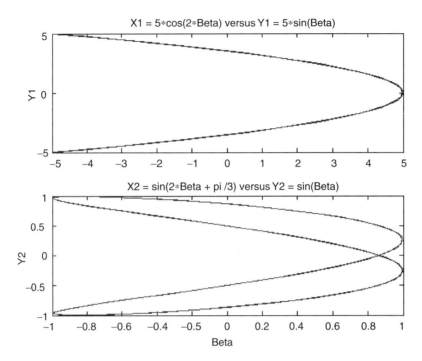

FIGURE 5.85
Plots of Example 5.9(1).

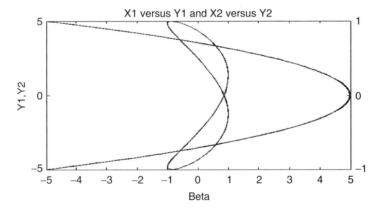

FIGURE 5.86
Plots of Example 5.9(2).

Observe that the plot of Figure 5.86 shows a clearer relation between the two plots illustrated in the separate plots of Figure 5.85.

Example 5.10

Create a 3-D plot from a 2-D, 11×11 matrix (in the xy plane) having ones along the main diagonals, zeros everywhere else, with the center element having a magnitude of 2. It is desired that the zx planes indicate the magnitudes of the elements position on the xy plane all having triangular shapes.

MATLAB Solution

```
>> format compact
>> M = eye(11);
>> N = fliplr(M);
>> addMN = M+N

addMN =
    1    0    0    0    0    0    0    0    0    0    1
    0    1    0    0    0    0    0    0    0    1    0
    0    0    1    0    0    0    0    0    1    0    0
    0    0    0    1    0    0    0    1    0    0    0
    0    0    0    0    1    0    1    0    0    0    0
    0    0    0    0    0    2    0    0    0    0    0
    0    0    0    0    1    0    1    0    0    0    0
    0    0    0    1    0    0    0    1    0    0    0
    0    0    1    0    0    0    0    0    1    0    0
    0    1    0    0    0    0    0    0    0    1    0
    1    0    0    0    0    0    0    0    0    0    1

>> mesh(addMN)
>> AX = [0 12 1 12 0 2]

AX =
    0    12    1    12    0    2

>> axis(AX)
>> xlabel('X'), ylabel('Y'), zlabel('Z')
>> title('3D Plot of Example 5.9')
```

The resulting graph is shown in Figure 5.87.

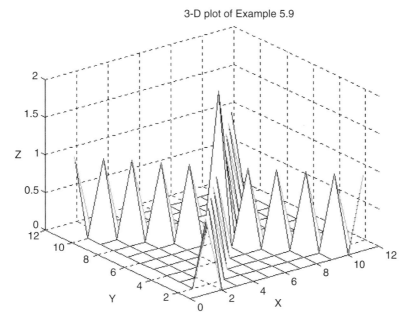

FIGURE 5.87
Three-dimensional plot of Example 5.10.

Example 5.11

Write a program that returns the plot of the 3-D trajectory of a particle defined by the following set of spatial equations:

$$x = r\,cos(2t)$$

$$y = r\,sin(2t)$$

$$z = t$$

where $r = e^{-t/7}$ over the range $0 \leq t \leq 12\pi$.

1. Show the 3-D view
2. Show the front and top views
3. Show the resulting view by using the command *view*(30, 120)

MATLAB Solution
```
>> T = linspace(0,12*pi,400);
>> Z =T;
>> R = exp(-T/7);
>> X = R.*cos(2*T);
>> Y= R.*sin(2*T);
>> subplot (2,2,1); plot3(X,Y,Z)
>> grid on
>> xlabel ('X'), ylabel('Y'), zlabel('Z')
>> title ('3D view for Example 5.11')
>> subplot(2,2,2); plot3(X,Y,Z); view(0,90)
>> grid on
>> xlabel ('X'),ylabel('Y')
>> title ('XY Plane Projection')
>> subplot
(2,2,3); plot3(X,Y,Z); view(0,0)
>> grid on
>> xlabel('X'),zlabel('Z')
>> title('XZ Plane Projection')
>> subplot(2,2,4); plot3(X,Y,Z); view(30,120)
>> grid on
>> xlabel('X'),ylabel('Y'),zlabel('Z')
>> title('3D Plot Using View (30,120)')
```

The resulting plots are shown in Figure 5.88.

Example 5.12

Write a program that returns the plot of the 3-D trajectory of a particle defined by the following set of spatial equations:

$$x = t$$

$$y = (-t + 10\pi)\,cos(t)$$

$$z = 3e^{0.2t} - 3$$

over the range $0 \leq t \leq 10$, and show the following views:

1. The 3-D view using *stem3*
2. The 3-D view using *plot3*
3. The *xy* view

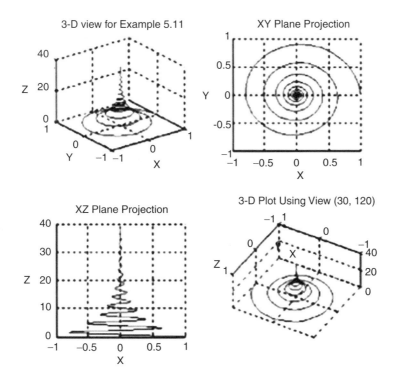

FIGURE 5.88
view plots of Example 5.11.

4. The *xz* view
5. The *zy* view

MATLAB Solution
```
>> format compact;
>> t = linspace(0,10*pi,200);
>> X = t;
>> Y=(-t+10*pi).*cos(t);
>> Z=3*exp(0.2*t)-3;
>> stem3(X,Y,Z)
>> xlabel('t'), ylabel('(-t+10*pi).*cos(t)')
>> zlabel('3*exp(0.2*t)-3'),title('Stem3(X,Y,Z), 3D View')
>> % the graph is shown in Figure 5.89
>> subplot(2,2,1); plot3(X,Y,Z)
>> grid on
>> xlabel('X'),ylabel('Y'),zlabel('Z')
>> title('Plot3(X,Y,Z), 3D View')
>> subplot(2,2,2); plot3(X,Y,Z); view(0,0)
>> xlabel('X'),zlabel('Z')
>> title('XZ View')
>> subplot(2,2,3); plot3(X,Y,Z); view(0,90)
>> xlabel('X'),ylabel('Y')
>> title('XY View')
>> subplot(2,2,4); plot3(X,Y,Z); view(90,0)
>> ylabel('Y'),zlabel('Z')
>> title('YZ View')
```

The resulting plots are shown in Figure 5.90.

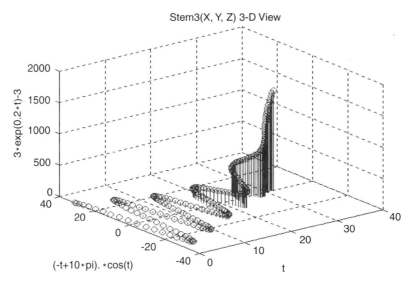

FIGURE 5.89
3-D *stem* plot of Example 5.12(1).

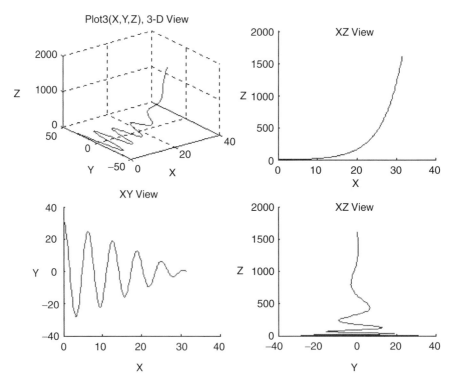

FIGURE 5.90
View plots of Example 5.12(2, 3, 4, and 5).

Example 5.13

Create a program that returns the polar plot of a rose in the polar coordinate system given by the following equations:

$$R_1 = 3\cos(n\beta) \text{ versus } \beta \quad \text{and} \quad R_2 = 3\sin(n\beta) \text{ versus } \beta$$

over the range $0 \leq \beta \leq 2\pi$, using 200 linearly spaced points.

Verify that the given set of equations returns a rose figure with $2n$ petals when n is even, and n petals when n is odd. Test the program and verify the preceding statements for $n = 4$ and $n = 5$.

MATLAB Solution
```
>> Beta = linspace(0,2*pi,200);
>> R1 = 3*cos(4*Beta);
>> subplot(2,2,1);polar(Beta,R1)
>> title('Rose Figure for R1=3*cos(4*Beta)')
>> R2 = 3*sin(4*Beta);
>> subplot(2,2,2); polar(Beta,R2)
>> title('Rose Figure for R2=3*sin(4*Beta)')
>> R3 = 3*cos(5*Beta);
>> subplot(2,2,3); polar(Beta,R3)
>> title('Rose Figure for R3=3*cos(5*Beta)')
>> R4 = 3*sin(5*Beta);
>> subplot(2,2,4); polar(Beta,R4)
>> title('Rose Figure for R4=3*sin(5*Beta)')
```

The resulting plots are shown in Figure 5.91.

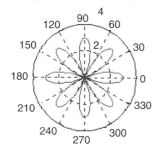
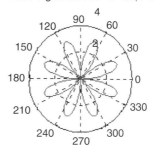
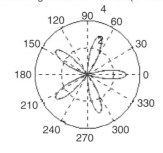
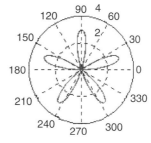

FIGURE 5.91
Plots of Example 5.13.

Example 5.14

On the basis of the earlier example, create the script file, *black_white_rose* that returns an eight-petal black rose and a five-petal white rose.

MATLAB Solution
```
% Script file: black _ white _ rose
clear; clf                       % clears variables and the figure window
T=linspace(0,2*pi,200);
subplot(1,2,1)
R=3*cos(4*T);
X=abs(R).*cos(T);
Y=abs(R).*sin(T);
fill(X,Y,'k')
axis('square')
title('8 petal black rose')
subplot(1,2,2)
R1=3*sin(5*T);
X1 = (R1).*cos(T);
Y1= (R1).*sin(T);
plot(X1,Y1);
axis('square')
title('5 petal white rose')
```

The script file, *black_white_rose* is executed, and the results are shown in Figure 5.92.

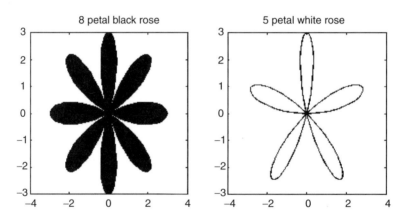

FIGURE 5.92
Plots of a black and white rose of Example 5.14.

Example 5.15

Write a program that returns the 3-D *mesh* and *surf* plots of the following function:

$$Z = sin(2x) \cdot cos(2y) \exp\left(\frac{-sqrt(x^2 + y^2)}{2}\right)$$

over the ranges $-\pi \le x \le \pi$ and $-\pi \le y \le \pi$.

MATLAB Solution
```
>> XY= linspace(-pi,pi,100);
>> YX =XY;
```

```
>> [X,Y] = meshgrid(XY,YX);
>> Z = sin(2.*Y).*cos(2.*X).*exp(-sqrt(X.^2+Y.^2)./2);
>> subplot (2,1,1);
>> mesh (X,Y,Z)
>> xlabel ('X'),ylabel('Y'),zlabel('Z')
>> title ('Mesh Plot')
>> subplot(2,1,2);
>> surf (X,Y,Z)
>> xlabel('X'),ylabel('Y'),zlabel('Z')
>> title('Surf Plot')
```

The resulting plots are shown in Figure 5.93.

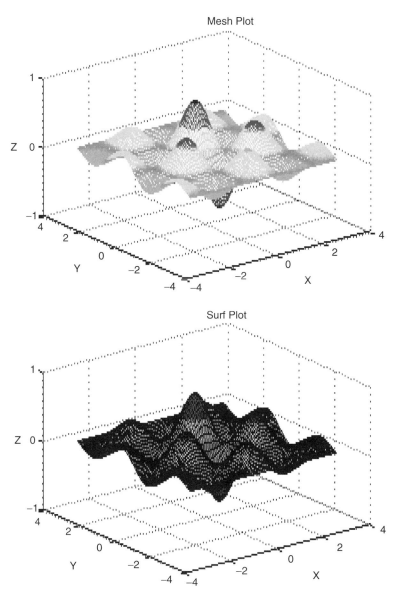

FIGURE 5.93
(See color insert following page 342.) 3-D *mesh* and *surf* plots of Example 5.15.

Example 5.16

Write a program that returns the *contour* and *view* plots with arguments *(0, 0), (90, 0), (−127.5, 0),* and *(−82.5, 0)* of the figure of Example 5.15.

MATLAB Solution
```
% Script file: cont _ view
XY = linspace (-pi,pi,100);
YX = XY;
[X,Y] = meshgrid(XY,YX);
Z = sin(2.*Y).*cos(2.*X).*exp(-sqrt(X.^2+Y.^2)./2);


Figure (1)                                    % contour plot
contour (X,Y,Z,13)
xlabel ('X'), ylabel ('Y'), zlabel ('Z')
title ('Contour plot')


figure (2)                                    % view plots
subplot (2,2,1)
surf (X,Y,Z)
shading faceted
view (0,0);
xlabel ('X'),ylabel ('Y'), zlabel ('Z')
subplot (2,2,2)
surf (X,Y,Z)
shading faceted
view (90,0)
xlabel ('X'),ylabel('Y'), zlabel('Z')
subplot (2,2,3)
surf (X,Y,Z)
shading faceted
view (-37.5-90,0)
xlabel ('X'),ylabel('Y'), zlabel('Z')
subplot (2,2,4)
surf (X,Y,Z)
shading faceted
view (-37.5-45,0)
xlabel ('X'),ylabel ('Y'), zlabel ('Z')
```

The resulting plots are shown in Figures 5.94 and 5.95.

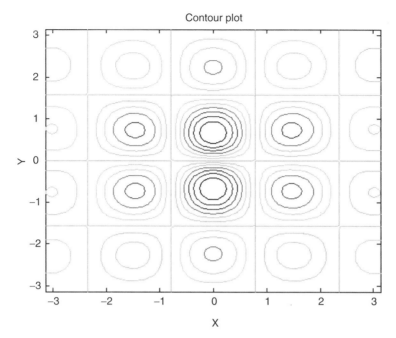

FIGURE 5.94
(See color insert following page 342.) *Contour* plot of Example 5.16.

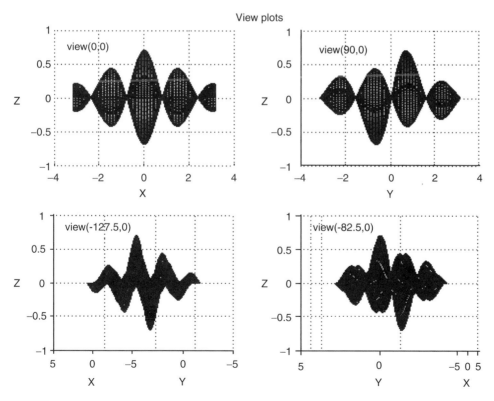

FIGURE 5.95
View plots of Example 5.16.

Example 5.17

Given the function $z = 16x^4 + 15x^2 - x + 6 - 2yx^2 + 15y^2 - y$, over the ranges of x and y given by $-3 \leq x \leq 3$ and $-3 \leq y \leq 9$. Create the script file, *fn_z* that returns the 3-D plots of z using the following commands:

1. *mesh*
2. *meshe*
3. *meshz*
4. *surf*
5. *surfc*
6. *waterfall*

MATLAB Solution

```
%Script file: fn _ z
XY=linspace(-3,3,19);
YX=linspace(-3,9,21);
[x,y]=meshgrid(XY,YX);
z=16*x.^4+15*x.^2-x+6-2*y.*x.^2+15*y.^2-y;

figure(1);
mesh(x,y,z);
box on; axis on;
xlabel('x-axis');ylabel('y-axis');zlabel('z-axis')
title('Plot using mesh')

figure(2)
meshc(x,y,z)
box on; axis on;
xlabel('x-axis');ylabel('y-axis');zlabel('z-axis')
title('Plot using meshc')

figure(3)
meshz(x,y,z)
box on; axis on;
xlabel('x-axis');ylabel('y-axis');zlabel('z-axis')
title('Plot using meshz')

figure(4);
surf(x,y,z);
box on; axis on;
xlabel('x-axis');
ylabel('y-axis');zlabel('z-axis')
title('Plot using surf')

figure(5)
surfc(x,y,z)
box on; axis on;
xlabel('x-axis');ylabel('y-axis');zlabel('z-axis')
title('Plot using surfc ')

figure(6)
waterfall(x,y,z)
box on; axis on;
xlabel('x-axis');ylabel('y-axis');zlabel('z-axis')
title('Plot using waterfall')
```

The script file, *fn_z* is executed and the resulting plots are shown in Figures 5.96 through 5.101.

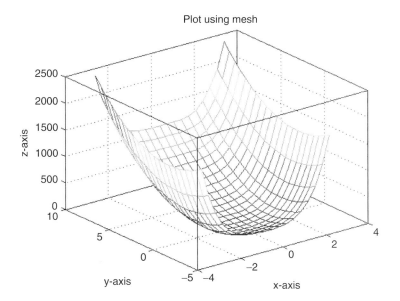

FIGURE 5.96
(See color insert following page 342.) 3-D plot using *mesh* of Example 5.17.

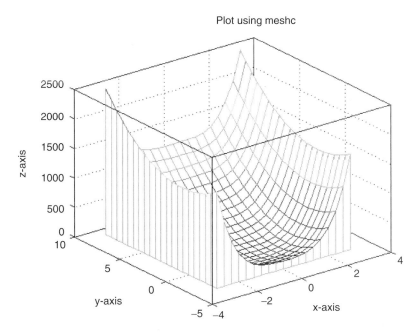

FIGURE 5.97
(See color insert following page 342.) 3-D plot using *meshc* of Example 5.17.

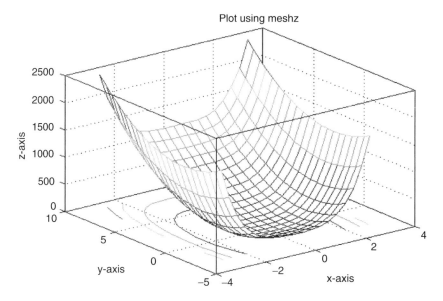

FIGURE 5.98
(See color insert following page 342.) 3-D plot using *meshz* of Example 5.17.

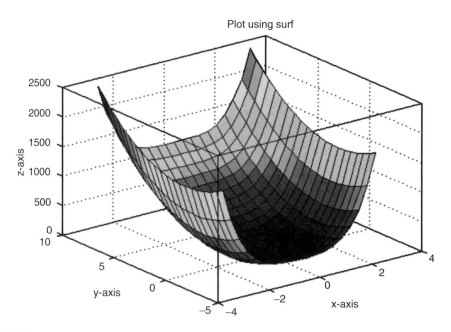

FIGURE 5.99
(See color insert following page 342.) 3-D plot using *surf* of Example 5.17.

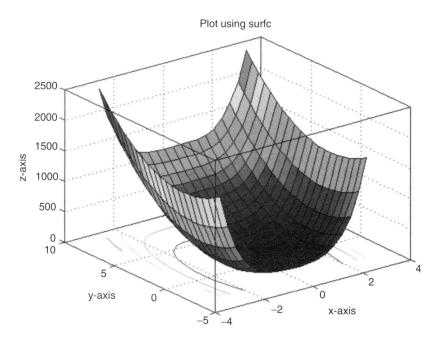

FIGURE 5.100
(See color insert following page 342.) 3-D plot using *surfc* of Example 5.17.

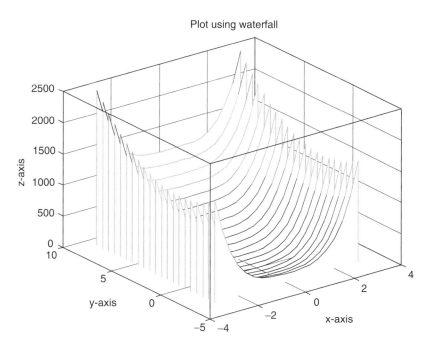

FIGURE 5.101
(See color insert following page 342.) 3-D plot using *waterfall* of Example 5.17.

Example 5.18

Create the script file, *ctl_sphere* that return the normalized shapes of: a *cone*, a *cylinder*, and a *sphere* using a purely mathematical approach (the MATLAB *cylinder* and *sphere* commands cannot be used in this example).

MATLAB Solution
```
% Script file: ctl _ sphere

figure(1)
[r,ang] = meshgrid(linspace(0,1,50),linspace(0,2*pi,50));
x = r.*cos(ang);
y = r.*sin(ang);
surf(x,y,r);axis equal

figure(2)
[a,b] = meshgrid(linspace(0,2*pi,50),-1:2:1);
x = cos(a);
y = sin(a);
surf(x,y,b)
axis equal

figure(3)
[c,d] = meshgrid(linspace(0,2*pi,50),linspace(0,pi,25));
x = cos(c).*sin(d);
y = sin(c).*sin(d);
z = cos(d);
surf(x,y,z)
axis equal
```

The resulting plots are shown in Figures 5.102 through 5.104.

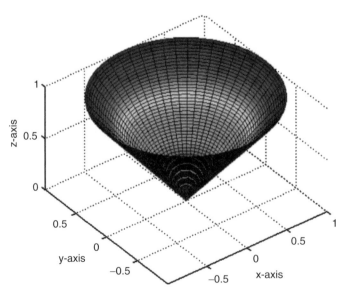

FIGURE 5.102
(See color insert following page 342.) 3-D plot of the cone of Example 5.18.

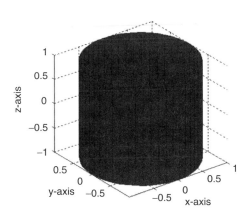

FIGURE 5.103
3-D plot of the cylinder of Example 5.18.

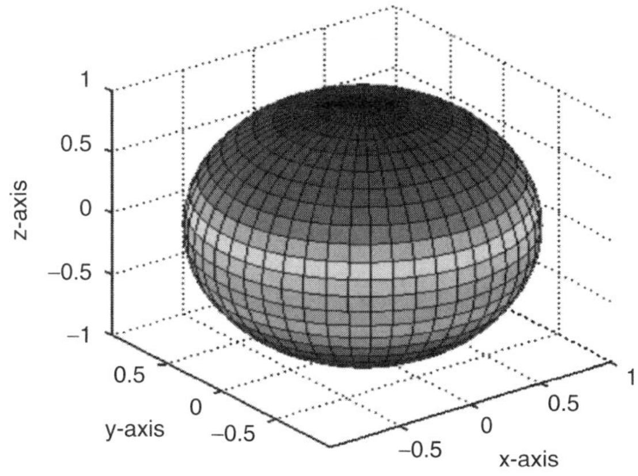

FIGURE 5.104
(See color insert following page 342.) 3-D plot of the sphere of Example 5.18.

5.5   Further Analysis

Q.5.1   Load and run the program of Example 5.1.

Q.5.2   Execute the first instruction without the semicolon. What is the objective and purpose of this instruction?

Q.5.3   Execute the second instruction without the semicolon. What is the purpose of the second instruction?

Q.5.4   With only the first and second instruction, is it possible to sketch *sin(x)* versus *x* by hand?

Q.5.5   Is it possible to create the four plots of Figures 5.64 through 5.68 using one *plot* instruction per figure? If it is, indicate how.

Q.5.6   Indicate how to move the legend to the upper-left corner of Figure 5.68.

Q.5.7   Identify the curves shown in Figure 5.67 using a text command.

Q.5.8   Is it possible to use the *stem* command when plotting more than one function? Test your answer.

Q.5.9   Is it possible to use the *stair* instruction when plotting more than one function? Test your answer.

Q.5.10   Load and run the program of Example 5.2.

Q.5.11   Rearrange the *subplots* shown in Figure 5.72 in at least two other ways.

Q.5.12   Rearrange the four *subplots* shown in Figure 5.72 in a way that each one occupies the whole figure window.

Q.5.13   Discuss the elements that distinguish each of the following families of curves:

   a.   Y1, Y2, and Y3

   b.   Y4, Y5, and Y6

   c.   Y7, Y8, and Y9

   d.   Y11, Y12, Y13, and Y14

   e.   Y15, Y16, Y17, and Y18

Q.5.14 Determine the frequencies, amplitudes, periods, and phase angle of each of the sinusoids used in the generation of the functions Y_n, for $n = 1, 2, 3, \ldots, 18$.

Q.5.15 Describe the shape of Figure 5.72.

Q.5.16 Describe the shape of Figure 5.74. Can Figure 5.74 give information about the relation of the input frequencies? Discuss.

Q.5.17 Run the portion of the program of Example 5.3 that returns Y_{1a}.

Q.5.18 Define the concept of even and odd functions, and identify which functions of Figure 5.75 are even and which ones are odd.

Q.5.19 Indicate if there is any relation between the exponent of x and the symmetry of the function Y_{1a}.

Q.5.20 Using the information given in Figure 5.75, sketch by hand the functions $Y1 = x + x^2$ and $Y2 = x + x^3$.

Q.5.21 Using the instruction *fplot*, obtain the plots of Q.5.20.

Q.5.22 Repeat Q.5.21 by using the instruction *ezplot*.

Q.5.23 Describe the effect of the coefficient b for the function Y_{2b}.

Q.5.24 Compare the function Y_{2b} with the function Y_{3c}.

Q.5.25 Describe the effect of the coefficient d in equation Y_{4d}.

Q.5.26 Describe the effect of the minus sign for x^2 in equation Y_{5e}.

Q.5.27 Compare equations Y_{4d} with Y_{5e}.

Q.5.28 Compare equations Y_{3c} with Y_{6f}.

Q.5.29 Discuss the effect of the coefficient g in equation Y_{7g}.

Q.5.30 Load and run the script file, *sin_x_over_x* of Example 5.4.

Q.5.31 Compare the graphs obtained using the instruction *fplot* with *ezplot*. Which graph provides a better resolution? Can you detect any differences?

Q.5.32 Compare the graphs obtained by using the instruction *plot* with *fplot*. Which graph provides a more reliable description of $sin(x)/x$?

Q.5.33 Load and run the script file, *three_circ* of Example 5.5.

Q.5.34 Using only one plot instruction, modify the program that returns a graph with the same three concentric circles.

Q.5.35 Modify the program to obtain 10 concentric circles with radius of $(0.8) * n$ for $n = 1, 2, 3, \ldots, 10$.

Q.5.36 Once the 10 concentric circles are obtained, modify the program that shades the center circle and all the even concentric strips.

Q.5.37 Load and run the program of Example 5.6.

Q.5.38 Compare the plots obtained in Figure 5.80. Clearly state the similarities, differences, and draw conclusions.

Q.5.39 State the equations of each plot in Figure 5.80.

Q.5.40 Discuss the effect of the sinusoid and its frequencies.

Q.5.41 Load and run the program of Example 5.7.

Q.5.42 Can you use *one polar* instruction to create Figure 5.81?

Q.5.43 Modify the program that replaces in the label of Figure 5.81, the word "Beta" by the Greek character "β."

Q.5.44 Modify the program that returns two shaded circles located on the x-axis.

Q.5.45 Load and run the script file, *ez_xsinx* of Example 5.8.

Q.5.46 Verify the solutions obtained outside $-0.15 \le x \le 0.15$ by direct substitution.

Q.5.47 Describe the curves for $-0.15 \le x \le 0.15$.

Q.5.48 Give reasons for the nature of the curves outside $-0.15 \le x \le 0.15$.

Q.5.49 Verify the solution obtained outside $-0.15 \le x \le 0.15$ by direct substitution.

Q.5.50 Run the script file, *num_xsinx* using *fplot*. Are the results obtained any better?

Q.5.51 Load and run the script file, *sin_cos* of Example 5.9.

Q.5.52 Draw a flowchart and describe the objective of each coded line of the program in the form of comments (%).

Q.5.53 What is the relation between the frequencies of x_1 and y_1?

Q.5.54 What is the relation between the frequencies of x_2 and y_2?

Q.5.55 Indicate how are the magnitudes related in Figures 5.85 and 5.86?

Q.5.56 What is the effect of the phase angle of x_2 on the plots?

Q.5.57 Load and run the program of Example 5.10.

Q.5.58 What is the objective and purpose of the variable *addMN*?

Q.5.59 What do the elements of *addMN* represent?

Q.5.60 Does the variable *addMN* define a plane? If so, what is the plane?

Q.5.61 Describe the purpose of the instruction *mesh(addMN)*.

Q.5.62 Define the purpose of variable *AX*.

Q.5.63 State the height of the center element of Figure 5.87 and identify its location in terms of its Cartesian coordinates.

Q.5.64 Describe how the centered element was created.

Q.5.65 Modify the label of Figure 5.87 by bold-style characters.

Q.5.66 Replace the axis labels X, Y, and Z by italic-style characters.

Q.5.67 Load and run the program of Example 5.11.

Q.5.68 What equation or equations best describe the top *view*?

Q.5.69 What equation or equations describe the front *view*?

Q.5.70 Describe the meaning of the command *view*(30, 120).

Q.5.71 Load and run the program of Example 5.12.

Q.5.72 Indicate the instruction that returns the analog 3-D *plot*.

Q.5.73 Indicate the instructions that return the discrete 3-D *plot*.

Q.5.74 Indicate the instructions that return the *xy*, *xz*, and *zy* views.

Q.5.75 Describe the argument of the *view* instruction.

Q.5.76 Replace the labels of Figure 5.89 by bold-style characters.

Q.5.77 Replace the equation on the axis to italic-style characters, the exponential exp(0.2t) to $e^{0.2t}$, and pi by the Greek character π.

Q.5.78 Load and run the program of Example 5.13.

Q.5.79 State the differences between the $cos(n * \beta)$ versus β and $sin(n * \beta)$ versus β plots in Figure 5.91.

Q.5.80 Load and run the script file, *black_white_rose* of Example 5.14.

Q.5.81   Replace the cosines by sines and rerun Example 5.14. Comment on the results.

Q.5.82   Rerun Example 5.14 to create a rose with red petals and a green background.

Q.5.83   Load and run the program of Example 5.15.

Q.5.84   State the purpose of the instruction *meshgrid*.

Q.5.85   State the purpose of the instruction *mesh*.

Q.5.86   Run the program of Example 5.15 by changing the *cos(2y)* by *sin(2y)* and compare the results.

Q.5.87   Load and run the script file, *con_view* of Example 5.16.

Q.5.88   Describe the *contour* command.

Q.5.89   Why is the response of the contour command a 2-D plot?

Q.5.90   What are the coordinates of the highest point of the plot of Figure 5.94?

Q.5.91   What are the coordinates of the lowest point of the plot of Figure 5.94?

Q.5.92   Discuss the meaning of the different colors.

Q.5.93   Estimate the number of peaks of the figure shown in Figure 5.94.

Q.5.94   Define the meaning and purpose of the *surf* command.

Q.5.95   Describe the command *view(0, 0)*.

Q.5.96   Explain why only five peaks are shown in Figure 5.95.

Q.5.97   What do the colors represent in the *view(0, 0)* command?

Q.5.98   How are the plots of *view(0, 0)* and *view(90, 0)* related?

Q.5.99   Identify at least three points (using the coordinates) that relate the plots obtained by *view(0, 0)* and *view(90, 0)*.

Q.5.100 Discuss why *view(-127, 5, 0)* is shown with a unique color.

Q.5.101 Discuss why *view(-87, 5, 0)* basically returns one-color figure.

Q.5.102 Discuss how *view(-127.5, 0)* and *view(-87.5, 0)* are related.

Q.5.103 Identify at least three points (using the coordinates) that relate *view(-127.5, 0)* with *view(90, 0)*.

Q.5.104 Load and run the script file, *fn_z* of Example 5.17.

Q.5.105 Compare the instructions *mesh* with *ezmesh*.

Q.5.106 Describe the command *shading*.

Q.5.107 Is it possible to use the shading instruction with the *mesh* instruction?

Q.5.108 What are the arguments of the *contour* instruction?

Q.5.109 Load and run the script file, *ctl_sphere* of Example 5.18.

Q.5.110 Define the variables *r* and *ang* used in the plot of Figure 5.102.

Q.5.111 State the variables that control the width, height, and length of the body of Figure 5.102.

Q.5.112 Label the variables that control the width, height, and length of the plot shown in Figure 5.103.

Q.5.113 List the variables that control the width, height, and length of the plot in Figure 5.103.

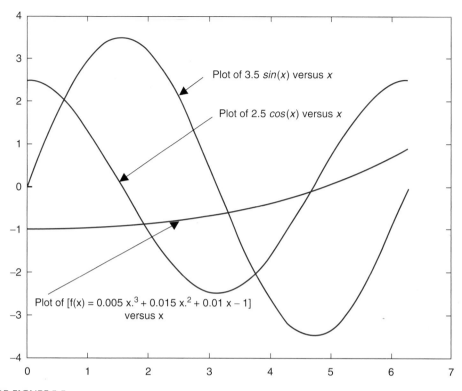

COLOR FIGURE 5.5
Plots of R.5.44.

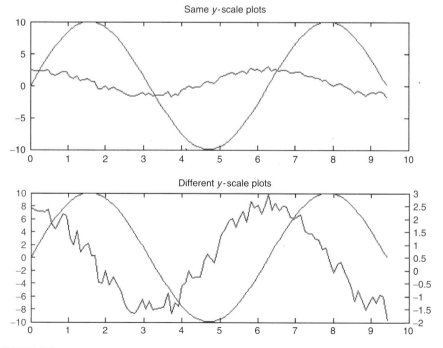

COLOR FIGURE 5.6
Plots of R.5.46 using one and two scales.

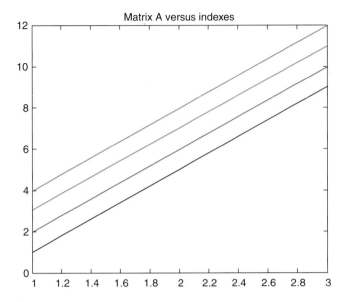

COLOR FIGURE 5.7
Plot of matrix A of R.5.50(a).

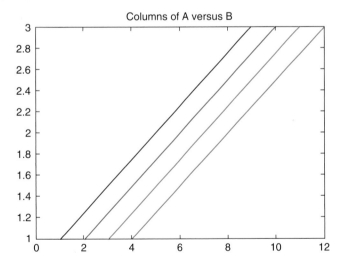

COLOR FIGURE 5.8
Plot of columns of matrix A versus B of R.5.50(b).

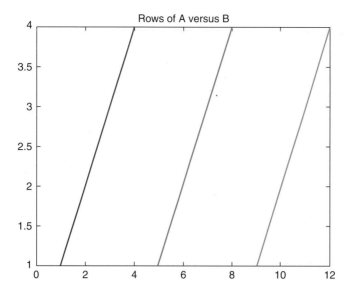

COLOR FIGURE 5.9
Plot of row of matrix A versus B of R.5.50(c).

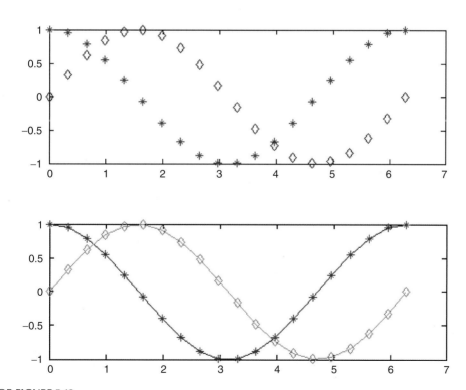

COLOR FIGURE 5.12
Plots with different markers of R.5.59.

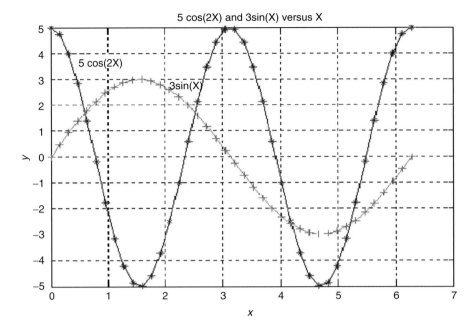

COLOR FIGURE 5.13
Plots with markers and text of R.5.71.

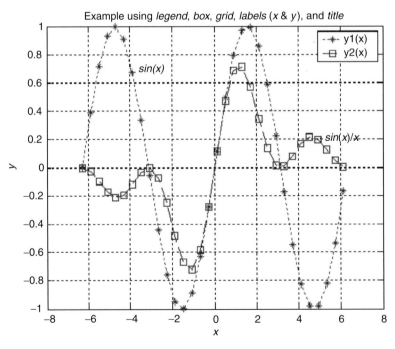

COLOR FIGURE 5.14
Plots with markers, text, and legend of R.5.73.

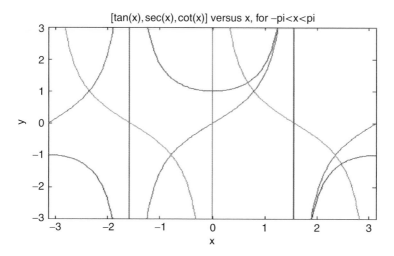

COLOR FIGURE 5.17
Multiple plots using *fplot* of R.5.90.

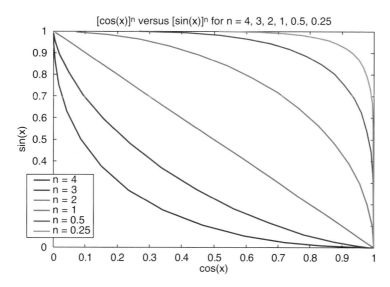

COLOR FIGURE 5.27
Plots of R.5.107(a).

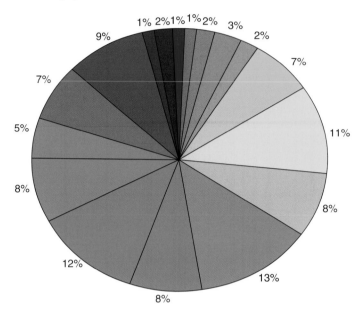

COLOR FIGURE 5.34
Pie plot of R.5.125.

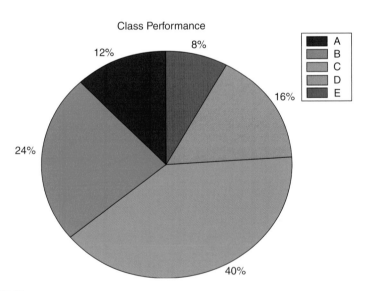

COLOR FIGURE 5.35
Pie plot of R.5.126.

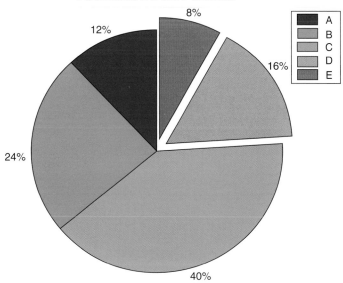

COLOR FIGURE 5.36
Detached pie plot of R.5.128.

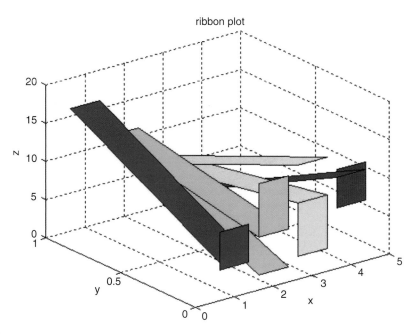

COLOR FIGURE 5.39
Ribbon plots of R.5.139.

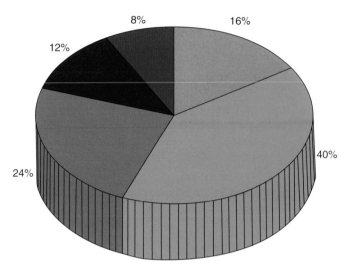

COLOR FIGURE 5.48
pie3 plot of R.5.150.

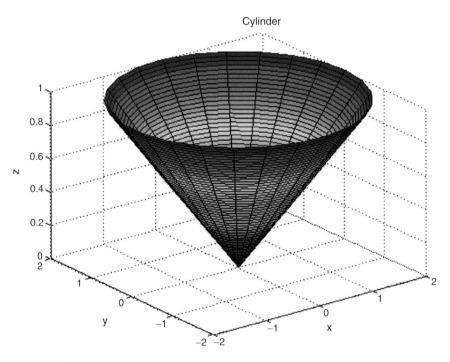

COLOR FIGURE 5.49
Plot of R.5.156.

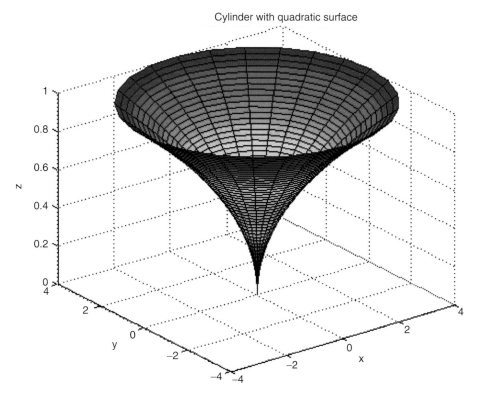

COLOR FIGURE 5.50
Plot of R.5.157.

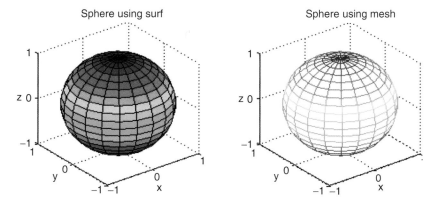

COLOR FIGURE 5.51
Plot of R.5.159.

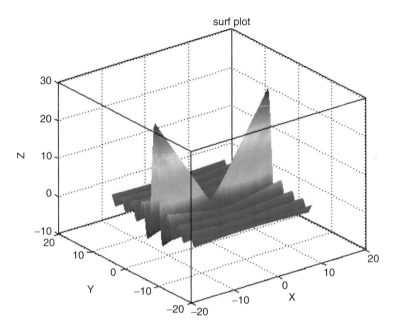

COLOR FIGURE 5.52
Surf plot of R.5.159(a).

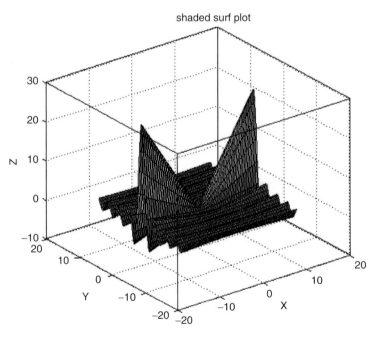

COLOR FIGURE 5.53
Shaded surf plot of R.5.159(a).

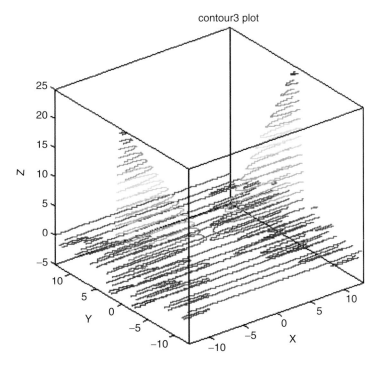

COLOR FIGURE 5.56
contour3 plot of R.5.159(c).

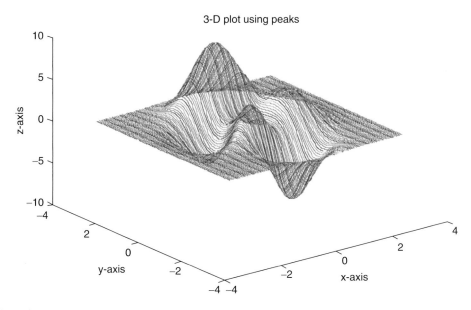

COLOR FIGURE 5.59
3-D plot of *peaks* of R.5.176.

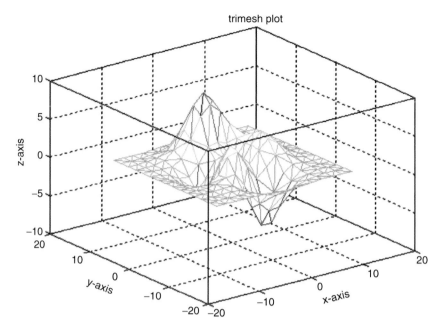

COLOR FIGURE 5.60
Plot of R.5.179(a).

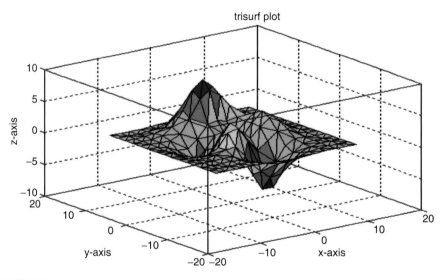

COLOR FIGURE 5.61
Plot of R.5.179(b).

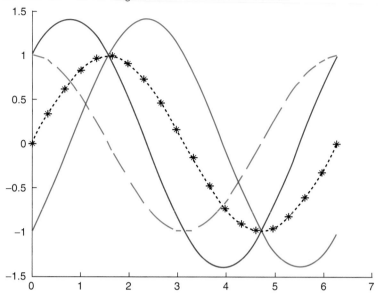

COLOR FIGURE 5.64
Trigonometric plots of Example 5.1 (*"Box Off"*).

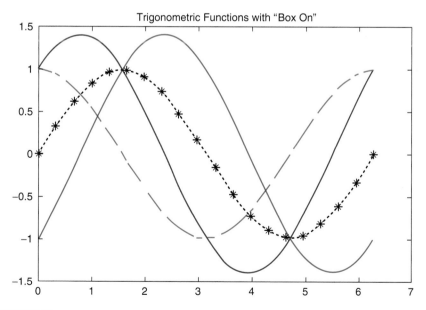

COLOR FIGURE 5.65
Trigonometric plots of Example 5.1 (*"Box On"*).

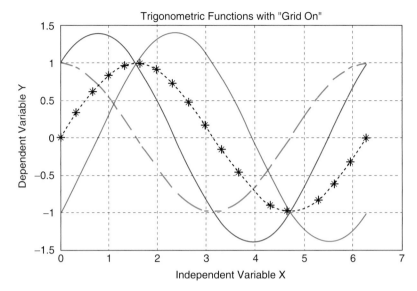

COLOR FIGURE 5.66
Trigonometric plots of Example 5.1 (*"grid on"*).

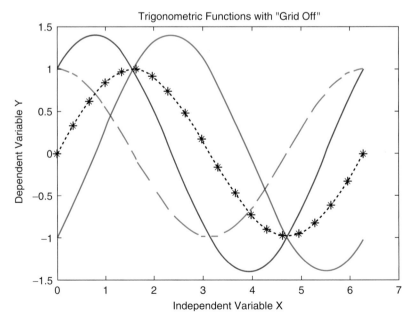

COLOR FIGURE 5.67
Trigonometric plots of Example 5.1 (*"grid off"*).

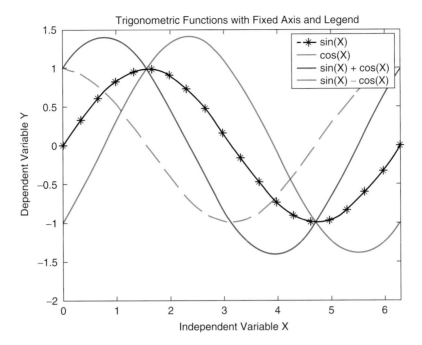

COLOR FIGURE 5.68
Trigonometric plots of Example 5.1 (*axis* and *legend*).

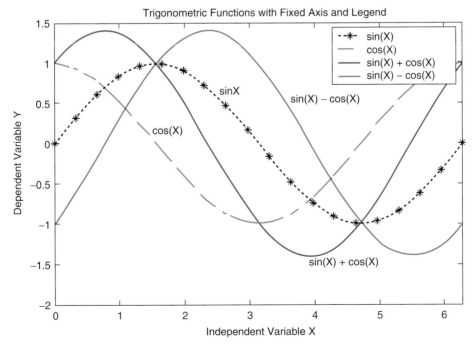

COLOR FIGURE 5.69
Trigonometric plots of Example 5.1 (*axis* and *legend*).

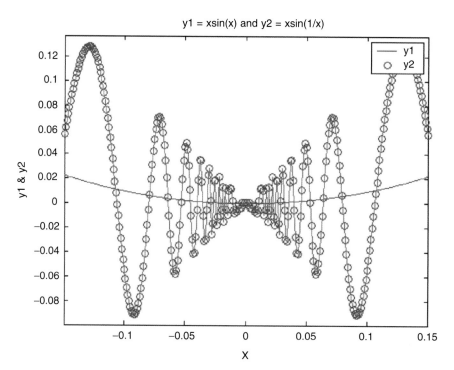

COLOR FIGURE 5.84
Numerical plot of Example 5.8 over $-0.15 \leq x \leq 0.15$.

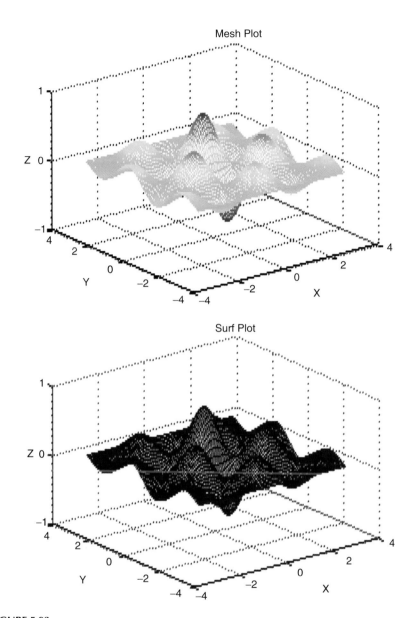

COLOR FIGURE 5.93
3-D *mesh* and *surf* plots of Example 5.15.

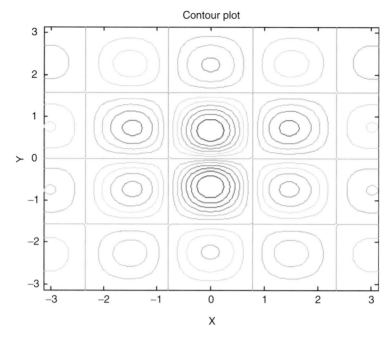

COLOR FIGURE 5.94
Contour plot of Example 5.16.

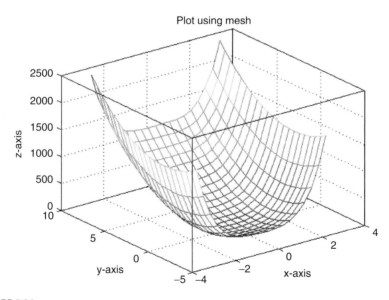

COLOR FIGURE 5.96
3-D plot using *mesh* of Example 5.17.

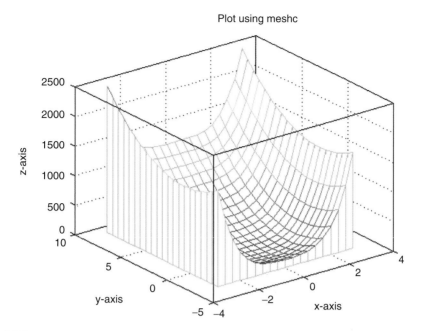

COLOR FIGURE 5.97
3-D plot using *meshc* of Example 5.17.

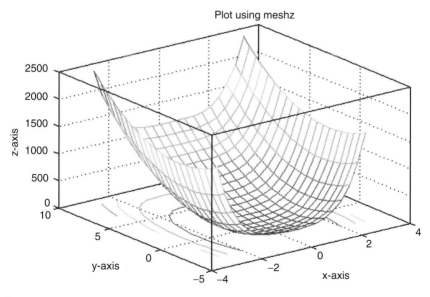

COLOR FIGURE 5.98
3-D plot using *meshz* of Example 5.17.

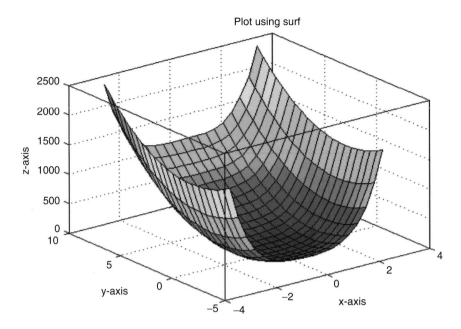

COLOR FIGURE 5.99
3-D plot using *surf* of Example 5.17.

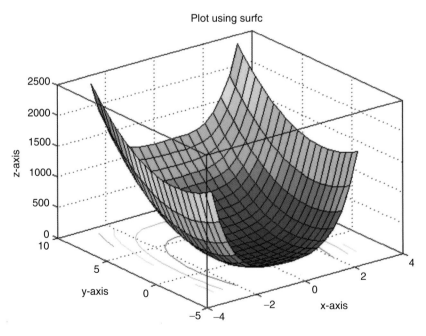

COLOR FIGURE 5.100
3-D plot using *surfc* of Example 5.17.

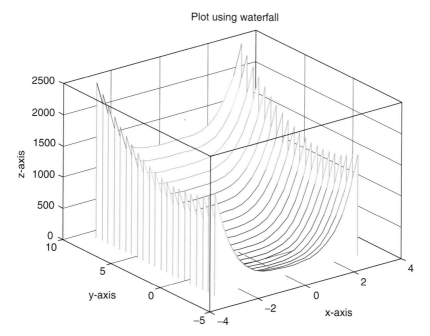

COLOR FIGURE 5.101
3-D plot using *waterfall* of Example 5.17.

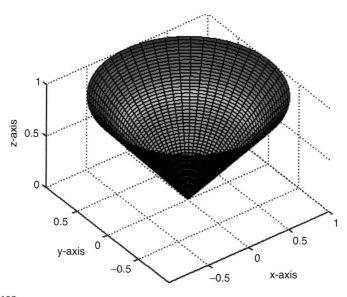

COLOR FIGURE 5.102
3-D plot of the cone of Example 5.18.

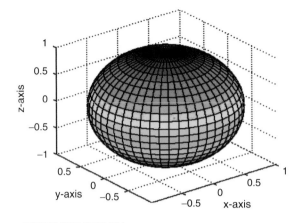

COLOR FIGURE 5.104
3-D plot of the sphere of Example 5.18.

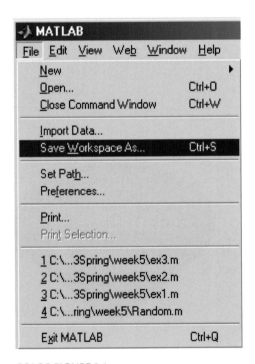

COLOR FIGURE 9.1
File menu.

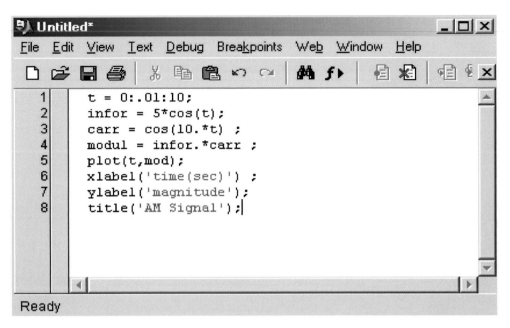

COLOR FIGURE 9.2
The edit window with the program just entered R.9.69.

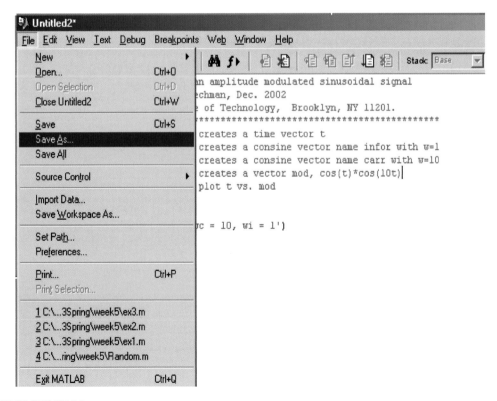

COLOR FIGURE 9.3
Saving the file.

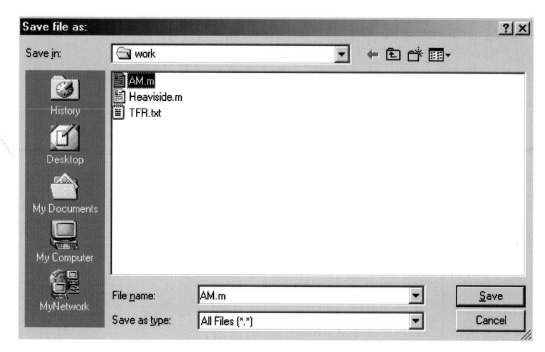

COLOR FIGURE 9.4
Saving the script file as *AM.m*.

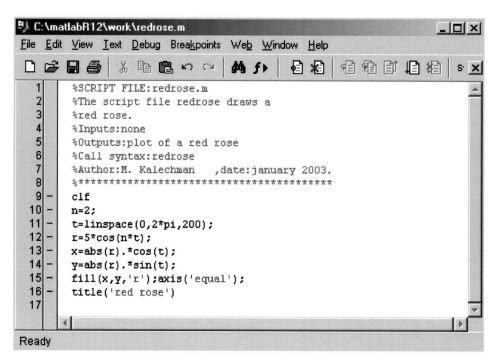

COLOR FIGURE 9.13
Edit window with the script file *redrose*.

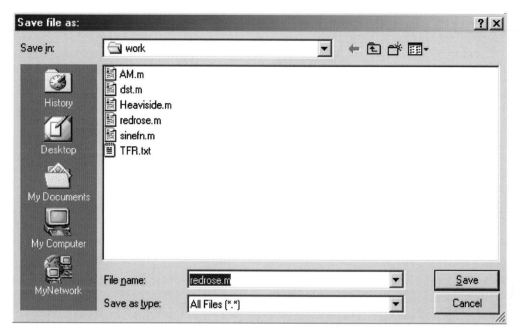

COLOR FIGURE 9.14
The script file *redrose* is stored in the folder *work*.

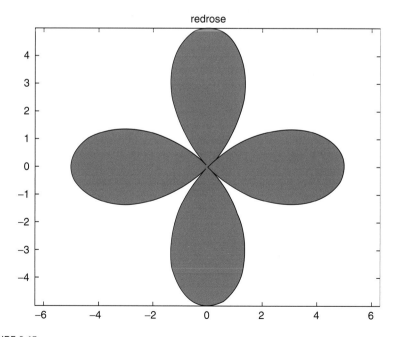

COLOR FIGURE 9.15
Plot of *redrose* of Example 9.1.

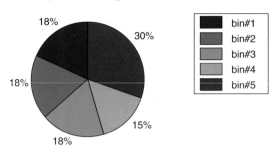

COLOR FIGURE 9.25
Pie plots of x_1 of Example 9.13.

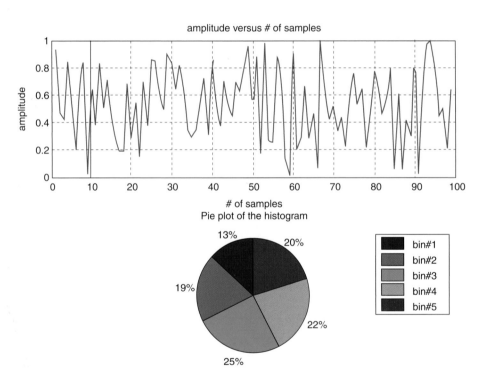

COLOR FIGURE 9.27
Pie plots for x_2 of Example 9.13.

5.6 Application Problems

P.5.1 Plot each of the points defined by the following Cartesian coordinates by hand and by using MATLAB: $<-3, 8>$, $<2, -9>$, $<-4.2, -10.3>$, and $<3, 3.3>$.

P.5.2 Let $x = [0:10\ 10:\ -1:0]$. Execute the command *plot (x)* and observe the resulting figure.

P.5.3 Duplicate the plot of P.5.2 by using the command *plot(x, y)*.

P.5.4 Let $x = [2:2:20]$ and $y = x.^2$. Execute the command *plot (x, y)* and observe the returning plot.

P.5.5 Replace the program of P.5.4 (consisting of three statements) using one command.

P.5.6 Write a script that returns the following plots over the domain $0 \le x \le 2\pi$ using 20 linearly spaced points:

1. $[y_1(x) = sin^2(x)]$ versus x
2. $[y_2(x) = cos^2(x)]$ versus x
3. $[y_3(x) = sin^2(x) + cos^2(x)]$ versus x

P.5.7 Rerun the programs of P.5.6 that returns the discrete plots using triangular markers.

P.5.8 Verify graphically that $sin(2x) = 2\ sin(x)\ cos(x)$ over the range $0 \le x \le 2\pi$ using 20 linearly spaced points.

P.5.9 Let $a = (n - 1)!$ and $b = (2\pi)^{0.5}\ n^{n-0.5}\ e^{-n}$. Write a program that returns the plots of a versus n and b versus n, and determine for what range of n, a constitutes a good approximation of b. This approximation is referred to as the Stirling's formula.

P.5.10 Solve the following equation graphically: $cos(2x) = cos^2(x) - sin^2(x)$ over the range $0 \le x \le 2\pi$. (Hint: plot $y_1(x) = cos(2x)$ and $y_2(x) = cos^2(x) - sin^2(x)$ and the intersection of $y_1(x)$ with $y_2(x)$ are the possible solutions.)

P.5.11 Estimate the solution for the following equation graphically: $sin(x) = cos(x)$ over the range $0 \le x \le 2\pi$. Use overlay plots.

P.5.12 Repeat problem P.5.11 for the following case: $sin(x) = 3\ cos(2x)$ over the range $0 \le x \le 2\pi$.

P.5.13 Estimate graphically the solution for the following equation: $5\ sin(t) = t$, for all t.

P.5.14 Write a program that returns the plot of the function $y = x^2$ over the domain $1 < x < 100$ using the following scales:

a. *Linear*

b. *Linear-log*

c. *Log-linear*

d. *Log-log*

P.5.15 Create a MATLAB program that returns the plot of a circle of radius $r = 2$, centered at $<x = 0, y = 0>$ using 50 linearly spaced points.

P.5.16 The equation of a circle in Cartesian coordinates centered at (x_0, y_0) is given by

$$x = x_0 + r\ cos(\theta) \quad \text{and} \quad y = y_0 + r\ sin(\theta)$$

a. Write a MATLAB script that returns a circle centered at $<1, 1>$ with a radius of 2.

b. On the same plot, add another circle centered at $<-2, 1>$ with a radius of 3 using the *hold on* command, and set the y-axis to be twice the size of the x-axis.

P.5.17 Create a MATLAB program that returns a table and the corresponding plot that relates degree Celsius (°C) to degree Fahrenheit (°F) by using the following conversion equation:

$$F = \frac{9}{5}C + 32$$

where F is in °F (degree Fahrenheit) and C is in °C (degree Celsius).

P.5.18 Plot the function $y(x) = -3x^2 + 6x + 1$ over the domain $-2 \le x \le 3$ and evaluate the following:

a. The y and the x intercepts (by letting $x = 0$ and solving for y, and by letting $y = 0$ and solving for x) by hand

b. Its maximum and minimum by hand

c. Repeat parts a and b using MATLAB

P.5.19 Repeat problem P.5.18 for the following function:

$$y = \frac{x}{(x-1)(x+1)}$$

over the domain $-1 \le x \le 1$.

P.5.20 Write a MATLAB script file that returns the plot of the following function:

$$y = (x - 2)^3$$

over the domain $8 \le x \le 2$.

P.5.21 Verify the following relations graphically over the range $0 \le x \le 2\pi$:

a. $(sin(3x) - sin(x))/(cos(3x) + cos(x)) = -2$

b. $cos(x) - cos^2(x/2) = -1$

c. $cot(x/2)\, tan(x/2) = 2\, csc(x)$

P.5.22 Solve the following equations over the domain $0 \le x \le 2\pi$:

a. $1 + cos(x) = 2\, cos(x/2)$

b. $cos(x) - tan(x) - sec(x) = 0$

c. $(sin(x)cot(x))/(cos(x)) = sin(x)$

d. $1 + tan(x) = 5\, sin(x)$

e. $sin^2(2x) - sin^2(x) = 1/2$

P.5.23 Solve the following systems of equations graphically:

a. $3x + 2y = 5$
   $-4x + 5y = 13$

b. $4y + 3x = 11$
   $-y + 3x = 1$

c. $x^2 + y^2 = 16$
   $4x - 3y = 0$

d. $y = 2x = -3$
   $x^2 + 2xy = -1$

e. $y = x$

$y = 5 \sin(2x) + 4$

f. $2cos^2(x) + cos(x) - 1 = 0$

P.5.24 Create a MATLAB script file that returns the plot of each equation in polar coordinates over the range $0 \leq \theta \leq 2\pi$.

a. $r_1 = 3/(1 - cos(\theta))$ (parabola)

b. $r_2 = 1/(4 - 3 cos(\theta))$ (ellipse)

c. $r_3 = 3/(2 - 3 cos(\theta))$ (hyperbola)

d. $r_4 = 3\sqrt{sin(2\theta)}$ (lemniscate equation—propeller shape)

e. $r_5 = 3e^{0.3}\theta$ (spiral equation)

f. $r_6 = 2 + 3 cos(\theta)$ (limaçons equation)

g. $r_7 = 3 sin(3\theta)$ (rose equation)

h. $r_8 = 3 cos(3\theta)$ (rose equation)

i. $r_9 = 3 + 3 cos(\theta)$ (cardioid's equation)

j. $r_{10} = 3\theta$ (Archimedean spiral equation)

P.5.25 Verify that the plot r versus θ over $0 \leq \theta \leq 2\pi$ using MATLAB returns the following:

a. A four-petal rose with petals at 0°, 90°, 180°, and 270°, If $r = 3 cos(2\theta)$

b. A four-petal rose with petals at 45°, 135°, −135°, and −45°, where $r = 3 sin(2\theta)$

c. The limniscate equation along the 0° and 180° diagonal, where $r = 3\sqrt{cos(2\theta)}$

d. The limniscate equation with the propellers at 45° and 225°, where $r = 3\sqrt{sin(2\theta)}$

P.5.26 Polar-to-rectangular conversion is given by the following equations:

$$x = r \, cos(\theta), \, y = r \, sin(\theta)$$

Rectangular-to-polar conversion is given by the following equations:

$$R = \pm\sqrt{x^2 + y^2}$$

$$sin \, \theta = \frac{\pm y}{\sqrt{x^2 + y^2}}$$

$$cos \, \theta = \frac{\pm x}{\sqrt{x^2 + y^2}}$$

$$tan \, \theta = \frac{\pm y}{x}$$

Plot the following equations in polar and rectangular coordinates:

a. $x^2 + y^2 = 16$ *(circle)*

b. $(x/16) + (y/4) = 1$ *(ellipse)*

c. $(x/16) - (y/4) = 1$ *(hyperbola)*

d. $r = 1/(4 - cos(\theta))$ *(ellipse)*

e. $r = 3/(1 - cos(\theta))$ *(parabola)*

P.5.27 The equation of a circle of radius r centered at the Cartesian coordinates $<x_0, y_0>$ is given by $(x - x_0)^2 + (y - y_0) = r^2$. Write a MATLAB script that returns the following plots:

a. A circle of radius 5, centered at $<1, -1>$

b. A circle of radius 4, centered at $<-1, 1>$

P.5.28 Write a program that returns a black circle of radius $r = 1$, centered at $<1, 1>$ (use the *fill* instruction).

P.5.29 Create the script and function files that returns a red circle, given the radius and the Cartesian coordinates of its center.

P.5.30 The following data represent the student enrollment in a given college per year:

Number of Students	311	413	503	562	651
Year	1993	1996	1999	2002	2006

Use MATLAB to construct *linear, stairs, bar,* and *stem* plots showing the enrollment trend.

P.5.31 The annual sales of a company are given as follows:

Sale in Thousands ($)	313	423	673	832	931
Year	1993	1996	1999	2002	2006

Use MATLAB and construct *linear, bar, stem,* and *stairs* graphs.

P.5.32 Write a program that returns the 3-D plot and xy, xz, and yz projections of the resulting figure for the following equation:

$$z = \frac{15}{3 + (x - 1)^2 + (y - 1)^2}$$

over the ranges $-5 \leq x \leq 5$ and $-5 \leq y \leq 5$.

P.5.33 Write a program that returns the 3-D mesh plots for the following functions:

a. $z = (-sin(2x^2 + 2y^2)^{1/2})/\sqrt{2x^2 + 2y^2}$ over the ranges $-10 \leq x \leq 10$ and $-10 \leq y \leq 10$

b. $z = (x - 1)^2 + (y - 2)^2 + xy$ over the ranges $-2 \leq x \leq 2$ and $-2 \leq y \leq 2$

c. $z = xe^{-\sqrt{(x^2-y^2)^2+x^2}}$ over the ranges $-3 \leq x \leq 3$ and $-3 \leq y \leq 3$

d. $z = sin(x) \cos xe^{((x-1)^2+(y-1)^2)}$ over the ranges $-10 \leq x \leq 10$ and $-10 \leq y \leq 10$

P.5.34 Write a program that returns the 3-D *meshgrid* and *mesh* plots for the following function:

$$z = \frac{2xy}{\sqrt{x^2 + y^2}}$$

over the ranges $1 \leq x \leq 4$ and $1 \leq y \leq 4$.

P.5.35 Write a program that returns the following plots using the commands indicated in the following for the function defined in P.5.34.

a. *waterfall*

b. *contour*

    c. *contour3*

    d. *plot3*

    e. *surf*

P.5.36 Write a program that returns the 3-D *contour* and *surface* plot of the following functions:

a. $z = (x - 1)^2 + (y - 2)^2 + xy$

b. $z = sin(x) \cdot sin(y) \cdot exp\left(\left(-\sqrt{2x^2 + 3y^2}\right)/5\right)$

c. $z = 3.5e^{-x^2} \cdot e^{-y^2}$ over the ranges $-5 \le x \le 5$ and $-5 \le y \le 5$

6

Complex Numbers

Imagination is more important than knowledge.

Albert Einstein

6.1 Introduction

The real number system consists of rational and irrational numbers that can be represented on a straight line called the real number line. Square roots of nonnegative real numbers may be represented on the real number line system, but not the square roots of negative numbers. Despite the fact that no negative number has a square root on the real number line system, it is still possible to develop algebraic expressions that contain such square roots.

There is no real number that multiplied by itself equals to -1, therefore the relation $(-1)^{1/2}$ has no real solution, and a simple equation such as $x = (-1)^{1/2}$ cannot be solved.

To find solutions to such equations the theory of complex numbers was developed. The square root of a negative real number is a pure imaginary number represented by the imaginary symbol i or j, where $\sqrt{-1} = j = i$. Then, the square root of any negative real number can be expressed in terms of i. Since i is used to denote electrical current, many technical books use j instead, to avoid confusion. For example, $\sqrt{-9} = \sqrt{9} \cdot \sqrt{-1} = 3j$. Imaginary numbers can be represented by a straight line called the imaginary number line. In general, a complex number z can be represented by two parts: a real and an imaginary.

A complex number can be represented as an ordered pair $z = (x, y)$, or more general as $z = x + jy$, where x and y are real numbers. The second form of writing complex number ($z = x + jy$) is more convenient to manipulate them in a computational environment. A complex number then can be represented in the real/imaginary coordinate system called the complex plane as a point, as shown in Figure 6.1. Thus, each point on the plane represents a complex number and conversely, each complex number represents a point on the plane. The collection of all these points constitutes the complex plane.

The complex plane consists of a Cartesian rectangular axes system, where the x-axis (horizontal) of the complex plane is referred as the real axis, where the real part is represented, whereas the y-axis (vertical) is referred as the imaginary axis, where the imaginary part is represented (with i as its unit). Observe that the real numbers are just a subset of the complex numbers, when the imaginary part is equal to zero. Much of modern mathematics is based on complex numbers, and they are used extensively in science and engineering. For example, in electrical circuit theory, when dealing with impedances, the real axis is generally associated with the resistance, whereas the imaginary axis is referred as the reactance axis.

Most standard MATLAB® algebraic manipulations defined for real numbers work with complex numbers. There are a few exceptions between real and complex numbers, such as

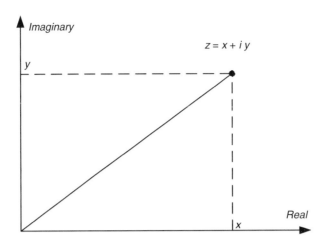

FIGURE 6.1
Complex plane.

the concept of equivalence. For the case of complex numbers, the only equivalent relation is the identity. Other relations such as greater than and smaller than have no meaning when dealing with complex numbers.

For example, two complex numbers $z_1 = x_1 + jy_1$ and $z_2 = x_2 + jy_2$ are equal, if and only if, $x_1 = x_2$ and $y_1 = y_2$. The concept of one complex number being greater than, or smaller than, another is meaningless.

The nature of complex numbers can best be illustrated and visualized by analyzing the following set of quadratic equations:

1. $y_1 = x^2 - 6x - 7$
2. $y_2 = x^2 - 6x + 9$
3. $y_3 = x^2 - 6x + 10$

(Observe that only the constant terms have been changed.)

Each one of the preceding equations has been solved graphically (Figure 6.2), and also numerically by the following program (by observing the respective plots and evaluating the respective roots of y_1, y_2, and y_3):

```
MATLAB Solution
>> X= [-2:0.05: 8];          % generates a 201 elements array X
>> Y1 = X.^2-6*X-7;          % generates a 201 elements array Y1
>> Y2 = X.^2-6*X+9;          % generates a 201 elements array Y2
>> Y3 = X.^2-6*X+10;         % generates a 201 elements array Y3
>> subplot(3,1,1)
>> plot (X,Y1)               % plots Y1 vs X
>> axis on, grid on
>> xlabel('X'), ylabel('Y1')
>> subplot(3,1,2)
>> plot (X,Y2)               % plots Y2 vs X
>> axis on, grid on
>> xlabel('X'), ylabel('Y2')
>> subplot(3,1,3)
>> plot (X,Y3)               % plots Y3 vs X
```

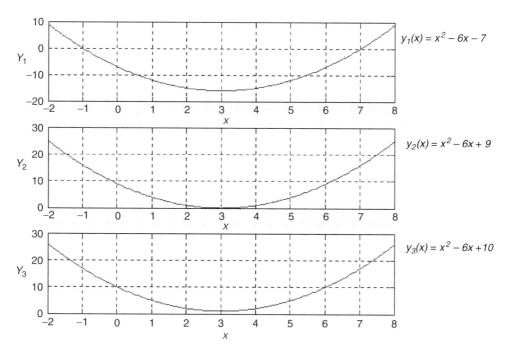

FIGURE 6.2
Plots of $y_1(x)$ versus x, $y_2(x)$ versus x, and $y_3(x)$ versus x.

```
>> axis on, grid on
>> xlabel('X'), ylabel('Y3')
>> P1 = [1 -6 -7];            % array of coefficients of Y1.
>> P2 = [1 -6 9];             % array of coefficients of Y2.
>> P3 = [1 -6 10];           % array of coefficients of Y3.
>> disp('The roots of Y1 are:');
>> roots(P1)                 % returns the roots of Y1

    The roots of y1(x) are:
        ans =
              7
             -1

>> disp('The roots of y2(x) are:');
>> roots(P2)                 % returns the roots of Y2

    The roots of y2(x) are:
        ans =
              3.0000 + 0.0000i
              3.0000 - 0.0000i

>> disp('The roots of y3(x) are:');
>> roots(P3)                 % returns the roots of Y3

    The roots of y3(x) are:
        ans =
              3.0000 + 1.0000i
              3.0000 - 1.0000i
```

From the plots shown in Figure 6.2 (and the preceding program), it can be seen that the roots of $y_1(x)$, $\{y_1(x) = 0 = x^2 - 6x - 7\}$ are $+7$ and -1. The roots of $y_2(x)$ are 3 and 3 (repeated roots); but the roots for $y_3(x)$ do not have a real solution because $y_3(x)$ does not intersect the x-axis. The roots of $y_3(x)$ can be evaluated analytically by using the standard quadratic formula

$$x_{1,2} = \left(\frac{-b \pm \sqrt{b^2 - 4ac}}{2a} \right)$$

Substituting the coefficients of $y_3(x)$ ($a = 1$, $b = -6$, and $c = 10$) in the preceding formula yields

$$x_{1,2} = \frac{6 \pm \sqrt{36 - 4(1)(10)}}{2}$$

$$x_{1,2} = \frac{6 \pm \sqrt{36 - 40}}{2}$$

$$x_{1,2} = \frac{6}{2} \pm \frac{\sqrt{-4}}{2}$$

$$x_1 = 3 + \sqrt{-1}$$

$$x_2 = 3 - \sqrt{-1}$$

Clearly, the roots of y_3 $(= 0)$ are complex.

Let us verify the solutions obtained by substituting $x_{1,2} = 3 \pm \sqrt{-1}$ in (equation) $y_3(x)$, yielding

$$(3 + i)^2 - 6(3 + i) + 10 = 9 + 6i - 1 - 18 - 6i + 10 = 0, \text{ and}$$

$$(3 - i)^2 - 6(3 - i) + 10 = 9 - 6i - 1 - 18 - 6i + 10 = 0$$

Clearly, the preceding substitutions indicate that indeed $3 \pm \sqrt{-1}$, or $3 \pm i$, are the roots of $y_3(x)$ and are complex.

Operations involving complex numbers, as well as complex functions, are extensively used in AC circuits (see Chapter 3 of the book entitled *Practical MATLAB® Applications for Engineers*), and in circuit analysis when using the Fourier or the Laplace operators (see Chapter 4 of the book entitled *Practical MATLAB® Applications for Engineers*). Owing to the importance of complex numbers, and complex functions in engineering and science, this chapter is dedicated to the algebra of complex number, manipulations, properties, and some applications (such as phasors). One of the nicest things about MATLAB is that complex numbers are treated in the same way as real numbers.

6.1.1 A Brief History

In older civilizations with strong foundations in phylosophy, logic and mathematics such as the Greeks, Arabs, and Babylonians, complex numbers did not exist or had no practical or physical meaning.

The theory of complex numbers was first seriously studied during the sixteenth century by two Italian mathematicians

Raffaele Bombelli (1501–1576)

Ferrari (1499–1557)

Bombelli called these new numbers *sophistic*. For centuries thereafter, mathematicians worked with complex numbers but without believing in their existence, or practical application. In later centuries starting around the 1600s, prominent figures such as *Leibnitz*, *Newton*, and *Bernoulli* saw the value of complex numbers as a means to explain their theories. The symbol *i*, which represents $\sqrt{-1}$, was first introduced by *Leonhard Euler* in 1777. The complex notation using *i* was first employed in 1777 by *Caspar Wessel* (1745–1818) and adopted in 1787 by *Jean Robert Ardon* (1787–1822). This representation became known for obvious reason as the *Ardon* representation or diagram. *Rene Descaters* (1637) first introduced formally the terms *real* and *imaginary*, and *Carl Frederick Gauss* (1777–1855) first used the terms *complex and Gaussian plane*; the modern terminology presently employed by engineers and scientists.

The complex numbers are the foundation of the theory of complex variables, first introduced by *Leonhard Euler* (1707–1783), who was born in Basel, Switzerland. If *Newton* is the greatest scientist of all times, then *Euler* is the greatest mathematician of all times. *Euler* published a total of 886 books and mathematical memoirs. This is an amazing accomplishment, even more amazing in light of the fact that *Euler* lost his vision in one eye in 1735, and became totally blind by 1771.

Like *Newton* and *Einstein*, *Euler* was not particularly brilliant as a child. At the University of Basel he studied theology, Greek, Latin, Hebrew (he was already fluent in French and German), physics, astronomy, medicine, and in particular mathematics, taught by a brilliant mathematician named *Jean Bernoulli* (1667–1748). The *Bernoullis* are considered the most distinguished family in the history of mathematics (Nicolaus III, Daniel I, Jean II, etc.).

Euler's work was extensive and brilliant, fundamental, and original, in particular, in the areas of calculus of variations and the theory of complex variables. He served as the chief mathematician at the Academy of Science at St. Petersburg, capital of Russia. Years later he was invited to serve at the Academy in Berlin, by Frederick II of Prussia, where he spent 25 years. Catherine II (the Great) of Russia instructed the Russian ambassador in Prussia, to return Euler to the Academy, by offering him the title of Director of the Academy, with a salary of 3000 rubles per year. In addition, his wife was to receive a pension of 1000 rubles per year in case of his death, and employment of his three sons (in the St. Petersburg area).

Euler left his mark in different scientific areas, including differential equations, number theory, geometry, probability, astronomy, strength of materials, mechanics, and hydrodynamics. In mathematics, Euler's contributions are many, and his name is often associated with some of his many discoveries that are commonly referred as

Euler's theorem on ..., Euler theorem of ..., Euler's identities, Euler's functions, Euler's proofs, Euler's coefficients, Euler's constant, etc.

Probably Euler's greatest contribution can be summarized by the following two equations:

a. $e^{j\varphi} = cos(\varphi) + j\,sin(\varphi)$

b. $e^{\pi j} + 1 = 0$

The above equations are complex, but in particular the second equation is considered by mathematicians as the most elegant equation ever written. This equation relates e and π, 1 (one), 0 (zero), and $\sqrt{-1} = j$, the most often used mathematical constants.

Laplace best described Euler's works and accomplishments by these words

Read Euler, read Euler, he is our master in everything.

6.2   Objectives

After completing this chapter the reader should be able to

- Enter manually complex numbers using MATLAB
- Assign values to a complex variable
- Perform arithmetic calculations using complex numbers such as addition, subtraction, multiplication, division, and exponentiation
- Determine the complex conjugate of a complex number or variable
- Convert a complex number from rectangular to trigonometric, exponential, or polar forms and vice versa
- Determine the real and imaginary parts of a complex number or expression
- Obtain the magnitude and angle of a complex number
- Know, understand, and use the DeMoivre theorem
- Calculate the complex roots, and be able to predict their locations and behavior
- Express a complex exponential as a complex number
- Express a sinusoidal function as a complex exponential
- Understand the concept of phasors, and know that a phasor is a short-hand notation or representation of a complex function
- Define the complex (Gaussian) plane
- Define the different coordinate systems such as Cartesian, polar, rectangular, and spherical
- Represent a complex number as a point on the complex plane
- Represent a complex number as a vector on the complex plane
- Understand the properties of a complex variable
- Manipulate complex numbers using MATLAB
- Create complex matrices and vectors
- Determine the transpose, inverse, and conjugate of a complex matrix
- Understand the meaning and concept of the principal value
- Use MATLAB to perform algebraic manipulations involving complex numbers or functions

6.3 Background

R.6.1 Any arbitrary complex number $z = a + ib$ can be represented as the sum of a real and an imaginary part, where a is the real part and b the imaginary part.

R.6.2 MATLAB stores the complex number $z = a + ib$ as two real numbers a and b.

R.6.3 MATLAB assumes that i and j represent $\sqrt{-1}$, unless i or j had been previously assigned a different value.

R.6.4 The following examples illustrate how MATLAB responds when the preassigned values of i and j ($\sqrt{-1}$) are used.

MATLAB Solution
```
>> sqrt(-1)

    ans =
          0 + 1.0000i
>> j

    ans =
          0 + 1.0000i
>> i*j

    ans =
          -1
```

R.6.5 A complex number $z = a + ib$ can be entered using MATLAB in three different ways, indicated as follows:

1. $z = a + bi$
2. $z = a + i*b$
3. $z = complex(a, b)$

where expression (1) is always complex, and expression (2) is complex, if i has not been assigned a value. MATLAB does not recognize z as complex, unless z is explicitly declared as complex ($z = a + bi$) (expression [3]).

R.6.6 For example, use MATLAB to enter the complex number $z = 1 + 2i$, in all possible ways.

MATLAB Solution
```
>> z1 = 1+2i

    z1 =
          1.0000 + 2.0000i
>> z2 = 1+2j

    z2 =
          1.0000 + 2.0000i
>> z3 = 1+j*2

    z3 =
          1.0000 + 2.0000i
>> z4 = 1+i*2

    z4 =
          1.0000 + 2.0000i
>> z5 = complex(1,2)

    z5 =
          1.0000 + 2.0000i
```

```
>> z6 = 1+i2
```

??? Undefined function or variable 'i2'.

R.6.7   The MATLAB command $C = complex(a, b)$ returns C, where *a* and *b* may be vectors, arrays, or matrices with identical sizes.

In the event that *b* is all zeros, C is complex with zeros as the imaginary part, unlike the result of the addition $a + 0i$, which returns a strictly real result.

The MATLAB command $C = complex(A)$, where *A* is a real matrix, returns the complex matrix C, with the matrix *A* as the real part, and the imaginary part comprises zeros.

R.6.8   For example, using MATLAB create the following complex sequence:

$$1 - 2i, 3 - 4i, 5 - 6i, 7 - 8i, ..., 11 - 12i$$

MATLAB Solution
```
>> a = 1:2:11;
>> b = 2:2:12
>> sequence = (complex(a,b))'

   sequence =
             1.0000 -  2.0000i
             3.0000 -  4.0000i
             5.0000 -  6.0000i
             7.0000 -  8.0000i
             9.0000 - 10.0000i
            11.0000 - 12.0000i
```

R.6.9   It is a recommended programming practice to reserve *i* and *j* exclusively to denote the imaginary part of a complex number ($\sqrt{-1}$), and avoid using *i* or *j* to define any other variable.

R.6.10   The standard operations defined for real numbers apply equally well for complex numbers.

For example, let $z = 1 + 2i$, use MATLAB and evaluate the following expressions:

a. $C1 = 1 + z * e^z$

b. $C2 = 3 - z^2 + 1/z * log(z)$

MATLAB Solution
```
>> z =1+2i;
>> C1 =1+z*exp(z)

   C1 =
         -5.0747 + 0.2093i
>> C2 = 3-z^2+1/z*log10(z)

   C2 =
          6.2622 -  4.0436i
```

R.6.11   The standard representation of z as $z = a + ib$ is called rectangular, binomial, or Cartesian form. z can be represented graphically as a point on the complex plane with the abscissa (horizontal axis) as the real axis and the ordinate (vertical axis) as the imaginary axis, as illustrated in Figure 6.3. Note that a complex number can also be considered as a vector.

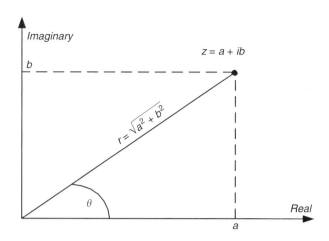

FIGURE 6.3
Plot of $z = a + ib$.

R.6.12 Clearly from Figure 6.3, $z = a + ib$, and z can also be represented as $z = r\cos(\theta) + ir\sin(\theta)$. This format of representing z is known as the trigonometric form, where

$$a = r\cos(\theta), \quad r = \sqrt{(a^2 + b^2)}$$

$$b = r\sin(\theta), \quad \text{and} \quad \theta = \tan^{-1}(b/a)$$

R.6.13 The variable r is referred as the absolute value of z ($|z|$), modulus z or magnitude of z, which represents the length of z from the origin $<x = 0, y = 0>$ to the terminal point: $<x = a$ and $y = b>$ and θ is often referred as the argument or (phase) angle.

R.6.14 Using Euler's relation, $e^{i\theta} = \cos\theta + i\sin\theta$, the trigonometric representation: $z = r\cos(\theta) + ir\sin(\theta)$ can also be expressed as $z = re^{i\theta}$, a format known as the exponential form of z.

R.6.15 The exponential form for z expressed in R.6.14 can be converted to the Steinmetz form (representing magnitude and phase). In the Steinmetz form, z is represented as $z = re^{i\theta} = r \angle \theta$. This last form is widely used in electrical circuit theory, and is referred as polar representation.

R.6.16 The conjugate of $z = a + ib$ is also a complex number represented in rectangular form as $z^* = a - ib$, if $z = a + ib$ (* denotes complex conjugate).

R.6.17 The conjugate z of the complex function expressed in exponential form as $z = re^{i\theta}$ is $z^* = re^{-i\theta}$.

R.6.18 The complex conjugate z^* of a complex number z is the image or projection of z with respect to the real axis, as indicated in Figure 6.4.

R.6.19 In summary, the complex conjugate of $z = a + ib$ can be expressed in different forms as $a - ib$, $re^{-i\theta}$, $r \angle -\theta$, or $r\cos\theta - ir\sin\theta$.

Thus, the complex conjugate of a complex number is obtained by reversing the sign of

1. The imaginary part when expressed in rectangular or trigonometric form

2. The angle in polar (Steinmetz) or exponential form

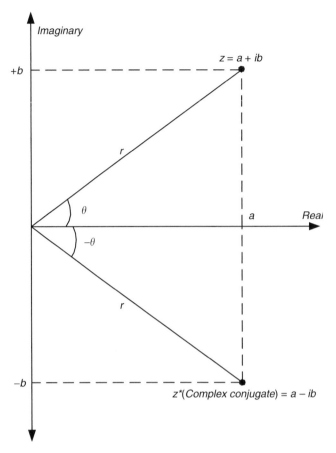

FIGURE 6.4
Plots of $z = a + ib$ and $z* = a - ib$.

R.6.20 Sinusoidal functions can be expressed as complex exponentials, by using Euler's equalities, indicated as follows:

$$sin(\theta) = \frac{e^{i\theta} - e^{-i\theta}}{2i}, \quad cos(\theta) = \frac{e^{i\theta} + e^{-i\theta}}{2}$$

R.6.21 The components of a sinusoid function can be expressed as two rotating vectors of the form e^{jwt} and e^{-jwt}, where w represents the angular rotational velocity.

R.6.22 For example, let the complex number z be given by $z = -6.7 + j\,8.43$. Transform the complex number z from rectangular to polar form.

ANALYTICAL Solution

$$z = -6.7 + 8.43j = (-6.7^2 + 8.43^2)^{1/2} \angle arc\,tan(8.3/-6.7)$$

$$z = 10.7682 \angle 128.4770° = 10.7682e^{j128.477°}$$

R.6.23 Let us explore the opposite relations. For example, express $z = 3e^{j\pi/4}$ in rectangular and polar forms.

ANALYTICAL Solution

The polar form of z is given by $z = 3 \angle [(\pi*180)/(4\pi)] = 3\angle 45°$

The rectangular form of z is $z = 3[\cos(45°) + \sin(45°)] = 2.1213 + 2.1213j = 2.1213(1 + j)$

R.6.24 Some useful and often used powers of i are presented as follows:

$$i^0 = (\sqrt{-1})^0 = 1$$

recall that $i = j$, for MATLAB

$$i^2 = (\sqrt{-1})^2 = -1$$

$$i^3 = (\sqrt{-1})^3 = (\sqrt{-1})^2 \cdot \sqrt{-1} = -j$$

$$i^4 = (\sqrt{-1})^4 = (\sqrt{-1})^2 \cdot (\sqrt{-1})^2 = (-1) \cdot (-1) = +1$$

$$i^5 = i^4 \cdot i = j$$

$$i^6 = i^4 \cdot i^2 = (1) \cdot (-1) = -1$$

$$i^7 = -i = -j$$

The powers of i are cyclic and follow the sequence $j, -1, -j, 1, j, -1, -j, 1, \ldots$ for the powers $1, 2, 3, 4, 5, 6, \ldots$.

What is $\sqrt{i}$ then?

ANALYTICAL Solution

$$\sqrt{i} = i^{1/2}$$

$$\sqrt{i} = (0 + i)^{1/2} = [\cos(\pi/2) + i\sin(\pi/2)]^{1/2}$$

$$\sqrt{i} = (1)^{1/2}[\cos(\pi/4) + i\sin(\pi/4)]$$

by the DeMoivre formula (see R.6.27 and R.6.28)

$$\sqrt{i} = \frac{\sqrt{2}}{2} + \frac{\sqrt{2}}{2}j$$

R.6.25 From R.6.24, it can be observed that any integer power of j takes one of four possible values: $j, -1, -j,$ and 1. It is useful to observe that any power of j evenly divided by four is equal to one.

For example, if $\quad i^n = 1, \qquad$ then $n = 4 \cdot k \qquad\qquad$ for $k = 1, 2, \ldots$
$\qquad\qquad\qquad\quad i^n = i, \qquad$ then $n = 4 \cdot k + 1$
$\qquad\qquad\qquad\quad i^n = -1, \qquad$ then $n = 4 \cdot k + 2$
$\qquad\qquad\qquad\quad i^n = -i, \qquad$ then $n = 4 \cdot k + 3$
$\qquad$ In general, $\quad i^n = i^R, \qquad$ where $n/4 = C + (R/4)$
$\qquad$ For example, $\quad i^{37} = (i^4)^9 i = i.$

R.6.26 The addition of two complex numbers z_1 and z_2 results in a new complex number consisting of adding the real and the imaginary parts separately of z_1 and z_2.

For example, let

$$z_i = a_1 + ib_1 \quad \text{and} \quad z_2 = a_2 + ib_2$$

then

$$z_1 + z_2 = (a_1 + a_2) + i(b_1 + b_2)$$

R.6.27 To subtract two complex numbers, z_1 from z_2, subtract the real and imaginary parts separately. For example,

$$z_1 - z_2 = (a_1 - a_2) + i(b_1 - b_2)$$

R.6.28 The addition and subtraction of two complex numbers can be performed conveniently, if both numbers are expressed in either rectangular or trigonometric form.

R.6.29 The product of two complex numbers z_1z_2, when expressed in rectangular form, is evaluated as follows:

$$z_1z_2 = (a_1 + ib_1)(a_2 + ib_2) = a_1a_2 + ia_1b_2 + ib_1a_2 + i^2b_1b_2$$

since $i^2 = (\sqrt{-1})^2 = -1$

$$z_1z_2 = (a_1a_2 - b_1b_2) + i(a_1b_2 + b_1a_2)$$

R.6.30 The product of two complex numbers z_1z_2, when expressed in trigonometric form, is shown as follows:

Let

$$z_1 = r_1 cos(\theta_1) + i r_1 sin(\theta_1)$$

and

$$z_2 = r_2 cos(\theta_2) + i r_2 sin(\theta_2)$$

then

$$z_1 \cdot z_2 = r_1 \cdot r_2 \left[cos(\theta_1 + \theta_2) + i sin(\theta_1 + \theta_2) \right]$$

or

$$z_1z_2 = r_1 \angle \theta_1 \cdot r_2 \angle \theta_2 = r_1r_2 \angle (\theta_1 + \theta_2)$$

R.6.31 The product of two complex numbers given by z_1 and z_2 can be conveniently evaluated when both numbers are expressed in exponential or polar form illustrated as follows:

$$z_1z_2 = (r_1e^{j\theta_1})(r_2e^{j\theta_2}) = r_1r_2e^{j(\theta_1+\theta_2)} = (r_1r_2)\left[cos(\theta_1 + \theta_2) + j sin(\theta_1 + \theta_2) \right]$$

R.6.32 The division of two complex numbers, z_1/z_2, expressed in rectangular form is accomplished by multiplying the numerator and denominator by the complex conjugate of the denominator (z_2), and simplifying the remaining expression.

The algebraic manipulations are indicated as follows:

$$\frac{z_1}{z_2} = \frac{a_1 + ib_1}{a_2 + ib_2} = \frac{(a_1 + ib_1)(a_2 - ib_2)}{(a_2 + ib_2)(a_2 - ib_2)}$$

$$\frac{z_1}{z_2} = \frac{(a_1 a_2 + b_1 b_2) + i(a_2 b_1 - b_2 a_1)}{a_2^2 + b_2^2}$$

$$\frac{z_1}{z_2} = \frac{(a_1 a_2 + b_1 b_2)}{a_2^2 + b_2^2} + \frac{i(a_2 b_1 - b_2 a_1)}{a_2^2 + b_2^2} \quad \text{(expressed in rectangular form)}$$

R.6.33 The division of two complex numbers, z_1/z_2, given in trigonometric form is evaluated as follows:

Let

$$z_1 = r_1 \cos(\theta_1) + i r_1 \sin(\theta_1)$$

and

$$z_2 = r_2 \cos(\theta_2) + i r_2 \sin(\theta_2)$$

then

$$z_1/z_2 = [r_1/r_2] [\cos(\theta_1 - \theta_2) + i \sin(\theta_1 - \theta_2)]$$

R.6.34 The division of two complex numbers, z_1/z_2, can be conveniently evaluated when both complex numbers are expressed in exponential or polar form indicated as follows:

$$\frac{z_1}{z_2} = \frac{r_1 e^{j\theta_1}}{r_2 e^{j\theta_2}} = \left(\frac{r_1}{r_2}\right) e^{j(\theta_1 - \theta_2)} = \left(\frac{r_1}{r_2}\right) \cdot \left[\cos(\theta_1 - \theta_2) + j \sin(\theta_1 - \theta_2)\right]$$

$$\frac{z_1}{z_2} = \frac{r_1 \angle \theta_1}{r_2 \angle \theta_2} = \left(\frac{r_1}{r_2}\right) \angle (\theta_1 - \theta_2)$$

R.6.35 The reciprocal of a complex number z is by definition 1 divided by z $(1/z = 1/re^{j\theta})$.

R.6.36 Since any complex number can be expressed in exponential form as $z = re^{i\theta}$ or $z = re^{(j\theta + 2\pi m)}$, hence $\sqrt[m]{z} = \sqrt[m]{r} \cdot e^{(i\theta + 2\pi \eta)/m}$. The m roots of a complex number can be obtained by assigning to n the values $n = 0, 1, 2, 3, \ldots, m - 1$, successively in the following equation:

$$z_m = \sqrt[m]{r} \left[\cos\left(\frac{\theta}{m} + \frac{2n\pi}{m}\right) + i \sin\left(\frac{\theta}{m} + \frac{2n\pi}{m}\right) \right]$$

where

$$\theta = \arctan\left(\frac{b}{a}\right), \quad \text{for } a > 0,$$

or

$$\theta = \arctan\left(\frac{b}{a} + \pi\right), \quad \text{for } a < 0$$

R.6.37 The DeMoivre* formula is employed when a complex number z is raised to a power n, where $n \geq 1$, illustrated as follows:

Let

$$z = |r| \cdot (\cos\theta + j\sin\theta)$$

then

$$z^n = |r|^n \cdot (\cos\theta + j\sin\theta)^n$$

and

$$z^n = |r|^n \cdot (\cos(n\theta) + j\sin(n\theta))$$

which is called the DeMoivre theorem.

The DeMoivre theorem was published in 1730 and was named after Abraham DeMoivre, but DeMoivre relations were known by many mathematicians as early as 1710.

For example, to evaluate the three cubic roots of 27 the following equation is used: $x = 27^{1/3}$ or $x^3 = 27 + j0$. Since $27^{1/3} = 3$, the three roots are located on a circle of radius 3, centered at the origin of the complex plane, and they are

a. $root\#1 = 3$

b. $root\#2 = 3[\cos(2\pi/3) + j\sin(2\pi/3)]$

c. $root\#3 = 3[\cos(-2\pi/3) + j\sin(-2\pi/3)]$

R.6.38 MATLAB returns only the principal value of the nth roots of a given complex number. The polynomial form illustrated in R.6.39 can be used to evaluate all the roots.

R.6.39 For example, evaluate using MATLAB the principal root as well as all the roots of $x = (-1)^{1/5}$, and verify the results obtained.

ANALYTICAL Solution

The 5th root of $x = (-1)^{1/5}$ can be evaluated by solving the following equation: $x^5 + 1 = 0$. This equation is converted into a polynomial MATLAB vector as $X = [1\ 0\ 0\ 0\ 0\ 1]$, and the roots are evaluated, indicated as follows (see Chapter 7 for more details about polynomials):

---

* Abraham DeMoivre (1667–1754) was a French Protestant exiled in London where he hoped to become a college professor. He was a friend of Isaac Newton and became a member of the *Royal Society of London*. He supported himself by solving problems related to games of chance and betting strategies. It is believed that he calculated the day of his own death. He died at the age of 87.

MATLAB Solution
```
>> x = (-1)^(1/5)              % MATLAB returns only the principal value

   x =
        0.8090 + 0.5878i
>> y = x^5                     % verifies the value of x

   y =
       -1.0000
>> X= [1 0 0 0 0 1];           % MATLAB polynomial vector for x^5+1 = 0
>> roots _ are = roots(X)      % evaluates the  five roots of -1

   roots _ are =
                -1.0000
                -0.3090 + 0.9511i
                -0.3090 - 0.9511i
                 0.8090 + 0.5878i
                 0.8090 - 0.5878i

>> roots _ 1 = roots _ are    % converts from column to a row vector

   roots _ 1 =
             -1.0000   -0.3090 - 0.9511i   -0.3090 + 0.9511i   0.8090 - 0.5878i
             0.8090 + 0.5878i

>> Checks _ results = roots _ 1.^5    % verifies the five roots

   Checks _ results =
                        -1.0000   -1.0000 + 0.0000i   -1.0000 - 0.0000i
                        -1.0000 - 0.0000i   -1.0000 + 0.0000i
```

R.6.40 The m roots of a complex number are cyclic in nature, and when graphed on the complex plane, the m roots of a complex number are equally spaced around a circle with radius $r^{1/m}$, and centered at the origin.

Whenever one root of a complex number is known, all the m roots can be evaluated and plotted on the complex plane. They are all located on the circle mentioned earlier and separated into m arcs of equal length.

R.6.41 The natural logarithm of a complex number can be evaluated, indicated as follows:
let

$$z = re^{i\theta} = re^{i(\theta+2\pi\eta)},$$

then

$$ln(z) = ln(re^{i\theta}) = ln(re^{i(\theta+2\pi\eta)}) = ln(r) + i(\theta + 2\pi n)$$

for any integer n.

The natural logarithm of a complex number is not unique and its principal value occurs at $n = 0$. The principal value is then given by

$$ln(z) = ln(r) + i\theta$$

R.6.42 For example, use MATLAB to evaluate $y = log(3 + 4i)$, and verify the result obtained.

MATLAB Solution
```
>> y = log(3+4i)

    y =
        1.6094 + 0.9273i

>> z = exp(y)                    % verify the result (value of y)

    z =
        3.0000 + 4.0000i
```

R.6.43 Some useful relations of the complex number $z = a + ib$, when expressed in rectangular form, are stated as follows:

$$z + z^* = 2a$$

$$z - z^* = 2ib$$

$$z \cdot z^* = a^2 + b^2 = |z|^2 = |z^*|^2$$

(recall that * denotes complex conjugate).

R.6.44 Some useful relations for the case of two complex numbers z_1 and z_2, when expressed in rectangular form, are stated as follows:

let

$$z_1 = a_1 + ib_1$$

and

$$z_2 = a_2 + ib_2$$

then

$$real(z_1) + real(z_2) = real(z_1 + z_2)$$

$$k\,real(z_1) = real(k\,z_1)$$

where k is a real number.

$$d/dt[real(z_1)] = real[d/dt(z_1)]$$

where d/dt means the derivative with respect to t.*

$$\int[real(z_1)]\,dt = real[\int(z_1)\,dt]$$

where $\int[\ ]\,dt$ means the integral with respect to t.

$$(z_1 + z_2)^* = z_1^* + z_2^*$$

(recall that * denotes complex conjugate).

$$(z_1 z_2)^* = z_1^* \cdot z_2^*$$

$$|z_1 + z_2| \le |z_1| + |z_2| \quad \text{(triangle inequality)}$$

$$\angle(z_1 \cdot z_2) = \angle z_1 + \angle z_2$$

$$\angle\left(\frac{z_1}{z_2}\right) = \angle(z_1) - \angle(z_2)$$

---

* See Chapter 7 for information about derivatives and integrals.

R.6.45 For example, use MATLAB to verify the following equation:

$$(z_1 z_2)^* = z_1^* \cdot z_2^*$$

(one of the equalities of R.6.44).

The preceding equality states that the complex conjugate of a product equals the products of the conjugates. The following MATLAB script file *conjugate prod* verifies that $(z_1 z_2)^* = z_1^* \cdot z_2^*$, for the following arbitrary complex numbers:

$$z_1 = 1 + 2i \quad \text{and} \quad z_2 = 3 + 4i$$

MATLAB Solution
```
%Script file: Conjugateprod
z1 = 1+2i;
z2 = 3+4i;
conj _ prodz1z2 = conj(z1*z2);
conj _ z1 = conj(z1);
conj _ z2 = conj(z2);
prod _ conjz1z2=conj _ z1*conj _ z2;
disp('************** RESULTS *************');
disp('***          conj(z1*z2) is :        ***');
conj _ prodz1z2
disp('***        conj(z1)*conj(z2) is :     ***');
prod _ conjz1z2
disp('************************************');
```

The script file: *Conjugateprod* is executed, and the results are shown as follows:
```
>> Conjugateprod
        **************** RESULTS ***************
        ***            conj(z1*z2) is :         ***
           conj _ prodz1z2 =
                        -5.0000 -10.0000i
        ***         conj(z1) *conj(z2) is :    ***
           prod _ conjz1z2 =
                        -5.0000 -10.0000i
        *****************************************
```

The results obtained clearly indicate that $(z_1 z_2)^*$ is indeed equal to $(z_1^* \cdot z_2^*)$.

R.6.46 Use MATLAB to verify the following relation:

$$(z_1 + z_2)^* = z_1^* + z_2^*$$

(one of the equalities of R.6.44).

The preceding equality states that the complex conjugate of a sum equals the sum of the conjugates. The following MATLAB script file verifies *conjugatesum* $(z_1 + z_2)^* = z_1^* + z_2^*$ for the same arbitrary complex numbers: given by $z_1 = 1 + 2i$ and $z_2 = 3 + 4i$.

MATLAB Solution
```
%Script file: Conjugatesum
z1 = 1+2i; z2 = 3+4i;
conj _ sumz1z2 = conj(z1+z2);
conj _ z1 = conj(z1); conj _ z2= conj(z2);
```

```
sum _ conz1z2 = conj _ z1 + conj _ z2;
disp('***************RESULTS***************');
disp('***        conj(z1+z2) is :       ***');
conj _ sumz1z2
disp('***  conj(z1)+conj(z2) is :  ***');
sum _ conz1z2
disp('************************************');
```

The script file *Conjugatesum* is executed below, and the results clearly indicate that indeed $(z_1 + z_2)^*$ is equal to $(z_1^* + z_2^*)$.

```
>> Conjugatesum

    ***************RESULTS*************
    ***          conj(z1+z2) is :         ***
      conj _ sumz1z2 =
                    4.0000 -  6.0000i
    ***     conj(z1)+conj(z2) is : ***
      sum _ conz1z2 =
                    4.0000 -  6.0000i
    **********************************
```

R.6.47 A complex function is a function whose argument (the independent variable) is a complex variable. For example,

$$f(z) = \frac{z^2 + 1}{z}$$

where $z = a + ib$, then

$$f(z) = f(a + ib) = \frac{(a + ib)^2 + 1}{a + ib} = \frac{(a^2 - b^2 + 1)a + 2ab^2}{a^2 + b} + i\frac{2a^2b - (a^2 - b^2 + 1)b}{a^2 + b^2}$$

R.6.48 Let $z = a + ib$, then the function $f(z) = ke^z$ is a complex exponential function.

R.6.49 Let us get some experience performing the basic operations such as addition, subtraction, multiplication, division, and exponentiation using complex numbers.

Let

$$z_1 = 8 + 10i$$

$$z_2 = 3 - 9i$$

$$z_3 = 5 - 12j$$

$$z_4 = 7 - i * 13$$

$$z_5 = 7 - i13$$

Perform the following commands using MATLAB:

1. Enter z_1, z_2, z_3, z_4, *and* z_5
2. *Sum_z_1z_2* $= z_1 + z_2$
3. *Prod_z_1z_2* $= z_1 * z_2$
4. *Div_z_1z_2* $= z_1/z_2$

5. $w = z_1 * z_2 + z_3 * z_4$

6. $v = z_1^2 + z_2^2 + z_3^2 + z_4^2$

MATLAB Solution

```
>> z1 = 8+10i                               % part (1)

   z1 =
        8.0000 +10.0000i
>> z2 = 3-9i

   z2 =
        3.0000 -  9.0000i
>> z3 = 5-12j

   z3 =
        5.0000 -12.0000i
>> z4 = 7-i*13

   z4 =
        7.0000 -13.0000i
>> z5 = 7-i13                               % wrong notation

   ??? Undefined function or variable 'i13'.
>> Sum_z1z2 = z1+z2                         % part (2)

   Sum_z1z2 =
                11.0000 + 1.0000i
>> Prod_z1z2 = z1*z2                        % part (3)

   Prod_z1z2 =
                1.1400e+002 -4.2000e+001i
>> Div_z1z2=z1/z2                           % part (4)

   Div_z1z2 =
                -0.7333 + 1.1333i
>> w = z1*z2+z3*z4                          % part (5)

   w =
        -7.0000e+000 -1.9100e+002i
>> v = z1^2+z2^2+z3^2+z4^2                  % part (6)

   v =
        -3.4700e+002 -1.9600e+002i
```

R.6.50 When complex numbers are entered in MATLAB within brackets, they become elements of a matrix.

R.6.51 When complex numbers are elements of a matrix, the matrix is referred as a complex matrix. Care must be taken when inputting complex numbers where blank spaces should be avoided, since blanks represent characters.

For example, let

$$x = 1 + 2j \quad \text{and} \quad y = 1 + 2j$$

blank space

then x is not equal to y.

R.6.52 The elements of a complex matrix can be entered in MATLAB by following the same rules defined for real matrices in Chapter 3. For example, let

$$A = \begin{bmatrix} 1+2j & 3-4j \\ 5 & 6-7j \end{bmatrix}$$

then the matrix A is entered using MATLAB syntax indicated as follows:

```
>> A = [1+2j 3-4j;5 6-7j];
```

R.6.53 The elements of a complex matrix can be entered in rectangular, exponential, or trigonometric form, but MATLAB always stores the elements in rectangular format.

R.6.54 Complex matrix and array operations use the same commands and follow the same rules as the ones defined for real matrices.

R.6.55 For example, let A and B be two complex matrices defined as follows:

$$A = \begin{bmatrix} [3e^{(j\pi/3)}] & 6\cos\left(\dfrac{\pi}{6}\right) + i6\sin\left(\dfrac{\pi}{6}\right) \\ 3+4i & 4.23e^{(-i\pi/18)} + 9 \end{bmatrix}$$

and

$$B = \begin{bmatrix} 5-9j & 5e^{j(\pi/3+\pi/5)} \\ (2-3j)^{3.3} & \log(6-8j) \end{bmatrix}$$

Create the script MATLAB file *Compmatop* that performs the following matrix operations:

1. Create the matrix A
2. Create the matrix B
3. $C = A + B$
4. $D = A * B$
5. $E = A. * B$
6. $F = \text{inv}(A)$
7. $G = F * A$
8. $H = A * F$
9. $I = A \wedge i$
10. $J = A. \wedge B$

MATLAB Solution
```
% Script file: Compmatop
A = [3*exp(pi/3*j) 6*cos(pi/6)+i*6*sin(pi/6);3+4i 4.23*exp(-i*pi/18)+9]
B = [5-9j 5*exp(pi/3*j+pi/5);(2-3j)^3.3 log(6-8j)]
C = A+B
D = A*B
E = A.*B
F = inv(A)
G = F*A
H = A*F
```

```
I = A^i
J = A.^B
```

The script file *Compmatop* is executed below and the results are shown as follows:

```
>> Compmatop
    A =
        1.5000 + 2.5981i     5.1962 + 3.0000i
        3.0000 + 4.0000i    13.1657 - 0.7345i
    B =
        5.0000 - 9.0000i     4.6861 + 8.1166i
       -68.5109 + 6.9866i     2.3026 - 0.9273i
    C =
        6.5000 - 6.4019i     9.8823 +11.1166i
       -65.5109 +10.9866i    15.4683 - 1.6618i
    D =
        1.0e+002 *
       -3.4607 - 1.6974i     0.0069 + 0.2644i
       -8.4587 + 1.3531i     0.1123 + 0.2919i
    E =
        1.0e+002 *
        0.3088 - 0.0051i     0.0000 + 0.5623i
       -2.3348 - 2.5308i     0.2963 - 0.1390i
    F =
        0.6976 - 0.1688i    -0.3077 - 0.1095i
       -0.2000 - 0.1846i     0.1059 + 0.1243i
    G =
        1.0000 - 0.0000i    -0.0000 - 0.0000i
        0.0000               1.0000
    H =
        1.0000 + 0.0000i    -0.0000 - 0.0000i
        0.0000 + 0.0000i     1.0000 - 0.0000i
    I =
        0.8065 - 0.0403i    -0.7942 - 0.2316i
       -0.5328 - 0.4375i    -0.6469 + 0.7881i
    J =
        1.0e+006
       -0.1832 + 3.0056i    -0.0000 - 0.0001i
       -0.0000 - 0.0000i    -0.0003 - 0.0002i
```

R.6.56 The prime operator (') on a complex matrix returns its conjugate transpose. Using the matrix A from R.6.52 as an example, perform the following command: $B = A'$. Then

$$B = \begin{bmatrix} 1 - 2j & 5 \\ 3 + 4j & 6 + 7j \end{bmatrix}$$

R.6.57 The *point transpose* (.') operation on a complex matrix returns the (unconjugate) transpose. Using the matrix A from R.6.52 as an example, perform the following operation: $C = A.'$. Then

$$C = \begin{bmatrix} 1 + 2j & 5 \\ 3 - 4j & 6 - 7j \end{bmatrix}$$

R.6.58 The MATLAB command *conj(z)* returns the complex conjugate of *z*. For example, let *z* = 1 − 2*i*, perform the following operation *conjz = conj(z)*

MATLAB Solution
```
>> z =1-2i

   z =
       1.0000 - 2.0000i
>> conjz = conj(z)

   conjz =
             1.0000 + 2.0000i
```

R.6.59 The MATLAB command *real(z)* returns the real part of *z*. For example, evaluate the real part of *z* defined in R.6.58.

MATLAB Solution
```
>> z = 3 + 4i;
>> realz = real(z)

   realz =
             3
```

R.6.60 The MATLAB command *imag(z)* returns the imaginary part of *z*.
       For example, let *z* = 3 + 4*i*, execute the following command *imagz = imag(z)* and observe the response.

MATLAB Solution
```
>> z = 3+4i;
>> imagz = imag(z)

   imagz =
             4
```

R.6.61 Let us illustrate some of the matrix concepts defined earlier, in this section. For example, let the complex matrix *A* be

$$A = \begin{bmatrix} [3e^{(j\pi/3)}] & 6\cos\left(\dfrac{\pi}{6}\right) + i\,6\sin\left(\dfrac{\pi}{6}\right) \\ \\ 3 + 4i & 4.23e^{(-i\pi/18)} + 9 \end{bmatrix}$$

Create the script file *Compmatrix* that performs the following operations:

1. *real_A = real(A)*
2. *imag_A = imag(A)*
3. *check_A = real_A + j*imag_A*, reconstructing the original matrix *A*

MATLAB Solution
```
% Script file: Compmatrix
A = [3*exp(pi/3*j) 6*cos(pi/6)+i*6*sin(pi/6);3+4i 4.23*exp(-i*pi/18)+9]
real _ A = real(A)
imag _ A = imag(A)
check _ A = real _ A+j*imag _ A
```

The script file *Compmatrix* is executed as follows, and the results indicate that indeed matrix *A* is equal to [*real(A)*] + [*j* * *imag(A)*].

```
>> Compmatrix
   A =
       1.5000 + 2.5981i    5.1962 + 3.0000i
       3.0000 + 4.0000i   13.1657 - 0.7345i
   real _ A =
               1.5000      5.1962
               3.0000     13.1657
   imag _ A =
               2.5981      3.0000
               4.0000     -0.7345
   check _ A =
               1.5000 + 2.5981i    5.1962 + 3.0000i
               3.0000 + 4.0000i   13.1657 - 0.7345i
```

Observe that each element of *A* is saved in rectangular form.

R.6.62 The MATLAB command *abs(z)* returns the absolute value of *z*. For example, let *z* = 3 + 4*i*, perform the following operation *absz* = *abs(z)*.

MATLAB Solution
```
>> z = 3+4i;
>> absz = abs(z) % note that absz = [absolute value of z] = sqrt (3^2+4^2)
   absz =
        5
```

R.6.63 Let us illustrate some of the matrix concepts presented earlier in this section. Let *A* and *B* be the two complex matrices defined as follows:

$$A = \begin{bmatrix} [3e^{(j\pi/3)}] & 6\cos\left(\dfrac{\pi}{6}\right) + i6\sin\left(\dfrac{\pi}{6}\right) \\ 3 + 4i & 4.23e^{(-i\pi/18)} + 9 \end{bmatrix}$$

and

$$B = \begin{bmatrix} 5 - 9j & 5e^{(j(\pi/3 + \pi/5))} \\ (2 - 3j)^{3.3} & \log(6 - 8j) \end{bmatrix}$$

Create the script file *com_matr_op* that performs the operations indicated as follows:

1. $A = [3 * exp(pi/3 * j) \quad 6 * cos(pi/6) + i * 6 * sin(pi/6); 3 + 4i \quad 4.23 * exp(-i * pi/18) + 9]$
2. $B = [5 - 9j \quad 5 * exp(pi/3 * j + pi/5); (2 - 3j)^\wedge 3.3 \quad log(6 - 8j)]$
3. $C = det(A)$
4. $D = conj(A)$
5. $E = A \wedge 2$
6. $F = A.^\wedge B$
7. $G = A'$
8. $H = A.'$
9. $I = [A\ B]$

10. *J = [A; B]*
11. *K = I(1, :)*
12. *L = J(:, 1)*
13. *M = eig(A)*

MATLAB Solution
```
% Script file : comp _ matr _ op
A = [3*exp(pi/3*j) 6*cos(pi/6)+i*6*sin(pi/6);3+4i 4.23*exp(-i*pi/18)+9]
B = [5-9j 5*exp(pi/3*j+pi/5);(2-3j)^3.3 log(6-8j)]
C = det(A)
D = conj(A)
E = A^2
F = A.^B
G = A'
H = A.'
I = [A B]
J = [A;B]
K = I(1,:)
L = J(:,1)
M = eig(A)
```

The script file *comp_matr_op* is executed below and the results are indicated as follows:

```
>>comp _ matr _ op
  A =                                    % A is defined as a 2 by
     1.5000 + 2.5981i    5.1962 + 3.0000i     2 complex matrix
     3.0000 + 4.0000i   13.1657 - 0.7345i  % observe that the
                                              elements of A are stored
                                          % in rectangular form
  B =                                    % B is defined as a 2 by
     5.0000 - 9.0000i    4.6861 + 8.1166i     2 complex matrix
   -68.5109 + 6.9866i    2.3026 - 0.9273i
  C =                                    % observe that C is the
    18.0685 + 3.3192i                       determinant of
                                          % matrix A, also complex
  D =                                    % D is the complex conj
     1.5000 - 2.5981i    5.1962 - 3.0000i     of matrix A
     3.0000 - 4.0000i   13.1657 + 0.7345i
  E =                                    % E is the matrix product
     1.0e+002                              [A*A]
    -0.0091 + 0.3758i    0.7061 + 0.5368i
     0.3654 + 0.6425i    1.7639 + 0.1044i
  F =                                    % F is A raised to B
     1.0e+006
    -0.1832 + 3.0056i   -0.0000 - 0.0001i
    -0.0000 - 0.0000i   -0.0003 - 0.0002i
  G =                                    % G is the transpose of A
     1.5000 - 2.5981i    3.0000 - 4.0000i
     5.1962 - 3.0000i   13.1657 + 0.7345i
```

```
H =                               % H is the un-conjugate
    1.5000 + 2.5981i    3.0000 + 4.0000i     transpose of A
    5.1962 + 3.0000i   13.1657 - 0.7345i
I =                               % I is A concatenated with
    1.5000 + 2.5981i  5.1962 + 3.0000i         matrix B (2x4matrix)
    5.0000 - 9.0000i  4.6861 + 8.1166i
    3.0000 + 4.0000i  13.1657 - 0.7345i
  -68.5109 + 6.9866i  2.3026 - 0.9273i
J =                               % J is matrix A followed
    1.5000 + 2.5981i    5.1962 + 3.0000i         by B (J is 4 x 2 matrix)
    3.0000 + 4.0000i   13.1657 - 0.7345i
    5.0000 - 9.0000i    4.6861 + 8.1166i
  -68.5109 + 6.9866i    2.3026 - 0.9273i
K =                               % K is the first row of
    1.5000 + 2.5981i  5.1962 + 3.0000i         matrix I
    5.0000 - 9.0000i  4.6861 + 8.1166i
L =                               % L is the first column of
    1.5000 + 2.5981i                    matrix  J
    3.0000 + 4.0000i
    5.0000 - 9.0000i
  -68.5109 + 6.9866i
M =                               % M are the eigenvalues of A
    1.3675 + 0.0646i
   13.2983 + 1.7989i
```

R.6.64 The MATLAB command *angle(z)* returns the value of the angle of the exponential or polar representation of z by evaluating the function $tan^{-1}[imag(z)/real(z)]$, or the MATLAB function *atan(imag(z), real(z))*, in radians, within the range $-\pi, +\pi$.

R.6.65 Complex data in polar form can be plotted using polar coordinates by employing the function *polar (alpha, r)*, where *alpha* is given by *angle(z)*, in rad and $r = abs(z)$.

   Observe that a point in the z-plane can be uniquely identified using polar coordinates by defining r and *alpha*. An arbitrary point in the z-plane can be represented by more than one pair of polar coordinates. For example, the polar coordinates $(9, 130°)$ and $(9, -230°)$ represent the same point in the z-plane.

R.6.66 Complex data can be represented as vectors, with an arrow drawn from the origin of the complex plane with length r, and an angle of $\Phi = arctan (b/a)$, by using the instruction *compass(z)*, where $z = a + jb$ and $r = \sqrt{a^2 + b^2}$. For example, let $z = 1 + 3i$, and perform then the command *compass(z)* and show the result.

MATLAB Solution
```
>> z =1+3i;
>> compass(z)              % returns the plot of Figure 6.5
```

R.6.67 Complex data can be shown as vectors by using the command *feather (z)*, or *feather (a, b)*. The argument of *feather*, z or $a + jb$, represents in the complex plane a directional arrow with a slope of b/a. For example, let $z_1 = 1 + i$ and $z_2 = -(1 + i)$; perform the instruction *feather (z)*, where $z = [z_1z_2]$, and show the resulting plot.

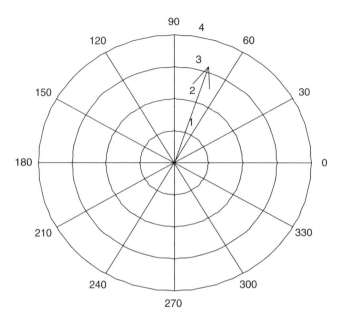

FIGURE 6.5
compass plots of z of R.6.66.

MATLAB Solution
```
>> z1 =1+i;
>> z2 = -(1+i);
>> z = [z1 z2];
>> feather (z)          % returns the plot shown in Figure 6.6
>> title ('plot using feather (z), z = ±(1+i)')
>> xlabel ('real');ylabel ('imaginary')
```

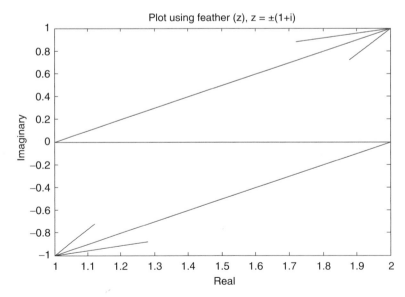

FIGURE 6.6
feather plots of z of R.6.67.

R.6.68 Two complex numbers $z_1 = a_1 + jb_1$ and $z_2 = a_2 + b_2$ are equal if and only if

$$a_1 = a_2 \quad \text{and} \quad b_1 = b_2$$

or

$$|z_1| = |z_2| \quad \text{and} \quad \angle z_1 = \angle z_2$$

Recall that the concept of comparing two complex numbers z_1 and z_2, and labeling one greater or smaller than the other is meaningless.

R.6.69 The phase angle of $z = a + jb$ is expressed as

$$\angle z = \tan^{-1}(b/a), \quad \text{for } a > 0$$

and

$$\angle z = \pi - \tan^{-1}(b/a), \quad \text{for } a < 0$$

R.6.70 The addition and subtraction of sinusoidal functions is frequently encountered in the physical sciences and engineering such as AC circuit analysis where the sinusoids have the same frequency, but different magnitudes and phase angles. One way to deal with this problem is by constructing the sinusoidal functions on the same set of axes and performing the required operation at every point along the abscissa.

This process is convenient if done by a computer, but is long and tedious if done by hand. A more efficient and convenient way, used extensively by engineers, is to use complex numbers to represent time sinusoidal functions using the polar or exponential complex form. A slightly modified polar form is commonly referred by engineers as a phasor representation. The conversion process is illustrated as follows. Let $f(t) = r \cos(wt + \theta)$, this function can be represented as

$$f(t) = real[r \; e^{j\theta} \; e^{jwt}]$$

where $F = real[r \; e^{j\theta}]$, or in short $F = r \; e^{j\theta}$, omitting the term *real*, but knowing that only the real part is considered.

If the time function is $f(t) = r \sin(wt + \theta)$, then $f(t)$ would be represented as

$$f(t) = imag[r \; e^{j\theta} \; e^{jwt}] \quad \text{or} \quad F = imag[r \; e^{j\theta}] \quad \text{or in short,} \quad F = r \; e^{j\theta}$$

This representation is frequently given in the compressed form as $r \angle \theta$ and is frequently referred as a phasor representation.

R.6.71 For example, the following time domain sinusoidal functions, shown in the left column, are converted to phasor representations in the right column.

Time Domain		Phasor Domain
$f_1(t) = 5.65 \sin(wt)$	$\longrightarrow$	$F_1 = 5.65 \angle 0°$
$f_2(t) = 9.13 \sin(wt + 35°)$	$\longrightarrow$	$F_2 = 9.13 \angle 35°$
$f_3(t) = 8.93 \cos(wt)$	$\longrightarrow$	$F_3 = 8.93 \angle 90°$

The preceding examples use the peak or maximum value of the time functions instead of the more frequently employed effective value (peak value times 0.707) in electrical AC circuit analysis by engineers and the angles are usually expressed in degrees, rarely in radians.

R.6.72 The MATLAB functions

$$[A, B, C] = cart2sph(x, y, z)$$

and

$$[D, E, F] = cart2pol(x, y, z)$$

transform the Cartesian coordinates into spherical, or polar coordinates (cylindrical), respectively, where A (azimuth), B (elevation), and D are angles expressed in radians; C and E represent the radius; and F represents the height.

Similarly, the instructions *pol2cart* and *sph2cart* convert polar or cylindrical coordinates to Cartesian.

For example, transform the 3D Cartesian point defined by

$$x = 1, y = 1, \text{ and } z = 1$$

into spherical and polar coordinates, and back to Cartesian coordinates as a check, using MATLAB.

MATLAB Solution
```
>> x =1;y =1;z =1;
>> [A,B,C] = cart2sph(x,y,z)              % spherical coordinate

      A =
          0.7854
      B =
          0.6155
      C =
          1.7321
>> [X,Y,Z] = sph2cart(A,B,C)              % Cartesian coordinates

      X =
          1
      Y =
          1.0000
      Z =
          1.0000
>> [D,E,F] = cart2pol(x,y,z)              % polar coordinates

      D =
          0.7854
      E =
          1.4142
      F =
          1
>> [X,Y,Z] = pol2cart(D,E,F)              % back to Cartesian coordinates

      X =
          1.0000
      Y =
          1
      Z =
          1
```

6.4 Examples

Example 6.1

Given the following two complex numbers:

$$z_1 = 1 + 2i$$

and

$$z_2 = 3 + 4i$$

Write a MATLAB program that performs the following operations:

a. $z_1 + z_2$

b. $z_1 - z_2$

c. $z_1 \cdot z_2$ (the . denotes product)

d. z_1/z_2

e. z_1^* and z_2^* (recall that the * denotes complex conjugate)

f. $real(z_1 \cdot z_2)$ and $imag(z_1 \cdot z_2)$

g. $real(z_1 + z_2)$ and $imag(z_1 + z_2)$

h. $v = real\ (z_1) + imag(z_1 * z_2)$

i. $w = imag\ (z_2) + i\ real(z_1 + z_2)$

j. $(z_1 \cdot z_2)^*$ and $(z_1 + z_2)^*$

k. *mag. of* $|z_1|$ and $|z_2|$

l. Phase angles of z_1 and z_2

MATLAB Solution

```
>> % MATLAB program that evaluates and illustrates the basic complex
                                                         operations
>> % for the following complex numbers:
>> z1 = 1+2i;
>> z2 = 3+4i;
>> sum = z1+z2                 % sum of z1+z2, part (a)

   sum =
         4.0000 + 6.0000i
>> dif = z1-z2                 % subtraction of z1-z2, part (b)
   dif =
         -2.0000 - 2.0000i
>> prod = z1*z2                % product of z1*z2, part (c)
   prod =
         -5.0000 +10.0000i
>> div = z1/z2                 % division of z1/z2, part (d)
   div =
         0.4400 + 0.0800i
>> z1conj = conj(z1)          % complex conjugate of z1 and z2, part (e)
   z1conj =
              1.0000 - 2.0000i
>> z2conj = conj(z2)
   z2conj =
              3.0000 - 4.0000i
```

```
>> realprod = real(prod)          % real and imaginary parts of the product
                                    z1*z2, part (f)
realprod =
            -5
>> imagprod = imag(prod)
imagprod =
           +10
>> realsum = real(sumcomp);       % part (g)
realsum =
             4
>> imasum = imag(sumcomp)
imagsum =
             6
>> v = real(z1) + i*imag(prod)    % create v, part (h)
v =
     1.0000 +10.0000i
>> w = imag(z2)+i*real(sum);      % part (i)
w =
     4.0000 + 4.0000i
>> conprod = conj(prod)           % complex conjugate of (z1.z2) and
                                    (z1+z2)*, part (j)
conprod =
           -5.0000 -10.0000i
>> consum = conj(sum)
consum =
           4.0000 - 6.0000i
>> magz1 = abs(z1)                % evaluates magnitudes of z1 and z2,
                                    part (k)
magz1 =
           2.2361
>> magz2 = abs(z2)                % angles are in radians
magz2 =
           5
>> angz1 = angle(z1)              % evaluates phase angles of z1 and z2,
                                    part (l)
angz1 =
           1.1071
>> angz2 = angle(z2)
angz2 =
           0.9273
```

Example 6.2

Given the exponential discrete complex sequence $y(n) = 3e^{zn}$, where $z = -1 + i(\pi/3)$; create the script MATLAB file *discrete* that returns the following plots:

a. *real(y)* versus n

b. *imag(y)* versus n

c. $\left[\sqrt{(real(y))^2 + (imaginary(y))^2} \right]$ versus n, over the range $0 \le n \le 12$, with regular spacing of $\Delta n = 0.2$

MATLAB Solution

```
% Script file: discrete
% generation of the complex exponential sequence y(n)
z = -1+i*(pi/3);
n = 0:0.2:12;
y = 3*exp(z*n);
a = real(y);
b = imag(y);
axis on ;

figure(1)
subplot(2,1,1)
stem(n,a)
xlabel('time index n'), ylabel('Amplitude')
title('Plot of real[3*exp(-1+i(pi/3))*n] vs. n')
grid on
subplot(2,1,2)
stem(n,b)
xlabel('time index n'),ylabel('Amplitude')
Title('Plot of imag [3*exp(-1+i(pi/3))*n] vs n')
grid on

figure(2)
stem(n,abs(y))
title('Plot of magnitude of [y] vs. n')
xlabel('time index n'),ylabel('Magnitude of [y]')
```

The script file *discrete* is executed and the results are shown in Figures 6.7 and 6.8.

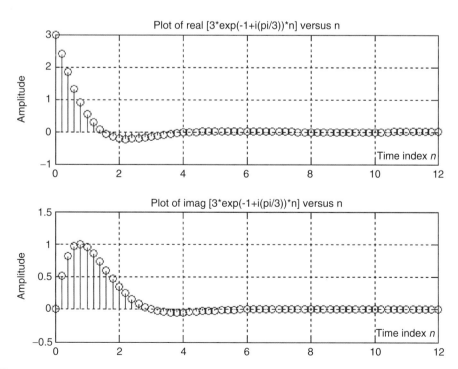

FIGURE 6.7
Discrete plots of Example 6.2(a and b).

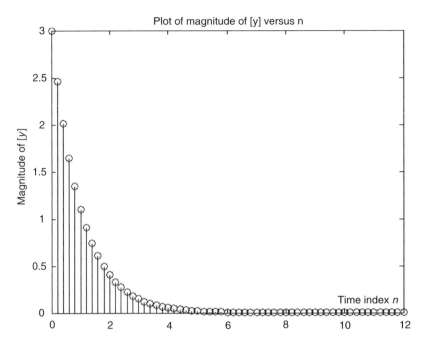

FIGURE 6.8
Discrete plot of Example 6.2(c).

Example 6.3

Write a program that plots the following functions:

$$y_1 = \frac{e^{jz} + e^{-jz}}{2} \quad \text{and} \quad y_2 = \frac{e^{jz} - e^{-jz}}{2j}$$

for $z = 2\pi n - \pi/4$ and $0 \leq n \leq 2$, with regular spacing of $\Delta n = 0.05$; verifying Euler's identities.

Recall that Euler's identities are

$$cos(z) = \frac{e^{jz} + e^{-jz}}{2} \quad \text{and} \quad sin(z) = \frac{e^{jz} - e^{-jz}}{2j}$$

MATLAB Solution

```
>>                                    % generation of a sinusoid using
                                        Euler's identities
>> n = 0:0.05:2;
>> z = 2*pi*n-pi/4;
>> y1 = 0.5*exp(i * z);
>> y = y1+ conj(y1);
>> y2 = -j.*(y1-conj(y1));
>> clf                                 % clears the figure window
>> subplot (1,2,1)
>> stem(n,y)                           % returns the discrete sinusoid
```

```
>> axis on;
>> axis ([0 2 -1.5 1.5])
>> grid on
>> xlabel ('Index n'),ylabel ('Amplitude')
>> title('Cosine Construction Using Eulers Identity')
>> subplot (1,2,2)
>> stem(n,y2)
>> xlabel ('Index n'), ylabel('Amplitude')
>> title ('Sine Construction Using Euler's Identity')
>> grid on
```

See Figure 6.9.

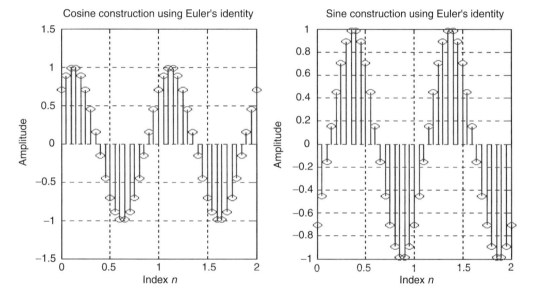

FIGURE 6.9
Plots of Example 6.3, verifying Euler's identity.

Example 6.4

Given the following complex numbers:

$$z_1 = 3 + i$$

and

$$z_2 = -4 + i2$$

Write a program that performs the following:

a. Represent z_1 and z_2 on the complex plane as points (indicated by * and +, respectively)

b. Represent z_1 and z_2 as vectors plotted in the complex plane (magnitude and angle)

c. Represent the vectors z_1, z_2, and $z = z_1 + z_2$, as plots in the polar coordinate system

d. Plot of [mag1] versus $\angle \beta/(4\pi)$, where [mag1] $= abs(z_1)(\beta/4\pi)$, over the range $0 \le \beta \le 4\pi$, in the polar coordinate system with regular spacing of 0.1

See Figure 6.10.

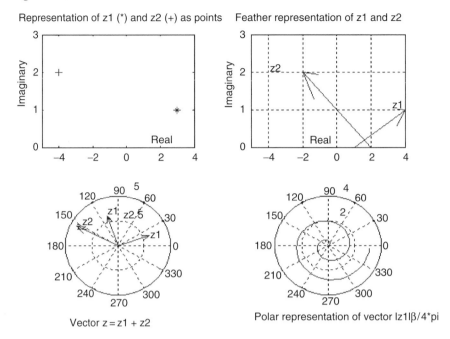

FIGURE 6.10
Plots of Example 6.4.

MATLAB Solution
```
>>                        % Figure 6.10 are the plots of Example 6.4
>> subplot(2,2,1)
>> plot(3,1,'*',-4,2,'+')   % char.* represents z1 and, + represents z2
>> xlabel('Real') ,ylabel('Imaginary')
>> title ('Representation of z1(*) and z2(+) as  points')
>> axis([-5 4 0 3])
>> z1=3+i;
>> z2 = -4+2i;
>> z = [z1 z2];
>> subplot(2,2,2)
>> feather (z)
>> axis ([-5 4 0 3])
>> grid on
>> text (real(z1), imag(z1), 'z1')
>> text (real(z2), imag(z2), 'z2')
>> xlabel ('Real'), ylabel('Imaginary')
>> title ('Feather Representation of z1 and z2')
>> z3 = z1+z2;
>> z4 = [z1 z2 z3];
>> subplot(2,2,3)
>> compass(z4)
>> text (real(z1),imag(z1),'z1')
>> text (real(z2),imag(z2),'z2')
>> text (real(z1+z2),imag(z1+z2),'z1+z2')
>> title('Vector z=z1+z2')
```

```
>> subplot (2,2,4)
>> beta = 0:0.1:4*pi;
>> mag1= abs(z1)*beta/(4*pi);
>> polar (beta,mag1);
>> title ('Polar representation of vector |z1|ß/(4*pi)')
```

Example 6.5

Given the following discrete complex function:

$$f(z) = \frac{z}{z-1}$$

where $z = x + 3j$, over the range $-10 \le x \le 10$, represented by regular linear spacing of 0.1.

    a. Obtain analytical expressions for $|f(z)|$ and $\angle f(z)$.
    b. Create the script file *Example65* that returns the plots of $|f(z)|$ versus x and $\angle f(z)$ versus x.

ANALYTICAL Solution

$$f(x+3j) = \frac{x+3j}{x+3j-1} = \frac{\sqrt{x^2+3^2}\angle \tan^{-1}(3/x)}{\sqrt{(x-1)^2+3^2}\angle \tan^{-1}(3/(x-1))}$$

The magnitude is given by

$$|f(z)| = \frac{\sqrt{x^2+3^2}}{\sqrt{(x-1)^2+3}}$$

The angle is given by

$$\angle f(z) = \angle \tan^{-1}(3/x) - \tan^{-1}(3/(x-1))$$

MATLAB Solution
```
% Script file: Example65
% The plots of Example 6.5 are shown in Figure 6.11
X=-10: .1: 10;
Z = X+3j; % Creates an array of complex elements
F = Z./(Z-1);
Subplot (2,1,1)
plot (X,abs(F))
xlabel ('x axis'), ylabel ('Magnitude')
title ('mag[f(z)] vs. x')
subplot (2,1,2)
angleF = angle(F);
plot (X,angleF)
xlabel ('x axis'), ylabel ('Angle in radians')
title ('angle[f(z)] vs. x')
```

The script file *Example65* is executed and the results are shown in Figure 6.11.

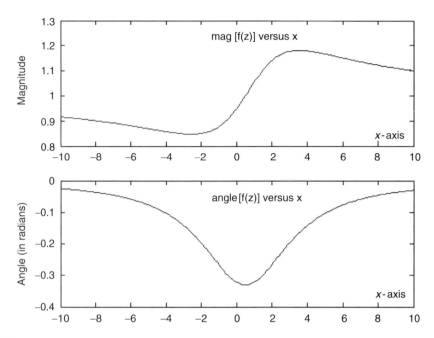

FIGURE 6.11
Plots of Example 6.5.

Example 6.6

Create the script file *Example66* that uses MATLAB to verify the following identity (Euler's identity):

$$e^{j\omega_o t} = cos(\omega_o t) + j \, sin(\omega_o t)$$

for any arbitrary value of ω_o. Let us choose $\omega_o = 2$ rad/s.

MATLAB Solution
```
%Script file: Example66
t =-pi:.1*pi:2*pi;
wo =2;
y = exp(j*wo.*t);
y1= cos(wo.*t); y2=sin(wo.*t);
realy = real(y);
imagy = imag(y);
subplot (2,1,1)
plot (t,realy,'o',t,y1)
legend ('real[y]', 'cos(2t)')
title ('real[exp(j*2*t)] vs t')
ylabel ('Amplitude of real[exp(j*2*t)]'), xlabel('time')
subplot (2,1,2)
plot (t,imag(y),'o',t,y2)
legend('imaginary[y]', 'sin(2t)')
title('imaginary [exp(j*2*t)]')
ylabel(' Amplitude of imaginary[(exp(j*2*t)]'),
xlabel('time')
```

The script file *Example66* is executed and the resulting plots are shown in Figure 6.12.

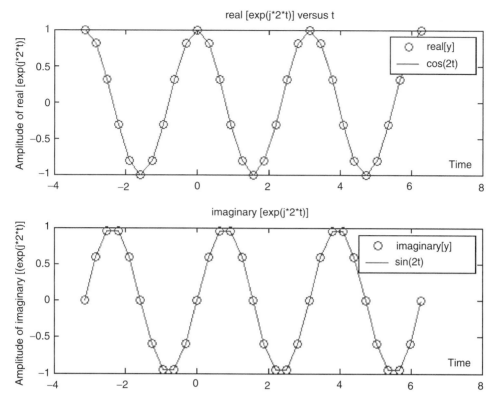

FIGURE 6.12
Plots of Example 6.6.

Example 6.7

Given the following function $f(t) = (1 - e^{0.1t}) [cos(2t) + j sin(2t)]$.
   Create the script file *Example67* that returns the following plots:

a. $|f(t)|$ versus t
b. *real[f(t)]* versus t
c. *imaginary[f(t)]* versus t
d. polar plot of {*angle[f(t)]*} versus *abs[f(t)]*

MATLAB Solution
```
%Script file: Example67
t=0:.1*pi:3*pi;
ft=(1-exp(.1.*t)).*(cos(2.*t)+j*sin(2.*t));
subplot(2,2,1);
plot(t,abs(ft));
title('abs [f(t)] vs. t'),
ylabel('magnitude [(f(t)]'); xlabel('time')
subplot(2,2,2);
plot(t,real(ft));
title('real[f(t)] vs. t'),
ylabel('Amplitude of real[f(t)]');xlabel('time')
subplot(2,2,3)
```

```
plot(t,imag(ft));
title('imag[f(t)] vs. t'), ylabel('Amplitude of imag [f(t)]');
xlabel('time')
subplot(2,2,4);polar(angle(ft), abs(ft))
title('Polar representation of |f(t)| vs. angle[(f(t)]')
```

The script file *Example67* is executed and the results are shown in Figure 6.13.

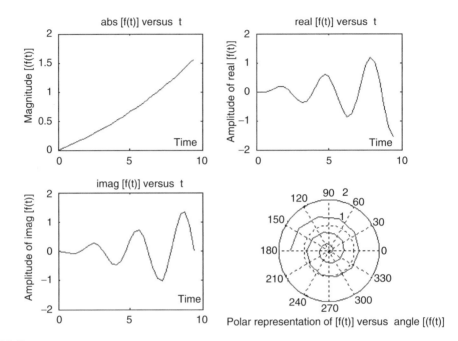

FIGURE 6.13
Plots of Example 6.7.

Example 6.8

Given the following time functions:

$$f_1(t) = 25\ sin(wt + 30°)$$

and

$$f_2(t) = 15\ sin(wt + 60°)$$

1. Determine by hand the sum of $[f_1(t) + f_2(t)]$, using complex algebra (phasors)
2. Use MATLAB as a calculator to evaluate $[f_1(t) + f_2(t)]$
3. Evaluate using complex exponentials the sum of $[f_1(t) + f_2(t)]$
4. Compare the answers obtained in parts: 1, 2, and 3
5. Obtain the plot in the time domain of $[f_1(t) + f_2(t)]$ versus t using complex algebra, assuming $w = \pi$
6. Plot the sum of $[f_1(t)\ plus\ f_2(t)]$ versus t, directly in the time domain, assuming $w = \pi$

ANALYTICAL Solution

Part (1)

Converting $f_1(t)$ and $f_2(t)$ from the time domain to the phasor domain yields

Time Domain	Phasor Domain
$f_1(t) = 25 \sin(\pi t + 30°)$ $\longrightarrow$	$F_1 = 25\angle 30°$
$f_2(t) = 15 \sin(\pi t + 60°)$ $\longrightarrow$	$F_2 = 15\angle 60°$

Convert F_1 and F_2 to rectangular form to perform the addition, obtaining the following relations:

$$F_1 = 25 \angle 30° = 25[\cos(30°) + j\sin(30°)] = 21.65 + j\,12.50$$

and

$$F_2 = 15 \angle 60° = 15[\cos(60°) + j\sin(60°)] = 7.5 + j\,13$$

finally

$$F = F_1 + F_2 = 29.15 + j\,25.50 = 38.72 \angle 41.72°$$

MATLAB Solution
```
% Script file: Example68
% sum of two sinusoids: f₁(t) = 25 sin (πt + 30°);
%    and f₂(t) = 15 sin (πt + 60°)
disp('Part (2)')
disp('Complex form using MATLAB as a calculator')
echo on
f1real = 25*cos(30*pi/180)
f1imag = 25*sin(30*pi/180)
f2real = 15*cos(60*pi/180)
f2imag =15*sin(60*pi/180)
freal = f1real+f2real
fimag = f1imag+f2imag
f = freal+j*fimag
fmag = abs(f)
fang = angle(f)*180/pi
echo off
% exponential complex form using MATLAB
f1 = 25*exp(j*30*pi/180);
f2 = 15*exp(j*60*pi/180);
fc = f1 + f2;
fcmag = abs(fc);
fcang = angle(fc)*180/pi;
disp('****************************************');
disp('*********Summary results *************');
disp('****************************************');
disp('**    MATLAB used as a calculator    **');
disp('**    the magnitude and phase        **');
```

```
disp('**    of   f1(t) + f2(t)  are :          **');
[fmag fang]
disp('**    MATLAB  phasor approach ;      **');
disp('**    the magnitude and phase        **');
disp('**    of f1(t) + f2(t) is:           **');
[fcmag fcang]
disp('*****************************************');
t = -2:.01:2;
fcc = fcmag*sin(pi.*t+fcang*pi/180);
subplot (2,1,1)
plot (t,fcc)
title ('Phasor domain [ f1+f2 ] vs t ')
ylabel ('Magnitude')
subplot (2,1,2)
f1input = 25*sin(pi.*t+30*pi/180);
f2input =15*sin(pi.*t+60*pi/180);
sum = f1input+f2input;
plot (t,sum)
title ('Time domain [ f1(t)+f2(t) ] vs t')
ylabel ('Magnitude')
xlabel ('time')
```

The script file *Example68* is executed below and the results are shown as follows:

```
>> Example68

Part (2)
Complex form using MATLAB as a calculator
f1real = 25*cos(30*pi/180)
f1real =
        21.6506
f1imag = 25*sin(30*pi/180)
f1imag =
        12.5000
f2real = 15*cos(60*pi/180)
f2real =
        7.5000
f2imag = 15*sin(60*pi/180)
f2imag =
        12.9904
freal = f1real+f2real
freal =
        29.1506
fimag = f1imag+f2imag
fimag =
        25.4904
f = freal+j*fimag
f =
    29.1506 +25.4904i
fmag =abs(f)
fmag =
        38.7236
fang =angle(f)*180/pi
fang =
        41.1676
```

```
echo off
**********************************************
***********Summary results ***************
**********************************************
**    MATLAB used as a calculator       **
**    the magnitude and phase           **
**    of f1(t) + f2(t) is :             **
      ans =
          38.7236    41.1676
**    MATLAB  phasor approach ;         **
**    the magnitude and phase           **
**    of f1(t) + f2(t) are :            **
      ans =
          38.7236    41.1676
**********************************************
```

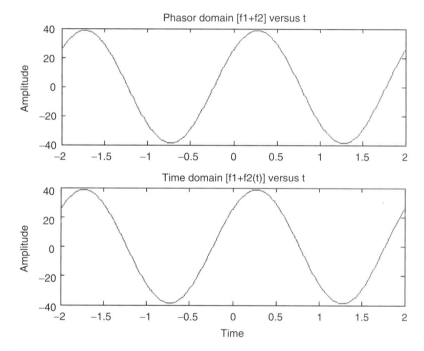

FIGURE 6.14
Plots of Example 6.8.

Clearly the results of parts 1, 2, and 3 are identical. Also observe that the phasor and the time domain approach return exactly the same plots, as shown in Figure 6.14.

Example 6.9

Using the time functions defined in Example 6.8:

$$f_1(t) = 25\,sin(wt + 30°)$$

and

$$f_2(t) = 15\,sin(wt + 60°)$$

1. Determine by hand the subtraction of $f_1(t) - f_2(t)$ using complex algebra (phasors)
2. Using MATLAB as a calculator evaluate $[f_1(t) - f_2(t)]$ using complex algebra
3. Use MATLAB complex exponentials to evaluate $[f_1(t) - f_2(t)]$
4. Compare the answers obtained in parts 1, 2, and 3
5. Plot in the time domain $[f_1(t) - f_2(t)]$ versus t, using the phasor approach, and plot $[f_1(t) - f_2(t)]$ versus t, directly in the time domain, and compare the results (assume $w = \pi$)

ANALYTICAL Solution

Part (1)

Converting $f_1(t)$ and $f_2(t)$ from the time (domain) to the phasor domain yields

$$f_1(t) = 25\,\sin(\pi t + 30°) \longrightarrow F_1 = 25 \angle 30°$$

$$f_2(t) = 15\,\sin(\pi t + 60°) \longrightarrow F_2 = 15 \angle 60°$$

And converting F_1 and F_2 to rectangular form to perform conveniently the subtraction yields

$$F_1 = 25 \angle 30° = 25[\cos(30°) + j\sin(30°)] = 21.65 + j\,12.50$$

and

$$F_2 = 15 \angle 30° = 15[\cos(60°) + j\sin(60°)] = 7.5 + j\,13$$

finally

$$F = F_1 - F_2 = 14.15 - j\,0.50 = 14.16 \angle -2°$$

MATLAB Solution
```
% Script file: Example69
% Subtraction of two sinusoids
disp ('complex form using MATLAB as a calculator')
echo on
f1real = 25*cos(30*pi/180)
f1imag = 25*sin(30*pi/180)
f2real =15*cos(60*pi/180)
f2imag =15*sin(60*pi/180)
freal = f1real-f2real
fimag = f1imag-f2imag
f = freal + j*fimag
fmag = abs(f)
fang =angle(f)*180/pi
echo off
% complex exponential  form
f1 =25*exp(j*30*pi/180);
f2 =15*exp(j*60*pi/180);
fc = f1-f2; fcmag = abs(fc);
fcang = angle(fc)*180/pi;
```

```
disp('*****************************************');
disp('**************R E S U L T S ***********');
disp('*****************************************');
disp('**    MATLAB used as a calculator    **');
disp('**    the magnitude and phase        **');
disp('**    of f1(t)-f2(t)  is:             **');
[fmag fang]
disp('**    MATLAB  phasor approach        **');
disp('**    the magnitude and phase        **');
disp('**    of f1(t)-f2(t)  are :           **');
[fcmag fcang]
disp('*****************************************');
t = -2:.01:2;
fcc = fcmag*sin(pi.*t+fcang*pi/180);
subplot (2,1,1)
plot (t,fcc);title('Phasor domain [ f1-f2 ] vs t ')
ylabel ('Amplitude');subplot(2,1,2);xlabel('time');
f1input = 25*sin(pi.*t+30*pi/180);
f2input =15*sin(pi.*t+60*pi/180);
diff = f1input-f2input;plot(t,diff);
title ('Time domain [ f1(t)-f2(t) ] vs t')
ylabel ('Amplitude');xlabel('time')
```

The script file *Example69* is executed below and the results are given as follows:

```
>> Example69

complex form using MATLAB as a calculator
f1real = 25*cos(30*pi/180)
f1real =
        21.6506
f1imag = 25*sin(30*pi/180)
f1imag =
        12.5000
f2real = 15*cos(60*pi/180)
f2real =
        7.5000
f2imag =15*sin(60*pi/180)
f2imag =
        12.990
freal = f1real-f2real
freal =
        14.1506
fimag =f1imag-f2imag
fimag =
        -0.4904
f = freal+j*fimag
f =
    14.1506 - 0.4904i
fmag =abs(f)
fmag =
        14.1591
fang =angle(f)*180/pi
fang =
        -1.9848
echo off
```

```
*****************************************
**************R E S U L T S ************
*****************************************
**    MATLAB used as a calculator     **
**    the magnitude and phase         **
**    of f1(t) - f2(t) is :           **
   ans =
        14.1591   -1.9848
**    MATLAB  phasor approach          **
**    the magnitude and phase         **
**    of f1(t)-f2(t) are :            **
   ans =
        14.1591-1.9848
*****************************************
```

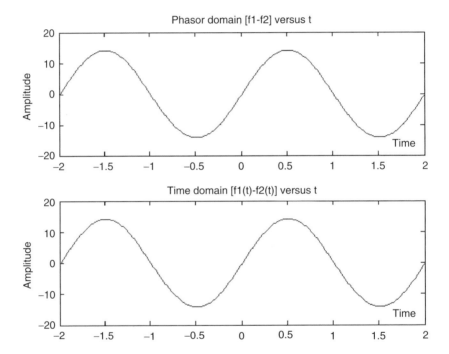

FIGURE 6.15
Plots of Example 6.9.

Clearly the results of parts 1, 2, and 3 are identical, and the corresponding plots are shown in Figure 6.15.

Example 6.10

Verify the following relation:

$$F_3 = F_1 + F_2$$

where

$$F_1 = 80 \angle -36.87°, F_2 = 60 \angle 53.13°, \text{ and } F_3 = 100 \angle 0°$$

a. In the phasor domain, using MATLAB complex algebra.

b. In the time domain plotting $F_1 + F_2$, converted from the phasors over the range $-2 \le t \le 2$, with linear spacing of $\Delta t = 0.01$.

c. On the same plot, obtain $[f_1(t) + f_2(t)]$ in time, phasor, and directly (F_3), over the range $-2 \le t \le 1.5$ (quarter cycle) with linear spacing of $\Delta t = 0.01$, and observe the three overlaying plots fully agree (assume $w = \pi$).

MATLAB Solution

```
% Script file: Example610
% verify
f1 = 80*exp(-j*36.87*pi/180);
f2 = 60*exp(j*53.13*pi/180);
fc = f1+f2;
fcmag = abs(fc);
fcang = angle(fc)*180/pi;
disp('***********************************************')
disp('**************R E S U L T S ******************')
disp('***********************************************')
disp('** MATLAB phasor approach                **')
disp('**    the magnitude and phase of F1+F2 are: **')
[fcmag fcang]
disp('***********************************************')
t =-2:.01:2;
fcc = fcmag*sin(pi.*t+fcang*pi/180);

figure(1)
subplot(2,1,1)
plot (t,fcc)
title ('Phasor domain [ f1+f2 ] vs. t ')
ylabel ('Amplitude'); xlabel('time');
subplot (2,1,2)
f1input = 80*sin(pi.*t-36.87*pi/180);
f2input = 60*sin(pi.*t+53.13*pi/180);
sum = f1input+f2input;
plot (t,sum)
title ('Time domain [ f1(t)+f2(t) ] vs. t')
ylabel ('Amplitude'); xlabel('time')

figure(2)
f3 = 100*sin(pi.*t);
plot (t,f3,'o',t,fcc,'h',t,sum,'+');
legend('direct', 'phasor add', 'time add')
title ('Plots of: [direct, phasor add, time add ]vs. t')
ylabel ('Amplitude');
xlabel ('time');axis([-2 -1.5 0 100])
```

The script file *Example610* is executed below, the results follow and their plots are shown in Figures 6.16 and 6.17.

```
>> Example610
***********************************************
```

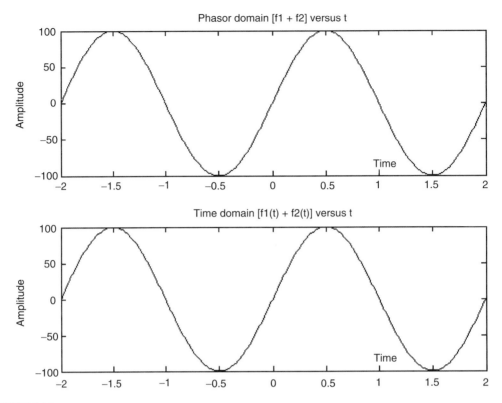

FIGURE 6.16
Plots of Example 6.10(b).

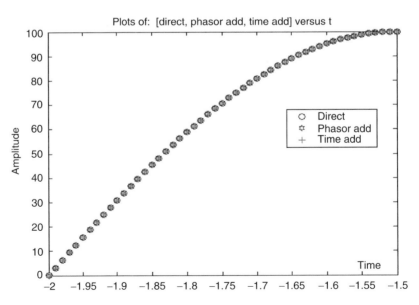

FIGURE 6.17
Plots of Example 6.10(c).

```
**************R  E  S  U  L  T  S  ***************
*********************************************
** MATLAB   phasor approach           **
** the magnitude and phase of F1+F2 are: **
     ans =
          100.0000    -0.0001
*********************************************
```

Example 6.11

Create the script file *Example611* that returns the following plots using the polar coordinate system of the following polar equations (Kay, 1994):

a. $r_1 = 1 + 2 \cos(nx)$, for $n = 1, 2, 3, 4$
b. $r_2 = 1 + 2 \sin(nx)$, for $n = 1, 2, 3, 4$
c. $r_3 = 2 \cos(nx)$, for $n = 1, 2, 3, 4$
d. $r_4 = x^2$, over the range $0 < x < 4\pi$

Observe the effects of the constant term, on the sinusoids (sines and cosines) and n.

MATLAB Solution
```
% Script file: Example611

figure(1)
subplot(2,2,1)
ezpolar('1+2*cos(x)',[0,2*pi])
subplot(2,2,2)
ezpolar('1+2*cos(2*x)',[0,2*pi])
subplot(2,2,3)
ezpolar('1+2*cos(3*x)',[0,2*pi])
subplot(2,2,4)
ezpolar('1+2*cos(4*x)',[0,2*pi])

figure(2)
subplot(2,2,1)
ezpolar('1+2*sin(x)',[0,2*pi])
subplot(2,2,2)
ezpolar('1+2*sin(2*x)',[0,2*pi])
subplot(2,2,3)
ezpolar('1+2*sin(3*x)',[0,2*pi])
subplot(2,2,4)
ezpolar('1+2*sin(4*x)',[0,2*pi])

figure(3)
subplot(2,2,1)
ezpolar('2*cos(x)',[0,2*pi])
subplot(2,2,2)
ezpolar('2*cos(2*x)',[0,2*pi])
subplot(2,2,3)
ezpolar('2*cos(3*x)',[0,2*pi])
subplot(2,2,4)
ezpolar('2*cos(4*x)',[0,2*pi])

figure(4)
ezpolar('(x)',[0,4*pi])
```

The script file *Example611* is executed and the resulting plots are shown in Figures 6.18 through 6.21.

Polar plots of $r1 = 1+2 \cos(nx)$ for $n = 1,2,3,4$

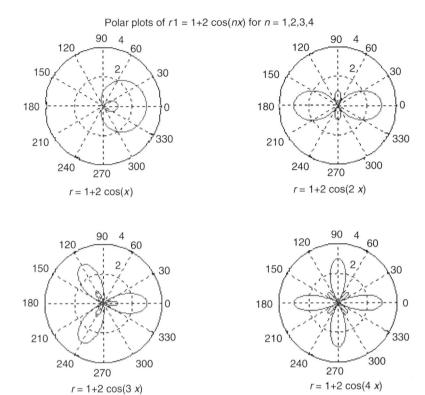

$r = 1+2 \cos(x)$

$r = 1+2 \cos(2x)$

$r = 1+2 \cos(3x)$

$r = 1+2 \cos(4x)$

FIGURE 6.18
Plots of Example 6.11(a).

Polar plots of $r2 = 1+2 \sin(nx)$ for $n = 1,2,3,4$

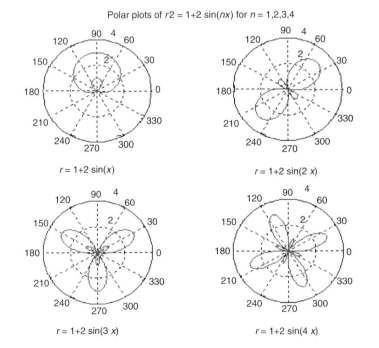

$r = 1+2 \sin(x)$

$r = 1+2 \sin(2x)$

$r = 1+2 \sin(3x)$

$r = 1+2 \sin(4x)$

FIGURE 6.19
Plots of Example 6.11(b).

Polar plots of $r3 = 2\cos(nx)$ for $n = 1,2,3,4$

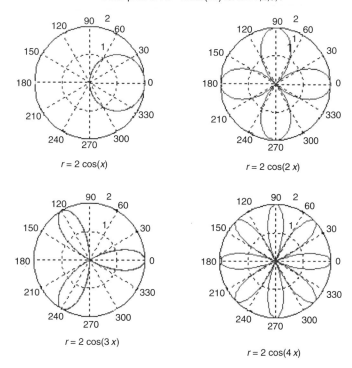

$r = 2\cos(x)$

$r = 2\cos(2x)$

$r = 2\cos(3x)$

$r = 2\cos(4x)$

FIGURE 6.20
Plots of Example 6.11(c).

Polar plot of $r4 = x^2$ for $0 < x < 4$ pi

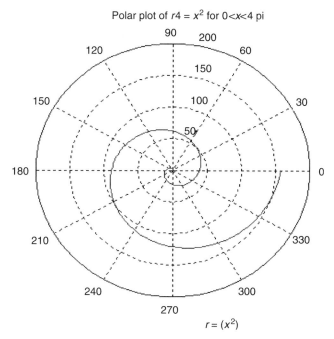

$r = (x^2)$

FIGURE 6.21
Plots of Example 6.11(d).

Example 6.12

Create the script file *Example612* that returns the plots on the complex plane of the roots of the following equations:

a. $x_1 = (-1)^{1/20}$
b. $x_2 = (-1)^{1/40}$

From the plots, verify that the roots are located on a circle with radius $r^{1/n}$, where $n = 20$ and 40, centered at the origin of the complex plane, separated by n equal arcs. Also verify for x_1 that the roots occur in complex conjugate pairs. It is left for the reader to verify that the roots for x_2 occur also in complex conjugate pairs.

The polynomial equations are given as follows:

a. $x_1^{20} + 1 = 0$
b. $x_2^{40} + 1 = 0$

MATLAB Solution
```
% Script file: Example612
X1 = [1 zeros(1,19) 1];
X2 = [1 zeros(1,39) 1];
roots _ X1=[roots(X1)]';


figure(1)
plot (real(roots _ X1),imag(roots _ X1),'o')
title('Plot of the roots of X1')
axis([-1.2 1.2 -1.2 1.2])
grid on
roots _ X2 = [roots(X2)]';


figure(2)
plot (real(roots _ X2),imag(roots _ X2),'o')
title ('Plot of the roots of X2')
axis ([-1.2 1.2 -1.2 1.2])
grid on


figure(3)
plot (real(roots _ X1),imag(roots _ X1),'o',real(roots _ X1),imag(roots _ X1))
title ('Plot connecting the roots of X1 ')
axis ([-1.2 1.2 -1.2 1.2])
grid on
```

The script file *Example612* is executed and the resulting plots are shown in Figures 6.22 through 6.24.

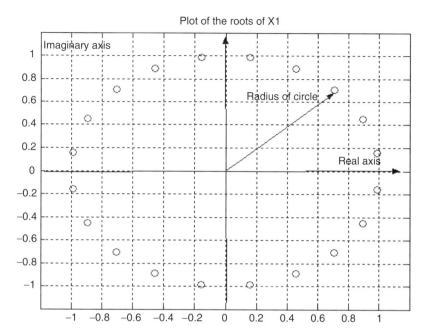

FIGURE 6.22
Plots of the roots of x_1 of Example 6.12.

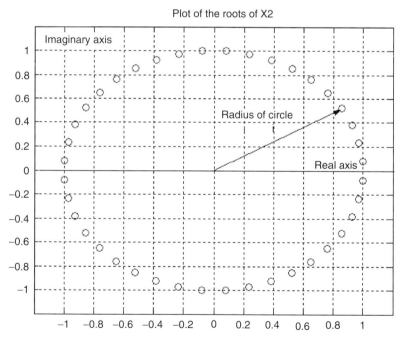

FIGURE 6.23
Plots of the roots of x_2 of Example 6.12.

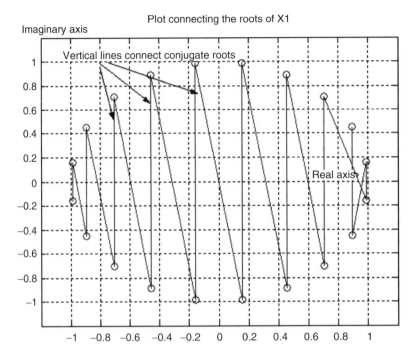

FIGURE 6.24
Complex conjugate plot of the roots of x_1 of Examples.

6.5 Further Analysis

Q.6.1 Load and run the program of Example 6.1.

Q.6.2 Define and discuss each of the following commands *conj*, *abs*, and *angle*.

Q.6.3 Evaluate by hand the following MATLAB instructions:

$conj(z_1)$

$conj(z_2)$

$abs(z_1)$

$abs(z_2)$

$angle(z_1)$

$angle(z_2)$

Q.6.4 Identify the variable names assigned to the commands defined in Q.6.3 and compare the MATLAB results with the manual results of Q.6.3.

Q.6.5 Modify and rerun Example 6.1 for the case $z_1 = 1 - i2$ and $z_2 = 3 - i4$.

Q.6.6 Repeat Q.6.5 for the case: $z_1 = +i2$ and $z_2 = +i4$.

Q.6.7 Repeat Q.6.5 for the case: $z_1 = 1$ and $z_2 = 3$.

Q.6.8 Define the parameters that control the magnitude, and phase of z_1^* and z_2^* (recall that * denotes complex conjugates).

Q.6.9 Define and discuss each of the following MATLAB commands: *real(z)* and *imag(z)*, when *z* is expressed in polar and exponential forms.

Q.6.10 Which parameters control the angle? Discuss how to obtain a maximum and minimum phase angle for z_1 and z_2.

Q.6.11 Modify Example 6.1 to display the plot $abs(z_1)$ versus $abs(z_2)$.

Q.6.12 Repeat Q.6.11 for the case of *compass(x, y)* and *feather(x, y)*, for $x = abs(z_1)$ and $y = abs(z_2)$.

Q.6.13 Load and run the script file *discrete* of Example 6.2.

Q.6.14 Which variables control the rate of growth of $y(n)$?

Q.6.15 Which variables control the amplitude of $y(n)$?

Q.6.16 What is the length of the sequence labeled y?

Q.6.17 Indicate how to change the length of the sequence label y to 100 elements, over the same range.

Q.6.18 Consider the following case: z is changed to $z = conj(z)$.

Will the magnitude change?

Will the phase change?

Q.6.19 Repeat Q.6.18 for the cases $z = conj(-z)$ and $z = -conj(z)$.

Q.6.20 Load and run the program of Example 6.3.

Q.6.21 Evaluate the frequencies of y_1 and y_2.

Q.6.22 What is the period and peak value for y_1 and y_2?

Q.6.23 Express the ys in term of a cosine and sine wave in a standard form as $y = cos(\omega t + \theta)$ and $y = sin(\omega t - \Phi)$

Q.6.24 Are the angles θ and Φ leading or lagging with respect to the reference wave $cos(\omega t)$? Discuss.

Q.6.25 Repeat questions Q.6.23 and Q.6.24, for the case $y = \dfrac{e^{jz} - e^{-jz}}{2j}$ and $z = 2\pi n + \pi/3$.

Q.6.26 Determine the length of the sequences y_1 and y_2 in Example 6.3, and indicate how they can be changed.

Q.6.27 Modify the equations for y_1 and y_2 of Example 6.3, when the frequency is $f = 0.4$ Hz and the amplitude is 2.

Q.6.28 Modify the equations for y_1 and y_2 in Example 6.3 for the case $f = 1.2$ Hz and unit amplitude.

Q.6.29 Modify the program of Example 6.3 for the case of $y1 = iy_1$ and $y2 = -iy_2$.

Q.6.30 Modify the program of Example 6.3 by replacing the instruction *stem(n, y)* with *bar(n, y)* and rerun the program. Compare and discuss the results.

Q.6.31 Repeat Q.6.30 for the case of *stairs(n, y)* and *plot(n, y)*.

Q.6.32 Load and run the program of Example 6.4.

Q.6.33 Analyze *subplot(2, 2, 1)* with *subplot(2, 2, 2)* of Figure 6.10. Is that what you expect? Explain and discuss.

Q.6.34 Define and discuss the command *polar*.

Q.6.35 Provide a practical example when the command polar can be of interest.

Q.6.36 Evaluate by hand $z = z_1 + z_2$, and compare it with the result obtained in *subplot(2, 2, 3)*.

Q.6.37 Describe the graph obtained in *subplot(2, 2, 4)* of Figure 6.9.

What variables affect the function growth?

What variables affect the displacement?

Q.6.38 Load and run the program of Example 6.5.

Q.6.39 What are the lengths of the sequences Z and F?

Q.6.40 Replace the *plot* command with the *stem* command and rerun the program.

Q.6.41 Repeat question Q.6.40 using the *stair* command.

Q.6.42 Modify the program of Example 6.5 to evaluate the maximum and minimum values of the magnitude of F, with respect to X.

Q.6.43 Repeat question Q.6.42 for the case of angle.

Q.6.44 Load and run the program of Example 6.6.

Q.6.45 Determine the amplitude, frequency, and phase of the function plotted on Figure 6.12 *subplot(2, 1, 1)*.

Q.6.46 Repeat question Q.6.45 for Figure 6.12 *subplot(2, 1, 2)*.

Q.6.47 Load and run the program of Example 6.7.

Q.6.48 What are the coefficients that control the magnitude of $f(t)$?

Q.6.49 Express analytically the real part of $f(t)$.

Q.6.50 Express analytically the imaginary part of $f(t)$.

Q.6.51 Rerun the program of Example 6.7 for $-\pi \le t \le 2\pi$, using 100 points.

Q.6.52 Use MATLAB to evaluate and plot *abs(f(t))* versus t, for $t < 0$.

Q.6.53 Modify the program to obtain the plot of *real [f(t)]* versus *imag[f(t)]*.

Q.6.54 Discuss and comment on how the plot of the *angle [f(t)]* versus *abs[f(t)]* would be affected for $t < 0$.

Q.6.55 Load and run the program of Example 6.8.

Q.6.56 What is a phasor and what variables define a phasor?

Q.6.57 Discuss when and why phasors are used in engineering problems.

Q.6.58 What is the main assumption when phasors are used?

Q.6.59 Describe the steps involved when evaluating *abs(z)* by hand, a calculator, or using MATLAB.

Q.6.60 Compare and discuss the time domain versus the phasor domain representation.

Q.6.61 If MATLAB is available, what would be the best way to evaluate the addition of Example 6.8?

Q.6.62 Load and run the program of Example 6.9.

Q.6.63 Describe the steps involved when evaluating *angle(z)* by hand, using a calculator, and by using MATLAB.

Q.6.64 Load and run the program of Example 6.10.

Q.6.65 Obtain plots of F_1 and F_2 as phasors.

Q.6.66 Observe that F_2 leads F_1. Evaluate the phase angle in degrees and radians between F_1 and F_2.

Q.6.67 Calculate by hand $F_1 + F_2$ and compare your result with F_3.

Q.6.68 Load and run the program of Example 6.11.

Q.6.69 Define what is a polar equation.

Q.6.70 Explain when and why it is convenient to use polar coordinates.

Q.6.71 Give a practical example when polar coordinates should be used.

Q.6.72 Modify and rerun your program by replacing *ezplot* by *plot* first, and then by *ezpolar*.

Q.6.73 What are the relations between n and the graphs obtained?

Q.6.74 What are the relations between the sine and cosine in polar coordinates for different values of n?

Q.6.75 Load and run Example 6.12.

Q.6.76 How would you calculate the arc length between the roots?

Q.6.77 How would you calculate the angle between the roots?

Q.6.78 What are the main values for the equations (a) and (b)?

Q.6.79 Plot the roots in polar coordinates.

6.6 Application Problems

P.6.1 Given the following quadratic equations:

    a. $x^2 - 4x - 5 = 0$

    b. $x^2 - 4x + 4 = 0$

    c. $x^2 - 4x + 13 = 0$

    Verify by hand and using MATLAB that the corresponding solutions are

    a. $x_1 = -1$ and $x_2 = 5$

    b. $x_1 = x_2 = 2$

    c. $x_1 = 2 + 3\sqrt{-1}$ and $x_2 = 2 - 3\sqrt{-1}$

P.6.2 Plot the quadratic equation given in P.6.1 and verify graphically the results obtained, if possible.

P.6.3 Verify analytically and by using MATLAB the following equalities:

$$i^{347} = i^3 = -i \qquad i^{943} = -i \qquad i^{-1134} = -1$$
$$i^{880} = -1 \qquad i^{-880} = 1$$

P.6.4 Let $z_1 = 2 - i$, $z_2 = -1 + 3i$, and $z_3 = -3 - 4i$.

    Verify which of the following relations are true:

    a. $z_1 + z_2 = 4 + 6i$

    b. $z_1 * z_2 = 1 + 7i$

    c. $z_1 i - (z_1 + z_2)^2 = -14 - 6i$

    d. $z_2^2 = 8 - 18i$

    e. $\sqrt{z_3} = \begin{cases} 1 - 2i \\ -1 + 2i \end{cases}$

P.6.5 Given $z_1 = 3 - 4i$ and $z_2 = 2 + i$. Verify that

$$\frac{z_1}{z_2} = \frac{2}{5} - \frac{11}{5}i$$

by hand and by using MATLAB.

P.6.6   Given $z_1 = 3 + i$ and $z_2 = 2 - 3i$. Verify the following equality:

$$|2z_1 - 5z_2| = \sqrt{305}$$

by hand and by using MATLAB.

P.6.7   Given $z_1 = 3 + i$ and $z_2 = -2 + 4i$. Verify that

$$\left| \frac{z_1 + z_2 + 1}{z_1 + z_2 + 1} \right| = \frac{3}{5}$$

P.6.8   Verify which of the following relations are true:
a. $(3 - 2i)^2 = -9 - 46i$
b. $\sqrt{3} - 2i = \sqrt{7}(\cos 310°89' + j \sin 310°89')$
c. $1 + \sqrt{3}i = (1 + \sqrt{3}) \cos 90°$
d. $1 - \sqrt{3}i = 2(\cos 300° + j \sin 300°)$
e. $-\sqrt{3}i = \sqrt{3}(\cos 180° + j \sin 180°)$
f. $\dfrac{(3 + i)(1 - i)^2}{(i - 3)(3 - i)} = \dfrac{13}{25} + \dfrac{9}{25} \cdot i$
g. $\dfrac{5(-3i)(1 + i)}{5 - \sqrt{5} \cdot i} = -3.62 - 1.38i$
h. $(1 + 5i)^{-3} = -0.004 + 0.06i$

P.6.9   Given the complex numbers A and B by

$$A = 2 + 2j \quad \text{and} \quad B = \sqrt{3} \angle 30°$$

perform the following operations:
a. $A + B$
b. $A * B$
c. A/B
d. A^2
e. $(A + B)^2$

P.6.10   Given the complex numbers $z_1 = 3 + 4j$ and $z_2 = 5 + j2$. Use MATLAB and evaluate the following:
a. $|z_1|$, $|z_2|$, and $|z|$, where $z = z_1/z_2$
b. $\angle z_1$, $\angle z_2$, and $\angle z$
c. $|z_1| \cdot |z_2|$
d. $|z_1| \cdot |z_2{}^*|$
e. $|z_1| \cdot |z_2{}^*|$
f. $|z_1{}^* \cdot z_2{}^*|$
g. $z_1 \cdot z_2$ (* denotes complex conjugate)

P.6.11 Let $z = 4 - 3j$, perform the following operations by hand and by using MATLAB:
   1. $sqrt(z) =$
   2. $log(z) =$
   3. $z^I =$
   4. $z^{3.25} =$
   5. $z*z' =$
   6. $z/z' =$

P.6.12 Verify analytically and by using MATLAB the following relation:

$$\frac{1}{j} = -j$$

P.6.13 Verify analytically and by using MATLAB that

$$e^{\pi\sqrt{-1}} = -1$$

P.6.14 Verify that

$$e^{j\varphi} = cos(\varphi) + jsin(\varphi), \quad \text{for } \varphi = 0, \frac{\pi}{3}, \frac{\pi}{2}, \text{and } \pi.$$

P.6.15 Express the complex functions $f(z)$ given below, where $z = 1 + 2j$ in terms of its real and imaginary parts
   a. $f(z) = z^2 + 2z + 5$
   b. $f(z) = (z + 1)(z - 2)/(z + 3)$
   c. $f(z) = (z + 1)/(z^2 + 2z + 5)$

P.6.16 The function $f(z)$ is defined by the following expression:

$$f(z) = 5(z + 2)(z^2 + 3)/(z - 1)$$

where $z = (2 + i)^2$
   Determine
   a. $|f(z)|$ and $\angle f(z)$
   b. $real[f(z)]$
   c. $imag[f(z)]$

P.6.17 Let $f(z) = 10e^{-zn}$ be a function of the complex variable $z = 1 - j$
   a. Express $f(z)$ in rectangular format for $n = 1$
   b. Plot $|f(z)|$ versus n
   c. Plot $|f(z)|$ versus $\angle f(z)$

P.6.18 Using MATLAB, obtain simplified expressions in rectangular and polar forms of the following equations:
   a. $z = (1 + 2j)(1 - 2j)e^{2j}$
   b. $z = (3 + 2j)(3 + 4j)/e^{2j}$
   c. $z = 4e^{-2j}(2 - 3j)/(\sqrt{3} \cdot e^{2j})$

P.6.19   Let $z = r(cos\theta + jsin\theta)$ be a complex function then the n roots of z are defined by

$$z_k = \sqrt[n]{r}\left(\cos\left(\frac{\theta}{n} + \frac{2\pi k}{n}\right) + j\sin\left(\frac{\theta}{n} + \frac{2\pi k}{n}\right)\right)$$

or if $z = r \angle \theta$, then

$$z_k = \sqrt[n]{r} \angle (\theta + n\,360°)/k, \quad \text{for } k = 0, 1, 2, ..., n - 1$$

Do you agree with this result? Solve then $z^3 = 1 + j$ using MATLAB and compare the result obtained with the formulas given above.

P.6.20   Repeat P.6.17 for the case $z = 3 + j4$.

P.6.21   The three roots of a complex number z are given as follows:

$$z_0 = \sqrt[3]{3}\,(\cos 40° + j\sin 40°)$$

$$z_1 = \sqrt[3]{3}\,(\cos 160° + j\sin 160°)$$

$$z_2 = \sqrt[3]{3}\,(\cos 280° + j\sin 280°)$$

Determine z in polar and rectangular forms.

P.6.22   The MacLaurin series expansions of $cos(z)$ and $sin(z)$ are given as follows:

$$\cos z = 1 - \frac{z^2}{2!} + \frac{z^4}{4!} - \frac{z^6}{6!}$$

$$\sin z = z - \frac{z^3}{3!} + \frac{z^5}{5!} - \frac{z^7}{7!}$$

Show by using MATLAB that $e^{jz} = cos\,z + j sin\,z$.

P.6.23   Using MATLAB prove the following relations:

$$z + z^* = 2a$$

$$z - z^* = -j2b$$

$$z * z^* = a^2 + b^2$$

where $z = a + jb$, for $a = 4$ and $b = -3$ (recall that z^* denotes complex conjugate).

P.6.24   Using MATLAB verify the following equalities:

a. $z = 2(\cos 30° + j\sin 30°) = 2\,[\frac{\sqrt{3}}{2} + 0.5i]$

b. $6 \angle 45° + 3 \angle 45° = 9 \angle 0°$

c. $6 \angle 0° - 7 \angle 180° = 13 \angle 0°$

P.6.25   Using MATLAB determine that if: $z_1 = r_1(cos\,\theta_1 + jsin\,\theta_1)$ and $z_2 = r_2(cos\,\theta_2 + jsin\,\theta_2)$, then

$$z_1 \cdot z_2 = r_1 r_2(\cos(\theta_1 + \theta_2) + j\sin(\theta_1 + \theta_2))$$

for $r_1 = 2, r_2 = 3, \theta_1 = \pm\pi/4$, and $\theta_2 = \pm\pi/3$.

P.6.26 Using z_1 and z_2 of P.6.25, prove

$$\frac{z_2}{z_1} = \frac{r_2}{r_1}(\cos(\theta_2 - \theta_1) + j\sin(\theta_2 - \theta_1))$$

P.6.27 The De Moivre theorem states that if $z = r(\cos\theta + j\sin\theta)$, then $z^n = r^n(\cos n\theta + j\sin n\theta)$. Write a MATLAB program that verifies the De Moivre theorem for the case when $z = 1 + 2j$, for $n = 2, 3,$ and 4.

P.6.28 Verify if the following relations are true:

a. $1 + j = \sqrt{2}e^{j45°}$

b. $(1 + j)^5 = -4 \cdot (1 + j)$

c. $(3 + 4j)^3 = -117 + 44j$

d. $8 \cdot (\cos 20° + j\sin 20°)^3 = 4 \cdot (1 + \sqrt{3}j)$

e. $(\cos 20° + j\sin 20°) \cdot (\cos(-80°) + j\sin(100°)) = \cos(280°) + j\sin(230°)$

f. $\dfrac{\cos 20° + j\sin 20°}{\cos(-80°) + j\sin 100°} = \cos 120° + j\sin 120°$

P.6.29 Given the following two complex matrices A and B:

$$A = \begin{bmatrix} 13 - j4 & -20 \\ -10 & 20 - 10\angle 45° \end{bmatrix}$$

$$B = \begin{bmatrix} 13 - j4 & -10 \\ -10 & 15 \end{bmatrix}$$

Write a MATLAB program that performs the following:

a. Input matrices A and B

b. Evaluate the determinant of the matrix A, $[det(A)]$, and show that it is equal to $156 \angle 247.4°$

c. Repeat point (b) for matrix B, and show that $det(B) = 112.3 \angle 32.4°$

d. Show then that $|A|/|B| = 1.39 \angle 279.7°$

P.6.30 Using MATLAB verify that $[A]/[B] = 3.33$, where

$$A = \begin{bmatrix} 5 + j5 & 55.8\angle - 17.4 \\ -j5 & 0 \end{bmatrix}$$

and

$$B = \begin{bmatrix} 5 + j5 & -j5 \\ -j5 & 8 + j8 \end{bmatrix}$$

P.6.31   Verify the following relations:

a. $\dfrac{(5 + j5)(15 + j5)}{5 + j5 + 15 + j5} = 4 + j3$

b. $5 \angle 30° \cdot \dfrac{5 + j5}{20 + j10} = 1.585 \angle 48.4°$

c. $5 \angle 30° \cdot \dfrac{5 + j5}{5 + j5 + 10} = 2.24 \angle 56.6°$

d. $\dfrac{12 + j24}{33 + j24} - \dfrac{(30 + j60) \cdot 20}{80 + j60} = 0.328 \angle 170.5°$

e. $\dfrac{15 - j6}{2 \cdot (4 - j)} = 1.94 - j0.265$

f. $\dfrac{(3 + 4i)}{(3 + 3i)} = 0.96 + 0.28i$

g. $(2 + 4i)^4 = -119 - 120i$

P.6.32   Using MATLAB show that $|e^{-jx}| = 1$ for any value of x. Test the preceding equality for following MATLAB sequence of numbers:

$$x = 0{:}.1\pi{:}2\pi$$

P.6.33   Using MATLAB evaluate

a. j^{-j}

b. j^j

P.6.34   Given the exponential function

$$y(t) = 3e^{-2t}(\cos 2t + i\sin 2t)$$

   Write a MATLAB program that generates the following plots:

a. $y(t)$ versus t, for $0 \le t \le 5$, using 100 points

b. $real[y(t)]$ versus t

c. $imag[y(t)]$ versus t

d. $real[y(t)]$ versus $imag[y(t)]$

e. $|y(t)|$ versus $\angle y(t)$ (use polar and plot commands)

P.6.35   Express the following time functions in phasor form:

a. $5[sin(wt - 35°)]$

b. $2.7^2 \cdot (0.34) \, sin(377t + 45°)$

c. $3.62 \, cos(wt + 60)$

d. $8.39 \, cos(377t - 123°)$

P.6.36   Express and plot in the time domain the following phasors as cosine waves, for a frequency of 60 Hz.

a. $A = \sqrt{3} \angle 30°$

b. $B = 5 \angle 135°$

c. $C = \sqrt{3}(3.1)^{2.23} \angle 58°$

d. $D = \dfrac{\sqrt{3} \, \sqrt[4]{8}}{\sqrt[3]{13}} \angle 35.8°$

P.6.37 Obtain plots of the sinusoidal expressions $f(t) = 15\sin(377t + 30°) - 22\sin(377t - 60°)$ by

a. Using the phasors representation

b. The time domain representation

P.6.38 Given $f_1(t) = 5\cos(\pi t/4 + 36°)$ and $f_2(t) = 7\cos(\pi t/4 - 30°)$. Evaluate $f_1(t) + f_2(t)$ and $f_1(t) - f_2(t)$ by

1. Hand

2. Using phasor representation

3. Using the time domain representation

P.6.39 Verify that the four roots of $x = (-1)^{1/4}$ are

$$x_1 = \frac{\sqrt{2}}{2} + j\frac{\sqrt{2}}{2}$$

$$x_2 = -\frac{\sqrt{2}}{2} + j\frac{\sqrt{2}}{2}$$

$$x_3 = -\frac{\sqrt{2}}{2} - j\frac{\sqrt{2}}{2}$$

$$x_4 = \frac{\sqrt{2}}{2} - j\frac{\sqrt{2}}{2}$$

Plot the roots of $x = (-1)^{1/4}$ on the complex plane.

7

Polynomials and Calculus, a Numerical and Symbolic Approach

> I keep the subject constantly before me and wait till the first dawning open little by little into full light.
>
> **Sir Isaac Newton**

7.1 Introduction

Polynomials are algebraic expressions consisting of the sum of one or more product terms. A function defined by a polynomial expression is referred to as a polynomial. In its simplest form, each of the product terms consists of a coefficient, and a variable of interest (x) raised to a nonnegative integer power. A polynomial {of one variable (x)} can in general be expressed as

$$f(x) = a_n x^n + a_{n-1} x^{n-1} + \cdots + a_1 x + a_0 = \sum_{1=0}^{n} a_1 x^1$$

where x is the variable, the a's ($a_n, a_{n-1}, a_{n-2}, \ldots, a_0$) are the coefficients assuming that $a_n \neq 0$, and the nonnegative integer n with the highest exponent defines the degree of the polynomial.

Some of the most frequently used polynomials are defined as follows:

$f(x) = 0,$	where $f(x)$ has no degree
$f(x) = a_0,$ where $a_0 \neq 0,$	and the degree of $f(x)$ is zero
$f(x) = a_1 x + a_0,$ where $a_1 \neq 0,$	the degree of $f(x)$ is 1, and defines a linear relation
$f(x) = a_2 x^2 + a_1 x + a_0,$ where $a_2 \neq 0,$	the degree of $f(x)$ is 2, and defines a quadratic relation
$f(x) = a_3 x^3 + a_2 x^2 + a_1 x + a_0,$ where $a_3 \neq 0,$	the degree of $f(x)$ is 3, and defines a cubic relation

Polynomials are extensively used in technology, engineering, and the sciences because they can best represent, or model a physical system. Polynomials are also easy to define and to evaluate because they involve the basic operations of additions, subtractions, and multiplications.

In engineering and science, it is often required to determine the roots of the polynomial equation defined by

$$f(x) = \sum_{m=0}^{n} a_m x^m = 0$$

where in general the coefficients a_m, for $m = 0, 1, 2, 3, \ldots, n$, may be complex.

An often-encountered polynomial in science and engineering is the quadratic, or second-order polynomial. It is well-known that the roots of the quadratic polynomial (equation of the form $a_2x^2 + a_1x + a_0 = 0$) are

$$x_{1,2} = \frac{-a_1}{2a_2} \pm \frac{\sqrt{a_1^2 - 4a_0a_2}}{2a_2}$$

For higher-order polynomials, such analytic expressions do not in general exist. Fortunately, cubic and higher-order polynomials occur infrequently in engineering and science, and when they do occur, they are difficult to solve.

Graphical means and trial and error were in general the preferred methods used in the past. Luckily, modern software packages such as MATLAB® provide simple and powerful functions to evaluate the roots (also called zeros) of a polynomial. For example, the MATLAB command *roots (P)* returns a column vector with the roots of the polynomial P, where P is defined by the order coefficients of the polynomial $f(x)$ as an array $P = [a_n, a_{n-1}, ..., a_0]$.

Polynomials as well as an extensive list of other functions can be converted into symbolic objects or expressions. Symbolic variables differ from the ordinary MATLAB variables in the sense that ordinary MATLAB variables are associated with numerical values, which must be defined and stored in the computer's memory before they are used, whereas symbolic variables are not associated with any numerical value, but rather indicate a qualitative relation among the variables.

MATLAB requires for that reason that symbolic variables be defined as symbolic (*syms*) before they are used.

Integration and differentiation are just two examples of symbolic applications defined as functions in the MATLAB Symbolic Toolbox. These types of operations can be performed using MATLAB with almost unlimited precision.

The MATLAB Symbolic Toolbox is a link between MATLAB and the symbolic algebra known as Maple. Integration and differentiation are part of a branch of mathematics known as calculus. Calculus was developed in the seventeenth century with the objective of providing a tool for solving problems involving motion such as velocity and acceleration. Calculus was first applied in physics, and since then calculus has been used in many fields of studies, specially involving the rate of change of a variable of interest. The concept of rate of change is used in diverse disciplines and applications; for example, in economics, as the rate of growth of the money supply (as an index of inflation); in chemistry, as the rate of change of a chemical reaction with respect to heat in electricity, as the rate of change of an electric charge with respect to time better known as electric current; in biology, as the rate of change of the hormones; in psychology, as the rate of change of the IQ over time, etc.

In the past centuries, many engineers and scientists studied the dynamic behavior of a wide range of physical systems, and in many cases, the studies showed how the rate of change between two or more variables interrelate. Those problems are best described in terms of differential equations (DEs), and MATLAB provide with a number of functions to evaluate their solution, if such solution exists.

A preferred first model used by engineers and educators, in many practical applications, is to describe a given system by means of a second-order DE. In the past, the techniques used to solve those equations consisted of using a series of approximations, transforms, or Green's functions.

Most practical systems, however, are described by higher-order DEs.

These equations can usually be transformed into sets of interrelated first-order DEs, called the state–space form, which can easily be solved using MATLAB, or be transformed into an equivalent discrete system and then be solved.

MATLAB provides two approaches to solve a DE or a system of simultaneous DEs.

1. A symbolic solution using the function *dsolve*
2. A numerical solution using a family of *ode* (ordinary differential equations) solvers (such as, *23*, *45*, and *116*)

Both methods will be presented and discussed in this chapter from a general system approach, as well as many other applications.

7.2 Objectives

After completing this chapter the reader should be able to

- Input polynomials using MATLAB
- Determine the degree of a polynomial
- Determine the roots of a polynomial
- Evaluate a polynomial for a specific value or range of values
- Perform polynomial algebra
- Obtain the transfer function of a system (division of polynomials)
- Plot the transfer function of a system
- Decompose a rational function into partial fractions expansion
- Provide a pole-zero diagram of a system transfer function
- Create symbolic variables and symbolic expressions
- Understand concepts and nomenclatures used in calculus (such as limits, continuous, delta, maximal, and minimal)
- Understand the concepts of integration and differentiation
- Use MATLAB numerical methods to evaluate integrals and derivatives
- Integrate and differentiate a variable, a constant, a trigonometric, a logarithmic, and an exponential function
- Understand the concept of a differential and a partial DE
- Understand concepts and nomenclatures used in DEs (such as ordinary, linear, and homogeneous)
- Identify and understand the concept of smoothness and continuity
- Understand that differentiation and integration are inverse operations
- Perform algebra and calculus operations on symbolic expressions
- State the solution of first-order linear DEs as the sum of two solutions (particular and general)
- Know that the general solution of a linear DE contains arbitrary constants

- Know that the particular solution contains no arbitrary constant
- Know that a unique solution can be calculated if initial or boundary conditions are given
- Solve a system of DEs using symbolic and numerical methods
- Understand the strength and weakness of the numerical and symbolic method used to solve DEs
- Understand the meaning and implication of the term stiff DEs
- Convert a high-order linear DEs into a set of first-order DEs
- State the state–space model description of a given linear time invariant system
- Convert the state–space equations into a transfer function, DE, and vice versa
- Discretize a continuous (linear time invariant) system
- Know and be able to use a number of functions available in the MATLAB Symbolic Toolbox to solve a variety of problems and applications

7.3   Background

R.7.1   A polynomial consists of two or more terms. For example, $y - x$, $x^2 + 2x + y$.

R.7.2   A monomial is an algebraic expression that consists of exactly one term, for example, $2x$, xy, $3x^2$. A polynomial is a monomial, or a sum of monomials.

R.7.3   A binomial is a polynomial that consists of exactly two terms, for example, $3x + 2y$, $xy - 3z$.

R.7.4   Polynomial functions are referred to as polynomials.

R.7.5   A polynomial in one variable (x) can be arranged in one of two ways: ascending or descending order.

R.7.6   A polynomial arranged in ascending order means that the power of the variable of interest (x) increase for each succeeding term, such as $f(x) = 3 + 2x + 4x^2 + 5x^3$.

R.7.7   A polynomial arranged in descending order means that the power of the variable of interest (x) decrease for each succeeding term, such as $f(x) = 5x^3 + 4x^2 + 2x + 3$.

R.7.8   The graph of a polynomial is characterized by being smooth and continuous. Smooth means no sharp variations, and continuous means no gaps.

R.7.9   When a polynomial $f(x)$ is equated to zero, $f(x) = 0$ is referred as the polynomial equation.

R.7.10   Let $f(x)$ be a polynomial with x as its variable, and let r be a number (real or complex) such that $f(r) = 0$; then r is called a zero, a root, or a solution of $f(x)$.

R.7.11   To solve a polynomial equation means to find the roots of the equation $f(x) = 0$, or decomposing $f(x)$ into its factors.

R.7.12   Let $f(x) = a_n x^n + a_{n-1} x^{n-1} + a_{n-2} x^{n-2} + \cdots + a_1 x + a_0 = 0$, and if $a_n \neq 0$, then $f(x)$ has exactly n roots, also referred to as zeros or solutions.

R.7.13   Let $f(x) = a_n x^n + a_{n-1} x^{n-1} + a_{n-2} x^{n-2} + \cdots + a_1 x + a_0 = 0$, where $a_n \neq 0$. Then the term $a_n x^n$ is called the leading term (the term with the highest degree) and the coefficient a_n constitutes the leading coefficient.

R.7.14 The degree of a polynomial is given by the monomial with the highest degree $a_n x^n$.

R.7.15 If $f(x)$ is divided by a factor $(x + r)$ until a constant remainder is obtained, the remainder is $f(-r)$. Observe then that $f(x) = (x + r) Q(x) + f(-r)$, where $Q(x)$ is a polynomial of degree $n - 1$, one less than $f(x)$.

R.7.16 Let $(x + r)$ be a factor of $f(x)$, then $f(-r) = 0$, and $-r$ is a root or solution of $f(x)$.

R.7.17 Let $c + bj$ be a (complex) root of $f(x)$ (where c and b are real numbers and $b \neq 0$), then $c - bj$, its complex conjugate, is also a root of $f(x)$.

R.7.18 Let n be an odd integer (where n represents the degree of $f(x)$, then $f(x)$ has at least one real root.

R.7.19 Let r be a root of $f(x)$, then $(x - r)$ is a factor of $f(x)$.

R.7.20 Recall that if $f(x) = ax^2 + bx + c$ (quadratic equation), then the two roots of $f(x)$, labeled x_1 and x_2 ($f(x_1) = f(x_2) = 0$) are

$$x_1 = \frac{-b + \sqrt{b^2 - 4ac}}{2a}$$

$$x_2 = \frac{-b - \sqrt{b^2 - 4ac}}{2a}$$

The roots of a quadratic equation can be real or complex numbers. If $b^2 - 4ac < 0$, then the roots are complex, otherwise they are real. If the roots of $f(x)$ are complex, then they occur as complex conjugate pairs, in either quadratic or higher degree polynomials.

R.7.21 Let $(x - r)$ be a factor of $f(x)$ and if that occurs more than once, then r is called a repeated root of $f(x)$.

R.7.22 Let $f(x)$ be a repeated root of multiplicity n_i, then $(x - r)^{ni}$ is a factor of $f(x)$. For example, the polynomial $f(x) = x^3 - 2x^2 - 7x - 4 = 0$, has two repeated roots at -1, and a single root at 4. Then $f(x) = (x + 1)^2(x - 4)$.

R.7.23 Every polynomial $f(x)$ of degree n can generally be transformed into n linear product terms (factors), times a constant C, as indicated in the following equation:

$$f(x) = a_n x^n + a_{n-1} x^{n-1} \ldots a_1 x + a_0 = \sum_{m=0}^{n} a_m x^m = C \prod_{m=1}^{n} (x - r_m) = 0$$

Observe that the polynomial equation $f(x) = 0$ has exactly n roots, labeled $r_1, r_2, \ldots, r_{n-1}, r_n$.[*]

R.7.24 Let a and b be two numbers, and if $f(a)$ is positive and $f(b)$ is negative, then the function $f(x)$ has at least one zero between $x = a$ and $x = b$.

R.7.25 A rational function can be constructed as a division of polynomials.

---

[*] Karl Friedrick Gauss (1777–1855) is credited with important contributions in the study of polynomials. At the young age of 22, in his doctoral dissertation, Gauss proved the fundamental theory in Algebra, which says that a polynomial of degree n has n roots. At 19, Gauss demonstrated that the 17-sided polygon can be constructed using the compass. This idea contradicted the belief supported by the mathematicians since the time of Euclid. From early times mathematicians believed that the only regular polygons that can be constructed with a compass were the triangle and the pentagon.

The general form is given as

$$H(x) = \frac{P(x)}{Q(x)}$$

where $P(x)$ and $Q(x)$ are polynomial functions, with $Q(x) \neq 0$ and no common factors. Ratios of integers are called rational numbers. By extension, ratios of polynomials are called rational functions.

R.7.26 The domain of a rational function consists of all the real numbers except those for which the denominator of $H(x)\{Q(x) = 0\}$ is zero. To sketch a rational function it helps to factor the numerator and denominator of $H(x)$. The x-intercept determines by the zeros of the numerator of $H(x)$ and the y-intercept by the value of the function at $x = 0$.

R.7.27 A short list of often-encountered polynomials and their decomposition into factors are given below:

$$(a + b)(a - b) = a^2 - b^2$$

$$(a \pm b)^2 = a^2 \pm 2ab + b^2$$

$$(a \pm b)^3 = a^3 \pm 3a^2b + 3ab^2 + b^3$$

$$a^3 \pm b^3 = (a \pm b)(a^2 \mp ab + b^2)$$

$$a^n + b^n = (a + b)(a^{n-1} - a^{n-2}b + \cdots + b^{n-1})$$

$$a^n - b^n = (a + b)(a^{n-1} + a^{n-2}b + \cdots + b^{n-1})$$

R.7.28 Descarte's rule of signs for the roots of a polynomial states the following:

Suppose $f(x)$ is a polynomial whose terms are arranged in descending powers of x, then the number of positive real roots of $f(x)$ cannot exceed the number of variations in the sign of the coefficients of the polynomial, and the number of negative roots cannot exceed the number of variations in the sign of the coefficients of $f(-x)$.

R.7.29 To add or subtract polynomials, first arrange them in like terms, and then add or subtract the corresponding coefficients separately. For example, let

$$f(x) = 3x^2 + 5x - 4 \quad \text{and} \quad g(x) = 5x^2 - 2x + 2$$

then

$$f(x) + g(x) = 8x^2 + 3x - 2 \quad \text{and} \quad f(x) - g(x) = -2x^2 + 7x - 6$$

R.7.30 To multiply two polynomials $f(x)$ by $g(x)$, each term of the polynomial $f(x)$ must be multiplied by each term of the polynomial $g(x)$, and then the final expression must be simplified. For example, using the polynomials of R.7.29, then

$$f(x) \cdot g(x) = (3x^2 + 5x - 4)(5x^2 - 2x + 2)$$

$$= 15x^4 - 6x^3 + 6x^2 + 25x^3 - 10x^2 + 10x - 20x^2 + 8x - 8$$

$$= 15x^4 + 19x^3 - 24x^2 + 18x - 8$$

R.7.31 To divide a polynomial $f(x)$ by a monomial $g(x)$, proceed in the following way: divide each term of the polynomial $f(x)$ by the monomial $g(x)$. For example, let

$$f(x) = 8x^3 + 4x^2 + 10x$$

and

$$g(x) = 2x$$

then

$$\frac{f(x)}{g(x)} = 4x^2 + 2x + 5$$

R.7.32 To divide a polynomial by another polynomial, first rearrange the two polynomials in descending order and then use long division, as illustrated in the following:

for example, let

$$f(x) = 8x^3 + 10x^2 + 6x + 4$$

and

$$g(x) = 2x + 2$$

then the division of $f(x)$ by $g(x)$ is accomplished by the process illustrated below, referred to as long division:

$$
\begin{array}{r}
4x^2 + x + 2 \\
2x + 2 \overline{)8x^3 + 10x^2 + 6x + 4} \\
\underline{-(8x^3 + 8x^2)} \\
2x^2 + 6x + 4 \\
\underline{-(2x^2 + 2x)} \\
4x + 4 \\
\underline{-(4x + 4)} \\
0
\end{array}
$$

Therefore,

$$\frac{f(x)}{g(x)} = \frac{8x^3 + 10x^2 + 6x + 4}{2x + 2} = 4x^2 + x + 2$$

R.7.33 To factor $f(x)$, means to find two or more polynomials $y(x)$ and $z(x)$, such that $f(x) = y(x) \cdot z(x)$.

For example, let $f(x) = x^2 - 9$, then $y(x) = (x + 3)$ and $z(x) = (x - 3)$ are factors of $f(x)$.

In a similar way, $x^2 - 5x - 14 = (x - 7) \cdot (x + 2)$ and $4x^3 + 4x^2 + x = x \cdot (2x + 1)^2$.

R.7.34 MATLAB inputs a polynomial of one variable x, as a row vector having as elements the coefficients of the polynomial, real or complex arranged in descending powers of x, placed in brackets. For example, let $y(x) = 3x^4 + 2x^3 + x^2 - x + 5$; then the Y MATLAB vector that represents the polynomial $y(x)$ is given by

```
>> Y = [3 2 1 -1 5];
```

In general, if $y(x) = a_n x^n + a_{n-1} x^{n-1} + a_{n-2} x^{n-2} + \cdots + a_1 x + a_0$, then the polynomial expressed as a MATLAB vector Y is given by

$$Y = [a_n \quad a_{n-1} \quad a_{n-2} \quad \cdots \quad a_1 \quad a_0]$$

R.7.35 When some coefficients of a polynomial are not present, then the missing coefficients are entered as zeros in the MATLAB vector representation. For example, let $y(x) = 8x^7 + 6x^6 + 3x^4 + x^2$, then the vector Y is given by

```
>> Y= [8 6 0 3 0 1 0 0]
```

Observe that the missing coefficients of $y(x)$ are a_5, a_3, a_1, and a_0 and are indicated by zeros in Y.

R.7.36 Let $p(x)$ be a polynomial of a single variable (x), defined by a row MATLAB vector P, where the coefficients of $p(x)$ are the order elements in P.

Then the MATLAB function $r = roots\ (P)$ returns the column vector r with the roots of the polynomial $p(x)$. Using the polynomial of R.7.35, $p(x) = 8x^7 + 6x^6 + 3x^4 + x^2$ as example the following commands return its roots.

```
>> P = [8 6 0 3 0 1 0 0];
>> r = roots(P)


   r =
          0
          0
          1.1246
          0.3594 + 0.4796i
          0.3594 - 0.4796i
          0.1720 + 0.5290i
          0.1720 - 0.5290i
```

R.7.37 Let the roots of the polynomial $p(x)$ be a column vector r; then the MATLAB command *poly (r)* returns the row vector P, with the coefficients of the polynomial $p(x)$ arranged in descending powers of x, as in the following illustration, using the vector r of R.7.36:

```
>> poly(r)

   ans =
         1.0000    0.7500   -0.0000    0.3750   -0.0000    0.1250    0    0
```

Observe that the preceding coefficients are the coefficients of $p(x)$, scaled by a factor of 8 (the command *poly* always returns a value of unity to the leading coefficient).

R.7.38 Let the polynomial $p(x)$ be defined by a row vector P, then the MATLAB function *polyval* (P, k) returns the polynomial $p(x)$ evaluated at $x = k$. For example, let $p(x) = \pi x^4 - \sqrt{7}\, x^3 + 5x - 1$, then the value of $p(x = 0)$ is evaluated by executing the following commands

```
>> P = [ pi - sqrt ( 7 ) 0 5 -1];
>> polyval (P,0)                    % returns p(x) for x = 0

    ans =
        -1
```

The MATLAB command $A = polyval(P, X)$, where X may be a matrix, returns the matrix A with the same size and shape of X, having as its elements the polynomial P, evaluated for each element (value) of X.

R.7.39 The MATLAB command $B = polyvalm(P, X)$, where X is a square matrix, returns B, representing the polynomial matrix evaluated for the matrix X. For example, evaluate the polynomial $p(X) = \pi X^4 - \sqrt{7}X^3 + 5X - 1$, for

$$X = \begin{bmatrix} 1 & 2 & 3 \\ 4 & 5 & 6 \\ 7 & 8 & 9 \end{bmatrix}$$

by using $polyval(P, X)$. Also evaluate $polyvalm(P, X)$ and observe the profound difference.

MATLAB Solution
```
>> P = [pi -sqrt(7) 0 5 -1];
>> X = [1 2 3; 4 5 6; 7 8 9];
>> A = polyval (P,X)

    A =
        1.0e+004 *
        0.0004    0.0038    0.0197
        0.0654    0.1657    0.3529
        0.6669    1.1552    1.8727

>> B = polyvalm (P,X)

    B =
        1.0e+005 *
        0.2252        0.2767        0.3281
        0.5099        0.6265        0.7431
        0.7946        0.9763        1.1581
```

R.7.40 Let us assume that it is desired to obtain a plot of the polynomial $p(x) = \pi x^4 - \sqrt{7}x^3 + 5x - 1$, over the range $-1 \le x \le +1$. The following program returns the plot of $p(x)$, defined by the vector P, using *101* points as shown in Figure 7.1:

MATLAB Solution
```
>> x = -1:2/100:1;
>> length(x)

    ans =
        101
```

```
>> P = [pi -sqrt (7) 0 5 -1];
>> Y = polyval(P,x);
>> plot (x,Y)
>> grid on
>> title ('Polynomial p(x) for -1 < x < 1')
>> xlabel ('variable x'), ylabel ('Amplitude of p(x)')
```

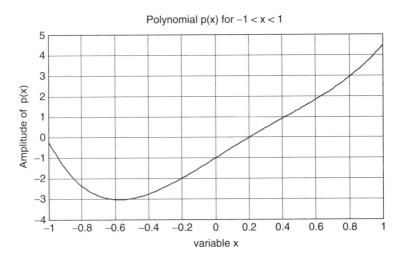

FIGURE 7.1
Plot of $p(x)$ of R.7.40.

R.7.41 Let the coefficients of the two polynomials P and Q be the row vectors defined by $P = [p_n \ p_{n-1} \ \ldots \ p_0]$ and $Q = [q_n \ q_{n-1} \ \ldots \ q_0]$. Recall that the addition $P + Q$ (or subtraction $P - Q$) is accomplished by adding the coefficients of like exponents of P and Q (or subtracting them).

R.7.42 MATLAB can perform the addition (or subtraction) of two polynomials represented by P and Q, only if the two (MATLAB) vectors (P and Q) have the same number of elements (length and size). For example, let

$$p(x) = 3x^4 + 2x^3 + x - 1$$

be represented by P, and

$$q(x) = 5x^3 - 2x^2 + 6$$

be represented by Q. Then the addition and subtraction of the polynomials $p(x)$ and $q(x)$ is indicated by the following program:

```
MATLAB Solution
>> P = [3 2 0 1 -1];
>> Q = [0 5 -2 0 6];
>> sum = P + Q                        % sum = p(x) + q(x)

    sum =
            3       7      -2       1       5
```

```
>> sub = P- Q                          % sub = p(x) - q(x)

   sub =
         3      -3      2      1      -7
```

Note that the vector's lengths are defined by the degree of the highest polynomial {$p(x)$ or $q(x)$}. Recall that the missing coefficients are entered as zeros.

R.7.43 The MATLAB function $M = conv(P, Q)$ returns the row vector M consisting of the coefficients of the product of the two polynomials, $p(x)$ by $q(x)$ represented as row vectors P and Q.

For example, let $p(x) = 3x + 2$ be represented by the variable P and $q(x) = 2x + 4$ be represented by the variable Q. Then the program that returns the product of $p(x)$ by $q(x)$ is illustrated as follows:

MATLAB Solution
```
>> P = [3 2];
>> Q = [2 4];
>> prod = conv(P,Q)

   prod =
         6     16      8
```

This result is interpreted as prod $(x) = p(x) \cdot q(x) = 6x^2 + 16x + 8$.

R.7.44 Observe that the product of more than two polynomials requires repeated use of the *conv* function. For example, let

$$p(x) = 2x^2 + 6x + 4$$

$$q(x) = -3x^2 + 7x - 5$$

$$y(x) = -6x^2 + 18$$

$$z(x) = 5x^3 + 3x + 2$$

Use MATLAB and evaluate the polynomial product consisting of [$p(x)$ $q(x)$ $y(x)$ $z(x)$].

MATLAB Solution
```
>> P = [2 6 4];
>> Q = [-3 7 -5];
>> Y = [-6 0 18];
>> Z = [5 0 3 2];
>> prodPQ = conv(P,Q)

   prodPQ =
          -6     -4     20     -2    -20

>> prodYZ = conv(Y,Z)

   prodYZ =
         -30      0     72    -12     54     36
```

```
>> prodPQYZ = conv(prodPQ, prodYZ)

   prodPQYZ =
           Columns 1 through 6
           180      120      -1032      -156      1764      -816
           Columns 7 through 10
           -480     852      -1152      -720
```

This result is interpreted as the product of

$$(p(x)\, q(x)\, y(x)\, z(x)) = 180x^9 + 120x^8 - 1032x^7 - 156x^6 + 1764x^5 - 816x^4 - 480x^3 \\ + 852x^2 - 1152x - 720$$

R.7.45 The MATLAB command $[D, R] = deconv\,(P_1, P_2)$ performs the following polynomial operation, $P_1/P_2 = D + R/P_2$, and returns D and R as row vectors, where D is the quotient polynomial and R is the residue or remainder polynomial. For example, let

$$p_1(x) = x^4 + 3x^3 + x^2 + 16x$$

and

$$p_2(x) = x^2 + 3x - 1$$

then, $p_1(x)/p_2(x)$ is evaluated by executing the following sequence of commands:

MATLAB Solution
```
>> P1 = [1 3 1 16 0];
>> P2 = [1 3 -1];
>> [D,R] = deconv(P1, P2)

       D =
             1      0      2
       R =
             0      0      0      10      2
```

The preceding result is interpreted as

$$(p_1(x))/(p_2(x)) = x^2 + 2 + ((10x + 2)/(x^2 + 3x - 1))$$

R.7.46 Let the set of order points $<x_i, y_i>$ be defined by the vectors X and Y. Then the MATLAB command $P = polyfit(X, Y, n)$ returns a row vector defining the polynomial P with $n + 1$ coefficients that is the best fit polynomial of degree n or smaller. The coefficients of P are returned as a row vector in descending powers of x.

R.7.47 Recall that a way to connect a given set of order points $<x_i, y_i>$ using a curve that appears to the eye to be smooth is by using the *spline* command. Polynomials of various degrees are used to interconnect polynomial points.

One of the most frequently used plotting techniques is the cubic *spline*. The cubic *spline* uses a cubic polynomial to connect adjacent data points. A general cubic polynomial $y = ax^3 + bx^2 + cx + d$ has four unknown coefficients a, b, c, and d. These unknowns are estimated by solving the following four equations for adjacent points K, $K + 1$ or $K - 1$.

a. Equation 1, data point K

b. Equation 2, data point $K + 1$ or $K - 1$

c. Equation 3, continuity of slopes dy/dx at joints

d. Equation 4, continuity of curvature dy^2/dx^2 at joints

For a set of n points, $n - 1$ cubic polynomials are evaluated having a total of $4(n - 1)$ unknowns. The points as well as the first and second derivatives* provide with $2(n - 1) + 2(n - 2)$ equations, respectively. Since $4(n - 1) = 2(n - 1) + 2(n - 2) + 2$, the extra two equations can be obtained from the data points at both ends. For additional details regarding this subject consult the MATLAB help file.

R.7.48 Let $H(x)$ be a rational function of the form $H(x) = P(x)/Q(x)$; then the partial fraction expansion can be accomplished by using the MATLAB function $[r, p, k] = residue(P, Q)$, where r are the partial fraction coefficients, p the roots of Q (also called poles), and k represents the gain or stand-alone term. For example, let

$$P(x) = 9x^3 + 8x^2 + 7x + 6$$

$$Q(x) = 5x^3 + 4x^2 + 3x + 2$$

Use MATLAB and obtain the partial fraction expansion of $H(x) = P(x)/Q(x)$.

MATLAB Solution
```
>> P = [9 8 7 6];
>> Q = [5 4 3 2];
>> [r,p,k] = residue(P,Q)

    r =
          0.0812 - 0.2848i
          0.0812 + 0.2848i
          0.3224

    p =
          0.0353 + 0.7397i
          0.0353 - 0.7397i
          0.7293

    k =
          1.8000
```

The results are interpreted as

$$\frac{P(x)}{Q(x)} = \frac{9x^3 + 8x^2 + 7x + 6}{5x^3 + 4x^2 + 3x + 2} = \frac{0.0812 - 0.2848i}{x - 0.0353 - 0.7397i} + \frac{0.0812 + 0.2848i}{x - 0.0353 + 0.7397i}$$

$$+ \frac{0.3224}{x - 0.7293} + 1.800$$

The *residue* command illustrated earlier in this chapter can be used if $Q(x)$ does not have multiple roots. If n multiple roots are present, the expansion would include terms such as $x - a \ldots (x - a)^n$, discussed in Chapter 4 of the book titled *Practical MATLAB® Applications for Engineers*.

R.7.49 The same MATLAB function $[P, Q] = residue(r, p, k)$ can be used to evaluate the numerator and the denominator of the polynomials of the rational function

---

* The concept of derivative is introduced later in this chapter. At this point, the reader can skip the theory and concentrate on how a function y is plotted using discrete points, and return later for more depth if desired.

$H(x) = P(x)/Q(x)$, where the input arguments are the column vector r (partial fraction coefficients), the column vector p {roots of $Q(x)$}, and gain k (stand-along term).

For example, using the residues, poles, and gain obtained in Example R.7.48 (where $P(x) = 9x^3 + 8x^2 + 7x + 6$, $Q(x) = 5x^3 + 4x^2 + 3x + 2$ and $H(x) = P(x)/Q(x)$) as input arguments, the original rational polynomials are reconstructed with the coefficients slightly modified (by a factor of 5), as indicated in the following:

```
>>  [P,Q] = residue(r,p,k)

     P =
         1.8000      1.6000      1.4000      1.2000
     Q =
         1.0000      0.8000      0.6000      0.4000
```

Observe that all the coefficients obtained are scaled by 5. Recall that MATLAB always returns the leading coefficient of $Q(x)$ set to 1.

R.7.50 The MATLAB command *residue* is used for continuous or analog functions. In electrical applications, for example, the independent variable is commonly frequency, expressed as w (Fourier) or s (Laplace) (see Chapter 4 of the book titled *Practical MATLAB® Applications for Engineers*). A slightly modified version is used for the case of discrete functions where the independent variable is labeled z. For example, let $H(z)$ be a discrete rational function of the form

$$H(z) = \frac{P(z)}{Q(z)} = \frac{z^2 + 2z + 5}{15z^3 + 13z^2 + 6z + 10}$$

where z^{-1} represents a unit delay (see Chapter 5 of the book titled *Practical MATLAB® Applications for Engineers* for additional information), then $[r, p, k] = residuez(P, Q)$ is the equivalent discrete version of the analog function $[r, p, k] = residues(P, Q)$.

R.7.51 The MATLAB function $[z, p, k] = tf2zp(P, Q)$ returns the zeros(z) and poles(p) as column vectors, as well as the gain constant k of the function $H(z)$. The input arguments are the row vectors P and Q containing the coefficients of the numerator and denominator of $H(z)$ arranged in descending power of z as indicated in the following:

$$H(z) = \frac{P(z)}{Q(z)} = \frac{p_0 + p_1 z^{-1} + p_2 z^{-2} + \cdots + p_n z^{-n}}{g_0 + g_1 z^{-1} + g_2 z^{-2} + \cdots + g_n z^{-n}} = K \frac{\prod_{i=1}^{n}(z - z_i)}{\prod_{i=1}^{n}(z - p_i)}$$

Observe that the degree of the numerator $P(z)$ is assumed to be equal to the degree of the denominator $Q(z)$. If the degrees of $P(z)$ and $Q(z)$ are not the same, then the missing coefficients of the polynomial with lower degree are entered as zeros. Note that $length(P) = length(Q)$.

R.7.52 The MATLAB function $[P, Q] = zp2tf(z, p, k)$, where the zeros (z), the *poles* (p), and the gains (k) are input as column vectors, MATLAB returns the rational function $H(z)$ in the form of two row vectors P and Q (numerator and denominator of $H(z)$).

R.7.53 The MATLAB function *zplane*(z, p) and *zplane*(P, Q) return a pole-zero plot, where either the z and the p arguments (*zeros* and *poles*) are entered as column vectors or P and Q are entered {numerator and denominator of $H(z)$} as row vectors arranged in descending powers of z.

R.7.54 The MATLAB function $[h, n] = impz(P, Q)$ divides the polynomial P by the polynomial Q resulting in a row vector h, with length n. This function is referred to in system applications as the impulse response (see Chapter 1 of the book titled *Practical MATLAB® Applications for Engineers*).

R.7.55　The rational function $H(x) = P(x)/Q(x)$ is often associated by engineers and scientists to what is referred the system transfer function, where the variable x is commonly replaced by s or w (Laplace or Fourier), where either s or w represents frequencies.

The transfer function assumes that the system is de-energized, meaning that all the initial conditions (ICs) are set to zero. The system transfer function can be determined from the system DE by taking the Laplace transformation and ignoring all the ICs. Assuming that the transfer function is known and given by $H(s) = P(s)/Q(s)$, then the system DEs can be obtained by replacing s by d/dt and $1/s$ by $\int dt$ (see Chapter 4 of the book titled *Practical MATLAB® Applications for Engineers*).

The concepts of integration, differentiation, as well as DEs are introduced later in this chapter. The system transfer function $H(s) = P(s)/Q(s)$ is used to evaluate the behavior, performance, and efficiency or gain of a given system, where $Q(s)$ represents the input to the system, and $P(s)$ represents its output. Frequently the output is labeled $Y(s)$ while the input is $X(s)$).

Frequency response plots are widely used in engineering to understand important characteristics about the system such as gain, attenuation, stability, and performance, which in the time domain are not evident.

R.7.56　The MATLAB function *freqs(P, Q, w)* or *freqresp(P, Q, w)* returns the analog transfer $H(s) = P(s)/Q(s)$ points evaluated over the range w, where P and Q are input as row vectors arranged in descending powers of s, where, in general, s is a complex variable of the form $s = \sigma + jw$. For example, let

$$H(s) = \frac{3s + 7}{s^2 - 2s - 3}$$

be the transfer function of a continuous time system. Then the sequence consisting of the following four MATLAB instructions:

```
>> P = [0 3 7];
>> Q = [1 -2 -3];
>> w = [0:1:10];
>> Hs = freqs (P,Q,w);
```

returns *11* points of the transfer function $H(s)$, over the domain of frequencies (s) from 0 to 10 rad/s in increments of 1 rad/s.

The command $[H, w] = freqs(P, Q)$, assigns automatically a set of 200 frequencies (to w) and returns the analog transfer function $H(s)$ for those frequencies. MATLAB automatically chooses and returns a range of frequencies that best describes the system.

R.7.57　Another useful MATLAB function is $[P, Q] = invfreqs (H, w, a, b)$, which returns the coefficients of P and Q, of order a and b, respectively, when the transfer function H is known, over the range w.

For the example used earlier, the *11* values assigned to H are used to obtain back the polynomials P and Q, respectively, as indicated in the following:

MATLAB Solution
```
>> [num, den] = invfreqs (Hs, w, 1, 2)

    num =
         3.0000      7.0000
    den =
         1.0000     -2.0000     -3.0000
```

R.7.58 The plots of $|H(s)|$ versus s (magnitude) and $\angle H(s)$ versus s (phase) (where $s = jw$) are referred to as Bode plots. Bode plots are used extensively in industry to describe the behavior of a system using experimental or analytical data.

Bode plots consist of two plots—a magnitude and a phase plot, which are used to convey important information about a given system. Both plots use the horizontal axis to represent frequencies, employing a linear or logarithmic scale.

The magnitude is usually expressed in dB {where $dB = 20log_{10} |H(jw)|$} or $|H(jw)|$ (unitless gain), whereas the phase plot is expressed in degrees (vertical axis).

The function $[mag, phase] = bode(P, Q, w)$ returns the values of magnitude and phase of $H(s) = P(s)/Q(s)$, over the range of frequencies specified by the row vector w, where P and Q are also specified as row vectors arranged in descending powers of s.

For example, let

$$H(s) = \frac{s}{s^2 + 2s + 1}$$

be the transfer function of a given analog system, then the program below illustrates the commands involved in obtaining the Bode plots of $H(s)$ over the range $0 \le w \le 10$.

MATLAB Solution
```
>> w = [.1: .1:10];
>> num = [1, 0];
>> den = [1 2 1];
>> [mag , phase] = Bode(num, den, w);
>> subplot (2 ,1, 1)
>> plot (w, mag);
>> xlabel('w'), ylabel('Magnitude')
>> title('Bode plots for 0 ≤ w ≤ 10') ;grid on
>> subplot (2 , 1, 2)
>> plot(w, phase) ; xlabel ('w'), ylabel('Phase') ; grid on
>> % the resulting plots are shown in Figure 7.2
```

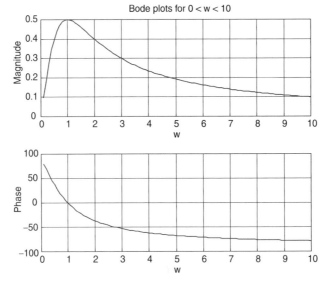

FIGURE 7.2
Bode plot of $H(s)$ of R.7.48 (given range s).

The function *Bode* can be used with the argument w (presented earlier), or without it, as illustrated in the following:

MATLAB Solution
```
>> num = [1 0];
>> den = [1 2 1];
>> [mag, phase, w] = Bode(num,den);
>> subplot (2,1,1)
>> plot (w,mag) ;grid on;
>> subplot (2,1,2)
>> title ('Bode plots with no input frequencies')
>> plot (w,phase)
>> grid on;          % the resulting  plots are shown in Figure 7.3
```

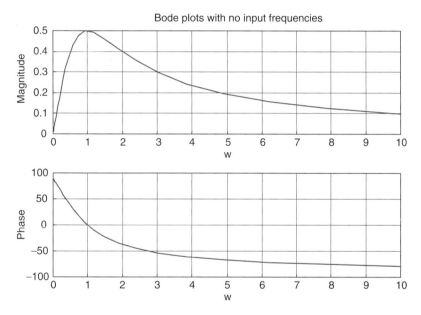

FIGURE 7.3
Bode plot of $H(s)$ of R.7.48 (no range s).

Observe that MATLAB assigns a range of key frequencies to w, when w is not specified and returns a column vector of $mag[H(w)]$ versus w, and *phase* $[H(w)]$ versus w.

Bode magnitude plots are also referred as Alpha plots, whereas Bode phase plots are also referred as Beta plots.

For the case of discrete systems, where the data is sampled with a sampling rate Ts, Bode uses z as the independent variable $[e^{jwTs}]$ to get a plot (see Chapter 5 of the book titled *Practical MATLAB® Applications for Engineers* for additional information).

The frequency response is plotted for frequencies smaller than the Nyquist frequency π/Ts, with a default value of 1 s (when Ts is not specified). This chapter deals exclusively with MATLAB functions and the reader should not be too concerned about the applications, which are revisited in later chapters.

R.7.59 Let $H(z) = P(z)/Q(z)$ be the discrete transfer function of a digital system, where $z = e^{-jWT}$; then the MATLAB function $[H, W] = freqz(P, Q, W)$, where P and Q are the coefficients of the polynomials $P(z)$ and $Q(z)$ (row vectors arranged in descending order of z) returns H as a function of W, where H is evaluated over the range of frequencies specified by W. For example, let

$$H(z) = \frac{0.37525 + 0.4235z^{-1} + 0.3725z^{-2}}{1 - 0.3592z^{-1} + 0.2092z^{-2}}$$

be a discrete system transfer function.

Then the following program returns 11 frequency points of the transfer function $H(z) = P(z)/Q(z)$:

MATLAB Solution
```
>> P = [0.3725     0.4235 0.725];
>> Q = [1    -0.359 0.2092];
>> W = 0: 0.2 :2 ;
>> [H, W]  = freqz(P,Q,W);
>> disp ('****************************')
>> disp ('     W            abs(H)')
>> disp ('****************************')
>> [H' W']
>> disp ('****************************')

****************************************
          W                    abs(H)
****************************************
    ans =
    1.7890                     0
    1.7358 + 0.4149i           0.2000
    1.5656 + 0.8207i           0.4000
    1.2515 + 1.1860i           0.6000
    0.7758 + 1.4333i           0.8000
    0.1923 + 1.4425i           1.0000
   -0.3224 + 1.1488i           1.2000
   -0.5775 + 0.6691i           1.4000
   -0.5547 + 0.2185i           1.6000
   -0.3767 - 0.0839i           1.8000
   -0.1601 - 0.2382i           2.0000
****************************************
```

The preceding table may be shown graphically. The following instructions return the magnitude plot:

```
>> plot (W,abs(H),'d',W,abs(H))
>> xlabel ('Discrete frequencies W'),
>> ylabel ('abs(H)'); title ('abs[H(z)] vs W')
>> grid on;          % the resulting plot is shown in Figure 7.4
```

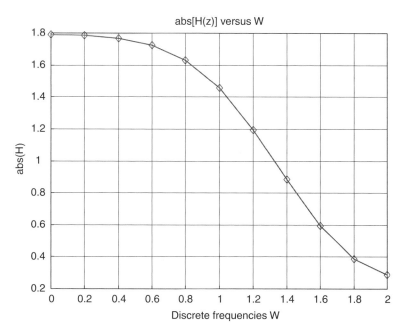

FIGURE 7.4
Magnitude plot of $H(z)$ of R.7.59.

R.7.60 The MATLAB function $y = filter(P, Q, x, IC)$ returns the discrete output y of a given discrete system transfer function $P(z)/Q(z)$, with input sequence x, given in the form of a row vector, where IC are the initial conditions specs of the system.

R.7.61 The MATLAB function $y = lsim(P, Q, x, t)$ returns the output y of an analog system, given the system continuous transfer function $P(s)/Q(s)$, for the input sequence x, over time t.

R.7.62 The following examples, given by the script file *cont_dis*, illustrate the use of the functions *filter* and *lsim*, when evaluating the output of a discrete system given by the discrete transfer function

$$H(z) = \frac{0.37525 + 0.4235z^{-1} + 0.3725z^{-2}}{1 - 0.3592z^{-1} + 0.2092z^{-2}}$$

with input $x(n) = 3.1 \cdot sin\,[6\pi(k-1)]$, for $0 \le n \le 101$, and the analog system given by the transfer function

$$H(s) = \frac{s}{s^2 + 2s + 1}$$

with input $x(t) = 5cos(2\pi t) + 10sin(6\pi t)$, over the range $0 \le t \le 10$. See Figure 7.5.

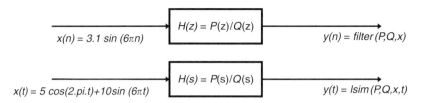

FIGURE 7.5
Discrete and analog systems representation of R.7.62.

```
MATLAB Solution
% Script file:cont _ dis
% discrete case
k = [0:100];
x = 3.1*sin(6*pi*(k-1));
P = [0.3725 0.4235 0.3725];
Q = [1 -0.3592 0.2092];

figure(1)
subplot (2, 1, 1)
stem (k, x), hold on;plot(k,x)
ylabel ('amplitude x(n)')
title ('Input   x(n) vs. n')
subplot(2, 1, 2)
y = filter(P,Q,x);
stem(k, y)
xlabel('index k')
ylabel ('amplitude y(n) ')
title ('Output using y = filter(P,Q,x)')
% The  input and output plots are shown in Figure 7.6
% as continuous functions
% analog case

figure(2)
t = 0:.1:10;
P = [1 0];Q = [1 2 1];
x = 5*cos(2*pi.*t)+10*sin(6*pi.*t);
y = lsim (P,Q,x,t);
subplot(2,1,1)
plot(t,x);
ylabel('amplitude x(t)') ;title('input x(t) vs. t')
subplot(2,1,2) ;plot(t,y);
xlabel('time (sec)') ; title('Output using y=lsim(P,Q,x,t)')
ylabel('amplitude y(t)')

% The input and output plots are shown in Figure 7.7
```

R.7.63 The MATLAB function $y = filtfilt(P, Q, x)$ returns the output sequence y for a given system transfer function given by P and Q, for the forward- and time-reversed input sequence x.

R.7.64 MATLAB allows variables to be used in mathematical expressions without assigning them numerical values. MATLAB calls this variables *symbolic*. For example, if the instruction $y = log(x)$ is entered, MATLAB probably responds with an error message, such as *undefined variable x*. But if x is declared as a symbolic object using the statement *sym('x')*, then MATLAB accepts the instruction $y = log(x)$ as symbolic,

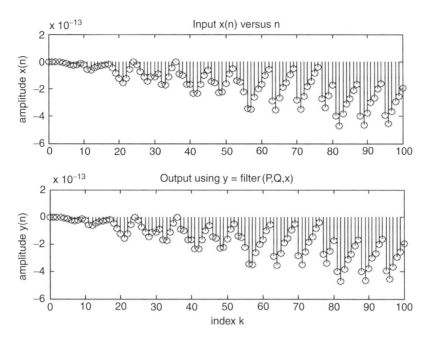

FIGURE 7.6
Discrete input and output plots of the system defined by $H(z)$.

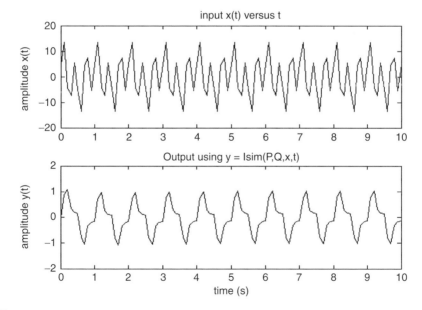

FIGURE 7.7
Continuous input and output plots of the system defined by $H(s)$.

and no numerical value is assigned to x. The symbolic expression y is then stored as a string.

Symbolic objects can be declared one at a time such as $x = sym('x')$, $y = sym('y')$, or they can be declared all at once by using the command *syms x y*, where the order is unimportant. The MATLAB Symbolic Toolbox consists of about 100 specialized functions, some of which will be presented later in this chapter. The standard numerical operators: $+$, $-$, $*$, $/$, and $\wedge$ can be used on symbolic variables to create symbolic algebraic expressions.

R.7.65 MATLAB can perform mathematical operations on symbolic expressions consisting of addition, subtraction, multiplication, division, simplification, integration, differentiation, factorization, solution of set of equations, and many other operations, some of which are presented below.

R.7.66 Symbolic expressions can be constructed using symbolic variables. For example, write a program that defines the following functions as symbolic objects:

a. $f_1 = x + 1$

b. $f_2 = cos(y)$

c. $f_3 = log10(z)$

d. $f_4 = 3x^2 + 4x + 16$

MATLAB Solution
```
>> sym x y z;                % declares x, y, and z as symbolic
>> f1 = x+1 ; f2 = cos(y);  f3 = log10 (z);
>> f4 = 3 *x^2 +4*x +16;
>>                           % f1, f2, f3 and f4  become sym objects
```

R.7.67 The MATLAB command *whos* can be used to check whether a given variable is symbolic, numerical (array), string or complex. For example, let

a. $A = 3$ (constant)

b. $B = \begin{bmatrix} 1 & 2 \\ 3 & 4 \end{bmatrix}$ (matrix)

c. $C = 3 - 4j$ (complex)

d. $D = $ "string sequence" (string)

e. $x, y,$ and z (symbolic)

Process each of the variables defined above then check the class of variables used by executing the command *whos*.

MATLAB Solution
```
>> A = 3;
>> B = [1 2; 3 4];
>> syms x y;
>> C =3-4*j;
>> D = 'string sequence';
>> z = 2*x+3*y;
>> whos
```

Name	Size	Bytes	Class
A	1x1	8	double array
B	2x2	32	double array
C	1x1	16	double array (complex)
D	1x15	30	char array
x	1x1	126	sym object
y	1x1	126	sym object
z	1x1	138	sym object

```
Grand total is 33 elements using 476 bytes
```

R.7.68 Let us construct the symbolic matrix *symmatrix*, by using the symbolic MATLAB variables x, y, and z, as indicated in the following:

$$sym\_matrix = \begin{bmatrix} x & 2*x & x^2 \\ y*x & 1/y+x & \log(y) \\ z*x & z*y & y*x \end{bmatrix}$$

MATLAB Solution
```
>> syms x y z;
>> sym _ matrix = [x   2*x   x^2; y*x   1/y+x   log10(y);z*x   z*y   x*y]

    sym _ matrix =
                    [x,   2*x,        x^2]
                    [y*x, 1/y+x, log(y)]
                    [ z*x, z*y,        y*x]
```

R.7.69 The elements of a symbolic matrix are identified by using the same rules defined for the case of numerical matrices (indexes of row by column). For example,

```
>> sym _ matrix(1,2),   % identifies the element located on row # 1, column # 2

    ans =
          2*x

>> sym _ matrix(2,:),                    % returns row # 2

    ans =
          [y*x,    1/y+x    log(y)]

>> sym _ matrix(:,3) ,                    % returns column # 3

    ans =
          [    x^2]
          [ log(y)]
          [    y*x]
```

R.7.70 The MATLAB command *findsym(y)* is used to identify or find the symbolic variables in a symbolic expression y, which is defined in the following:

For example, let

```
>> a = 1;
>> y = sym ('3*X^2+4*Y+Z+a');    % then
>> findsym(y)
```

returns a list of all the symbolic variables that make up y, as indicated in the following:

```
    ans =
          X, Y, Z, a
```

R.7.71 The command *findsym(y, n)* returns only the n symbolic variables that are closer to x.

R.7.72 Standard algebraic operations can be performed on symbolic expressions such as addition, subtraction, multiplication, division, and exponentiation employing the same rules defined for the numerical case.

For example, define the following expressions as symbolic objects:

a. $y_1 = 2x + 3y + 4z$

b. $y_2 = 5x + 6y + 7z$

c. $y_3 = 8x + 9y$

d. $y_4 = (1/(x + 2)(x + 1)$

e. $y_5 = (x + 1)(x + 3)$

and perform the following symbolic operations:

i. $y_6 = y_1 + y_2$

ii. $y_7 = y_1 - y_3$

iii. $y_8 = y_6 + y_4$

iv. $y_9 = y_{6*2}$

MATLAB Solution
```
>> y1 = sym('2 * x + 3 * y + 4 * z')

   y1 =
        2 * x + 3 * y + 4 * z

>> y2 = sym('5 * x + 6 * y + 7 * z')

   y2 =
        5 * x + 6 * y + 7 * z

>> y3 = sym('8 * x + 9 * y')

   y3 =
        8 * x + 9 * y

>> y4 = sym('1/(x+2) * (x+1)')

   y4 =
        1/(x+2) * (x+1)

>> y5 = sym('(x+1) * (x+3)')

   y5 =
        (x+1) * (x+3)

>> y6 = y1 + y2

   y6 =
        7*x+9*y+11*z

>> y7 = y1 - y3

   y7 =
        6*x-6*y+4*z
```

```
>> y8 = y6 * y4

   y8 =
          (7*x+9*y+11*z)/(x+2)*(x+1)

>> y9 = y6 * 2

   y9 =
          14*x+18*y+22*z
```

R.7.73 Algebraic operations such as addition, subtraction, multiplication, and division using the symbolic expressions f_1 and f_2 can also be accomplished by using the following MATLAB symbolic instructions:

symadd(f_1, f_2)	returns the symbolic addition of f_1 and f_2
symsub(f_1, f_2)	returns the symbolic subtraction of f_2 from f_1
symmul(f_1, f_2)	returns the symbolic multiplication of f_1 by f_2
symdiv(f_1, f_2)	returns the symbolic division of f_1 by f_2

R.7.74 The MATLAB command *pretty(f)* returns the symbolic expression f in typeset format.

For example, let $y_{10} = 2.y_5$, where $y_5 = (x + 1)(x + 3)$.

Use symbolic techniques to evaluate and express y_{10}.

MATLAB Solution
```
>> y5 = sym('(x+1) * (x+3)');
>> y10 = pretty (y5 * 2)


   y10 =
           2x² + 8x + 6
```

R.7.75 The MATLAB command *subs* (y, y_1, y_f) returns the expression for y in which the symbolic variable y_1 is substituted by the symbolic variable y_f. For example, let $y = 3x + 4y$ be a symbolic expression. Write a short MATLAB program that substitutes in y x by z.

MATLAB Solution
```
>> syms x y z
>> y = 3*x + 4*y;
>> subxbyz = subs(y,x,z)

   subxbyz =
              3*z+4*y
```

R.7.76 The MATLAB command $y = sym('y(x)')$ defines y as a function of x. For example, use MATLAB to define $y(x)$ and create the function $g\_x = y(x) - 2 * y(x - 1) + 3 * y(x - 2)$.

MATLAB Solution
```
>> syms x
>> y = sym('y(x)');
>> g _ x = subs(y,x,x) - 2* subs(y,x,x-1) + 3*subs(y,x,x-2)
```

```
g _ x =
        y(x)  -  2*y(x-1)  +  3*y(x-2)
```

R.7.77 A two symbolic variables substitution x, z by p, q in the symbolic expression y can be accomplished by the following command: *subs(y, {'x', 'z'}, {'p', 'q'})*.

R.7.78 For example, let $y = 2x^2 + log(z^2)$ be a symbolic expression. Write a program that defines y, and then substitutes x by p and z by q.

MATLAB Solution
```
>> syms x y z p q
>> y=2*x^2+3*log(z^2);
>> subsxzbypq = subs(y,{'x','z'},{'p','q'})


   subsxzbypq =
            2*p^2+3*log(q^2)
```

R.7.79 The MATLAB function *compose(f, y)* returns the function $f(y(x))$.

R.7.80 For example, let $f(x) = 1 + 2x$ and $y(x) = log(x)$. Use MATLAB to obtain an expression for $f(y)$.

MATLAB Solution
```
>> f=sym('1+2*x');
>> y=sym('log(x)');
>> comfy = compose(f,y)


   comfy =
            1+2*log(x)
```

R.7.81 The symbolic MATLAB function $H = symdiv(A, B)$ returns the symbolic function $H = A/B$, where A and B consist of two symbolic polynomials. Observe that the *symdiv* command can be used to define a symbolic system transfer function expression where $(A = P)$ and $(B = Q)$.

R.7.82 For example, let $A = 2x^3 + 3x^2 + 2x - 10$ and $B = 5x^4 + 4x^3 + 3x^2 + 2x - 25$ be two symbolic polynomial expressions. Use MATLAB to create the symbolic expression $H = A/B$.

MATLAB Solution
```
>> A = sym('2 * x^3 + 3 * x^2 + 2*x - 10');
>> B = sym('5 * x^4 + 4 * x^3 + 3 * x^2 + 2*x - 25');
>> H = symdiv(A, B)


   H =
        (2*x^3+3*x^2+2*x-10)/(5*x^4+4*x^3+3*x^2+2*x-25)
```

R.7.83 The MATLAB command *[num, den] = numden(H)* returns the numerator *num* and the denominator *den*, respectively, of the symbolic expression H. For example, using the result obtained in R.7.82, then *[num, den] = numden(H)* returns

```
num = 2 * x^3 + 3 * x^2 + 2*x - 10
den = 5 * x^4 + 4 * x^3 + 3 * x^2 + 2*x - 25
```

R.7.84 Let y be a symbolic expression. Then the MATLAB functions *simple(y)* and *simplify(y)*, returns the simplified expressions for y. The difference between the functions *simple* and *simplify* is that the command *simple* attempts to simplify y using a variety of methods and displays the various results, whereas *simplify* returns a unique simplified expression for y.

R.7.85 For example, use the commands *simple* and *simplify* on the following symbolic function

$$f = \sqrt{\cos(a.t)^2 + (j.\sin(a.t))^2}$$

MATLAB Solution
```
>> syms a t
>> f = sqrt (cos(a*t)^2 + (j*sin(a*t))^2);
>> simple(f)

    simplify:
    (2*cos(a*t)^2-1)^(1/2)
    radsimp:
    (cos(a*t)^2-sin(a*t)^2)^(1/2)
    combine(trig):
    cos(2*a*t)^(1/2)
    factor:
    (-(sin(a*t)-cos(a*t))*(sin(a*t)+cos(a*t)))^(1/2)
    expand:
    (cos(a*t)^2-sin(a*t)^2)^(1/2)
    combine:
    cos(2*a*t)^(1/2)
    convert(exp):
    ((1/2*exp(i*a*t)+1/2/exp(i*a*t))^2+1/4*(exp(i*a*t)-
    1/exp(i*a*t))^2)^(1/2)
    convert(sincos):
    (cos(a*t)^2-sin(a*t)^2)^(1/2)
    convert(tan):
    ((1-tan(1/2*a*t)^2)^2/(1+tan(1/2*a*t)^2)^2-
    4*tan(1/2*a*t)^2/(1+tan(1/2*a*t)^2)^2)^(1/2)
    collect(t):
    (cos(a*t)^2-sin(a*t)^2)^(1/2)
    ans =
    cos(2*a*t)^(1/2)

>> simplify(f)

    ans =
            (2*cos(a*t)^2-1)^(1/2)
```

R.7.86 The command *[r, how] = simple(f)* returns r the compact form for f, and *how*, represents a string sequence defining the algorithm used to obtain r.

R.7.87 For example, use the command *[r, how] = simple(f)* for $f = \sqrt{\cos(a.t)^2 + (j.\sin(a.t))^2}$ and observe its response.

MATLAB Solution
```
>> syms a t
>> f = sqrt (cos(a*t)^2 + (j*sin(a*t))^2);
>> [r,how] = simple(f)
```

```
    r =
        cos(2*a*t)^(1/2)
    how =
        combine
```

R.7.88 The MATLAB command *collect(y)* returns *y* in which like terms are collected. The more general function *collect(y, z)* returns *y*, in which the coefficients dependent on the symbolic variable *z* are collected.

R.7.89 For example, let $y = 3x + 2y - 2x - 2y + 2$ be a symbolic expression. Use MATLAB to

a. Create *y*

b. Substitute *x* by π in *y*

c. Collect the like terms in *y*

d. Simplify (*y*)

For each case observe and verify the MATLAB results.

MATLAB Solution
```
>> syms x y
>> y = 3*x+2*y-2*x-2*y+2;
>> sub = subs (y,x,pi)

    sub =
            5.1416

>> collect(y)

    ans =
            x+2

>> simplify(y)

    ans =
            x+2
```

R.7.90 The MATLAB function *factor(y)* returns the symbolic expression *y*, in terms of its factors. For example, let $y = x^2 - x - 2$ be a symbolic expression. Use MATLAB and decompose *y* into its factors.

MATLAB Solution
```
>> y = sym ('x^2-x-2');
>> factor(y)

    ans =
            (x+1)*(x-2)
```

R.7.91 The MATLAB function *factorial(n)* returns the value of $y = prod(1:n)$. For example, evaluate $y = 9!$ by symbolic and numerical means.

MATLAB Solution
```
>> factorial(9)

    ans =
            362880
```

```
>> prod(1:9)

    ans =
          362880
```

Note that the numerical and symbolic results fully agree.

R.7.92 The MATLAB function *expand(y)* returns the symbolic expanded expression for *y*.

R.7.93 For example, let $y_1 = (x + a + b)^3$ and $y_2 = cos(a + b)$ be symbolic expressions. Use MATLAB to obtain the expanded version for

a. $y_1 = (x + a + b)^3$

b. $y_2 = cos(a + b)$

MATLAB Solution
```
>> syms x a b;
>> y1 = (x+a+b)^3

    y1 =
       (x+a+b)^3

>> expy1 = expand(y1)

    expy1 =
            x^3+3*x^2*a+3*x^2*b+3*x*a^2+6*x*a*b+3*x*b^2+
            a^3+3*a^2*b+3*a*b^2+b^3

>> y2 = cos(a+b);
>> expye = expand(y2)

    expye =
            cos(a)*cos(b)-sin(a)*sin(b)
```

R.7.94 The MATLAB function *numsp = sym2poly(sp)* returns the symbolic polynomial *sp(x)*, converted into a numerical polynomial consisting of a row vector with its coefficients arranged in descending powers of *x*.

R.7.95 For example, let $sp(x) = 5x^4 + 4x^3 + 3x^2 + 2x + 1$ be a symbolic polynomial.
Use MATLAB to convert the symbolic polynomial *sp* into a numerical vector polynomial *numsp*.

MATLAB Solution
```
>> syms x
>> sp = sym('5 * x^4 + 4 * x^3 + 3 * x^2 + 2*x + 1');
>> numsp = sym2poly(sp)

    numsp =
             5     4     3     2     1
```

R.7.96 The MATLAB function *sp = poly2sym(P)* returns the numerical row vector *P* consisting of the coefficients of the polynomial *p(x)* converted into a symbolic polynomial *sp(x)*.

R.7.97   For example, let $p(x) = x^4 + 2x^3 + 3x^2 + 4x + 5$ be represented by the row vector $P = [1\ 2\ 3\ 4\ 5]$. Use MATLAB to convert the numerical polynomial P, into the symbolic polynomial sp.

MATLAB Solution
```
>> P = [1 2 3 4 5] ;      % corresponds to p(x) = x^4+2*x^3+3*x^2+4*x+5
>> sp = poly2sym(P)       % returns the symbolic polynomial sp(x)


    sp =
         x^4+2*x^3+3*x^2+4*x+5
```

R.7.98   The MATLAB function *horner(sp)* returns the symbolic expression *sp* in a nested format.

R.7.99   For example, use MATLAB to transform the symbolic object $sp(x) = x^4 - 4x^2 + 10x - 30$, into a nested format.

MATLAB Solution
```
>> sp = sym('x^4-4*x^2+10*x-30');
>> hor _ for = horner (sp)


    hor _ for =
               30+(10+(-4+x^2)*x)*x
```

R.7.100   Let x be the independent variable of f, and let x_1 and x_2 be two numerical values of x. Then subtracting x_2 from x_1 is called the increment of x or delta x expressed as Δx.

R.7.101   Recall that a tangent is the straight line which touches the (curve) function defined by $f(x)$ at a point.

R.7.102   A straight line that passes through the (two) points $x = a$ and $x = b$ of $f(x)$ that are near one another on a continuous curve, separated by a Δx, is referred to as secant line. Its slope is given by

$$slope[a,b] = \frac{f(x + \Delta x) - f(x)}{\Delta x}$$

Notice that when $\Delta x = 0$, then the slope is undefined. In the limit as Δx approaches zero, the secant line becomes a tangent, and the two points a and b merge into one.

R.7.103   The independent variable x of $f(x)$ is said to have a constant value l as limit, when the successive values of x are such that $|l - x|$ ultimately become and remain less than any preassigned positive number. The notation used to define the limit of a function $f(x)$ is $\lim_{x \to a} [f(x)] = l$.

The preceding expression says that the limit of $f(x)$ as x approaches a equals l. x will be very close to a, but x will not be equal to a, and $f(x)$ will get closer to the number l as x gets closer to a. The limit of $f(x)$ for a particular value $x = a$ describes the behavior of $f(x)$ at the point a. The concept of limit is one of the basic concepts of calculus, and is generally difficult to visualize, understand, and apply.

R.7.104 The following relations are used to define and summarize some frequently used properties associated with limits. Let us assume that $\lim\limits_{x \to a}[f(x)]$ and $\lim\limits_{x \to a}[g(x)]$ exist and are finite, then

a. $\lim\limits_{x \to a}[f(x) \pm g(x)] = \lim\limits_{x \to a}[f(x)] \pm \lim\limits_{x \to a}[g(x)]$

b. $\lim\limits_{x \to a}[f(x) \cdot g(x)] = [\lim\limits_{x \to a}[f(x)]][\lim\limits_{x \to a}[g(x)]]$

c. $\lim\limits_{x \to a}\left[\dfrac{f(x)}{g(x)}\right] = \lim\limits_{x \to a}[f(x)]/\lim\limits_{x \to a}[g(x)]$ if $\lim\limits_{x \to a}[g(x)] \neq 0$

d. $\lim\limits_{x \to a}\left[\sqrt[n]{f(x)}\right] = \sqrt[n]{\lim\limits_{x \to a}[f(x)]}$, where n is a positive integer

e. $\lim\limits_{x \to a}\left[f(x)\right]^n = \left\{\lim\limits_{x \to a}[f(x)]\right\}^n$, for any positive integer n

f. $\lim\limits_{x \to a}[A \cdot f(x)] = A \cdot \lim\limits_{x \to a}[f(x)]$, where A is a constant

g. If $\lim\limits_{x \to a}[f(x)] = f(a)$, then the function $f(x)$ is said to be continuous at the point $x = a$

R.7.105 The symbolic MATLAB command $F = limit(f, x, a)$ returns F, the numerical value of the limit of $f(x)$ as x approaches a {expressed as $F = \lim\limits_{x \to a}[f(x)]$}. The default value for a is zero.

R.7.106 For example, use MATLAB and evaluate the limits of the following expressions:

a. $F_1 = \lim\limits_{t \to a}\left[\dfrac{\sin(a.t)}{a.t}\right]$

b. $F_2 = \lim\limits_{t \to a}\left[\dfrac{1}{t + a}\right]$

c. $F_3 = \lim\limits_{t \to 0}\left[\dfrac{1}{t + a}\right]$

d. $F_4 = \lim\limits_{n \to \infty}\left[\left(1 + \dfrac{1}{n}\right)^2\right]$

e. $F_5 = \lim\limits_{n \to \infty}\left[\dfrac{1}{n} + a\right]$

f. $F_6 = \lim\limits_{t \to \infty}\left[\dfrac{1}{t + a}\right]$

MATLAB Solution
```
>> syms a t n
>> F1= limit(sin(a*t)/(a*t))

    F1 =
         1

>> F2 = limit(1/(t+a),t,a)

    F2 =
         1/2a

>> F3 = limit(1/(t+a),t,0)

    F3 =
         1/a

>> F4 = limit((1+1/n)^n, n, inf)

    F4 =
         exp(1)

>> F5= limit((1/n+a),n, inf)

    F5 =
         a

>> F6 = limit(1/(t+a),t,inf)

    F6 =
         0
```

R.7.107 Let $y = f(x)$, and let $\lim_{\Delta x \to 0}[\Delta y/\Delta x]$ exist in the limit. Then this limiting process is known as the derivative of y with respect to x.

R.7.108 The derivative of $y = f(x)$ with respect to x is defined as the slope of the tangential line at any point of y, often referred as the *slope function*, given by

$$\frac{dy}{dx} = f'(x) = \lim_{\Delta x \to 0}\left[\frac{\Delta y}{\Delta x}\right] = \lim_{\Delta x \to 0}\left\{\frac{f(x + \Delta x) - f(x)}{\Delta x}\right\}$$

Since the concept of derivative is so important in science and engineering, let us explore its definition and meaning.

Let $y = f(x)$, then $y + \Delta y = f(x + \Delta x)$; increment both sides by Δ, then $\Delta y = f(x + \Delta x) - y$, or $\Delta y = f(x + \Delta x) - f(x)$. Then

$$\frac{\Delta y}{\Delta x} = \frac{f(x + \Delta x) - f(x)}{\Delta x}$$

dividing both sides of the equation by Δx, and taking the limit as x approaches zero yields

$$\lim_{\Delta x \to 0}\left[\frac{\Delta y}{\Delta x}\right] = \lim_{\Delta x \to 0}\left[\frac{f(x + \Delta x) - f(x)}{\Delta x}\right]$$

Recall that the preceding relation was defined as the derivative, given by

$$\frac{df(x)}{dx} = \lim_{\Delta x \to 0} \left[\frac{\Delta y}{\Delta x} \right]$$

R.7.109   The process of evaluating the derivative is called differentiation. Let $y = f(x)$, then the first derivative corresponds to the slope of the tangent line at any point of y, and the equation $dy/dx = 0$ is often used to determine the locations of the maxima and minima of y, and it is probably one of the most important applications of differential calculus. Observe that when the function $f(x)$ is at its maximum or minimum, its slope is zero, because the tangent line at y is parallel to the abscissa.

R.7.110   Let's get an insight into the process of differentiation, for example, for an economist the variable of interest can represent an investment, and the economist may be interested in maximizing profits and minimizing costs.

R.7.111   In the physical sciences, a distance may be expressed as a function of the independent variable time (t), denoted by $f(t)$ (in meters). Then $v(t) = (df(t))/dt$, represents velocity (in meter per second), and $a(t) = (d^2[f(t)])/dt^2$ represents acceleration (in meter per second square). Or let $w(t)$ represent energy (in joules), then $p(t) = (d[w(t)])/dt$ represents power (in watts or joules per second).

R.7.112   The first derivative of $y(x)$ can be used to predict the behavior of $y(x)$ by means of a new function that represents the slope of $y(x)$ with respect to x that indicates the rate of change. The units of the derivative are then the unit of the dependent variable y divided by the unit of the independent variable x.

For example, let $y = f(x)$, then the first derivative can be used to determine if y tends to increase or decrease, as indicated in the following:

a. If $dy/dx > 0$ on an interval x, then the function $f(x)$ increases, otherwise.
b. If $dy/dx < 0$, then $f(x)$ decreases on an interval x.
   If $y(x)$ does not change, then the derivative is zero.
   The first derivative of $y(x)$ is expressed using the following standard notation:

$$\frac{dy}{dx} = y' = f'(x)$$

The notation used to denote the second derivative of $y(x)$ and its meaning is given in the following:

$$\frac{d^2y}{dx^2} = \frac{d}{dx}\left[\frac{dy}{dx}\right]$$

$$y'' = f''(x) = \frac{f'(x + \Delta x) - f'(x)}{\Delta x} \approx \frac{f(x + 2\,\Delta x) - f(x + \Delta x) + f(x)}{(\Delta x)^2}$$

R.7.113   The second derivative is often used to test the concavity of $y = f(x)$, over a particular interval of x as indicated in the following:

a. If $f''(x) > 0$, for any point on an interval of x, then $f(x)$ is concave up (otherwise).
b. If $f''(x) < 0$, for any point on an interval of x, then $f(x)$ is concave down.

R.7.114   The MATLAB function $derivP = polyder(P)$, where P is a row vector consisting of the coefficients of the polynomial $p(x)$, expressed in descending powers of x returns *derivP*, the coefficients of the expression $d[p(x)]/dx$, as a row vector arranged in descending powers of x.

R.7.115 For example, let $p(x) = 3x^4 + 2x^3 - 4x^2 + 7x + 21$. Then use MATLAB to evaluate $d[p(x)]/dx$.

MATLAB Solution
```
>> P = [3   2   -4   7   21];
>> derivP = polyder(P)

    derivP =
            12      6      -8      7
```

The preceding result is interpreted as $d[p(x)]/dx = 12x^3 + 6x^2 - 8x + 7$.

R.7.116 The command *polyder* can be used to evaluate the derivatives of either a product or a quotient of polynomials (entered as vectors), as indicated in the following:

$$derivProd = polyder\,(P1, P2) = \frac{d[P1 * P2]}{dx}$$

and

$$[der\_P1,\, der\_P2] = polyder\,(P1,\, P2)$$

where

$$\frac{d[P1/P2]}{dx} = \frac{der\_P1}{der\_P2}$$

R.7.117 For example, let $p_1(x) = 1x^4 + 2x^3 - 3x^2 + 5x + 7$ and $p_2(x) = 3x^4 + 6x^3 - 4x^2 + 9x + 13$. Then use MATLAB and evaluate the following:

a. $\dfrac{d[P1 * P2]}{dx}$

b. $\dfrac{d[P1/P2]}{dx}$

MATLAB Solution
```
>> P1 = [1 2 -3 5 7];
>> P2 = [3 6 -4 9 13];
>> derP1xP2 = polyder(P1,P2)

    derP1xP2 =
            24      84      -6      -10      376      63      -44      128


>> [derP1,derP2] = polyder (P1,P2)

    derP1 =
            10      -8      -56      -55      -22      2
    derP2 =
            9       36      12      6      202      84      -23      234      169
```

The preceding results are interpreted as

$$\frac{d[P1 * P2]}{dx} = 24x^7 + 84x^6 - 6x^5 - 10x^4 + 376x^3 + 63x^2 - 44x + 128$$

$$\frac{d[P1/P2]}{dx} = \frac{10x^5 - 8x^4 - 56x^3 - 55x^2 - 22x + 2}{9x^8 + 36x^7 + 12x^6 + 6x^5 + 202x^4 + 84x^3 - 23x^2 + 234x + 169}$$

R.7.118 Let $y_1 = f_1(x)$ and $y_2 = f_2(x)$ be two functions of the independent variable x, and let c be an arbitrary constant. Then the relations in the following summarize the rules and relations frequently used in the evaluation of derivatives:

a. $\dfrac{d}{dx}(c) = 0$

b. $\dfrac{d}{dx}[c \cdot f_1(x)] = c\dfrac{d}{dx}f_1(x)$

c. $\dfrac{d}{dx}[f_1(x) \pm f_2(x)] = \dfrac{df_1(x)}{dx} \pm \dfrac{df_2(x)}{dx}$

d. $\dfrac{d}{dx}[f_1(x) \cdot f_2(x)] = f_1(x)\dfrac{d}{dx}f_2(x) + f_2(x)\dfrac{d}{dx}f_1(x)$

e. $\dfrac{d}{dx}\left[\dfrac{f_1(x)}{f_2(x)}\right] = \dfrac{f_2(x)(d/dx)f_1(x) - f_1(x)(d/dx)f_2(x)}{f_2^2(x)}$

f. $\dfrac{d}{dx}[x^c] = cx^{c-1}$

g. $\dfrac{d}{dx}f_1^c(x) = cf_1^{c-1}(x)\dfrac{d}{dx}[f_1(x)]$

h. $\dfrac{d}{dx}(\ln(x)) = \dfrac{1}{x}$

i. $\dfrac{d}{dx}[e^{cx}] = ce^{cx}$

j. $\dfrac{df(x)}{dx} = \dfrac{df(x)}{du} \cdot \dfrac{du}{dx}$ (chain rule)

R.7.119 A list of the derivatives of the standard trigonometric functions are given below:

a. $\dfrac{d}{dx}[\sin(x)] = \cos(x)$

b. $\dfrac{d}{dx}[\cos(x)] = -\sin(x)$

c. $\dfrac{d}{dx}[\tan(x)] = \sec^2(x) = \dfrac{1}{\cos^2(x)}$

d. $\dfrac{d}{dx}[\cot(x)] = -\csc(x)$

e. $\dfrac{d}{dx}[\sec(x)] = [\sec(x)][\tan(x)]$

f. $\dfrac{d}{dx}[\csc(x)] = [-\csc(x)][\cot(x)]$

g. $\dfrac{d}{dx}[\arcsin(x)] = \dfrac{1}{\sqrt{1-x^2}}$

h. $\dfrac{d}{dx}[\arctan(x)] = \dfrac{1}{1+x^2}$

i. $\dfrac{d}{dx}[\operatorname{arcsec}(x)] = \dfrac{1}{x\sqrt{1-x^2}}$

j. $\dfrac{d}{dx}[\arccos(x)] = -\dfrac{d}{dx}[\arcsin(x)]$

k. $\dfrac{d}{dx}[\operatorname{arccot}(x)] = -\dfrac{d}{dx}[\arctan(x)]$

l. $\dfrac{d}{dx}[\operatorname{arccsc}(x)] = -\dfrac{d}{dx}[\operatorname{arcsec}(x)]$

R.7.120   The examples below illustrate the evaluation process of the derivatives when performed by hand for each of the following expressions:

a. $y_1(t) = \dfrac{d}{dt}[3\cos(6t)] \quad$ at $\quad t = \pi/18$

b. $y_2(t) = \dfrac{d}{dt}[3/\sin^2(t)]$

c. $y_3(t) = \dfrac{d}{dt}\left[3e^{4t^5}\right]$

d. $y_4(t) = \dfrac{d}{dt}[2e^{-3t}\sin(5t)]$

e. $y_5(t) = \dfrac{d}{dt}[3\ln(t-5)]$

f. $y_6(t) = \dfrac{d}{dt}[t\ln(t)]$

ANLYTICAL Solutions

a. $y_1(t) = \dfrac{d}{dt}[3\cos(6t)] = -3 * 6 * \sin(6t)\Big|_{t=\pi/18}$

$\qquad = -18\sin(6 * \pi/18) = -18\sin(\pi/3)$

$\qquad = -18\sin(60°) = -18(0.886) = -15.588$

b. $y_2(t) = \dfrac{d}{dt}\left[\dfrac{3}{\sin^2(t)}\right] = 3\dfrac{d}{dt}\left[\sin^{-2}(t)\right]$

$\qquad = 3 * (-2) * \sin^{-3}(t) * \dfrac{d}{dt}[\sin(t)]$

$\qquad = -6\cos(t)/\sin^3(t)$

c. $y_3(t) = \dfrac{d}{dt}\left[3e^{4t^5}\right] = 3e^{4t^5}\dfrac{d}{dt}\left[4t^5\right]$

$\quad y_3(t) = 3e^{4t^5}\left[20t^4\right]$

$\qquad = 60 * t^4 * e^{4t^5}$

d. $y_4(t) = \dfrac{d}{dt}\left[2e^{-3t} \cdot \sin(5t)\right] = 2e^{-3t}\dfrac{d}{dt}[\sin(5t)] + 2\sin(5t)\dfrac{d}{dt}\left[e^{-3t}\right]$

$\quad y_4(t) = 2e^{-3t}[5\cos(5t)] + 2\sin(5t)\left[-3e^{-3t}\right]$

$\quad y_4(t) = 10e^{-3t}\cos(5t) - 6e^{-3t}\sin(5t)$

$\qquad = e^{-3t}[10\cos(5t) - 6\sin(5t)]$

e. $y_5(t) = \dfrac{d}{dt}[3\ln(t-5)] = \dfrac{3}{t-5}\dfrac{d}{dt}(t-5)$

$\qquad = \dfrac{3}{t-5}$

f. $y_6(t) = \dfrac{d}{dt}[t\ln(t)] = t\dfrac{d}{dt}[\ln(t)] + \ln(t)\dfrac{d}{dt}[t]$

$\qquad = t\left(\dfrac{1}{t}\right) + \ln(t) = 1 + \ln(t)$

R.7.121 The process, called differentiation of $f(x)$ with respect to x, is the process that returns the expression for $(df(x))/dx$. Integration is the inverse of differentiation. The notation used to denote integration is $\int f(x)\,dx$ that reads the integral of $f(x)$ with respect to x.

R.7.122 Let $y_1 = f_1(x)$ and $y_2 = f_2(x)$ be two functions of the independent variable x, and let b and c be arbitrary constants, where $b \neq 0$.

Then the relations below summarize the rules and properties frequently used in the evaluation of integrals by hand:

a. $\int dx = x + c$

b. $\int \left[f_1(x) \pm f_2(x)\right]dx = \int f_1(x)dx \pm \int f_2(x)dx$

c. $\int b\,dx = bx + c$

d. $\int x^b\,dx = \dfrac{x^{b+1}}{b+1} + c \quad b \neq -1$

e. $\int \dfrac{dx}{x} = \ln |x| + c$

f. $\int e^{bx}\,dx = \dfrac{1}{b} e^{bx} + c$

R.7.123 A list of integrals of the standard trigonometric functions are given in the following:

a. $\int \sin(x)\,dx = -\cos(x) + c$

b. $\int \cos(x)\,dx = \sin(x) + c$

c. $\int \tan(x)\,dx = \ln|\sec(x)| + c$

d. $\int \cot(x)\,dx = \ln|\sin(x)| + c$

e. $\int \sec(x)\,dx = \ln|\sec(x) + \tan(x)| + c$

f. $\int \csc(x)\,dx = \ln|\csc(x) - \cot(x)| + c$

g. $\int \sinh(x)\,dx = \cosh(x) + c$

h. $\int \cosh(x)\,dx = \sinh(x) + c$

i. $\int \dfrac{dx}{1 + x^2} = \arctan(x) + c$

j. $\int \dfrac{dx}{\sqrt{1 + x^2}} = \arcsin(x) + c$

k. $\int \dfrac{dx}{x\sqrt{x^2 - 1}} = \text{arcsec}(x) + c$

R.7.124 The term definite integral defines the expression, when the limits of integration are over the range a and b. The notation used is indicated by

$$\int_a^b f(x)\,dx = \phi(b) - \phi(a)$$

where $\phi(b) = \int f(x)\,dx$ at $x = b$ and $\phi(a) = \int f(x)\,dx$ at $x = a$.

R.7.125 The indefinite integral is defined by $\int f(x)\,dx = \phi(x) + c$, with no assigned limits of integration.

R.7.126 Observe that some integrals may be undefined depending on the limits of integration. For example,

$$\int \frac{dx}{x} = \ln|x| + c$$

is undefined for $x = 0$.

These integrals are called *improper,* and the points where the integral becomes undefined are called *singularities.*

R.7.127 The examples below illustrate the integration process performed by hand for the following expressions:

a. $\int_0^1 6\sin(5\pi t)\,dt =$

b. $\int \left[\dfrac{1}{3 - 5t}\right] dt =$

ANALYTICAL Solutions

a. $\int_0^1 6\sin(5\pi t)\,dt = \left.\dfrac{-6}{5\pi}\cos(5\pi t)\right|_0^1$

$$= \frac{-6}{5\pi}[\cos(5\pi) - \cos(0)]$$

$$= \frac{-6}{5\pi}[-1 - 1] = \frac{12}{5\pi}$$

b. $\int \left[\dfrac{1}{3 - 5t}\right] dt = \dfrac{1}{-5}\int\left[\dfrac{1}{3 - 5t}\right]d(-5t)$

$$= -\frac{1}{5}\ln(3 - 5t) + c$$

R.7.128 In the physical sciences, the concept of integration is used in a variety of applications such as the evaluations of planar areas, lengths, areas of surfaces, and volumes.

The area of a continuous function $y = f(x)$, over the range $a \le x \le b$, can be evaluated (approximated) discretely by the following equation:

$$\sum_{k=0}^{n} f(x_k)\,\Delta x \quad \text{where} \quad \Delta x = \frac{b - a}{n}$$

Of course, the approximation improves if n increases (in the limit as n approaches infinity), implying that Δx decreases.

Recall that the areas above the x-axis are considered positive, whereas areas under the x-axis are negative. In systems, integration is used to determine average and *rms* or effective values, *energies, voltages,* and *currents.*

For many functions the integral can be evaluated analytically, but for many other functions the integral cannot be accomplished analytically, and requires numerical or symbolic approximations.

R.7.129 The following examples, chosen from the physical sciences illustrate various applications where integration is employed:

Let $y = f(x)$ be defined in the interval of interest between $x = a$ and $x = b$. Then,

a. The area of $y = f(x)$, over the range $y|_{x=b} \leq y \leq y|_{x=b}$, is given by

$$\int_a^b f(x)dx$$

b. The surface area obtained by rotating $f(x)$ about the x-axis is given by

$$2\pi \int_a^b f(x)\left[\sqrt{1 + \left(\frac{df(x)}{dx}\right)^2}\right]dx$$

c. The volume obtained by rotating the function $f(x)$ about the x-axis, over the range $x = a$ and $x = b$, is given by*

$$\pi \int_a^b \left[f(x)\right]^2 dx$$

d. The length of the curve defined by $f(x)$ between $x = a$ and $x = b$ is given by (Linderburg, 1982)

$$\int_a^b \sqrt{1 + \left(\frac{df(x)}{dx}\right)^2} dx$$

R.7.130 The MATLAB function *diff(sp)* returns the derivative of the symbolic expression *sp* with respect to the default variable x. The most frequent MATLAB differentiation commands are defined below for the general symbolic object $sp = f(x, z)$.

Newer versions of MATLAB use x as the default variable, when present, or the variable that is closest to x.

a. $diff(sp) = \dfrac{d(sp)}{dx}$

b. $diff(sp,'z') = \dfrac{d}{dz}(sp)$

c. $diff(sp,n) = \dfrac{d^n}{dx^n}(sp)$

d. $diff(sp,'z',n) = \dfrac{d^n}{dz^n}(sp)$

---

* For surfaces of revolution, and double and triple integrals consult Jensen (2000).

The reader should not confuse the symbolic command *diff* with the MATLAB numerical command *differ* = *diff(Y)*, where *Y* is a vector that represents the points of *y(x)* for a specified range of *x*. Recall that MATLAB returns in this case the vector *differ* consisting of the differences of adjacent elements in *differ* = [(x_2 − *x*) (x_3 − x_2) (x_4 − x_3) ... (x_n − x_{n-1})] (see Chapter 3).

The [(*dy(x)*)/*dx*] expressed as the MATLAB variable *dydx* can be evaluated by numerical means by the code line given by

```
>> dydx = diff(Y)./(deltax*ones(1,length(x)-1))
```

where deltax represents the step size over *x*.

R.7.131 For example, let *y* = x^2, over the range $0 \le x \le 3$. Use MATLAB to evaluate numerically (*dy(x)*)/*dx*, by using the MATLAB command *diff* with a step size $\Delta x = 0.5$.

```
MATLAB Solution
>> x = 0:.5:3

    x =
        0      0.5000     1.0000     1.5000     2.0000     2.5000     3.0000

>> y = x.^2

    y =   .
        0      0.2500     1.0000     2.2500     4.0000     6.2500     9.0000

>> diffy = diff(y)

    diffy =
           0.2500     0.7500     1.2500     1.7500     2.2500     2.7500

>> dxdy = diffy./(ones(1,length(x)-1)*0.5)

    dxdy =
           0.5000     1.5000     2.5000     3.5000     4.5000     5.5000
```

Clearly, *dxdy* = *y*′ represents the equation of a straight line with *y*′ = slope = change(*y*)/change(*x*) = 1/0.5 = 2.

R.7.132 The integration process can be accomplished by *numerical* or *symbolic* means.

Numerical integration for the case of a polynomial can be accomplished by using the function *P* = *polyint(P1, k)* and the inverse function given by *P1* = *polyder(P)*, where *k* represents the constant term of *P*. Recall that *P* is a vector that consists of the coefficients of *p(x)* in descending powers of *x*.

R.7.133 For example, let *p(x)* = x^4 + $2x^3$ − $3x^2$ + 5*x* + 7. Use MATLAB and evaluate by numerical means

a. $\dfrac{dp(x)}{dx} =$

b. $\displaystyle\int \left[\dfrac{dp(x)}{dx}\right] dx =$

MATLAB Solution
```
>>P = [1 2 -3 5 7];
>> der _ P = polyder(P,k)


    der _ P =
           4      6     -6      5


>> int _ of _ der _ P = polyint(der _ P,7)


    int _ of _ der _ P =
                   1     2     -3     5     7
```

R.7.134 Numerical integration can also be approximated by using the MATLAB function
area = trapz(x, Y)* (see Chapter 3 for additional details), where *Y* is an array that
represents *Y = f(x)*, the values of *f(x)* over the domain specified by the array *x*. There
are other simple numerical ways to evaluate areas such as *area = sum(Y) * deltax*,
or *area = trapz(Y) * deltax*. The area under *y(x)* as a function of the independent
variable *x*, also referred as the *running integral*, can be approximated by the following MATLAB instructions:

a. *run_int = cumtrapz(Y) * deltax*

b. *run_int = cumsum(Y) * deltax*

where *deltax* is the step size used to generate *x*. In general, to improve accuracy the
domain of *x* should include a relatively large number of elements.

R.7.135 There are other MATLAB integration solvers such as (Jensen, 2000)

a. *quad('f', a, b, tol, trace)*

b. *quad1('f', a, b, tol, trace)*

c. *quad8('f'.a, b, tol, trace)*

that accept directly the function *f* as a symbolic object, and returns

$$\int_a^b f(x)dx$$

where *tol* is an optional parameter that represents the error tolerance (the default
value is 10^{-3}), and *trace* is a scalar optional parameter used to control the display
of the intermediate results.

The difference between *quad*, *quad1*, and *quad8* is that the first uses the Simpson's rule,[†] the second uses the Lobatto's algorithm (not used in newer MATLAB
versions), and the third uses the Newton–Cotes' algorithm.[‡]

---

* The trapezoidal rule approximates the area under *y = f(x)* by dividing *x* into Δ*x* subintervals, connected by
straight lines, and adding the areas of the subintervals. Clearly, as the number of subintervals increases and
approaches infinity, the piecewise straight lines better approximate *f(x)*.
† The Simpson's rule uses a quadratic polynomial approximation to *f(x)* over adjacent pairs of subintervals.
‡ The Newton–Cotes formulas use the Simpson's rule, but approximate *f(x)* by a higher degree polynomial
through the given number of points (*quad8* uses a polynomial approximation of order 8).

R.7.136 Examples of the use of the *quad* family, abbreviation for quadrature* are illustrated
as follows. Evaluate

$$y(x) = \int_0^{\pi/4} \cos(x)dx$$

by using *quad, quad8, trapz, cumtrapz,* and the *sum.* Observe, compare, and verify
the accuracy of each MATLAB solution.

MATLAB Solution
```
>> area1quad = quad('cos',0.0,pi/4)

    areaquad =
            0.7071

>> areaquad8=quad8('cos',0.0,pi/4)

    areaquad8 =
            0.7071

>> clear
>> x = 0:.01:pi/4;
>> y = cos(x);
>> areatrap1= trapz(x,y)

    areatrap1 =
            0.7033

>> areatrap2 = trapz(y)*0.01

    areatrap2=
            0.7033

>> areasum = sum(y)*0.01

    areasum =
            0.7118

>> areacumtra = cumtrapz(y)*0.01                % cumulative evaluations

areacumtra =
Columns 1 through 8
0    0.0100    0.0200    0.0300    0.0400    0.0500    0.0600    0.0699

Columns 9 through 16
0.0799    0.0899    0.0998    ......        ..........        ...............

Columns 73 through 79
0.6594    0.6669    0.6743    0.6816    0.6889    0.6961    0.7033
```

---

* Old term used to evaluate areas.

R.7.137 The MATLAB symbolic function *int(sp)* returns the indefinite symbolic integral of *sp* with respect to the default variable *x*. The most frequent MATLAB integration commands are defined below for the symbolic object *sp* = *f*(*x*, *z*):

a. $\text{int}(sp) = \int f(x,z)dx$

b. $\text{int}(sp, a, b) = \int_a^b f(x,z)dx$

c. $\text{int}(sp, 'z') = \int f(x,z)dz$

d. $\text{int}(sp, 'z', a, b) = \int_a^b f(x,z)dz$

R.7.138 Examples illustrating the *integration* and *differentiation* processes are presented below using the following expression:

$$y_1(x) = 2x^3 + 3x^2 + 4x - 5 \quad \text{and} \quad y_2(x) = \sin(x) + (1/4)x^2$$

Evaluate by hand and by using MATLAB the following expressions:

a. $\int y_1 dx =$

b. $\int_1^2 y_1 dx =$

c. $\int y_2 dx =$

d. $\int_0^\pi y_2 dx =$

e. $\dfrac{d(y_1)}{dx} =$

f. $\dfrac{d(y_2)}{dx} =$

g. $\dfrac{d^5(y_2)}{dx^5} =$

ANALYTICAL Solutions

a. $\int y_1 dx = 0.5x^4 + x^3 + 2x^2 - 5x$

Let us evaluate the preceding expression in the interval $x = 0.1$ and $x = 0.2$.

$\int_{.1}^{.2} y_1 dx = 0.5\big[(0.2)^4 - (0.1)^4\big] + \big[(0.2)^3 - (0.1)^3\big] + 2\big[(0.2)^2 - (0.1)^2\big] - 5[0.2 - 0.1]$

b. $= -\dfrac{1729}{4000} = -.4323$

c. $\int y_2 dx = \int (\sin(x) + (1/4)x^2)dx = -\cos(x) + (1/12)x^3$

d. $\int_0^\pi y_2 dx = -[\cos(\pi) - \cos(0)] + \dfrac{1}{12}(\pi^2 - 0^2) = 4.584$

e. $d(y_1)/dx = 6x^2 + 6x + 4$

f. $\dfrac{d(y_2)}{dx} = \cos(x) + \left(\dfrac{1}{2}\right)x$

g. $\dfrac{d^5(y_2)}{dx^5} = \cos(x)$

MATLAB Solutions

```
>> syms x
>> y1 = 2*x^3+3*x^2+4*x-5;
>> inty1= int(y1)                        % integrate y1

    inty1 =
            1/2*x^4+x^3+2*x^2-5*x

>> inty1lim = int(y1,.1,.2)              % evaluate integral y1 between
                                           x = 0.1 to x = 0.2

    inty1lim =
                -1729/4000

>> double(inty1lim);                     % converts inty1lim into double
                                           precision

    ans =
          -0.4323

>> y2 = sin(x) + (1/4)*x^2;
>> inty2 = int(y2)                       % integral of y2

    inty2 =
            -cos(x)+1/12*x^3

>> inty2lim = int(y2,0,pi) ;             % integral of y2 between x=0
                                           and x=pi

    inty2lim =
                2+1/12*pi^3

>> double(inty2lim)

    ans =
          4.5839

>> dify1= diff(y1)                       % dy1(x)/dx

    dify1 =
            6*x^2+6*x+4

>> dify2 = diff(y2)                      % dy2(x)/dx

    dify2 =
            cos(x)+1/2*x
```

```
>> dif5y2 = diff(y2,5)                      %[fifth derivative of y2 with
                                              respect to x

    dif5y2 =
            cos(x)
```

Observe that the analytical results fully agree with the MATLAB results.

R.7.139 The MATLAB function *sinint(x)* frequently encountered in engineering and technology is referred as the sampling function denoted by *(Sa)* returns the integral

$$\int_0^x \frac{\sin(y)}{y}\,dy$$

where *x* may represent a constant, or a matrix.

R.7.140 For example, evaluate the following expressions using the MATLAB *sinint* command:

a. $\int_0^3 \frac{\sin(y)}{y}\,dy$

b. $\int_0^1 \frac{\sin y}{y}\,dy, \int_0^\pi \frac{\sin(y)}{y}\,dy, \int_0^{e^{\wedge 2}} \frac{\sin(y)}{y}\,dy,$ and $\int_0^{1/5} \frac{\sin(y)}{y}\,dy$

by employing a matrix approach.

MATLAB Solution
```
>> syms y1 y2 a
>> y1= sinint(3)                                          % part(a)

    y1 =
            1.8487

>> a = [1 pi exp(2) 1/5];                                 % part(b)
>> y2 = sinint(a)

    y2 =
            0.9461      1.8519      1.4970      0.1996
```

R.7.141 The function *double(c)* converts the symbolic object *c* (constants, scalar, or matrix) into a double precision floating point variable.

R.7.142 For example, let $y = x^3 + 2x^2 + x - 15$. Use MATLAB to evaluate the following:

a. $y_1 = y(x = 1)$

b. Convert y_1 into a floating point variable (y_2)

c. $y_3 = y_2^2$

d. Verify the class of variables employed in this example

MATLAB Solution
```
>> syms x y y1
>> y = x^3+2*x^2+x-15;
>> y1 = subs(y,x,1)
```

```
    y1 =
          -11

>> y2=double(y1)

    y2 =
          -11

>> y3= y2^2

    y3 =
          121

>> whos

            Name        Size        Bytes            Class
            x           1x1         126           sym object
            y           1x1         152           sym object
            y1          1x1           8           double array
            y2          1x1           8           double array
            y3          1x1           8           double array
            Grand total is 20 elements using 302 bytes
```

R.7.143 The MATLAB function *vpa(k, d)*, which stands for variable precision arithmetic, returns *k* with *d* digits of accuracy where *k* may be a constant or a matrix. For example, evaluate

a. e (natural logarithm) to 25 digits of accuracy

b. $\pi/2$ to 30 digits of accuracy

c. The 3 × 3, Hilbert matrix to six digits of accuracy

MATLAB Solution
```
>> f1= vpa(exp(1), 25)

    f1 =
          2.718281828459045534884808

>> f2 = vpa(pi/2, 30)

    f2 =
          1.57079632679489661923132169164

>> A= vpa(hilb(3), 6 )

    A =
          [       1., .500000, .333333]
          [ .500000, .333333, .250000]
          [ .333333, .250000, .200000]
```

R.7.144 The MATLAB function *digit(d)* defines the precision used to perform symbolic operations using variable precision arithmetic. The default precision is set to 32 digits for the *rpa* command.

R.7.145 For example, use MATLAB to express the variables *x*, *z*, and *w*, defined below:

   a. $x = 1/3$ using the default and 32 digits

   b. $z = e$ using the default and 20 digits

   c. $w = \pi$ using the default and 25 digits

MATLAB Solution
```
>> syms v w x y
>> x =1/3

    x =
        0.3333

>> v = vpa(x)                                  % part(a)

    v =
        .33333333333333333333333333333333333

>> z = exp(1)

    z =
        2.7183

>> digits(20)                                  % part(b)
>> z = vpa(z)

    z=
        2.7182818284590455349

>> digits(25)
>> w = pi

    w =
        3.1416

>> w = vpa(pi)                                 % part(c)

    ww =
        3.141592653589793238462643
```

R.7.146 The MATLAB function *taylor(sp, n)* returns the symbolic object *sp*, using an *n* term Taylor (Maclarin) polynomial series approximation.

R.7.147 For example, approximate cos(*x*) by a Taylor's series, using six and eight terms (Lindfield, 2000).

MATLAB Solution
```
>> syms y1 x
>> y1 = taylor(cos(x),6)                       % six term approximation

    y1 =
        1-1/2*x^2+1/24*x^4
```

```
>> y2 = taylor(exp(cos(x)),8) )              % eight term approximation

   y2 =
       exp(1)-1/2*exp(1)*x^2+1/6*exp(1)*x^31/720*exp(1)*x^6
```

R.7.148 The symbolic function $sum\_n = symsum(f(x), a, b)$ returns sum_n, which consist of the sum of the sequence of elements defined by $f(x)$ over the range n given by a, $a + 1, a + 2, ..., b$ for $b > a$.

R.7.149 For example, use MATLAB to evaluate the sum of $f(x)$, as defined in the following expression:

$$f(x) = \sum_{m=1}^{n} x^m \quad \text{for } x = 0.5$$

for

a. $n = 20$

b. $n = 5$

c. $6 \leq n \leq 20$ in two ways

   i. [part a] – [part b]

   ii. Direct evaluation

MATLAB Solution
```
>> syms n sum20 sum5 sumdif
>> sum20 = symsum(.5^n, 1, 20)              % sum of first 20 terms

   sum20 =
           1048575/1048576

>> sum5 = symsum(.5^n, 1, 5)                % sum of first 5 terms

   sum5 =
           31/32

>> sumdif = sum20-sum5                      % sum over terms 6 to
                                              20 / part(c1)

   sumdif =
           32767/1048576

>> sumdifsym = symsum(.5^n, 6, 20) ;        % direct evaluation /
                                              part(c2)

   sumdifsym =
           32767/1048576
```

R.7.150 The commands defined for numerical matrices in Chapter 3, such as *det, inv, eig,* and *trace,* work equally well for symbolic matrices.

R.7.151 For example, let *A* be a 4 × 4 symbolic matrix, defined in the following expression
with elements *a*, *b*, *c*, and *d*:

$$A = \begin{bmatrix} a & b & c & d \\ b & c & d & a \\ c & d & a & d \\ d & a & b & c \end{bmatrix}$$

Use MATLAB and obtain symbolic expressions for

a. B = *det(A)*

b. C = *inv(A)*

c. D = *eig(A)*

d. E = *diag(A)*

e. F = *trace(A)*

MATLAB Solution
```
>> syms a b c d A
>> A = [a b c d ;b c d a ;c d a d ;d a b c]


   A =
        [ a, b, c, d]
        [ b, c, d, a]
        [ c, d, a, d]
        [ d, a, b, c]


>> B = det(A)                        % part (a)


   B =
        2*a^2*c^2-2*a*c*d*b-4*a*d^2*c+3*d*a^2*b+a^2*d^2-a^4-
        2*b^2*a*c+d*b^3+3*b*d*c^2-d^2*b^2-c^4-d^3*b+d^2*c^2+d^4


>> C = inv(A)                        % part(b)


   C =
        (-a*c^2+c*d*b+d^2*c-d*a*b-a*d^2+a^3)/(-2*a^2*c^2+2*a*c*d*b+4*a*d
        ^2*c-3*d*a^2*b-a^2*d^2+a^4+2*b^2*a*c-d*b^3- ...


>> D = eig(A)                        % part (c)


   D =
        RootOf( _Z^4+(-2*c-2*a)* _Z^3+(-b^2-2*d^2-d*b+4*a*c)* _
        Z^2+(d^2*c-2*a^2*c-2*a*c^2+2*c^3+a*d^2+2*a^3+c*b^2
        -2*d*a*b-2*c*d*b+a*b^2)* _Z+2*a^2*c^2-2*a*c*d*b-4*a*d^2*c+3
        *d*a^2*b+a^2*d^2-a^4-2*b^2*a*c+d*b^3+3*b*d*c^2-d^2*b^2-c^4-
        d^3*b+d^2*c^2+d^4)
```

```
>> E = diag(A)                        % part (d)

    E =
        [ a]
        [ c]
        [ a]
        [ c]

>> F = trace(A)                       % part (e)

    F =
        2*a+2*c
```

R.7.152 The MATLAB symbolic functions $y = solve(eq1)$, or $solve(eq1, eq2, eq3, …)$ returns the symbolic solution of an equation ($eqs1$) or the system of equations given by $eq1$, $eq2, …, eqn$.

R.7.153 For example, solve the following equation, or sets of equations given below by using the MATLAB symbolic solver:

a. $x^2 = 9$

   $x - 0.5 * y + 1.5 * z = 5$

b. $6 * x + 4 * y - 2 * z = 10$

   $x - y + z = -1$

MATLAB Solution
```
>> y1 = sym('x^2-9');                 % equation x^2-9=0
>> y2 = sym('x-0.5*y+1.5*z-5');       % equationx-0.5*y+1.5*z-5=0
>> y3 = sym('6*x+4*y-2*z-10');        % equation 6*x+4*y-2*z-10=0
>> y4 =   sym('-x-y+z+1');            % equation -x-y+z+1=0
>> x = solve(y1)                      % solution of part(a)

    x =
        [ 3]
        [ -3]

>> [x,y,z]=solve(y2,y3,y4)            % solutions for part (b)

    x =
        5.
    y =
        -6.
    z =
        -2.

>> % part a can also be solved by the following commands:
>> syms x
>> x = solve('x^2-9=0')

    x =
        [ 3]
        [ -3]
```

R.7.154 When more than one solution satisfies a given equation, MATLAB returns the solution that is closest to zero.

R.7.155 For example, use MATLAB to solve the following equation: $sin^2(x) = cos^2(x)$.

```
MATLAB Solution
>> y = sym('sin(x)^2-cos(x)^2')

    y =
        sin(x)^2-cos(x)^2

>> x = solve(y) ;                          % returns the closest solution
                                             with respect to x= 0

    x =
        [  1/4*pi  ]
        [ -1/4*pi  ]
```

R.7.156 When a system of equations consists of more equations than unknowns, MATLAB returns a warning message. On the other hand, when a system consists of more unknowns than equations, MATLAB treats the first alphabetic variable (s) as a constant and returns the solution in term (s) of that variable (s).

R.7.157 MATLAB is also capable of solving DEs.

A DE is an equation that involves derivatives or integrals. A DE is a mathematical relation between the variable x (where $y = y(x)$) and y and its derivatives with respect to x. This relation can best be stated mathematically by

$$\frac{d^n y}{dx^n} = f(x, y, y', y'', \ldots, y^{n-1})$$

Recall that y is the dependent variable, x is the independent variable, and y^n denotes y differentiated n times with respect to x.

At least one derivative or integral must be present in an equation to make that equation a DE.

R.7.158 Given a DE, the problem consists of finding the function or set of functions $y(x)$ that satisfies the equation $f(x, y, y', y'', \ldots, y^{n-1}y^n) = 0$.

The set of functions $\{y(x)\}$ is referred to as the solutions of the DE.

R.7.159 For example, let

$$\frac{d^2 y(x)}{dx^2} - 16y(x) = 0$$

be a DE, then let us assume that the solution is of the form $y(x) = C_1 e^{4x} + C_2 e^{-4x}$. Then,

$$y'(x) = 4C_1 e^{4x} - 4C_2 e^{-4x}$$

and

$$y''(x) = 16C_1 e^{4x} + 16C_2 e^{-4x}$$

Substituting the preceding derivatives in the DE yields

$$16C_1 e^{4x} + 16C_2 e^{-4x} - 16C_1 e^{4x} - 16C_2 e^{-4x} = 0$$

verifying that the resulting relation is indeed an equation, and its solution is $y(x) = C_1 e^{4x} + C_2 e^{-4x}$. Observe in general that if $y(x)$ is substituted in a given DE and the resulting expression is an identity over x in some interval, then $y(x)$ is the solution of the DE. The constants C_1 and C_2 are referred to as the boundary conditions of the DE.

R.7.160   The order of the DE is given by its highest derivative of the dependent variable y, with respect to the independent variable x.

R.7.161   For example, a first-order DE contains only the first derivative of y with respect to x. The equation

$$\frac{d^2 y(x)}{dx^2} - 16y(x) = 0$$

is an example of a second-order DE.

R.7.162   A DE is linear if the dependent variable y is raised to the first power. For example,

$$3\frac{d^2 y}{dx^2} + 2t\frac{dy}{dx} + y\sin(x) = \cos^2(x)$$

is a linear, second-order DE.

R.7.163   The term ordinary DE refers to a DE where only the derivatives are functions of one variable. The general form is

$$k_n \frac{d^n y}{dx^n} + k_{n-1} \frac{d^{n-1} y}{dx^{n-1}} \cdots k_0 y = f(y, x)$$

Observe that the solution of an ordinary DE is a function of one variable.
When the solution is a function of more than one variable, the derivatives are then called partial derivatives, and the equation is referred as a partial DE.
Only certain types of ordinary DEs will be considered in this chapter.

R.7.164   DEs have in general an infinite number of solutions.
A unique solution $y(x)$ can be evaluated when ICs (or boundaries) are specified, such as $y(x = a) = ya$, as well as the range of x for which the solution $y(x)$ holds.

R.7.165   A second-order DE is defined in general by the following relation:

$$a\frac{d^2 y}{dx^2} + b\frac{dy}{dx} + cy = d$$

where a, b, c, and d can be constants or functions of the independent variable x.

R.7.166 When the equation in R.7.165 is equated to zero ($d = 0$), then the equation is called homogeneous, otherwise it is referred as inhomogeneous.

R.7.167 A DE with ICs is referred to as an initial value problem.

In general, an n-order DE needs n ICs, where the ICs are given by the dependent variable y specified at particular values of x, the independent variable, and its $n - 1$ derivatives of y with respect to x, specified at particular values of x.*

R.7.168 Ordinary DEs that are of initial value are often encountered in real world problems in the physical sciences and engineering such as in electrical circuits, electronics, mechanics, heat transfer, and dynamics.

R.7.169 Since linear DEs constitute an important part of real world problems, particular emphasis is given in its treatment. The solution of linear DEs consists of the sum of two solutions called (Stanley, 2005)

a. *The particular solution* (denoted by y_1)

b. *The general solution* (denoted by y_2)

R.7.170 The particular solution of the homogeneous equation usually has the form of the right-hand side of the DE, and the general solution involves exponential functions.

R.7.171 An often-encountered DE in the sciences and engineering is

$$\frac{dy(t)}{dt} = ky(t)$$

Its solution is $y(t) = Ae^{kt}$, a growing or decaying exponential depending on if $k > 0$ or $k < 0$, where A is an arbitrary constant that satisfies a given boundary condition.

This DE models situations as diverse as the growth of the world population, the growth of the economy, or the charging or discharging of a capacitor.

R.7.172 To gain some experience in solving DE, four examples of analytical solutions of DEs are illustrated as follows:

The steps involved in arriving at the solutions are indicated and hopefully can be followed by the reader.

Example 1

Solve the following DE:

$$\frac{dy(t)}{dt} = 4(y(t) + 3)$$

with the following condition $y = 5$ at $t = 0$, commonly expressed as $y(0) = 5$, or $y_0 = 5$.

---

* In many practical problems, the independent variable is time, denoted by t.

ANALYTICAL Solution

$$\frac{dy(t)}{dt} = 4y(t) + 12$$

$$\frac{dy(t)}{dt} - 4y(t) = 12$$

then the particular solution (y_1) is a constant. Then,

$$\frac{dy(t)}{dt} = 0 \quad \text{and} \quad y_1 = -\frac{12}{4} = -3$$

The general solution (y_2) involves exponentials of the form $y_2 = Ae^{st}$, where the coefficient of the exponential is evaluated by first replacing $\frac{d}{dt}$ by s and the constant term by zero, obtaining in this way what is called the auxiliary equation $s - 4 = 0$.
  Solving for s, yield $s = 4$. The solution is then given by $y = y_1 + y_2 = -3 + Ae^{4t}$.
  The given ICs yield

$$y|_{t=0} = 5 = -3 + Ae^{4(0)}$$

$$5 = -3 + A$$

$$A = 5 + 3 = 8$$

The complete solution $y(t)$ is therefore $y(t) = -3 + 8e^{4t}$
  This solution can be verified by substituting $y(t)$ into the given DE. Then

$$\frac{d}{dt}[-3 + 8e^{4t}] = 4[-3 + 8e^{4t}] + 12$$

Performing the differentiation yields $32e^{4t} = -12 + 32e^{4t} + 12$, verifying in this way that indeed $y(t)$ is the solution of the given DE, satisfying the given IC.

Example 2

Solve analytically the following DE:

$$\frac{dy(t)}{dt} = \frac{1}{3 + t^2}$$

with the IC $y = 7$ at $t = 0$, or $y(0) = 7$.

ANALYTICAL Solution

DEs can be solved in some cases using simple algebraic manipulations, and by applying calculus concepts, as in the following illustration:

Solving for dy, and then integrating yields,

$$dy = \left[\frac{1}{3 + t^2}\right]dt \quad \text{(separation of variables)}$$

$$\int dy = \int \left[\frac{1}{3+t^2} \right] dt$$

$$y(t) = \left(1/\sqrt{3} \right) \tan^{-1}\left(t/\sqrt{3} \right) + k$$

The preceding solution for the given IC yields

$$y\big|_{t-0} = 7 = tan^{-1}(0) + k, \quad \text{therefore} \quad k = 7$$

and

$$y(t) = \left(1/\sqrt{3} \right) \tan^{-1}\left(t/\sqrt{3} \right) + 7$$

Example 3

Obtain the particular and general solutions for the following second-order DE, given by

$$\frac{d^2 y(t)}{dt^2} + \frac{dy(t)}{dt} - 2y(t) = 10$$

ANALYTICAL Solution

Since, $dy/dt = 0$, then the particular solution is $y_1 = -10/2 = -5$, and the general solution y_2 is obtained by evaluating the auxiliary equation given by $s^2 + s - 2 = 0$, replacing $\frac{d}{dt} = s$. Then $(s-1)(s+2) = 0$, and $s = +1$, and $s = -2$ are its roots. Thus the general solution y_2 is of the form $y_2(t) = Ae^t + Be^{-2t}$ and the complete solution is then given by

$$y(t) = y_1(t) + y_2(t)$$

$$y(t) = -5 + Ae^t + Be^{-2t}$$

The constants A and B can be evaluated if boundary conditions are known or given.

Example 4

Solve the following DE by separation of variables:

$$\frac{dy(t)}{dt} = 4t\sqrt{1 - y^2}$$

ANALYTICAL Solution

$$\frac{dy(t)}{\sqrt{1 - y^2}} = 4t\,dt$$

$$\int \left[\frac{1}{\sqrt{1 - y^2}} \right] dy = \int [4t]dt + c$$

$$sin^{-1}(y) = 2t^2 + c$$

$$y = sin(2t^2 + c)$$

R.7.173 The MATLAB functions *ode23* and *ode45* are used in this book to illustrate the steps followed in the numerical solution of either one, or a set of ordinary DEs using the variable step Runge–Kutta method (approximations using second/third for *ode23* or fourth/fifth order for *ode45*). MATLAB has, in addition, a library of other ordinary DEs solvers, used for particular cases, such as *ode113*, *ode15s*, *ode23s*, *ode23t*, and *odets*. The implementation and usage of the *ode* solvers are presented in the next section.

R.7.174 The MATLAB functions

$$[y, t] = ode23('difeq,' tin, tfin, yic, tol, trace)$$

$$[y, t] = ode45('difeq,' tin, tfin, yic, tol, trace)$$

are used to solve ordinary couple DEs using numerical techniques, where "*difeq*," is the given DE defined as a string, contained in an M-function file specially created for this purpose. The solution *y(t)* is evaluated over the range *tin* < *t* < *tfin*, with an optional accuracy given by (tolerance) *tol* with MATLAB default values of 10^{-3} for *ode23* and 10^{-6} for *ode45*. The optional argument *trace* can be nonzero, in which case the intermediate results are displayed. The tolerance (*tol*) and other parameters can be specified by the *odeset* function (see script file R_6_176). The functions *ode23* and *ode45* are very similar.

The only difference is that *ode45* is more accurate but much slower than *ode23*. *yic* represent the ICs specified as a column vector. MATLAB has a number of solvers for ordinary DEs. The preferred function is *ode45* and is the one that usually provides satisfactory results. Other MATLAB numerical solvers of DEs are

ode113, *ode15s*, *ode23s*, *ode23t*, and *ode23tb*.

The MATLAB solvers

ode15s, *ode23t*, *ode23s*, *ode23tb*

are particularly useful when the ordinary DE is *stiff*.

The purpose of all the numerical solvers is to find approximations to almost any system of DEs. The syntax is very similar for all the *ode* solvers, and by learning one we learn how to use any of them.

For additional information about any of the numerical solvers and how to use them, check the online help.

R.7.175 A *stiff* DE is an equation that affects unequally different time intervals, and any time scale cannot accurately reflect and plot its behavior.

For example, the following DE can be considered *stiff*:

$$\frac{d^2y(t)}{dt^2} + \frac{dy(t)}{dt} + 10,000 = 0$$

The general solution is of the form $y(t) = Ae^{-t} + Be^{-10,000t}$.

To give physical meaning to the preceding equation, assume that *t* represents time in seconds. Observe that Ae^{-t} has an effective range over $0\,\text{s} \leq t \leq 5\text{s}$, whereas $Be^{-10,000t}$ has an effective range over, $0\,\text{s} \leq t \leq 0.005$ s.

Clearly, a time interval of 5 s should be appropriate to observe the contributions of each exponential. Then, any time scale plot of $Be^{-10,000t}$ would appear as a disturbance, because a change of $Be^{-10,000t}$ from its maximum to its minimum would take a very small interval of time, requiring a large number of small steps in its evaluation.

R.7.176 For example, consider the first-order, linear, ordinary DE

$$\frac{dy}{dt} = 4t - 2y$$

with IC given by $y(0) = 150$.

The analytical solution is then given by $y(t) = 151e^{-2t} + 2t - 1$.

$$\frac{dy}{dt} = 4t - 2y$$

is solved later by creating the script file R_6_176 using the functions *ode23, ode45,* and *ode113*. The analytical solution is plotted in Figure 7.8, over the range $0 \le t \le 4$ and $0 \le t \le 30$.

The reader can compare the various solutions that are illustrated graphically and observe that they are not exactly equal. Observe that for the domain shown, between 0 and 30 the lengths of the solutions presented are not equal (40 versus 105). Furthermore, observe that the solution using *ode45* represents a much better approximation *t* than the solution employing *ode23*, but is equivalent to *ode113*.

Recall that to use the functions *ode23, ode45,* or *ode113* (or any *ode* solver), the DE must first be defined in a function file,* named f in this example, and given as follows:

$$\text{function dery} = f(t, y)$$

$$\text{dery} = 4 * t - 2 * y$$

The functions *ode23, ode45,* and *ode113* are called by the script file R_6_176, with an initial and final time of 0 and 30, and IC $y(0) = 150$.

MATLAB Solution
```
% Script file:R _ 6 _ 176
format compact;
[t,y1] = ode23('f',[0,30],150);
[t,y2] = ode45('f',[0,30],150);
lengtht = length(t); lengthy1=length(y1); lengthy2 = length(y2);

figure (1)
subplot (2,1,1); x = linspace(0,30,40);
plot (x,y1,'*'); xlabel ('t'); grid on;
title ('Solution using ode23,for 40 points for 0<t<30');
subplot (2,1,2);
plot (t(1:40),y2(1:40),'d');grid on;xlabel ('t(time)');
title ('Solution using ode45,for first 40 points for 0<t<30');

figure(2)
sol = 151*exp(-2.*t)+2.*t-1;
subplot(2,1,1);
plot(t(1:40),sol(1:40),'+') ; xlabel('t'); grid on;
title('Analytic solution ,for  first 40 points for 0<t<4.1');
subplot(2,1,2);
plot(t,sol,'+');grid on;
title('Analytic solution ,for first 105 points for 0<t<30');
xlabel ('t(time)');
```

---

* Function files are treated with details in Chapter 9, but were briefly introduced in Chapter 1. At this point, just by observing the format and structure of the function file the reader can get a clear idea of how to use the *ode* solver.

```
figure(3)
subplot(2,1,1);
plot(t,y2,'^');
title ('Solution using ode45,for 105 points for 0<t<30');
grid on; xlabel('t');
subplot(2,1,2);[t,y3]=ode113('f',[0 30],150);
plot(t,y3,'o')
title('Solution using ode113,for 105 points');
grid on; xlabel('t(time)');
```

Back in the command window the function file *R_6_176* is executed, and the solutions in the form of plots are shown in Figures 7.8 through 7.10.

```
>> R _ 6 _ 167

    lenhtt =
             105
    lenhty1 =
             40
    lenhty2=
             105


K>>return
K>>return
```

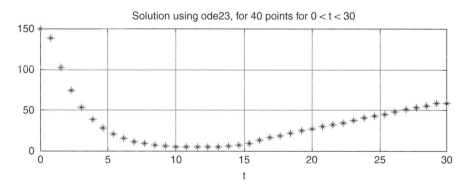

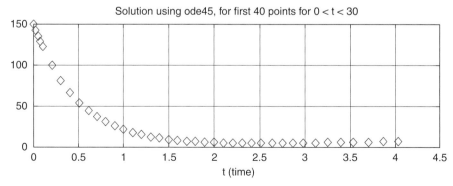

FIGURE 7.8
Solution using *ode* of R.7.176.

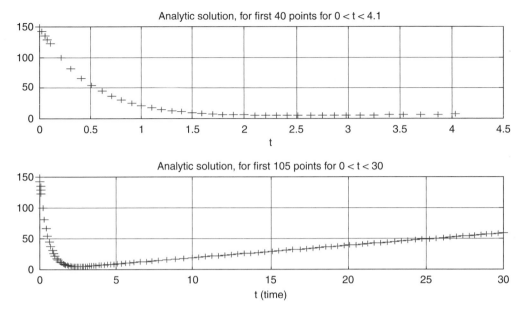

FIGURE 7.9
Analytical solution of R.7.176.

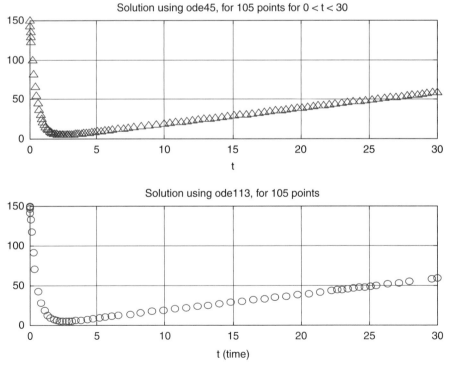

FIGURE 7.10
Solutions using *ode45* and *ode113* of R.7.176.

R.7.177 The MATLAB Symbolic function

$$dsolve('difeq1', 'difeq2', 'difeq3', \ldots 'IC1', 'IC2', 'IC3', \ldots)$$

is used to solve up to 10 DEs given by (*difeq1, difeq2, …, difeq10*) with the corresponding ICs, provided that the total number of arguments do not exit 12 (Polking and Arnold, 2004).

The function *dsolve* returns a solution if one exists. If no solution exceed, then the numerical approach using an *ode* solver, such as *ode23, ode45, or ode113*, is the choice left.

The arguments *difeq1, difeq2, …, difeq10* must be specified using standard symbolic notation where the character *D* denotes the first derivative. Similarly, *D2*, *D3*, …, *Dn* represent higher derivatives, assuming that the dependent variable is *t*.

Observe that *dsolve* requires that the DEs be entered as strings, which means inside single quotes.

The following are syntax examples used by *dsolve*:

a. The second derivative is denoted by

$$D2y = \frac{d^2y}{dt^2}$$

b. The IC is denoted by $y(0) = a$.

c. The first derivative of y with respect to t evaluated at $t = 0$ is denoted by $Dy(0) = b$, where a and b can be either constants or symbolic variables.

R.7.178 To illustrate the use of the function *dsolve*, the following three examples are presented and solved:

a. Example 1: $dy/dt = -3y$, with $y(0) = 1$

b. Example 2: $(1 + t^3)(dy/dt) + 3ty = \cos(t)$, and *simplify* the result

c. Example 3: $dy/dt = 4t - 2y$, with $y(0) = 150$, for $0 \le t \le 4$, and obtain the plot $y(t)$ versus t

Example 1

Solve the following DE:

$$\frac{dy}{dt} = -3y$$

with the IC $y(0) = 1$.

MATLAB Solution
```
>> y = dsolve('Dy=-3*y,'y(0)=1')

    y =
            exp(-3*t)
```

A more complex example, where the steady-state response (solution) is determined (when no ICs are given) is illustrated in Example 2.

Example 2

Solve the following DE:

Let

$$(1+t^3)\frac{dy}{dt} + 3ty = \cos(t)$$

and simplify the solution obtained.

MATLAB Solution
```
>> syms t y
>> solution = dsolve('(1+t^3)*Dy+3*t*y=cos(t)')

   solution =
      (exp(-3^(1/2)*atan(2/3*3^(1/2)*t-1/3*3^(1/2)))*Int(cos(t)*(t^2-
      +1)^(1/2)*exp(3^(1/2)*atan(2/3*3^(1/2)*t-
      1/3*3^(1/2)))/(t^4+t^3+t+1),t)*t+
      exp(-3^(1/2)*atan(2/3*3^(1/2)*t-1/3*3^(1/2)))*Int(cos(t)*(t^2-
      t+1)^(1/2)*exp(3^(1/2)*atan(2/3*3^(1/2)*t-
      1/3*3^(1/2)))/(t^4+t^3+t+1),t)+exp(-3^(1/2)*atan(2/3*3^(1/2)*t-
      1/3*3^(1/2)))*C1*t+exp(-3^(1/2)*atan(2/3*3^(1/2)*t-
      1/3*3^(1/2)))*C1)/(t^2-t+1)^(1/2)

>> simplify (solution)

   ans =
      exp(-3^(1/2)*atan(2/3*3^(1/2)*t-1/3*3^(1/2)))*(Int(cos(t)*(t^2-
      t+1)^(1/2)*exp(3^(1/2)*atan(2/3*3^(1/2)*t-
      1/3*3^(1/2)))/(t^4+t^3+t+1),t)*t+Int(cos(t)*(t^2-
      t+1)^(1/2)*exp(3^(1/2)*atan(2/3*3^(1/2)*t-
      1/3*3^(1/2)))/(t^4+t^3+t+1),t)+C1*t+C1)/(t^2-t+1)^(1/2)
```

Observe that in many cases the solution obtained using MATLAB is long and complex, and given in terms of arbitrary constant note that, Example 3, the first-order, linear, ordinary DE presented in R.7.176 is revisited. Recall that the DE that was solved using the *ode* solver (numerical) is now solved using the command *dsolve* (symbolic) below.

Example 3

Solve and plot the following DE:

$$\frac{dy}{dt} = 4t - 2y$$

with $y(0) = 150$, over the range $0 \le t \le 4$.

MATLAB Solution
```
>> syms t y
>> y = dsolve('Dy = 4*t - 2*y', 'y(0) = 150')

    y =
        2*t-1+151*exp(-2*t)
>> ezplot (y)                      % returns the plot of the sym object y
>> axis([0 4 0 200])
>> xlabel('t (time)'); ylabel('y(t)');
```

The plot of the solution $y(t)$ versus t is shown in Figure 7.11.
Note that the symbolic solution fully agrees with the analytical solution of R.7.176.

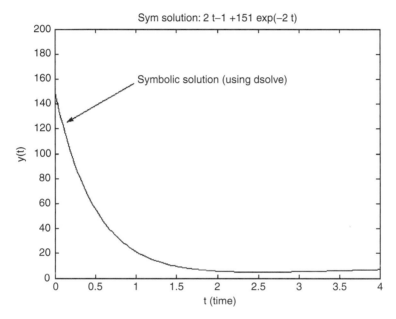

FIGURE 7.11
Symbolic solution of R.7.181.

R.7.179 The command *class (y)* returns the class of y as either a symbolic or a numerical variable. An alternate way to determine the class is by using the command *whos*.

R.7.180 The *ode* commands can be used to solve first-order DEs. If a given DE is of second- or higher-order, then the equation must first be transformed into a set of first-order DEs, a technique referred as the state–space, state variable, or Cauchy form.

R.7.181 To discuss the state–space approach, and the solution of a system of DEs, let us get back to the model of a system.

Recall that any linear system can be described in terms of an n-order DE, in the time domain, or the equivalent relation called the transfer function in the frequency domain, given by

$$H(s) = \frac{Y(s)}{X(s)}$$

with all the ICs set to zero (*IC's* = 0).

R.7.182 The transfer functions, as well as the system DE, gives a quantitative and qualitative relation between the system input(s) and output(s). This relationship results essentially from the elimination of all internal system variables.

The block box system variables can be evaluated from experimentation and measurements, without any knowledge of the internal structure of the system.

In practical systems, the internal structure is known and the variables that affect its performance can be observed and analyzed.

R.7.183 The state model, on the other hand, utilizes some of the system variables, which ordinarily are not included in the transfer function. These additional variables permit the resolution of a system into a set of first-order subsystems, or analytically in a set of first-order DEs.

For example, the state model of an n-order system can be decomposed in a set of n first-order subsystems, involving the n state–space variables.

R.7.184 Recall that the input–output description of a system is appropriate only if the system is initially relaxed. If the system is not initially relaxed, then the output depends also on the system's ICs. The set of ICs are referred to as a state. A state of a system can be defined as the set of variables such that if the output is known at a time t_0, and all inputs and ICs are known for $t > t_0$, then its output can be determined uniquely for any $t > t_0$.

R.7.185 The growing interest in system optimization has led to the extensive use of state–space equations in a variety of disciplines. The concept of state is a universally accepted concept that is equally useful when applied to any type of dynamic system such as economical, political, epidemiological, educational, ecological, and meteorological just to mention a few.[*]

R.7.186 Let v be the system state variable, $x(t)$ the input, and $y(t)$ the system output. Then the state variable system description is given by the following set of equations:

$$\dot{v} = Av(t) + Bx(t)$$

and the system output equation given by

$$y(t) = Cv(t) + Dx(t)$$

where in general A, B, C, and D are nxn, nxp, qxn, and qxp matrices, respectively.

R.7.187 A sufficient condition for the state–space equations to have a unique solution is that every element of A, B, C, and D should be a continuous function of t over the range $-\infty$ to ∞.

If A, B, C, and D are time dependent, then the set of equations describes a linear time-varying dynamic system. If, on the other hand, A, B, C, and D are independent of t, then the system is a linear-invariant dynamic system.

---

[*] Under the supervision of Prof. John R. Raggazzini of Columbia University (1950), Rudolf E. Kalman, a graduate student. They became the leading advocate of the decomposition of a system into a state–space structure. The work by Zadeh and Desoer, *Linear Systems Theory, the State Space Approach*, published in 1963, and reprinted in 1979, was in general well received by the electrical engineering community, and became the standard approach in the field of control theory.

R.7.188 Let the following general n-order system DE be given by

$$\sum_{k=0}^{n} a_k \frac{d^k y(t)}{dt^k} = b_0 x(t) \quad \text{with } a_n = 1$$

where $x(t)$ is the system input and $y(t)$ is the system output. Then, the preceding dynamic system can be represented by n first-order DEs, by assigning the (state) variables $v_1(t)$, $v_2(t)$, ..., $v_k(t)$ as indicated in the following relations:

$$v_1(t) = y(t)$$

$$v_2(t) = \frac{dy(t)}{dt} = \dot{y}$$

$$v_3(t) = \frac{d^2 y(t)}{dt^2} = \ddot{y}$$

$$\ldots$$

$$v_{n-1}(t) = \frac{d^{n-2} y(t)}{dt^{n-2}}$$

$$\frac{dv_n(t)}{dt} = -a_0 v_1(t) - a_1 v_2(t) - a_2 v_3(t) + \cdots - a_{n-1} v_n(t) + b_0 x(t)$$

The preceding set of equations can be written in matrix form as

$$
\begin{bmatrix} \dot{v} \\ \dot{v}_2 \\ \dot{v}_3 \\ \ldots \\ \dot{v}_{n-1} \\ \dot{v}_n \end{bmatrix} =
\begin{bmatrix}
0 & 1 & 0 & \ldots & 0 & 0 \\
0 & 0 & 1 & 0 & \ldots & 0 \\
0 & 0 & 0 & 1 & \ldots & 0 \\
0 & \ldots & \ldots & \ldots & 1 & 0 \\
0 & 0 & 0 & \ldots & \ldots & 1 \\
-a_0 & -a_1 & -a_2 & \ldots & \ldots & -a_{n-1}
\end{bmatrix}
\begin{bmatrix} v_1 \\ v_2 \\ v_3 \\ \ldots \\ v_n \end{bmatrix} +
\begin{bmatrix} 0 \\ 0 \\ 0 \\ 0 \\ \ldots \\ b_0 \end{bmatrix} x(t)
$$

where

$$
A = \begin{bmatrix}
0 & 1 & 0 & \ldots & 0 & 0 \\
0 & 0 & 1 & 0 & \ldots & 0 \\
0 & 0 & 0 & 1 & \ldots & 0 \\
0 & \ldots & \ldots & \ldots & 1 & 0 \\
0 & 0 & 0 & \ldots & \ldots & 1 \\
-a_0 & -a_1 & -a_2 & \ldots & \ldots & -a_{n-1}
\end{bmatrix} \quad \text{and} \quad B = \begin{bmatrix} 0 \\ 0 \\ 0 \\ 0 \\ \ldots \\ b_0 \end{bmatrix}
$$

and the output matrix equation becomes

$$y(t) = \begin{bmatrix} 1 & 0 & 0 & \cdots & \cdots & 0 \end{bmatrix} \begin{bmatrix} v_1 \\ v_2 \\ v_3 \\ \cdots \\ v_{n-1} \\ v_n \end{bmatrix} + [0][x(t)]$$

where $C = [1 \quad 0 \quad 0 \quad \cdots \quad \cdots \quad 0]$ and $D = [0]$.

For simplicity, the discussions in this text will be limited to the time-unvarying systems with constant coefficients, because they constitute the vast majority of practical cases.

R.7.189 For example, let

$$7\frac{d^2y(t)}{dt^2} + 3\frac{dy(t)}{dt} + 4y = 5\sin(t)$$

be a system equation.

Then, obtain the system's state–space matrix equation.

```
Analytical Solution
```

$$\begin{bmatrix} \dot{v}_1 \\ \dot{v}_2 \end{bmatrix} = \begin{bmatrix} 0 & 1 \\ -4/7 & -3/7 \end{bmatrix} \begin{bmatrix} v_1 \\ v_2 \end{bmatrix} + \begin{bmatrix} 0 \\ 1/7 \end{bmatrix} [5\sin(t)]$$

and

$$y(t) = \begin{bmatrix} 1 & 0 \end{bmatrix} \begin{bmatrix} v_1 \\ v_2 \end{bmatrix} + [0][5\sin(t)]$$

R.7.190 The MATLAB command $[y, v] = lsim(A, B, C, D, x, t, Vo)$ returns the system output y and the state variables v, given the system state–space matrices A, B, C, D, and input x defined over an interval t, with the ICs given by Vo, assuming that the system is time invariant.

R.7.191 The MATLAB command $lsim(A, B, C, D, x, t, Vo)$ returns the plot $y(t)$ versus t.

Color, line style, and marker can be used to define the responses, when dealing with multiple systems in which case $lsim(sys1, 'r', sys2, 'y--', sys3, \ldots, x, t)$.

R.7.192 The MATLAB command $initial\ (A, B, C, D)$ returns the plot of the free response of the linear, time-invariant (LTI) system defined by the state–space equations (A, B, C, D).

R.7.193 For example, create the script file $lsim\_plots$ that returns the plot of $y(t)$ versus t of the system equation

$$7\frac{d^2y(t)}{dt^2} + 3\frac{dy(t)}{dt} + 4y = 5\sin(t)$$

for the following cases:

a. With no IC

b. With IC = [2 ; 3]

c. The free response

MATLAB Solution
```
% Script file: lsim_plots
A= [0 1;-4/7 -3/7]; B = [0;1/7]; C =[1 0]; D = [0];
t =linspace(0,6,500);
subplot(3,1,1)
x = 5*sin(t);
lsim (A,B,C,D,x,t);
subplot (3,1,2)
Vo = [2;3];lsim(A,B,C,D,x,t,Vo)
subplot (3,1,3)
initial(A,B,C,D,Vo) % Figure 7.12
```

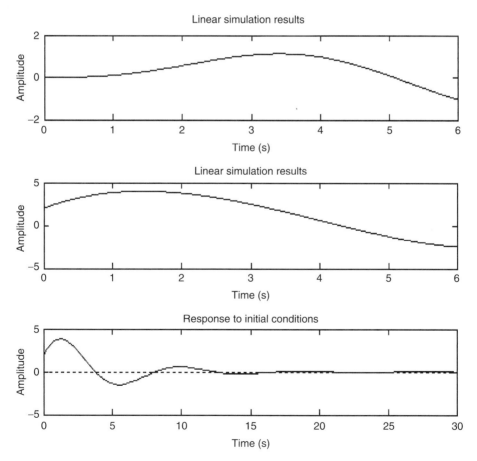

FIGURE 7.12
Solutions of R.7.193(a, b, and c).

R.7.194   The MATLAB command $[Y, X] = tf2ss(A, B, C, D)$ returns the system transfer function $H(s) = Y(s)/X(s)$ as vectors (or polynomial) containing the coefficients of the Y (numerator) and X (denominator) in descending powers of s, given the state–space matrices A, B, C, and D.

R.7.195   The MATLAB command $[A, B, C, D] = tf2ss(Y, X)$ returns the state–space matrix representation of the system given by

$$\dot{v} = Av + Bx$$

$$y = Cv + Dx$$

where the transfer function is $H(s) = Y(s)/X(s)$.

This command also works for discrete systems. For discrete-time transfer functions, the length of the numerator and denominator are made equal to ensure correct results (pad with zeros).

R.7.196   For example, use MATLAB and obtain the transfer-function of the system defined by the DE is given by

$$7\frac{d^2y(t)}{dt^2} + 3\frac{dy(t)}{dt} + 4y = x(t)$$

Since the state–space matrices are already known, let us verify the system transfer function as illustrated in the following:

MATLAB Solution
```
>> A = [0 1;-4/7 -3/7]; B=[0;1/7]; C = [1 0]; D=[0];
>> [Y,X] = ss2tf (A,B,C,D)

    Y =
         0            0          0.1429
    X =
         1.0000       0.4286     0.5714
```

Observe that the transfer function can be calculated directly from the DE as

$$Y(S)(7s^2 + 3s + 4) = X(s)$$

or

$$\frac{Y(s)}{X(s)} = \frac{1}{7s^2 + 3s + 4}$$

The MATLAB command *ss2tf* returns the leading denominator (X) coefficient set to one (observe that all the coefficients are divided by 7).

Let us verify the transfer function by getting the ss matrices.

```
>> [A,B,C,D] = tf2ss(Y,X)

    A =
         -0.4286      -0.5714
          1.0000       0
```

```
    B =
        1
        0
    C =
        0       0.1429
    D =
        0
```

Note that the matrices obtained represent an equivalent system to the one in P.7.189.

R.7.197 The MATLAB command *tf(indep, depend)* returns the system transfer function, where *indep* represents the coefficients of the DE arranged in descending order of the independent variable (*x*) and *depend* represents the coefficients of the DE arranged in descending order of the dependent variable (*y*).

For example, the command *tf* is used to obtain the transfer function for the system DE given by

$$7\frac{d^2y(t)}{dt^2} + 3\frac{dy(t)}{dt} + 4y = x(t)$$

where *indep* = [1] and *depend* = [7 3 4], then

```
>> tf([1],[7 3 4])
    Transfer function:
       1
    ---------------
    7 s^2 + 3 s + 4
```

R.7.198 Clearly, the state variable model of a given system is not unique.

Instead of having a model of a given LTI system based directly on its transfer function in rational fraction form, it may instead be modeled upon the partial fraction expansion of the transfer function. Each partial fraction component is then considered separately as a subsystem and the system response *y* is the addition of all the subsystems, which may be considered as connected in parallel.

R.7.199 The MATLAB command [*Ad, Bd, Cdd, ICd*] = *c2d(A, B, C, D, Ts)* returns the discrete-time space–state system description given the continuous time state–space model, including the discrete ICs *ICd*, using the sampling rate *Ts*.

The discretization method employed by MATLAB is an option chosen from the following:

'zoh'	zero-order hold on the inputs
'foh'	linear interpolation of inputs (triangle approximation)
'tustin'	bilinear approximation
'prewarp'	with frequency prewarping

The default option is *'zoh'*.

R.7.200 For example, using the state–space continuous matrices *A* and *B* from the system given by the DE

$$7\frac{d^2y(t)}{dt^2} + 3\frac{dy(t)}{dt} + 4y = x(t)$$

obtain the discrete state–space matrices using the following sampling times: Ts = 1, 0.5, and 0.25, and *zoh* as the discretization option.

MATLAB Solution
```
>> A= [0 1;-4/7 -3/7], B = [0;1/7]

    A =
                0        1.0000
          -0.5714      -0.4286
    B =
                0
          0.1429

>> [Ad1,Bd1] = c2d(A,B,1)

    Ad1 =
          0.7624       0.7383
         -0.4219       0.4460
    Bd1 =
          0.0594
          0.1055

>> [Ad05,Bd05] = c2d(A,B,0.5)

    Ad05 =
          0.9342       0.4394
         -0.2511       0.7459
    Bd05 =
          0.0165
          0.0628

>> [Ad025,Bd025] = c2d(A,B,0.25)

    Ad025 =
          0.9828       0.2357
         -0.1347       0.8818
    Bd025 =
          0.0043
          0.0337
```

R.7.201 The MATLAB command [*A, B*] = *d2c(Ad, Bd, Ts*] returns the continuous-time space–state system description, given the discrete time state–space model discretized with a sampling rate of *Ts*. The conversion method employed by MATLAB is an option that can be chosen from the following:

'zoh'	zero-order hold on the inputs
'tustin'	bilinear approximation
'prewarp'	with frequency prewarping

The default option is *'zoh'*.

R.7.202 For example, using the discrete state–space matrices Ad025 and Bd025, with $Ts = 0.25$, obtained in R.7.200 from the continuous system given by the DE

$$7\frac{d^2y(t)}{dt^2} + 3\frac{dy(t)}{dt} + 4y = x(t)$$

is used to get back the continuous state–space matrices A and B.

MATLAB Solution
```
>> Ad025

    Ad025 =
                0.9828      0.2357
               -0.1347      0.8818

>> Bd025

    Bd025 =
                0.0043
                0.0337

>> [A,B] = d2c(Ad025,Bd025,0.25)

    A =
            0.0000      1.0000
           -0.5714     -0.4286
    B =
           -0.0000
            0.1429
```

Observe that continuous matrices fully agree with the matrices of R.7.200.

R.7.203 The MATLAB command $[Add, Bdd] = d2d(Ad, Bd, Ts]$ returns a resample-time space–state discrete equivalent system model discretized with a resampling rate of Ts.

R.7.204 The MATLAB command $[y, v] = dlsim(Ad, Bd, Cd, Dd, x.V_o)$ or $[y,v] = dlsim(Y, X, x, V_o)$ returns the system output y and state variables v, given the discrete system state–space matrices Ad, Bd, Cd, Dd or the coefficients of the transfer function, given by Y and X (functions of z), and input x, with the ICs V_o.

R.7.205 The MATLAB command $dlsim(Ad, Bd, Cd, Dd, x, Vo)$ or $lsim(num, den, x)$ returns the time response plot. Color, line style, and the choice of markers are options that can define the responses when dealing with multiple systems, in which case $dlsim(sys1, 'r', sys2, 'y--', sys3, ..., x)$.

The discrete system simulation corresponds to the state–space model given by the difference equations

$$v[n + 1] = Av[n] + Bx[n]$$

$$y[n] = Cx[n] + Dx[n]$$

If $lengh(Y) = lengh(x)$, then $dlsin(Y, X, x)$ is equivalent to $filter(num, den, x)$.

R.7.206 Recall that MATLAB offers a number of symbolic plotting commands (see Chapter 5) that returns the plots of symbolic objects. These commands have the prefix *ez*, such as

> *ezcontour*
>
> *ezcontourf*
>
> *ezmesh*
>
> *ezmeshc*
>
> *ezplot*
>
> *ezplot3*
>
> *ezpolar*
>
> *ezsurf*
>
> *ezsurfc*

For any additional information regarding any of the preceding functions, use the help online command.

R.7.207 Recall that *ezplot* was introduced and used in earlier chapters.

For example, let us quickly review *ezplot* by plotting the function $y(x) = (x - 1)^3 + 2$, over the range $0.5 \le x \le 2.5$.

MATLAB Solution
```
>> y = sym('(x-1)^3+2');
>> ezplot(y), grid on
>> axis([-0.5 2.5 0 5])
>> xlabel('X'), ylabel('Y'), title('Y VS X')
>> % the resulting plot is shown in Figure 7.13
```

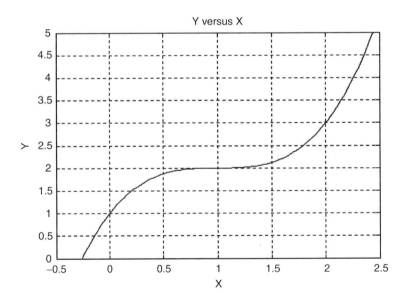

FIGURE 7.13
ezplot of R.7.207.

R.7.208 Let $y = f(x)$, then the MATLAB function *fmin('yy', x1, x2)* returns the minimum of $f(x)$ over the range $x1 \leq x \leq x2$, where the function y is defined by the function file yy.

The newer versions of MATLAB replaces *fmin* by *fminbnd('yy', x1, x2)*.

R.7.209 For example, use MATLAB to determine the local minimum of $y(x) = cos(2*x)$, over the range $0.5 \leq x \leq 1.0$, using the function *fmin*, with the following arguments:

a. y is specified as a function file

b. y is specified as a string

MATLAB Solution
The function y is defined by the function file yy:

```
        function y = yy(x)
        y = cos(2*x);
```

Back in the command window, the following commands are executed

```
>> x = fmin('yy', .5, 1.0)              % using the function file yy


    x =
        0.9999


>> x = fmin('cos(2*x)', .5, 1.0)        % y is a string


    x =
        0.9999
```

R.7.210 The MATLAB function *eval('ya')*, where *ya* is a string representing a polynomial or any arbitrary function returns the numerical value for *ya*.

R.7.211 For example, the symbolic command *eval ('ya=6 * sin(3 * pi+pi/3)')* is equivalent of executing the numerical command *ya=6 * sin(3 * pi+pi/3)*, as in the following illustration:

MATLAB Solution
```
>> eval('ya = 6*sin(3*pi+pi/3)')         % symbolic evaluation

        ya =
             -5.1962

>> ya = 6*sin(3*pi+pi/3)                  % numerical evaluation

        ya =
             -5.1962
```

R.7.212 The function *yzero = fzero('y', a)* returns the zero of the expression y, closest to a, where y can be a polynomial or any other function. The general format of the function is given by *yzero = fzero('y', a, tol, trace)*. Recall that the optional specs *tol* and *trace* were defined for the *ode* commands (R.7.174). This function is particularly useful to solve nonlinear equations involving one variable.

R.7.213 For example, solve the following equation $sin(x) = cos(x)$, for the value of x closest to *1*.

MATLAB Solution
```
Let us define the nonlinear equation in the function file: funct.m,
indicated as follows:

        function y = funct(x)
        y = sin(x) -cos(x);

Then, back in the command window the following command is executed.

>> x1 = fzero('funct', 1)      % returns the solution closest to x =1

   x1 =
        0.7854
```

R.7.214 Since symbolic functions were introduced and used in this chapter, it is appropriate to say a few words about other MATLAB symbolic functions not used yet.

The MATLAB Symbolic Toolbox provides access to a number of specialized functions used in engineering and technology. Some of these specialized functions, because of their importance, constitute the topics of portions or entire chapters in this book such as

- *Dirac-delta*, and *Heaviside* function, in Chapter 1 of the book titled *Practical MATLAB® Applications for Engineers*

- Laplace and Fourier transforms, in Chapter 4 of the book titled *Practical MATLAB® Applications for Engineers*

- Ztransforms and Fast Fourier Transforms, in Chapter 5 of the book titled *Practical MATLAB® Applications for Engineers*

R.7.215 The MATLAB Symbolic Toolbox also provides access to a number of Maple functions.

These functions can be evaluated numerically by using the command *mfun*. The syntax is *mfun('f', a1, a2,....an)*, where *f* is the name of a Maple function, and the *a*'s are the numerical quantities assigned to *f*.

The Maple functions are not MATLAB functions, because they are not defined in standard M files, and the MATLAB help command cannot be used to get any information about them. In general, Maple functions are not available in the student edition of MATLAB.

R.7.216 A list of Maple functions available in MATLAB can be obtained by using the *help* command followed by *mfunlist*. A partial list of the Maple functions are given as follows:

bernoulli	*(Bernoulli numbers/Bernoulli polynomial)*
Bessel	*(Bessel function of the first kind)*
Beta	*(Beta function)*
Binomial	*(Binomial coefficients)*
erfc	*(Error function)*
erf	*(Error function)*
euler	*(Euler numbers/Euler polynomials)*
Fresnel	*(Fresnel cosine integral)*

Gamma	*(Gamma function)*
harmonic	*(Harmonic function)*
chi	*(Hyperbolic cosine integral)*
shi	*(Hyperbolic sine integral)*
w	*(Lambert's w function)*
zeta	*(Riemann/zeta function)*
T	*(Chebyshev)*
H	*(Hermite)*
P	*(Jacobi)*
L	*(Laguerre)*
P	*(Legendre)*

7.4 Examples

Example 7.1

The polynomial $p(x)$ is the product of two polynomials, $p_1(x)$ and $p_2(x)$, given by

$$p_1(x) = x^2 + 2x + 8 \quad \text{and} \quad p_2(x) = x^2 + 15x + 25$$

Write a MATLAB program that returns

1. The coefficients of $p(x)$
2. The polynomial $p(x)$
3. The degree of $p(x)$
4. The roots of $p(x)$, $p_1(x)$, and $p_2(x)$

MATLAB Solution

```
>> clc                            % clears the command window
>> format compact                 % suppresses extra line-feeds
>> P1 = [1  2  8] ;               %   defines p₁(x)
>> P2 = [1 15 25] ;               %   defines p₂(x)
>> coeff = conv(P1, P2)           % product of  P1*P2
>> disp ('The coefficients of p(x) are:');   disp(coeff)

        The coefficients of  p(x) are:
        coeff =
                1    17    63    170    200

>> syms x
>> px = poly2sym(coeff)
>> disp('The polynomial p(x) is:');disp(px)

        The polynomial p(x) is:
        px =
             x^4 + 17*x^3 + 62*x^2 + 170*x + 200
```

```
>> D = length(P)-1;
>> disp('The degree of the polynomial p(x) is :'); disp(D)

        The degree of the polynomial p(x) is :
                                                4

>> R1= roots(P1);
>> R2 = roots(P2);
>> R3 = roots(P);
>> disp ('The roots of p1(x) are:');  disp (R1)

        The roots of p1(x) are :
                            1.0000 + 2.6458i
                            1.0000 -  2.6458I

>> disp ('The roots of p2(x) are :'); disp(R2)

        The roots of p2(x) are :
                            -13.0902
                            -1.9098

>> disp('The roots of p(x) are:'); disp(R3)

        The roots of p(x) are :
                            -13.0902
                            -1.0000 + 2.6458i
                            -1.0000 - 2.6458i
                            -1.9098
```

Example 7.2

Let $p_1(x) = x + 3$, $p_2(x) = x^2 + 3x + 14$, $p_3(x) = 10$, $p_4(x) = x^3 + 2x^2 + 8x + 4$, and $p(x) = p_1(x) * p_2(x) + p_3(x) * p_4(x)$.

Write a MATLAB program that returns the coefficients of the polynomial $p(x)$ as well as the explicit polynomial $p(x)$.

```
MATLAB Solution
>> P1 = [1 3];
>> P2 = [1 3 14];
>> P3 = 10;
>> P4 = [1 2 8 4];
>> % Obtain the partial products of P12 = P1*P2 and P34 = P3*P4
>> P12 = conv ( P1, P2)

    P12 =
         1    6   23   42

>> P34 = conv (P3,P4)

    P34 =
        10   20   80   40

>> % determine the length of these polynomials for compatibility,
>> % and add zero when required
```

```
>> L1 = length(P12)

  L1 =
        4

>> L2 = length (P34)

  L2 =
        4

>> L= [L1 L2]

  L =
        4        4

>> M = max(L)

  M =
        4

>> % Obtain the coefficients of the polynomial P=P12+P34.
>> P = [zeros(1,M-L1),P12]+[zeros(1,M-L2),P34];
>> disp ('The coefficients of the resulting polynomial p(x) = p₁(x) *
                                        p₂(x) + p₃(x) * p₄(x) are:');
>> disp (P)
```

The coefficients of the resulting polynomial p(x) = p₁(x) * p₂(x) + p₃(x) * p₄(x) are:

                    11   26   103   82

```
>> px = poly2sym(P)
>> disp ('The polynomial p(x) = p₁(x) * p₂(x) + p₃(x) * p₄(x)  is:');
disp(px)
```

The polynomial p(x) = p₁(x) * p₂(x) + p₃(x) * p₄(x) is:

```
  px =
        11*x^3+26*x^2+103*x+82
```

Example 7.3

Analyze the discrete system shown in Figure 7.14 where $x(n) = 5 \cos(0.3 * 2 * \pi * n/256) + 3 \sin(0.8 * 2 * \pi * n/256)$; for $n = 0, 1, 2, 3, \ldots, 600$, and the transfer function of each box is given by

$$H_1(z) = \frac{1 - 5z^{-1}}{4 + 2z^{-1} + 2z^{-2}}$$

$$H_2(z) = \frac{3 + 2z^{-1} + 3z^{-2}}{4 + 2z^{-1} + 2z^{-2}}$$

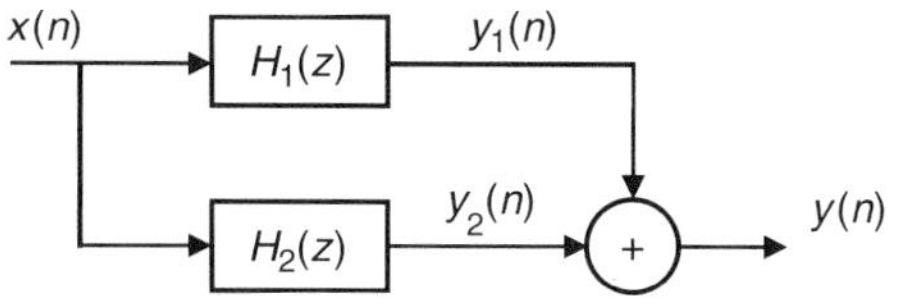

FIGURE 7.14
Block box of the discrete system of Example 7.3.

Create the script file Example 7.3 that returns the following plots:

   a. *x(n)* versus *n*
   b. $y_1(n)$ versus *n*
   c. $y_2(n)$ versus *n*
   d. *y(n)* versus *n*
   e. *Repeat parts 2, 3, and 4 if a transient is observed*

MATLAB Solution
```
% Script file:Example73.m
n = 0:600;
x1= 5*cos(.3*2*pi*n/256);
x2 = 3*sin(.8*2*pi*n/256);
x = x1+x2; p1=[1 -5];
q1 = [4 2 2]; p2=[3 2 3];
y1 = filter(p1,q1,x); y2=filter(p2,q1,x);

figure(1)
subplot (2,2,1);
stem(n,x); ylabel ('Amplitude')
title('Input sequence x(n)');grid on;
subplot(2,2,2);
stem(n,y1); ylabel('Amplitude')
title ('Output sequence y1(n)'); grid on;
subplot (2,2,3);
stem(n,y2); xlabel('index n')
ylabel ('Amplitude') ;title('Output sequence y2(n)');
grid on;
subplot (2,2,4);
y = y1+y2; stem(n,y);
xlabel ('index n')
ylabel ('Amplitude') ; title ('Output sequence  y(n) =y1(n)+y2(n)');
grid on;                                  %  plots are shown in Figure 7.15

figure(2)
nn = 0:20;x1=5*cos(.3*2*pi*nn/256);
x2 = 3*sin(.8*2*pi*nn/256);
xx = x1+x2; p1=[1 -5];
q1 = [4 2 2]; p2=[3 2 3];
y11 = filter(p1,q1,xx); y22=filter(p2,q1,xx);
subplot(3,1,1);
stem (nn,y11);hold on; plot(nn,y11);
ylabel('y1(n)')
title ('Output sequences y1(n),y2(n) & y(n), for 0<n<20');
subplot (3,1,2);
stem (nn,y22);hold on; plot(nn,y22);
ylabel ('y2(n)') ;
subplot (3,1,3);
yy = y11+y22;
stem(nn,yy);hold on; plot(nn,yy);
ylabel ('y(n)') ;xlabel('index n')     %  plots are shown in Figure 7.16
```

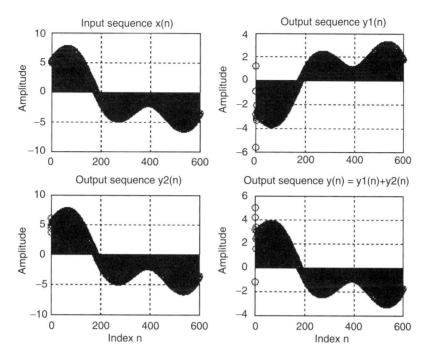

FIGURE 7.15
Plots of Example 7.3(a, b, c, and d).

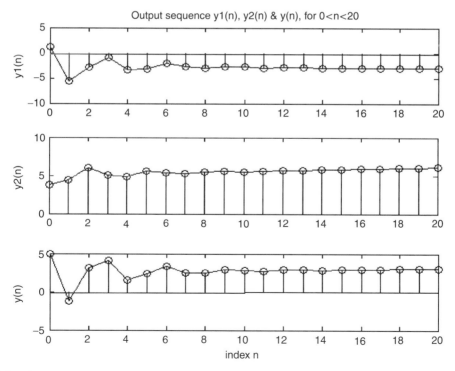

FFIGURE 7.16
Plots of Example 7.3(e).

Example 7.4

Given the following system discrete transfer function:

$$H(z) = \frac{3z^4 + 10z^3 + 5z^2 + 30z + 41}{8z^4 + 3z^3 - 12z^2 + 18z - 10}$$

Write a MATLAB program that performs the following:

1. Input $H(z)$ as two row vectors pH and qH, the numerator and denominator of $H(z)$
2. Returns the poles, zeros, and gain of $H(z)$
3. Returns the magnitudes of the poles of $H(z)$
4. Returns the plot of the poles of $H(z)$, and evaluates if the poles are inside the unit circle. (If they are inside the unit circle, then the system is stable, otherwise the system is unstable.)
5. From the poles, zeros, and gain of part 2, reconstruct the transfer function $H(z)$

MATLAB Solution
```
>> pH = [3 10 5 30 41];                          % numerator of H(z)
>> qH = [8 3 -12 18 -10];                         % denominator of H(z)
>> [z,p,k] = tf2zp(pH,qH);
>> disp('The zeros of H(z) are at :'), disp(z)

        The zeros of H(z) are at :
                                3.3626
                                0.6287 + 1.7071i
                                0.6287 - 1.7071i
                                1.2281

>> disp ('The poles of H(z) are at :'), disp (p)

        The poles of H(z) are at :
                                1.9301
                                0.4178 + 0.8518i
                                0.4178 - 0.8518i
                                0.7195

>> disp ('The gain of H(z) is :'), disp(k)

        The gain of H(z) is :
                                0.3750

>> magpole = abs(p);
>> disp('The magnitude of the poles are:'), disp(magpole)

        The magnitude of the poles are:
                                1.9301
                                0.9488
                                0.9488
                                0.7195

>> zplane(pH,qH);                                 % zplane plot shown in Figure 7.17
>> title('Plot of poles and zeros')
>> grid on ;
```

```
>> [a,b] = zp2tf(z,p,k);
>> disp ('The coefficients of the  numerator of H(z) are:'),
>> disp(a)
```

     The coefficients of the numerator of H(z) are:
                    0.3750   1.2500   0.6250   3.7500   5.1250

```
>> disp('The coefficients of the  denominator  of H(z) are:')
>> disp(b)
```

     The coefficients of the denominator of H(z) are :
                    1.0000   0.3750   -1.5000   2.2500   -1.2500

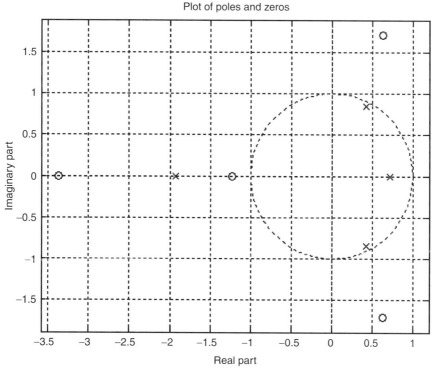

FIGURE 7.17
Plot of poles and zeros of Example 7.4.

Observe that one pole $<x = -1.9301, y = 0>$ is located outside the unit circle. Then the system is unstable. Observe also that the function *zp2tf* returns the numerator and denominator of the transfer function by a scaling factor of 8 (recall that MATLAB sets the leading coefficient of $H(z)$ to 1).

Example 7.5

Using the transfer function of Example 7.4, obtain the following:

a. The partial fraction expansion of $H(z) = P(z)/Q(z)$
b. Verify the partial fraction expansion of part a, by reconstructing the transfer function $H(z) = P(z)/Q(z)$, from the partial fraction expansion (by a factor of 8)

c. The first five coefficients of the impulse response using the *impz* function[*]
d. Verify the first three coefficients of the impulse response of part c using the long division
e. The plot of the *impulse response* versus *n*, obtained in part c

MATLAB Solution
```
>> pH = [3 10 5 30 41];
>> qH = [8 3 -12 18 -10];
>> [r,p,k] = residuez(pH,qH);
>> disp('The partial fraction residues of H(z) are:'), disp( r)
```

```
     The partial fraction residues of H(z) are:
                                        -0.1118
                                        -0.5049 + 1.5915i
                                        -0.5049 - 1.5915i
                                        -5.5966
```

```
>> disp ('The poles of H(z) are:'), disp(p)
```

```
     The poles of H(z) are :
                            -1.9301
                            -0.4178 + 0.8518i
                            -0.4178 - 0.8518i
                            -0.7195
```

```
>> disp ('The constant K is :'), disp(k)
```

```
     The constant K is :
                         4.1000
```

```
>> [A,B] = residuez (r',p',k');
>> disp (' The coefficients of the numerator  of H(z) are :'), disp(A)
```

```
     The coefficients of the numerator of H(z) are :
               0.3750      1.2500      0.6250      3.7500      5.1250
```

```
>> disp ('The coefficients of the denominator of H(z) are:'),disp(B)
```

```
     The coefficients of the denominator of H(z) are:
               1.0000      0.3750     -1.5000      2.2500     -1.2500
```

```
>> L= 5;
>> [y,n] = impz (pH,qH,L);
>> disp ('The impulse response output sequence is:'), disp(y) % part(c)
```

```
     The impulse response output sequence is:
                                        0.3750
                                        1.1094
                                        0.7715
                                        4.2810
                                        2.6495
```

---

[*] The impulse–response is the response of the system to an input consisting of a single one followed by zeros.

```
>> [A,B]= deconv(pH,qH);
>> disp ('The residue1 is :'), disp(A)
```

   The residue1 is:
      0.3750

```
>> disp ('The quotient's coefficients are:'), disp(B)
```

   The quotient's coefficients are:
     0 8.8750 9.5000 23.2500 44.7500

```
>> B(1)=[] ; B(5) = 0;
>> [C, D] = deconv(B, qH);
>> disp('The residue2 is:');disp ( C )
```

   The residue2 is:
      1.1050

```
>> D(1) = []; D(5) = 0;
>> [E, F] = deconv(C, qH);
>> disp('The residue3 is :');disp(E)
```

   The residue3 is:
      0.7715

```
>> % observe that the residues obtained correspond to the
                              coefficients of impz
>> stem(n, y)
>> title('impz of H(z) of Example 7.4')
>> grid on;
>> xlabel('index n'); ylabel('magnitude of impz');        % impulse plot
                                                          Figure 7.18
```

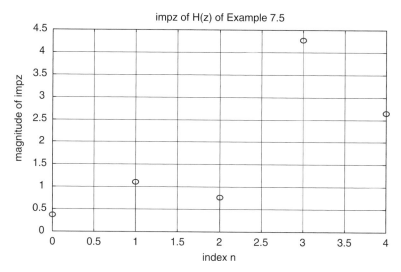

FIGURE 7.18
Discrete system impulse–response plot of Example 7.5.

TABLE 7.1

Cartesian Coordinate Points of Example 7.6

x	−2	−1	0	1	2	3
y	−11	−6	−7	−8	3	14

Example 7.6

Given the Cartesian coordinate points, shown in Table 7.1.

Write a MATLAB program that approximates the six (6) given points by a linear, quadratic, and cubic polynomial, and for each case indicate the points as well as the corresponding approximation.

Show also the *spline* function approximation plot that illustrates the best possible fit for the given data (six points).

MATLAB Solution
```
>> x = -2:3;
>> y = [-11 -6 -7 -8 3 14];                    % < x,y >  points
>> subplot (2,2,1)
>> plot (x,y,'*'), xlabel ('x'), ylabel('y'),
>> title ('Input data'), grid on
>> p1= polyfit(x,y,1)                          % linear approximation

   p1 =
        4.3143   -4.6571

>> x1 = linspace (-2,3,100);
>> pa1 = polyval (p1,x1);
>> subplot (2,2,2)
>> plot (x,y,'*',x1,pa1); grid on ; xlabel ('x'); ylabel ('y') ;
>> title ('First Order Approximation')
>> p2 = polyfit (x,y,2)                         % quadratic approximation.

   p2 =
        1.3929    2.9214   -8.3714

>> pa2 = polyval(p2,x1);
>> subplot (2,2,3)
>> plot (x,y,'*',x1,pa2), grid on, xlabel('x'),ylabel('y')
>> title ('Quadratic Approximation')
>> p3 = polyfit(x,y,3)                          % cubic approximation.

   p3 =
        0.6111    0.4762    0.2937   -6.9048

>> pa3 = polyval(p3,x1);subplot(2,2,4)
>> plot(x,y,'*',x1,pa3),grid on, xlabel('x'), ylabel('y')
>> title('Cubic Approximation')                % Figure 7.19.
>> xx =-2:.1:3;
>> yy = spline(x,y,xx);
>> clf;plot (x,y,'*', xx, yy), grid on
>> xlabel ('x'), ylabel('y'), title('Spline Fit')
>> % the spline approximation plot is shown in Figure 7.20
```

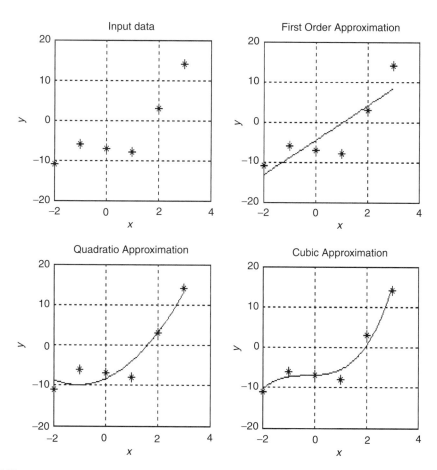

FIGURE 7.19
Approximation plots of the points of Example 7.6 using *polyval*.

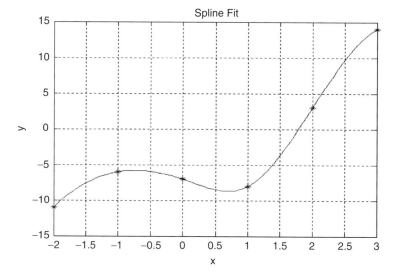

FIGURE 7.20
Approximation plots of the points of Example 7.6 using *spline*.

Example 7.7

Given the following polynomials:

$$p_1(s) = 3x^4 + 2x^3 + x^2 + 4x + 5$$

and

$$p_2(s) = 5x^4 - x^3 - 3x^2 + x - 1$$

a. Perform the following MATLAB symbolic operations and observe the respective responses
   i. $p_1 + p_2$
   ii. $p_1 - p_2$
   iii. $p_1 * p_2$
   iv. p_1/p_2
   v. $p_1 \wedge 3$
   vi. *pretty* (p_1/p_2)
b. Evaluate or determine using MATLAB:
   i. The symbolic variable used in p_1
   ii. The nested form of p_1
   iii. The symbolic polynomials p_1 and p_2 converted to a numerical polynomial (vector)
   iv. The numerical representation of the polynomial that results from the symbolic product of p_1 with p_2
   v. The numerical polynomial p_1*p_2 and convert the product to symbolic
   vi. Factor the product of part v, and verify that the factors are p_1 and p_2
   vii. $(dp_1(x))/dx$ and $(dp_2(x))/dx$
   viii. $(d^3p_1(x))/dx^3$ and $(d^3p_2(x))\,dx^3$
   ix. The expression for $\int p_1(x)\,dx$
   x. The numerical value for $\int_1^2 p_1(x)\,dx$
   xi. Identify the symbolic variables using the *class* command for $p_1(x)$, $p_2(x)$, $(d/dx)[p_1(x)]$, and $\int p_1(x)\,dx$
   xii. Identify all the variables used (symbolic or numerical), using the command *whos*
c. Create the following plots, over the range $-6 \le x \le 6$:
   i. $p_1(x)$ versus x
   ii. $p_2(x)$ versus x
   iii. $[p_1(x) * p_2(x)]$ versus x (symbolic)
   iv. $[p_1(x) * p_2(x)]$ versus x (numerical)
   v. $\left[\int p_1(x)\,dx\right]$ versus x (symbolic)
   vi. $[(dp_1(x))/dx]$ versus x (symbolic)

MATLAB Solution
```
>> p1 = sym('3*x^4+2*x^3+x^2+4*x+5');    % symbolic polynomials p1 and p2
>> p2 = sym('5*x^4-x^3-3*x^2+x-1');
>> %                     part(a)
>> Symb _ sum = symadd(p1,p2)

   Symb _ sum =
                 8*x^4+x^3-2*x^2+5*x+4

>> Symb _ sub = symsub(p1,p2)

   Symb _ sub =
                 -2*x^4+3*x^3+4*x^2+3*x+6
```

```
>> Symb _ prod = symmul(p1,p2)

   Symb _ prod =
                (3*x^4+2*x^3+x^2+4*x+5)*(5*x^4-x^3-3*x^2+x-1)

>>            % observe that the product is indicated but not performed
>> Symb _ div = symdiv (p1,p2)

   Symb _ div =
                (3*x^4+2*x^3+x^2+4*x+5)/(5*x^4-x^3-3*x^2+x-1)

>> psqr = p1^3;
>> pretty(psqr)

              4       3     2               3
          (3 x   + 2 x   + x   + 4 x + 5)

>> pretty(Symb _ div)

              4       3     2
          3 x   + 2 x   + x   + 4 x + 5
          ---------------------------------
              4       3     2
          5 x   - x   - 3 x   + x - 1

>> %        part (b)
>> findsym ( p1)          %  sym variables used in p1

      a ns =
            x

>> nestp1= horner(p1)    % p1 expressed in nested form

   nestp1 =
             5+(4+(1+(2+3*x)*x)*x)*x

>> p11 = sym2poly(p1)    % convert the sym polynomial p1 to numerical

   p11 =
          3      2      1      4      5

>> p22 = sym2poly(p2)    % converts symbolic polynomial p2 to numerical

   p22 =
          5     -1     -3      1     -1

>> p33 = conv(p11, p22)  % numerical coef of the product of   (p1*p2)

   p33 =
          15      7     -6     16     17     -18    -12      1     -5

>> p33sym = poly2sym(p33)          % converts p1*p2 to symbolic

   p33sym =
             15*x^8+7*x^7-6*x^6+16*x^5+17*x^4-18*x^3-12*x^2+x-5

>> factor (p33sym)

   ans =
          (3*x^4+2*x^3+x^2+4*x+5)*(5*x^4-x^3-3*x^2+x-1)

>> poly2sym (p11)                  % observe that MATLAB returns p1
```

```
    ans =
          3*x^4+2*x^3+x^2+4*x+5

>> poly2sym(p22)                          % observe that MATLAB   returns p2

    ans =
          5*x^4-x^3-3*x^2+x-1

>> dp1 = diff(p1)                         % dp1/dx

    dp1 =
          12*x^3+6*x^2+2*x+4

>> dp2 = diff(p2)                         % dp2/dx

    dp2 =
          20*x^3-3*x^2-6*x+1

>> dp13 = diff(p1,3)

    dp13 =
          72*x+12

>> dp23 = diff(p2,3)

    dp23 =
          120*x-6

>> intp1= int(p1)                         % integral of p1

    intp1 =
          3/5*x^5+1/2*x^4+1/3*x^3+2*x^2+5*x

>> intp112 = int (p1,1,2)                 % integral of p1 from x=1 to x=2

    intp112 =
              1183/30

>> class p1, p2, intp1, dp1               % check class of p1, p2, intp1, dp1
                                          are   symbolic

    ans =
          char
     p1 =
          3*x^4+2*x^3+x^2+4*x+5
     p2 =
          5*x^4-x^3-3*x^2+x-1
   intp1 =
          3/5*x^5+1/2*x^4+1/3*x^3+2*x^2+5*x
    dp1 =
          12*x^3+6*x^2+2*x+4

>> whos

          Name          Size      Bytes        Class
          Symb _ sum    1x1        166         sym object
          ans           1x4          8         char array
          dp1           1x1        160         sym object
```

dp13	1x1	138	sym object
dp2	1x1	160	sym object
dp23	1x1	138	sym object
intp1	1x1	190	sym object
intp112	1x1	138	sym object
estp1	1x1	170	sym object
p1	1x1	166	sym object
p11	1x5	40	double array
p2	1x1	162	sym object
p22	1x5	40	double array
p33	1x9	72	double array
p33sym	1x1	224	sym object
psqr	1x1	174	sym object
rootsp1	4x1	5664	sym object
symb _ div	1x1	214	sym object
symb _ prod	1x1	214	sym object
symb _ sub	1x1	172	sym object

Grand total is 3037 elements using 8410 bytes

```
>> % part ( c )

>> figure(1)
>> subplot(2,1,1)
>> ezplot(p1)                              %  plot of  p1( x )
>> xlabel('x'), ylabel('p1(x)');
>> title('p1(x) vs. x'), grid on;
>> subplot(2,1,2)
>> ezplot(p2);                             % plot of  p2(x)
>> xlabel('x'), ylabel('p2(x)'),
>> title('p2(x) vs x');
>> grid on;
>> % Figure 7.21

>> figure(2)
>> subplot(2,1,1)
>> ezplot(symb _ prod); grid on;
>> xlabel('x'), ylabel(' mag. syms(p1(x)*p2(x))  ')
>> title('symbolic product[ p1(x)*p2(x)] vs x');
>> subplot(2, 1, 2);
>> x = -6:0.1:6;
>> y = polyval(p33, x);
>> plot(x, y)
>> grid on;
>> title('numerical product [p1(x)*p2(x)] vs x');
>> xlabel('x'); ylabel('mag[(p1(x)*p2(x)]');
>> % Figure 7.22

>> figure(3)
>> subplot(2,1,1)
>> ezplot(intp1);    % plots integral of p1(x)
>> xlabel('x'), ylabel('mag.int[p1(x)]dx')
>> grid on;
>> title('integral[p1(x)]dx vs. x')
>> subplot(2, 1, 2);
>> ezplot(dp1)   % plots dp1/dx
```

```
>> xlabel('x'); ylabel('mag. [dp1(x)/dx]');
>> title('[dp1(x)/dx] vs. x');
>>  grid on;
>>  % Figure 7.23
```

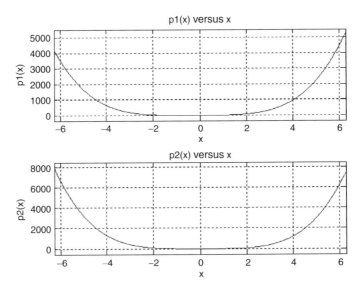

FIGURE 7.21
Plots of $p_1(x)$ and $p_2(x)$ of Example 7.7.

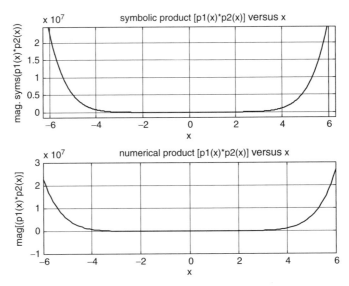

FIGURE 7.22
Plots of the products of $p_1(x) * p_2(x)$ of Example 7.7.

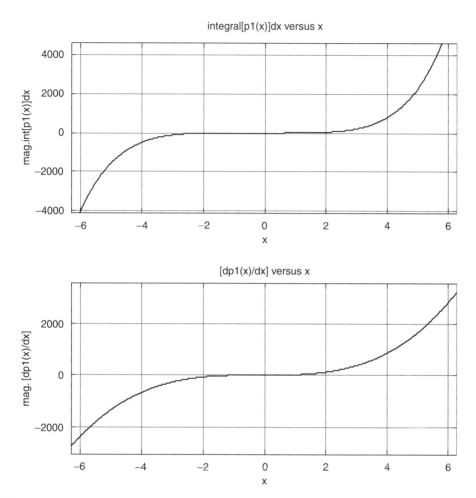

FIGURE 7.23
Plots of the integration and differentiation of $p_1(x)$ of Example 7.7.

Example 7.8

The function $y(x) = x^2$ over the range $1 \le x \le 2$ is shown in Figure 7.24.
Write a MATLAB program that returns

1. The shaded area
2. The surface of the area obtained by rotating $y(x) = x^2$ about the x-axis (for $1 \le x \le 2$)
3. The volume of the body shown in Figure 7.25
4. The length of the curve $y = x^2$, over the range $1 \le x \le 2$*

---

* The equations of the area, surface area, volume, and length are given in R.7.129.

MATLAB Solution
```
>> y = sym('x^2');
>> area = int(y,1,2);% area is the integral(shaded area in Figure7.24)
>> disp('The shaded area is ='); disp(area)
```

The shaded area is =

                                      7/3

```
>> dy = diff(y);
>> y1 = (1+dy^2)^0.5;
>> y2 = y*y1;
>> integ = int(y2,1,2);
>> A= sym2poly(integ);
>> surf = 2*pi*A;
>> disp('The surface area is:'); disp(surf)
```

The surface area is:

                                     49.4162

```
>> p = y^2;
>> I = pi*int(p,1,2);
>> volume = sym2poly(I);
>> disp ('The volume is:');disp(volume)
```

The volume is :

                                    19.4779

```
>> C = int(y1,1,2);
>> L = sym2poly ( C );
>> disp ('The length is:'); disp(L)
```

The length is:

                                      3.1678

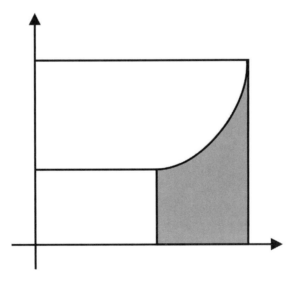

FIGURE 7.24
Sketch of $y(x) = x^2$ over the range $1 \leq x \leq 2$ of Example 7.8.

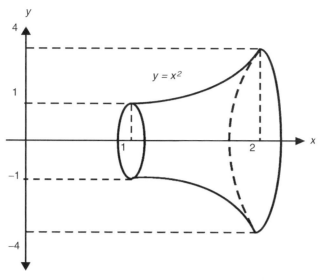

FIGURE 7.25
Surface sketch of Example 7.8.

Example 7.9

Let's revisit the often encountered problem of solving a set of linear independent equations.

For example let

$$x + 2y + z = 0 \tag{7.1}$$

$$2x - y + z = 5 \tag{7.2}$$

$$4x + 2y + 5z = 6 \tag{7.3}$$

a. Solve the set of equations by hand, and show that the solutions are $x = 2$, $y = -1$, and $z = 0$.
b. Solve the preceding set of equations using MATLAB matrix algebra.
c. Solve the same set of equations using MATLAB symbolic techniques.

ANALYTICAL Solution

By hand

$$
\begin{array}{llll}
 & x + 2y + z = & 0 & (7.1) \\
- & 2x - y + z = & 5 & -(7.2) \\
\hline
 & -x + 3y \quad\;\; = & -5 & (7.4)
\end{array}
$$

$(x + 2y + z) * 5 = 5x + 10y + 5z = 0$ (Equation 7.1 multiplied by 5)

$$5x + 10y + 5z = \quad 0 \qquad \text{(Equation 7.1 multiplied by 5)}$$

$$\underline{\; 4x + \; 2y + 5z = -6} \qquad \text{(7.3)}$$

$$x + \; 8y \qquad = -6 \qquad \text{(7.5)}$$

$$-x + 3y = -5 \qquad \text{(7.4)}$$

$$\underline{+ \quad x + 8y = -6} \qquad +\text{(7.5)}$$

$$11y = -11$$

and $\qquad y = -1$

Substituting $y = -1$ in Equation 7.4,

$$-x + 3\,(-1) = -5$$

$$-\,x - 3 = -5$$

$$x + 3 = +5$$

$$x = 5 - 3 = 2$$

$$x = 2$$

Substituting $x = 2$ and $y = -1$ in Equation 7.1,

$$2 + 2\,(-1) + z = 0$$

$$2 - 2 + z = 0$$

$$z = 0$$

MATLAB Solution
```
>> % part (b),using matrix algebra
>> A= [1 2 1;2 -1 1;4 2 5]                    % coefficients of the set of
                                                equations

   A =
        1      2      1
        2     -1      1
        4      2      5

>> B = [0; 5; 6]

   B =
        0
        5
        6

>> Solution = inv(A)*B;
>> disp('The matrix solution ( part b) for x, y, and z are:');
>> disp(Solution)

        The matrix solution ( part b) for x, y, and z are:
                                               2.0000
                                             - 1.0000
                                               0

>> % Solution ( c ), using Symbolic Expressions
>> eq1 = sym('x+2*y+z');
>> eq2 = sym('2*x-y+z-5');
```

```
>> eq3 = sym('4*x+2*y+5*z-6');
>> [x,y,z] = solve(eq1,eq2,eq3);
>> disp ('The sym solution(part  c) for x, y, and z are:');disp(x;y;z)
```

$$\text{The sym solution (part c) for x, y, and z are:}$$

```
                                                     x =
                                                        2
                                                     y =
                                                        -1
                                                     z =
                                                        0
```

Example 7.10

Let the set of first-order DEs be given by

$$\frac{dy_1(t)}{dt} = -2y_1(t) - y_2^3(t)$$

$$\frac{dy_2(t)}{dt} = y_1(t)e^{(2-t)} + 2y_2(t)$$

with the following ICs: $y_1(0) = 0$, $y_2(0) = 3$, over the range $0 \le t \le 2$.

Create the script file *Example710* that returns the solution of the system of DEs, by using the numerical solvers

1. *ode23*
2. *ode45*

MATLAB Solution
```
%function file that defines the differential equation
function y1y2 = fy1y2(t, y)
y1y2 = [-2*y(1)-y(2)^3; y(1)*exp(2-t)+2*y(2)];

% main program
% Script file:Example710
y0 = [0; 3];
[t, y1] = ode23('fy1y2', [0 2], y0);
subplot(2, 1, 1)
plot(t, y1(:, 1), '*')
grid on
title('Solution y1 using ode 23')                % Figure 7.26
subplot(2, 1, 2)
plot(t, y1(:, 2), 'd')
title('Solution y2 using ode 23')
grid on ;
keyboard;
subplot(2, 1, 1)
[t, y2] = ode45('fy1y2', [0 2], y0);
plot(t, y2(:, 1), '*');
grid on;
title('Solution y1 using ode 45')                % Figure 7.27
subplot(2, 1, 2)
plot(t, y2(:, 2), 'd')
grid on;
title('Solution y2 using ode 45')
```

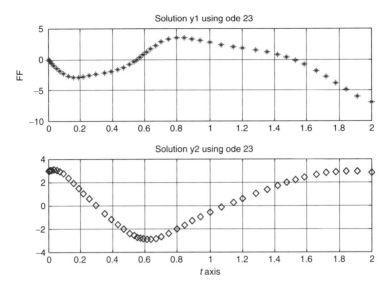

FIGURE 7.26
Plots of the solutions of Example 7.10 using *ode23*.

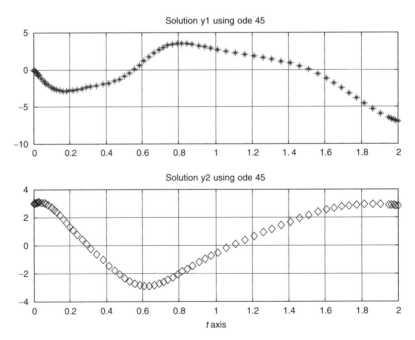

FIGURE 7.27
Plots of the solutions of Example 7.10 using *ode45*.

Example 7.11

Create the MATLAB script file *diff_eqs* that returns the solutions for *x(t)* and *y(t)*, for the following set of DEs:

$$\frac{dx(t)}{dt} = -2y(t) - 3x(t)$$

$$\frac{dy(t)}{dt} = -3y(t) + 5x(t)$$

given the following ICs: *x*(0) = 1 and *y*(0) = 2, over the range $-5 \le t \le -3$. Use the solver *dsolve* and obtain plots of

1. *x(t)* versus *t*
2. *y(t)* versus *t*

MATLAB Solution
```
% Script file: diff _ eqs
disp('***************************************************')
disp('The solution of the equations : dx/dt =-2*y-3*x,
  and dy/dt=-3*y+5*x')
disp('with the initial conditions: x(0)=1,y(0)=2,are given by ')
[x,y] = dsolve('Dx=-2*y-3*x,Dy=-3*y+5*x','x(0)=1,y(0)=2')
disp('The pretty x(t) is given by')
pretty(x)
disp('The pretty y(t) is given by')
pretty(y)
disp('***************************************************')
subplot(2,1,1);
ezplot(x)
title('plot of x(t) vs t using dsolve')
axis([-5 -3 -5e5 9e5])
subplot(2,1,2)
ezplot(y)
title('plot of y(t) vs t using dsolve')
xlabel('t (time)')
axis([-5 -3 -6e5 6e5])
% the plots of x(t) vs.t and y(t) vs. t, are shown in Figure 7.28
```

Back in the command window, the file *diff_eqs* is executed and the results are shown as follows:

```
>> diff _ eqs


***************************************************
The solution of the equations: dx/dt =-2*y-3*x, and dy/dt=-3*y+5*x
with the initial conditions: x(0) =1, y(0)=2,are given by
    x =
       1/5*exp(-3*t)*(5*cos(t*10^(1/2))- 2*10^(1/2)*sin(t*10^(1/2)))
    y =
       1/2*exp(-3*t)*(10^(1/2)*sin(t*10^(1/2)) +4*cos(t*10^(1/2)))
```

```
The pretty x(t) is given by
                1/2              1/2                1/2
   - 1/5 exp(-3 t) (-5 cos(t 10   ) + 2 10    sin(t 10   ))
The pretty y(t) is given by
             1/2           1/2                      1/2
   1/2 exp(-3 t) (10    sin(t 10   ) + 4 cos(t 10   ))
         ************************************************
```

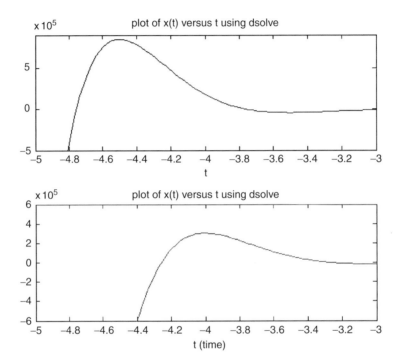

FIGURE 7.28
Plots of *x(t)* and *y(t)* of Example 7.11.

Example 7.12

Let

$$\frac{d}{dt}[y(t)] + 2y(t) - 4t = 0$$

with the IC given by $y(0) = 10$.

Create the script file *Example712* that performs the following:

1. Solves the preceding DE using the symbolic solver *dsolve*
2. Verifies using MATLAB the solution obtained in part 1
3. Obtains the general solution of the given DE

4. Verify the solution using MATLAB from the given IC (initial conditions)
5. The plot of (the solution) *y(t)* versus *t*, over the range $0 \le t \le 5$

MATLAB Solution

```
%Script file: Example712
syms y y1 t
disp ('*****************************************************')
disp ('The solution of the equations : dy/dt+2*y-4*t=0')
disp ('with the initial conditions: y(0)=10 is :')
y = dsolve('Dy+2*y-4*t=0','y(0)=10','t')
% verifies the solution
disp('Evaluate: verify = dy/dt +2*y-4*t, yields')
verify = diff(y,t)+2*y-4*t
% solves the given equation for the general solution
clear ;
disp ('The general solution is:');
y1= dsolve('Dy1+2*y1-4*t=0','t')
% solves for the initial conditions
y=10; t=0;
disp('The calculated y(0)=C1, where:')
C1 = subs(y)
Disp ('*****************************************************')
y = dsolve('Dy+2*y-4*t=0','y(0)=10','t');
ezplot (y,[0,5])
title ('Solution of y(t),with y(0)=10, for 0<t<5 ')
ylabel (' y(t)'); xlabel ('t') ; grid on
```

Back in the command window, the script file *Example712* is executed, and the results are shown in the following (Figure 7.29):

```
>> Example712
*****************************************************
   The solution of the equations : dy/dt+2*y-4*t=0
   with the initial conditions: y(0)=10  is:

   Y =
     2*t-1+11*exp(-2*t)

   Evaluate: verify = dy/dt+2*y-4*t, yields
              verify =
                      0

      The general solution is:
                  y1 =
                      2*t-1+exp(-2*t)*C1
      The calculated y(0)=C1, where:
                  C1 =
                       10
                  y =
                      2*t-1+11*exp(-2*t)
*****************************************************
```

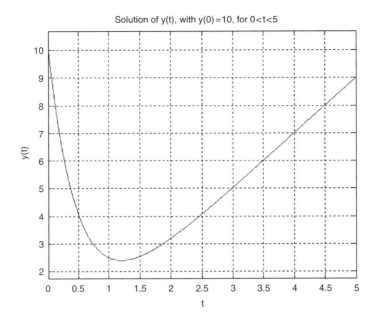

FIGURE 7.29
Plots of $y(t)$ of Example 7.12.

7.5 Further Analysis

Q.7.1   Load and run Example 7.1.

Q.7.2   Run Example 7.1 without the semicolons, and place descriptive comments (%) at the end of each line.

Q.7.3   What are the degrees of p_1, p_2, and p, and how are those polynomials related to the MATLAB variables used?

Q.7.4   Discuss the purpose of variable D.

Q.7.5   Are the roots of p related to the roots of p_1 and p_2?

Q.7.6   What is in general the relation between the degree of a polynomial and its vector length representation?

Q.7.7   Discuss the relation between the degrees of p_1 and p_2, and $p_1 * p_2$.

Q.7.8   What is the relation between the degrees of p_1, p_2, and the MATLAB command length of the *conv(p1, p2)*?

Q.7.9   Load and run Example 7.2.

Q.7.10 Define and discuss the purpose of the statement L = [L1 L2].

Q.7.11 What is the MATLAB variable that defines the polynomial p, and why zeros are added in defining the array P?

Q.7.12 Load and run Example 7.3.

Q.7.13 Evaluate the number of points used for each plot of Figures 7.15 and 7.16.

Q.7.14 Modify and rerun Example 7.3 for the case $H_{eq}(z) = H_1(z) + H_2(z)$, as indicated in the block diagram of Figure 7.30.

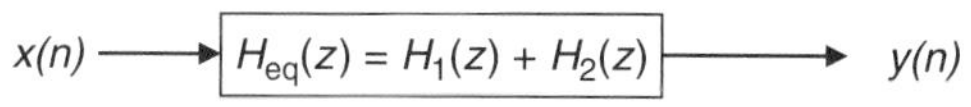

$$x(n) \longrightarrow \boxed{H_{eq}(z) = H_1(z) + H_2(z)} \longrightarrow y(n)$$

FIGURE 7.30
Block diagram of Q.7.14.

Q.7.15 Compare the output y obtained in Q.7.14 with the result obtained for Example 7.3 ($y(n)$), shown in Figure 7.15.

Q.7.16 Compare the outputs y_1 and y_2 shown in Figure 7.15. What are the main similarities and differences?

Q.7.17 Compare the transfer function $H_1(z)$ with $H_2(z)$, and state similarities and differences.

Q.7.18 Is it possible to approximate $H_1(z)$ in terms of $H_2(z)$?

   If it is, how can it be done?

Q.7.19 Modify and rerun Example 7.3 for the case $H(z) = H_1(z) * H_2(z)$.

Q.7.20 Is it possible to predict the system output of Q.7.19 based on the results obtained in the Example 7.3?

Q.7.21 Modify script file Example73.m to generate the outputs y_1, y_2, and y for $n = 50, 100, 300, 500$, and 1000.

Q.7.22 Load and run Example 7.4.

Q.7.23 Rerun Example 7.4 without the semicolons, and place brief descriptive comment (%) at the end of each line of the program.

Q.7.24 What is the purpose of the command *tf2zp(pH, qH)*?

Q.7.25 Analyze the function *magpole = abs(p)*, and evaluate by hand the values for *magpole*.

Q.7.26 Describe the function *zplane(pH, qH)*.

Q.7.27 Discuss the objective and meaning of the graph shown in Figure 7.17 (*z-plane*).

Q.7.28 Draw and estimate the numerical values of *magpole* (magnitude and direction) shown in Figure 7.17.

Q.7.29 Compare the function *residuez* with *zp2tf*.

Q.7.30 Modify and rerun Example 7.4 using the function *roots* to determine the poles and zeros of $H(z)$.

Q.7.31 What is the relation between the stability of a discrete system and the unit circle in the z-plane?

Q.7.32 What is meant by a stable system?

Q.7.33 Are the poles of $H(z)$ inside the unit circle? Indicate which ones are inside and which ones are outside. Discuss why they are relevant and their implications.

Q.7.34 Load and run Example 7.5.

Q.7.35 Discuss the purpose of the function *residuez*.

Q.7.36 What is meant by *residues* of $H(z)$?

Q.7.37 Describe the purpose of the function *impz(p, q, n)*.

Q.7.38 Describe the purpose of the function *deconv(p, q)* and its corresponding output.

Q.7.39 Load and run Example 7.6.

Q.7.40 Determine the total number of points used in the *subplot(2, 2, 1)*.

Q.7.41 Determine the number of points used in the *subplot(2, 2, 2)*.

Q.7.42 Describe the approximations used in Example 7.6.

Q.7.43 Define the linear, quadratic, and cubic polynomial approximations, and indicate which one is the "best" approximation.

Q.7.44 What is meant by "best" approximation?

Q.7.45 Describe and define variable p_1.

Q.7.46 Describe and define variable p_2.

Q.7.47 Describe the relations between the variables p_1, p_2, and p_3.

Q.7.48 Load and run Example 7.7.

Q.7.49 Indicate the similarities and differences between the following set of instructions:
  a. *psums = p1 + p2* and *psum = symadd(p1, p2)*
  b. *psubs = p1 − p2* and *psub = symsub(p1, p2)*
  c. pdivs = *deconv(p1, p2)* and *pdiv = symdiv(p1, p2)*
  d. pprods = *conv(p1, p2)* and *pprod = symmul(p1, p2)*

Q.7.50 Indicate if the instruction *symdiv(p1/p2)* returns the result you expected.

Q.7.51 Compare and indicate the differences between the MATLAB functions *roots, factor*, and *solve*.

Q.7.52 Indicate and discuss the purpose(s) of converting a symbolic polynomial expression into a row numerical array.

Q.7.53 Define and discuss the main purpose of integration.

Q.7.54 Repeat question Q.7.53 for the case of differentiation.

Q.7.55 What is the purpose of using a symbolic variable, and when should a symbolic variable be used?

Q.7.56 Load and run Example 7.8.

Q.7.57 What is meant by "surface area"?

Q.7.58 Repeat question Q.7.57 for "length of a curve."

Q.7.59 Replace variable y by x and x by y, and rerun Example 7.8. Indicate the outputs and discuss their meaning and implication.

Q.7.60 Load and run Example 7.9.

Q.7.61 Compare the analytical solutions with the MATLAB solutions.

Q.7.62 Load and run the couple first-order DEs of Example 7.10.

Q.7.63 Determine the total number of variables used to define the set of DEs.

Q.7.64 Determine and discuss the number of ICs required to obtain a unique solution. Generalize for any given set of different equations.

Q.7.65 Compare the solutions shown in Figures 7.26 and 7.27.

Q.7.66 Obtain error plots of y_1 and y_2, between *ode23* and *ode45*.

Q.7.67 Solve the given set of DEs using the numerical solvers *ode113* and *ode23tb*.

Q.7.68 Compare the solutions obtained.

Q.7.69 Load and run Example 7.11.

Q.7.70 Rerun Example 7.11 using numerical techniques and compare the result with the symbolic solutions.

Q.7.71 Load and run Example 7.12.

Q.7.72 Rerun Example 7.12 using *ode* solvers numerical techniques and compare the result with the symbolic solution.

Q.7.73 Obtain an error plot by comparing the numerical solution with the system solution.

7.6 Application Problems

P.7.1 Given the polynomial $y = x^5 + x^4 + x^2 - 2x + 2$, write a MATLAB program that
  a. Defines 100 Cartesian rectangular coordinate points for y over the range $-2 \leq x \leq 2$
  b. Determines the roots of y
  c. Returns the plot $y(x)$ versus x over the following ranges:
     i. $-2 \leq x \leq +2$
     ii. $-5 \leq x \leq +10$
  d. Use the function *polyfit* to approximate y with the polynomials of degree n, for $n = 1, 2, 3, 4$, and 5
  e. Returns the plots for each approximation and compares each approximation at the points $<x(1), y(1)> <x(10), y(10)>, <x(20), y(20)>, <x(30), y(30)>, <x(40), y(40)>,$ and $<x(100), y(100)>$
  f. Using the roots (obtained in part b), reconstruct the polynomial y
  g. Verify if the roots of $y(x)$ are

$$x_1 = -1.8182$$
$$x_{2,3} = -0.2775 \pm 1.206i$$
$$x_{4,5} = 0.6866 \pm 0.4966i$$

P.7.2 Let $y(x) = x^3 + 2x^2 - 5x - 6$.
  a. Verify by hand that the roots of y are $x = -1, -3$, and $+2$
  b. Write a MATLAB program that returns the roots of y by employing symbolic and numerical techniques, and compare the results obtained
  c. Verify the results of part b by using Descarte's rule of signs

P.7.3 The polynomials $y(x)$ and $z(x)$ are given as follows:

$$y(x) = -4x^7 + x^3 - x^2 - 1$$
$$z(x) = -2x^6 + x^4 + 5$$

Write a MATLAB program that performs the following operations:
  a. $A = y + z$
  b. $B = y - z$
  c. $C = z - y$
  d. $D = y * z$
  e. $E = y/z$

  f. The roots of y and z

  g. The roots of D

  h. Compare and discuss the results obtained in parts f and g

P.7.4   Let $y(x) = x^8 + 7x^7 - 5x^3 + 2x - 2$.

  Write a MATLAB program that returns

  a. The remainder when $y(x)$ is divided by $x + 1$

  b. The value of y evaluated at $x = -1, +1$, and 0

P.7.5   Given $y(x) = x^5 + 3x^4 - 2k + 2$, determine the numerical value for k such that $x - 2$ is a factor of y.

P.7.6   Verify using MATLAB that $(x + a)$ is a factor of $x^{2n} - a^{2n}$ for any arbitrary n. Test your program for $a = 3$ and $n = 2$.

P.7.7   Verify using MATLAB that $(x - a)$ is a factor of $x^n - a^n$ for any natural n. Test your program for $a = 5$ and $n = 10$.

P.7.8   Given $y(x) = x^4 + x^3 - 13x^2 - x + 12$.

  a. Verify using MATLAB that all the roots of y are in the range $-1 \le x \le 3$

  b. Plot $y(x)$ versus x, over the range $-1 \le x \le 3$, and indicate with a "*" the location of all the roots.

P.7.9   Given the polynomial $y(x) = 3x^4 - 16x^3 + 18x^2 - 12x - 24$.

  a. Verify that $x = 2$, is the root of y with multiplicity 3

  b. Verify that the gain is 3

  c. Verify that the forth root is -1

P.7.10   Construct using MATLAB a polynomial having the following roots:

$$-2, +2, -1, +1, \pm 3i$$

P.7.11   Construct a fourth-degree polynomial having the following roots:

$$-2 \text{ with a multiplicity of } 3, \text{ and } 0 \text{ with multiplicity } 1.$$

P.7.12   Construct a polynomial having the following roots:

$$-1 \pm i, 2 \pm i, \text{ and } 3 \pm \sqrt{2}$$

P.7.13   Given the xy Cartesian rectangular coordinate points shown in Table 7.2, determine and plot the polynomial function that best approximates y in terms of x, and evaluates the largest error over the range $-3 \le x \le 4$ at the given points.

TABLE 7.2

Cartesian Coordinates for P.7.13

x	-3	-2	-1	0	1	2	3	4
y	0	-3	1	0	-1	-0	-2	-4

P.7.14 Given the xy Cartesian rectangular coordinate points shown in Table 7.3.
  a. Approximate y with linear, quadratic, and cubic equations
  b. Which approximation of part (a) is the "best"?
  c. Determine the roots of y for the best approximation
  d. Using the cubic approximation, plot $y(x)$ versus x as a continuous function over the domain $-6 \leq x \leq 1$ using 100 points, and indicate on the plot the data points as well as the approximation.

TABLE 7.3

Cartesian Coordinates for P.7.14

x	-3	$\sqrt{2}$	$-\sqrt{2}$	-6	-1	2	-2	1
y	0	1.5	1	0	-2	1	2	-4

P.7.15 Repeat P.7.14 for the set of points shown in Table 7.4.

TABLE 7.4

Cartesian Coordinates for P.7.15

x	0	1	-1	2	-2
y	1	-1	3	3	-1

P.7.16 Given the following analog transfer function:

$$H(s) = \frac{3s + 7}{s^2 - 2s - 3}$$

Use MATLAB to
  a. Determine the poles, zeros, and gain of $H(s)$
  b. Decompose $H(s)$ into partial fractions expansion
  c. From the poles, zeros, and gain reconstruct $H(s)$
  d. From the partial fractions expansion reconstruct $H(s)$
  e. Obtain the plot $H(s)$ versus s, over the range $0 \leq s \leq 10$, using 100 points
  f. Obtain the following plots:
    i. $H_1(s) = 4/(s-3)$ versus s
    ii. $H_2(s) = -1/(s + 1)$ versus s
    iii. $[H_1(s) + H_2(s)]$ versus s
  g. Compare the plot s of $[H_1(s) + H_2(s)]$ versus s, with $H(s)$ versus s. Discuss.

P.7.17 Given the following analog transfer function:

$$H(s) = \frac{5s^2 - 15s - 11}{(s + 1)(s - 2)^2} \frac{5s^2 - 15s - 11}{(s + 1)(s - 1)^2}$$

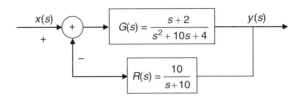

FIGURE 7.31
System diagram of P.7.18.

Write a MATLAB program that

a. Proves that the partial fraction expansion of $H(s)$ is given by

$$H(s) = \frac{-1/3}{(s+1)} + \frac{-7}{(s-2)^3} + \frac{4}{(s-2)^2} + \frac{-1/3}{s-2}$$

b. Plots $H(s)$ versus s, and each one of the terms of the partial fraction expansion of $H(s)$

c. Determines the poles, zeros, and residues of $H(s)$

d. Evaluates $H(s)$ for $s = 0$, $s = 5$, $s = 10$, and $s = 100$

P.7.18   The block box diagram of a feedback control system is shown in Figure 7.31. The transfer function of the preceding system is

$$H(s) = \frac{y(s)}{x(s)} = \frac{G(s)}{1 + R(s) * G(s)}$$

Use MATLAB to evaluate the following:

a. The coefficient of the numerator and denominator of $H(s)$

b. The poles, zeros, and gain of $H(s)$

c. Analyze the stability of the system (for a system to be stable the poles must be on the left half of the complex plane, which means that the real part of the poles must be negative) by plotting the poles of $H(s)$

d. The system transfer function by using the command *tf*

e. The system state–space equations

P.7.19   For the functions defined in the following:

$$f(x) = 1 + 3sin(x)$$

$$y(x) = \frac{x}{x+1} + 2$$

$$z(x) = log(x) + 3\,tan(x)$$

a. Create the symbolic expressions for f, y, and z

b. Determine the expression $y(f(x))$

c. Verify part b

d. Evaluate $f + y$

e. Evaluate $f * y * z$

f. Determine the expressions of the numerator and denominator of part e

g. Obtain an expression for $df(x)/dx$

h. Evaluate $dy(x)/dx$, at $x = 0$

i. Evaluate $\int_0^1 y(x)\,dx$

j. Obtain an expression for $dz(x)/dx$

k. Evaluate $\int_0^{2\pi} f(x)\,dx$

l. Verify the result obtained in part k by numerical means

P.7.20 Given the following set of linear equations:

$$3x + 4y = 18$$
$$x - 3y = -7$$

a. Verify by hand calculations that $x = 2$, and $y = 3$ are the solutions

b. Solve graphically the given set of equations over the range $0 \le x \le 5$

c. Use symbolic techniques to evaluate the solution of the given equations

d. Repeat part c, using matrix algebra

P.7.21 Repeat P.7.20 b, c, and d for the following set of equations:

$$2x - 5y = -11$$
$$x + y = 5$$

P.7.22 Given the following set of equations:

$$2x^2 - 3xy + y^2 = 15$$
$$x^2 - 2xy + y^2 = 9$$
$$x = -2xy$$

solve for x and y, and verify that $x = \pm 2$ and $y = \pm 1$.

P.7.23 Use MATLAB to verify the following identities:

a. $16 - x^4 = (4 + x^2) \cdot (2 + x) \cdot (2 - x)$

b. $x^4 - 36 = (x^2 + 6) \cdot (x^2 - 6)$

c. $6x^3 + 24x^2 - 72x = 6x \cdot (x - 2) \cdot (x + 6)$

d. $7x^3 - 49x^2 - 420 = 7x \cdot (x + 5) \cdot (x - 12)$

e. $2x^2 - 315x - 5500 = (2x - 275) \cdot (x - 20)$

P.7.24 The following program corresponds to the analysis of a discrete system. The resulting plots are shown in Figure 7.32.

```
>> n = 0:0.2:200;
>> x1= 5*cos(5*pi*n/256);
>> x2 = 3*cos(50*pi*n/256);
>> x = x1+x2;
>> p1= [0.6 0.3 0.8];
>> y1 = filter(p1,1,x);
>> p2 = [0.5 0.55 0.48];
>> q2 = [1 -0.6 0.48];
>> y2 = filter(p2,q2,x);
>> y = y1+y2;
>> subplot(2,2,1)
>> plot(n,x), grid on
>> ylabel('Amplitude'), title('Input sequence')
```

```
>> subplot(2,2,2)
>> plot(n,y1), grid on, ylabel('Magnitude'), title('Output y1')
>> subplot(2,2,3)
>> plot(n,y2), grid on, xlabel('Index n')
>> ylabel('Magnitude'), title('Output y2')
>> subplot(2,2,4)
>> plot(n,y), grid on, xlabel('Index n')
>> ylabel('Magnitude'), title('Output y')
```

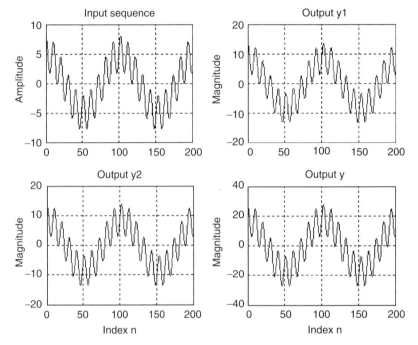

FIGURE 7.32
The system input and output plots of P.7.24.

    a. Draw a block box diagram representation of the system, and clearly indicate inputs, outputs, and transfer function

    b. Draw a flowchart of the program

    c. What are the main frequencies present in the system?

    d. What does the system do?

    e. Do you agree with the choice of variables and commands used?

P.7.25   a. Express the following series using symbolic commands.

$$sum1 = 1 + \frac{1}{2} + \frac{1}{3} + \frac{1}{4} + \cdots + \frac{1}{n}$$

$$sum2 = 1^{-2} + 2^{-2} + 3^{-2} + 4^{-2} + 5^{-2} + \cdots + n^{-2}$$

$$sum3 = exp(-1) + exp(-2) + exp(-3) + \cdots + exp(-n)$$

$$*sum4 = 1 + \frac{2}{2!} + \frac{3}{3!} + \frac{4}{4!} + L + \frac{n^*}{n!}$$

$$sum5 = 0.3 + \frac{(-0.3)^2}{1!} + \frac{(-0.3)^3}{2!} + \frac{(-0.3)^4}{3!} + \frac{(-0.3)^5}{4!} + \frac{(-0.3)^6}{5!} \ldots \frac{(-0.3)^n}{n!}$$

b. Obtain for each series a numerical value for the first 20 terms

c. Evaluate the sums of the preceding series over the range $15 \le n \le 25$

P.7.26 Find the derivatives with respect to t, for each of the following functions:

a. $y_1(t) = 2t + 2$

b. $y_2(t) = 3t^2 + 2t + 5$

c. $y_3(t) = t^4 + log(t) + 4$

P.7.27 Find the value of the slope or the tangent line for the following functions:

a. $y_1(t) = 3t^2 + 2t - 7$, at $t = 0$

b. $y_2(t) = 3t^3 + 2t^2 - 6t + 3$, at $t = 1.2$

P.7.28 Obtain the expressions of the derivative with respect to t, of the following functions:

a. $f_1(t) = cos(t) - tan(t)$

b. $f_2(t) = cos(t) - sin(t)$

c. $f_3(t) = e^{(sin(t))}$

d. $f_4(t) = tan(\sqrt{t^2 - 1})$

e. $f_4(t) = 3cos(3t + pi/3)$

P.7.29 Evaluate by hand and by using MATLAB the following limits:

a. $F_1 = \lim_{t \to a} [t + 2] =$

b. $F_2 = \lim_{t \to a} \left[t^3 + 2t^2 3t - 4 \right] =$

c. $F_3 = \lim_{t \to a} \left[\frac{t^2 - 9}{t + 3} \right] =$

P.7.30 Perform the following integrals and plot the results over the range $0 \le t \le 3$, for each of the following expressions:

a. $\int t^3 \, dt$

b. $\int 3t^2 \, dt$

c. $\int \sqrt{(1 - t^2)} \, dt$

d. $\int log_2(t) \, dt$

---

* The gamma function may be used, defined by $gamma(x) = 1*2*3*4\ldots(x - 1) = (x - 1)!$

P.7.31 Evaluate by hand and by using MATLAB, the following definite integrals:

a. $\int_0^4 t^3 dt$

b. $\int_{-10}^1 (t^2 - 2t + 3)dt$

c. $\int_{-1}^2 \sqrt{t}\, dt$

d. $\int_0^2 \cos^2(2t)dt$

P.7.32 Given the functions

$$g(x) = x^5 + 4x^3 + 2x^2 + 3x - 5$$

and

$$y(x) = x^3 + 3x^2 + x - 4$$

use MATLAB and evaluate the following expressions:

a. $g(x) * y(x)$
b. $g(x)/y(x)$
c. $g(x)|_{x=pi}$
d. $\int_0^1 g(x)\, dx$
e. $\int_{-1}^0 y(x)\, dx$
f. $(d/dx)g(x)$
g. $(d/dx)y(x)$
h. $(d^3/dx^3)y(x)$

P.7.33 Use MATLAB (symbolic) to obtain the series expansion of the following expressions:

a. $tan(2x)$
b. $cos(3x)$
c. $sin^{-1}(4x)$

P.7.34 Simplify the resulting expressions obtained in parts a, b, and c of P.7.33 and arrange the expressions in a nested format.

P.7.35 Given the symbolic matrix A, shown as follows:

$$A = \begin{bmatrix} 1 & a & b \\ a^2 & b^2 & a*b \\ 1/b & 1/b & a/b \end{bmatrix}$$

evaluate the following:

a. A^{-1}
b. $det(A)$
c. $eig(A)$
d. $diag(A)$

P.7.36 Verify that the general solution $[y_2(t)]$ for the DE

$$\frac{d^2y}{dx^2} = 2e^t - 3t^2 \quad \text{is} \quad y_2(t) = 2e^t - (t^4/4) + At + B$$

P.7.37 Given the following DE:

$$\frac{d^2}{dt^2}[y(t)] + 2\frac{d}{dt}[y(t)] + 5y(t) = 0 \quad \text{with IC given by } y(0) = 1 \quad \dot{y}(0) = 2$$

a. Verify that the roots of the characteristic equation are $s_{1,2} = -1 \pm 2i$
b. Verify that the general solution is of the form $y(t) = Ae^{-t}\cos(2t) + Be^{-x}\sin(2t)$
c. Solve for the constants A and B

P.7.38 Given the following DE:

$$\frac{d^2}{dt^2}[y(t)] + \frac{d}{dt}[y(t)] - 2y(t) = 0$$

a. Verify that the roots of the characteristic equation are $s_1 = -1$ and $s_2 = 2$
b. Verify that the general solution is of the form $y(t) = Ae^{-t} + Be^{2t}$
c. Solve for the constants A and B if $y(0) = 2$ and $\dot{y}(0) = 3$

P.7.39 Show that the general solution $[y_2(t)]$ of the DE

$$\frac{d^2}{dt^2}[y(t)] - 16y(t) = 0 \quad \text{is} \quad y_2(t) = Ae^{4t} + Be^{-4t}$$

P.7.40 Obtain the analytic and the MATLAB solutions for the following DE:

$$\frac{d^2}{dt^2}[y(t)] - \frac{d}{dt}[y(t)] - 5y(t) = 24$$

P.7.41 Find the complete solution to the following DE:

$$\frac{dy}{dt} + \frac{5}{3}y = 10$$

with the IC given by $y = 5$, at $t = 0$.

P.7.42 Find the general solutions for the following DEs:
a. $(t+3)\,dy = y^2\,dt$
b. $\frac{d^2y}{dx^2} + 25\ \ y = 0$

P.7.43 Solve the following DE:

$$\frac{dy(t)}{dt} - y(t) = 2te^{2t}$$

with the IC given by $y(0) = 3$

a. Using a numerical approach

b. Using a symbolic approach

c. Plot the solutions of parts a and b, over the range $0 \le t \le 2$

d. Compare the solutions of parts a with part b by means of an error plot

P.7.44 Let

$$2\frac{d}{dt}y(t) - 3y(t) - 9t = 0$$

with the IC given by $y(0) = 20$.

a. Verify by hand and by using MATLAB that the solution of the preceding DE is $y(t) = -44e^{1.5t} - 3t$

b. Solve the given DE by using the following solvers: *ode23*, *ode45*, and *ode113*, and plot the solutions over the range $0 \le t \le 2$

c. Repeat part b, using the symbolic function *dsolve*

d. Compare the result of parts b with c

P.7.45 Let

$$20\frac{d^2}{dt^2}y(t) + 6.3\frac{d}{dt}y(t) + 7.2y = 5\sin(3t)$$

be a system DE.

a. Identify the input and output system variables

b. Obtain the state–space system equations

c. Obtain its system transfer function

d. Obtain its discrete state–space model, for $Ts = 0.5, 1.0$, and 1.5

e. Obtain the impulse time plot response for the continuous and discrete systems

P.7.46 Repeat problem P.7.45 for the following system equation:

$$6\frac{d^3}{dt^3}y(t) + 3\frac{d^2}{dt^2}y(t) + 18\frac{d}{dt}y + 100y(t) = 3 + 4\cos(2t)$$

8

Decisions and Relations

Good judgment comes from experience. And where does experience come from? Experience comes from bad judgment.

Mark Twain

8.1 Introduction

MATLAB® executes the instructions in the same sequence as they are input. The first instruction is executed first, the second instruction is executed second, and so on, and the last instruction is executed last. In this chapter, MATLAB commands, which will allow a program to change the normal execution sequence, are introduced. The flow control in a MATLAB program can be altered by relations, logical expressions, and branching instructions. As a result, a computer program can execute different sequences of instructions depending on the condition set. Relational and logical operations allow the comparison of variables, and based on their results the transfer and selection of a pathway to satisfy a particular condition can be accomplished. For example, if statement A is true, then the program executes the instructions B, C, and D, but if not, then the program executes instructions E, F, and G instead.

Recall that MATLAB returns a 1 (one) if a statement is found to be true, otherwise MATLAB returns a 0 (zero). MATLAB supports the traditional branching commands universally used in other high-level programming languages (such as Fortran and Basic) as well as the standard logical operations that may be used to control the flow of the instructions that make up a program.

8.2 Objectives

After completing this chapter, the reader should be able to

- Know the meaning and syntax of the standard relations
- Know the meaning and syntax of standard logical operations
- Know the hierarchy of logical, relational, and arithmetic operations
- Perform relational and logical algebra on vectors, matrices, and scalars
- Work with and evaluate logical variables, logical relations, and logical expressions
- Know the meaning and syntax of the conditional branching instruction such as *for-end, while-end, if-end,* and *switch-end*

- Compare strings with substrings
- Locate and replace a string in a string by another string
- Use the experience gained with the decision and relation commands to solve a variety of problems in diverse areas such as economics, mathematics, and engineering

8.3 Background

R.8.1 In addition to the arithmetic operations, MATLAB also allows the use of the standard mathematical relations such as *larger than, equal to*, etc.
The relational characters are defined in Table 8.1.

R.8.2 When a comparison is executed using relational operators the possible outcomes are 1 if the relation is true and 0 otherwise (if false).

R.8.3 The MATLAB relational operations follow the syntax $C = A$ *relation* B, where C takes the value of 1 or 0, depending on the outcome of the condition set by the *relation*. In its simplest form, the arguments for A and B can be constants or arrays.

R.8.4 If the relational arguments consist of two arrays A and B, then A and B must necessarily have the same size. If A is an array and B is a scalar, the *relation* is used to compare each element of A with B, and MATLAB returns the binary matrix C indicating the result of the relation, where C has the same dimension as A.

R.8.5 For example, let the relational arguments be $A = 5$ and $B = 1$, then Table 8.2 shows the commands and its corresponding output C for $C = A$ *relation* B.

R.8.6 Examples showing relational commands for the case where the arguments, A and B are vectors, are illustrated in Table 8.3, for $A = [0\ 1\ 2\ 3]$ and $B = [-1\ 2\ 1\ 3]$.

TABLE 8.1

Algebraic Relational Characters

Relational Symbol	Description
>	Greater than
<	Less than
==	Equal to
>=	Greater than or equal to
<=	Less than or equal to
~==	Not equal to

TABLE 8.2

Relations Involving Constants

Input	Output
C = 5>1	C = 1
C = 5==1	C = 0
C = 5<1	C = 0
C = 5~=1	C = 1
C = 5>=1	C = 1
C = 5<=1	C = 0

TABLE 8.3

Relations Involving Vectors

Input	Output
C = A>B	C = [1 0 1 0]
C = A==B	C = [0 0 0 1]
C = A<B	C = [0 1 0 0]
C = A~=B	C = [1 1 1 0]
C = A>=B	C = [1 0 1 1]
C = A<=B	C = [0 1 0 1]

R.8.7 Relational operations can also be used to compare two matrices provided they have the same dimensions. For example, let

$$A = \begin{bmatrix} 1 & 2 \\ 3 & 4 \end{bmatrix} \quad \text{and} \quad B = \begin{bmatrix} 1 & 0 \\ 5 & 2 \end{bmatrix}$$

Write a short program that returns

a. The matrix C with the information that indicates if the elements in A are greater than the corresponding elements in B.

b. The matrix D with the information that indicates if the elements in A are equal to the corresponding elements in B. Observe that the relations are performed on the individual elements of A and B according to their location, that is, $A(n, m)$ with respect to $B(n, m)$, for all possible n and m ($n = 1, 2...$ and $m = 1, 2...$).

MATLAB Solution
```
>> A = [1 2; 3 4];
>> B = [1 0; 5 2];
>> C = A>B

  C =
        0    1
        0    1

>> D = A==B

  D =
        1    0
        0    0
```

R.8.8 Recall that relational operators can also be used to compare an array with a scalar. For example, let

$$A = \begin{bmatrix} 1 & 2 \\ 3 & 4 \end{bmatrix}$$

Then write a program that returns

a. The matrix B with the information that indicates if each element of A is greater than 2

b. The matrix C with the information that indicates if each element of A is negative

c. The matrix D with the information that indicates if each element of A is equal to 2

MATLAB Solution
```
>> A = [1 2; 3 4];
>> B = A>2              % part (a)

  B =
        0    0
        1    1
```

```
>> C = A<0              % part (b)


   C =
        0    0
        0    0

>> D = A==2             % part(c)


   D =
        0    1
        0    0
```

R.8.9   Let us consider now an example using vectors. Let

$$A = [-5 \quad -4 \quad -3 \quad -2 \quad -1 \quad 0 \quad 1 \quad 2 \quad 3 \quad 4 \quad 5]$$

and

$$B = [-1 \quad 1 \quad -1 \quad 1 \quad -1 \quad 1 \quad -1 \quad 1 \quad -1 \quad 1 \quad -1]$$

Write a MATLAB program that returns

a. An array C with the information indicating if the elements in A are greater than the elements in B
b. An array D with the information indicating if the elements in A are equal to the corresponding elements in B
c. An array E with the information indicating if the elements in A are smaller than the elements in B
d. An array F with the information indicating if the elements in B are positive
e. An array G with the information indicating if the elements in A are greater or equal to 5

MATLAB Solution
```
>> A = -5:5

   A =
       -5  -4  -3  -2  -1   0   1   2   3   4   5

>> B = (-1).^A

   B =
       -1   1  -1   1  -1   1  -1   1  -1   1  -1

>> C = A>B              % part (a)

   C =
        0   0   0   0   0   0   1   1   1   1   1

>> D = A==B             % part (b)

   D =
        0   0   0   0   1   0   0   0   0   0   0
```

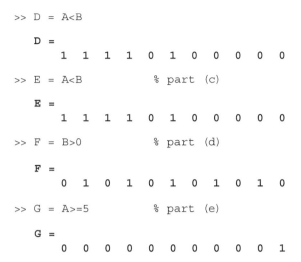

```
>> D = A<B

   D =
      1  1  1  1  0  1  0  0  0  0  0

>> E = A<B              % part (c)

   E =
      1  1  1  1  0  1  0  0  0  0  0

>> F = B>0              % part (d)

   F =
      0  1  0  1  0  1  0  1  0  1  0

>> G = A>=5             % part (e)

   G =
      0  0  0  0  0  0  0  0  0  0  1
```

R.8.10 It is possible to combine arithmetic and operational relations in one compact format.

R.8.11 For example, analyze the following program:

```
>> A = 1:10;
>> B = (-2) .^A;
>> C = B>=A;
>> result1 = A-C

   result1 =
            1  1  3  3  5  5  7  7  9  9
```

The last two lines of the preceding program can be compressed into one coded line by combining the arithmetic and relational operators as indicated in the following by the variable *result2*.

```
>> result2 = A- (B>=A)

   result2 =
            1  1  3  3  5  5  7  7  9  9
```

R.8.12 Relational operators can also be used when the arguments are characters.
    For example, write a program that compares whether the character *A* is greater than character *B*.

MATLAB Solution
```
>> X = 'A';
>> Y = 'B';
>> Z = X>Y
   Z =
       0
```

MATLAB performs a quantitative relation by converting the characters *A* and *B* into its *ASCII* code* representation.

---

* Recall that the ASCII code converts a character into a binary coded sequence. See Chapter 3 for additional information about the ASCII code.

R.8.13 MATLAB can be used to compare strings of characters. The MATLAB function $C = strcmp(x, y)$ compares the characters of the string x with the corresponding characters of the string y and returns the binary vector C in which the elements $C(n) = 1$ *(true)* if the element $x(n)$ is identical to the element $y(n)$, for $n = 1, 2,...$ *length(x)*, otherwise the element $C(n) = 0$ *(false)*. This function is case sensitive. Recall also that a blank (or a space) in *ASCII* is a character, like any other character.

R.8.14 The MATLAB function *stcmpi(x, y)* is similar to *strcmp(x, y)* but is not case sensitive.

R.8.15 The MATLAB function $C = strncmp(x, y, n)$ compares only the first n characters of the strings x with the corresponding elements of the string y and returns the element $C(n) = 1$, if $x(n) = y(n)$, otherwise MATLAB returns the element $C(n) = 0$, for all possible values of n. This function is case sensitive.

R.8.16 The MATLAB function *strncmpi(x, y, n)* is similar to *strcmp(x, y, n)* function, but is not case sensitive. The following examples illustrate the different modalities of the comparison commands just presented.

R.8.17 Let string A = 'Matlab' and string B = 'MATLAB.'
Write a set of MATLAB commands that compares the strings A with B using

a. *strcmp*

b. *strcmpi*

c. *strncmp* for the first 2 characters, $n = 2$

d. *strncmpi* for the first 2 characters, $n = 2$

MATLAB Solution
```
>> A = 'Matlab';
>> B = 'MATLAB';
>> C = strcmp(A,B),      % part (a) , case sensitive

   C =
       0

>> D = strcmpi(A,B),     % part(b), not-case sensitive

   D =
       1

>> E = strncmp(A,B,2),   % part(c), compares the first 2 characters,
   >>                    % case sensitive

   E =
       0

>> F = strncmpi(A,B,2),  % part(d), compares the first 2 characters,
   >>                    % not-case sensitive

   F =
       1
```

R.8.18 Two strings x and y can be compared by using the relational symbols defined in Table 8.1. For example, perform the following comparisons using the strings A and B defined in R.8.17 and the appropriate relational statements:

a. $A = B$

b. $A > B$

Observe and verify their responses.

```
MATLAB Solution
>> A = 'MatlaB';
>> B = 'MATLAB';      % part (a)
>> C = A==B           % observe that the relational ==is case sensitive

   C =
      1  0  0  0  0  1

>> D = A>B                  % part (b), according to the ASCII values for
>>                          % A(i) and  B(i),   for i=1,2,3,4,....6

   D =
       0  1  1  1  1  0
```

R.8.19 The MATLAB command $B = lower(A)$ assigns to the variable B consisting of the corresponding uppercase characters of A converted to lowercase characters, whereas all the lowercase characters of A remain unchanged.

R.8.20 The MATLAB command $B = upper(A)$ assigns to the variable B consisting of the lowercase characters of A converted to uppercase characters, whereas all the uppercase characters of A remain unchanged.

R.8.21 For example, let $A = 'MaTlaB'$ and $B = 'mAtLaB.'$
    Use MATLAB and perform the following:

a. Convert string A to uppercase characters

b. Convert string B to lowercase characters

c. Verify if $upper(A)$ is equal to $upper(lower(B))$

```
MATLAB Solution
>> A ='MaTlaB' ,   B ='mAtLaB'

   A =
       MaTlaB
   B =
       MAtLaB

>> UPPER _ CASE = upper(A)               % part (a)

   UPPER _ CASE =
                  MATLAB

>> lower _ case = lower(B)               % part (b)

   lower _ case =
                  matlab

>> C = UPPER _ CASE==upper(lower _ case)  % part (c)

   C =
       1  1  1  1  1  1
```

R.8.22 The MATLAB function *findstr (string, 'find')* returns the positions of the substring *'find'* given the string *string*. For example, identify the locations of the single character *'a'* in the string *'Matlab.'*

MATLAB Solution
```
>> X = 'Matlab';
>> locations _ of _ a = findstr(X,'a')

   locations _ of _ a =
                    2    5
```

R.8.23 The MATLAB function *newstring = strrep ('a', 'b', 'c')* replaces the string *'b'* in string *'a'* by string *'c.'*

R.8.24 For example, replace the string *'atla'* by the string *'ATLA'* in the string *'Matlab.'*

MATLAB Solution
```
>> X='Matlab';
>> Y= strrep (X,'atla','ATLA')

   Y =
      MATLAb
```

R.8.25 MATLAB uses the standard logical operators of *AND, OR, NOT,* and *X-OR* on arrays.
The syntax is defined in Table 8.4.

R.8.26 Recall that logic variables are binary consisting of ones and zero. Logic variables may be connected using logical operators forming logic expressions. Recall that logical as well as relational operators return a *1* when true and *0* otherwise.

R.8.27 Logical operations can be performed on logical variables and relational expressions. Logic expressions can be best defined by means of a truth table.

R.8.28 A truth table is a table that relates any possible input to its corresponding output. A truth table is an exhaustive form of defining an expression, relation, or a function output in terms of its inputs.

R.8.29 Table 8.5 shows a summary of the truth tables of the standard logical MATLAB relations. Recall that the input variables *A* and *B* are binary in the sense that they are either zero or not zero, where the nonzero elements are indicated by X in Table 8.5.

R.8.30 The command *C = A&B* returns the array *C* {where the *size(A) = size(B) = size(C)*} with entries of ones when both *A* and *B* are nonzero elements, and zeros when either *A, B* or both *A* and *B* are zero.

R.8.31 The command *C = A|B* returns the array *C* {where *size(A) = size(B) = size(C)*} with entries of ones when either *A, B* or both *A* and *B* are nonzero, and zero when both *A* and *B* are zero.

TABLE 8.4

Standard Logical Operators

Symbol	Description of Operation
&	AND
|	OR
~	NOT
Xor	EXCLUSIVE OR

TABLE 8.5

Truth Table of the Standard Logical Operators*

Logical Variables		A and B	A or B	A xor B	Not A	Not B
A	**B**	**A&B**	**A | B**	**xor (A, B)**	**~A**	**~B**
0	0	0	0	0	1	1
0	X	0	1	1	1	0
X	0	0	1	1	0	1
X	X	1	1	0	0	0

* The character X indicates a nonzero element.

TABLE 8.6

Logical Responses for A = [2 0 −1 5], and B = [1 0 0 6]

Input	Output
C = A&B	C = [1 0 0 1]
C = A|B	C = [1 0 1 1]
C = ~(A&B)	C = [0 1 1 0]
C = xor(A,B)	C = [0 0 1 0]
C = ~A	C = [0 1 0 0]

R.8.32 The command $C = \sim A$ returns the array C, with the dimension of A, with entries of ones when the elements of A are zeros, and zeros when the elements of A are nonzero.

R.8.33 The command $C = xor\ (A, B)$ returns the array C, with the dimension of A (or B) $\{size(A) = size(B)\}$, with entries of ones when the elements of A and B are not equal (0 and X or X and 0), and zeros when both the elements of A and B are equal (either 2 zeros or 2 Xs).

R.8.34 Examples of logical operations for the case when the arguments are vectors are illustrated in Table 8.6. Let $A = [2\ 0 - 1\ 5]$ and $B = [1\ 0\ 0\ 6]$. Then, MATLAB returns the array C after executing the commands indicated in the first column of the Table 8.6.

R.8.35 Logical operations can also be used when the arguments do not have the same dimensions, such as an array and a scalar.

R.8.36 For example, let

$$A = \begin{bmatrix} 0 & 1 \\ 2 & 3 \end{bmatrix} \quad \text{and} \quad B = 1$$

Execute the instructions indicated as follows and observe and verify their responses

a. $C = A \& B$

b. $D = A/B$

MATLAB Solution
```
>> A = [0 1;2 3];
>> B = 1;
>> C = A&B                        % part (a)

   C =
      0  1
      1  1
```

```
>> D = A|B                          % part (b)

D =
    1   1
    1   1
```

R.8.37 Observe that logical operations always return a binary vector, matrix, or scalar (consisting of 1's and 0's). Observe also that the input arguments may not necessarily be binary. Note that an array A may be connected by a logical operator to a constant c resulting in an array output with the same dimensions of A in which the logical operator of each element of A is evaluated with respect to c.

R.8.38 For example, let $A = 1{:}10$ and $B = (-2).^A$.

Write and execute the MATLAB statements that return

a. An array C that indicates the locations where $A > 6$ and $B < 3$

b. An array D that indicates the locations where $A > 6$ or $B < 3$

MATLAB Solution
```
>> A = 1:10

A =
    1   2   3   4   5   6   7   8   9   10

>> B = (-2).^A

B =
   -2    4   -8   16  -32   64  -128  256  -512  1024

>> C = (A>6)&(B<3)

C =
    0   0   0   0   0   0   1   0   1   0

>> D = (A>6)|(B<3)

D =
    1   0   1   0   1   0   1   1   1   1
```

R.8.39 Relational operators have a higher order of precedence than logical operators.

The hierarchy of arithmetic operations have been presented in Chapter 2. These rules need to be extended at this point to include the relational and logical operations. Table 8.7 summarizes the hierarchy of the arithmetic, relational, and logical operators.

Parentheses have the highest precedence of all the operators, and they can change the hierarchy of the operators. When a set of operators have the same hierarchy, then the hierarchy of the operators is from left to right.

R.8.40 Let us illustrate the logical and relational examples using vectors and arrays. For example, let $A = [2\ 0\ -1\ 5]$ and $B = [1\ 0\ 0\ 6]$. Verify the responses indicated by C if the input commands are given in the first column of Table 8.8.

TABLE 8.7

Operational Hierarchy

Operators	Hierarchy
Arithmetic operators	^ , '
	* , /
	+ , −
Relational operators	==
	~=
	>=
	<
	<=
	>
Logical operators	~
	&
	xor

TABLE 8.8

Examples for A = [2 0 −1 5], and B = [1 0 0 6]

Input	Output
C = A>B	C = [1 0 0 0]
C = A==B	C = [0 1 0 0]
C = (A>B)&(A==B)	C = [0 0 0 0]
C = (A>B)|(A==B)	C = [1 1 0 0]
C = A<B	C = [0 0 1 1]
C = (A<B)|(A==B)	C = [0 1 1 1]
C = xor(A, B)	C = [0 0 1 0]
C = xor((A>B), (A==B))	C = [1 1 0 0]
C = A&B	C = [1 0 0 1]
C = ~(A&B)	C = [0 1 1 0]
C = ~A&B	C = [0 0 0 0]

R.8.41 Additional examples using relational and logical operations on arrays are illustrated as follows:

let

$$A = \begin{bmatrix} 1 & 0 \\ 2 & 5 \end{bmatrix} \quad \text{and} \quad B = \begin{bmatrix} 1 & 2 \\ 1 & 6 \end{bmatrix}$$

Execute, observe, and verify the responses of each of the following commands:

a. $C = A>B$

b. $D = A==B$

c. $E = \sim(A==B)$

d. $F = A\&B$

e. $G = A|B$

f. $H = xor(A, B)$

g. $I = (A>B)|(A==B)$

MATLAB Solution

```
>> format compact
>> A = [1 0; 2 5]

   A =
        1    0
        2    5

>> B = [1 2; 1 6]

   B =
        1    2
        1    6
```

```
>> C = A>B                          % part (a)

   C =
        0   0
        1   0

>> D = A==B                         % part (b)

   D =
      1   0
      0   0

>> E =-(A==B)                       % part (c)

   E =
     -1   0
      0   0

>> F = A&B                          % part (d)

   F =
      1   0
      1   1

>> G = A|B                          % part (e)

   G =
      1   1
      1   1

>> H = xor(A,B)                     % part (f)

   H =
      0   1
      0   0

>> I = (A>B)|(A==B)                 % part (g)

      I =
              1       0
              1       0
```

R.8.42 MATLAB provides, besides the standard logical relations, an additional number of built-in logical functions (some were already introduced in previous chapters).
A brief summary of additional logical functions are presented in Table 8.9.

R.8.43 Let us consider an additional example. Let

$$A = \begin{bmatrix} 0 & 1 & 2 \\ 4 & 5 & 6 \\ -1 & -2 & -3 \end{bmatrix}$$

$$X = [-3 \quad -2 \quad -1 \quad 0 \quad 1 \quad 2 \quad 3 \quad \text{.......} \quad 8 \quad 9 \quad 10], \quad Y = []$$

and

$$\text{String} = \text{'ABC'}$$

TABLE 8.9

Built-In Logical Functions

Function	Description
all(x)	Returns a 1 if all the elements in the vector x are positive ($x > 0$) and 0 otherwise
any(x)	Returns a 1 if any of the elements of the array x are nonzero
exist(x)	Returns a 1 if the variable x exists
Isnumeric(x)	Returns a 1 if x is a numeric array, 0 otherwise
ischar(x)	Returns a 1 if x is a character array, 0 otherwise
isempty(A)	Returns a 1 if A is an empty matrix, 0 otherwise
isnan(A)	Returns a matrix with ones where the elements of the matrix A are *NaN*, and zero otherwise
find(A)	Returns the nonzero elements of the matrix A. This function was defined and used in Chapter 3
finite(A)	Returns a matrix with ones when the elements of A are finite and zero otherwise
isfinite(A)	Returns a 1 if all the elements of the matrix A are finite
isinf(A)	Returns a 1 or 0 for each element of the matrix A; 1 if the element is infinite, 0 otherwise
isreal(a)	Returns a 1 if a is real, 0 otherwise
isimag(a)	Returns a 1 if a is imaginary, 0 otherwise
isletter(a)	Returns a 1 if a is a letter, 0 otherwise
isspace(a)	Returns a 1 if a is blank, tab, or new line, 0 otherwise

Execute, observe, and verify the responses after executing the following MATLAB built-in logical functions:

a. *all(X)*

b. *all(A)*

c. *all(Y)*

d. *C = ones(1, 14)./X; any(X)*

e. *isnumeric(A)*

f. *isempty(A)*

g. *isempty(Y)*

h. *isinf(A)*

i. *isfinite(A)*

j. *isinf(C)*

k. *find(A)*

l. *ischar(String)*

MATLAB Solution

```
>> format compact.
>> A = [0 1 2; 4 5 6; -1 -2 -3]

    A =
         0     1     2
         4     5     6
        -1    -2    -3

>> X = -3:10

    X =
      -3  -2  -1   0   1   2   3   4   5   6   7   8   9  10
```

```
>> Y= []

   Y =
        []

>> String ='ABC'

   String =
            ABC

>> all(X)                                  % part (a)

   ans =
        0

>> all(A)                                  % part (b)

   ans =
        0       1       1

>> all(Y)                                  % part (c)

   ans =
        1

>> C = ones(1,14)./X                       % part (d)

Warning: Divide by zero.
     C =
-0.3333    -0.5000    -1.0000       Inf    1.0000    0.5000    0.3333
 0.2500     0.2000     0.1667    0.1429    0.1250    0.1111    0.1000

>> any(X)                                  % observe that one element
                                              of X is nonzero

   ans =
        1

>> isnumeric(A)                            % part (e)

   ans =
        1

>> isempty(A)                              % part (f)

   ans =
        0

>> isempty(Y)                              % part (g)

   ans =
        1
```

```
>> isinf(A)                              % part (h)

    ans =
              0      0      0
              0      0      0
              0      0      0

>> isfinite(A)                           % part (i)

    ans =
              1      1      1
              1      1      1
              1      1      1

>> isinf(C)                              % part (j)

    ans =
              0  0  0  1  0  0  0  0  0  0  0  0  0  0

>> find(A)                               % part(k)

    ans =
              2
              3
              4
              5
              6
              7
              8
              9

>> ischar(String)                        % part (l)

    ans =
              1
```

R.8.44 The relational and logical operations are frequently used to set up a conditional statement. A conditional statement is an instruction that sets up a condition, and based on its outcome, decides the correct program path to follow, such as

if <condition> is true, then execute the sequence of commands *B, C, and D*

if <condition> is not true, then execute the sequence of commands *E, F, and G*

The flowchart shown in Figure 8.1 illustrates the decision-making action and the two distinct paths.

R.8.45 The flow-control path of a program can be altered by each of the following four conditional MATLAB commands:

a. the *if-end*

b. the *for-end*

c. the *while-end*

d. the *switch-end*

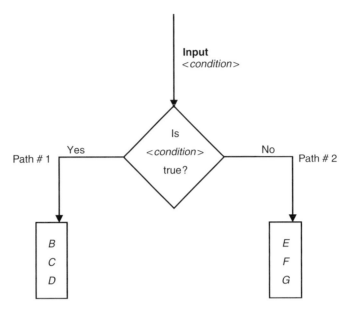

FIGURE 8.1
Decision-making flowchart representation.

The decision-making *conditions* set by using any of the conditional commands will alter the normal sequential flow of the program. The syntax and control command mechanisms are presented and discussed next.

R.8.46 The syntax and format of the simplest form of the *if-end* statement is as follows:

```
if          <condition>
            <statements>
end
```

Only one decision-making *<condition>* is used, and if this *<condition>* is *true*, then the *<statements>* are executed, followed by the *end.*

However, if the *<condition>* is *false*, then the *<statements>* are not executed, followed by the *end.*

The flowchart shown in Figure 8.2 graphically illustrates the decision-making action with the corresponding path.

R.8.47 The *if-end* statement can be expanded to include two different paths by executing two different sets of statements based on a single decision-making condition.

The syntax and format are as follows:

```
if          <condition>
            <statements _ 1>
else
            <statements _ 2>
end
```

Meaning that *if <condition>* is *true*, then *<statements_1>* is executed followed by the *end* (exit); but *if <condition>* is not true, then *<statements_2>* is executed followed by the *end* (exit). The flowchart shown in Figure 8.3 illustrates the decision-making action resulting in distinct execution paths.

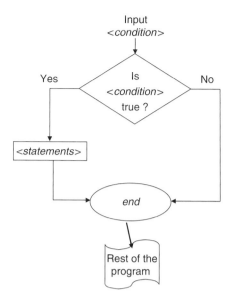

FIGURE 8.2
Decision-making flowchart representation of the *if-end* statement.

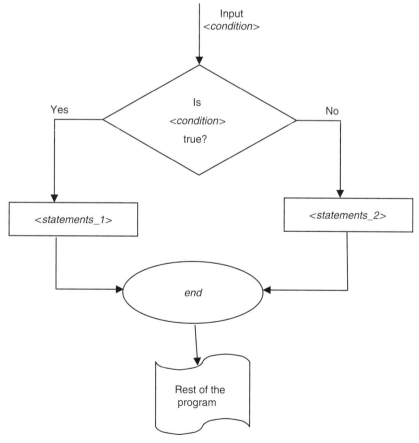

FIGURE 8.3
Decision-making flowchart of the *if-end* (with *else*) statement.

R.8.48 A multiple path can be set based on multiple decision-making conditions by using the general form of the *if-end* statement. The syntax and format are as follows:

```
if              <condition _ 1>
                <statements _ 1>
elseif          <condition _ 2>
                <statements _ 2>
elseif          <condition _ 3>
                <statements _ 3>


elseif          <condition _ n-1>
                <statements _ n-1>
else
                <statements _ n>
end
```

Meaning that *if* <condition_1> is *true*, then the <statements_1> are executed followed by the *end* (exit); but if <condition_1> is not *true*, then MATLAB checks if <condition_2> is *true*; if it is *true*, then <statements_2 > are executed, followed by the *end* (exit), otherwise the <condition_3> is tested, and so on; if none of the $n-1$ conditions are *true*, then MATLAB executes <statements_n> followed by the *end* (exit).

R.8.49 Note that the multiple conditions format based on the *elseif* <condition_k> for $k = 1$, $2, 3, ..., n-1$ provide n alternate paths based on the n conditions.

Observe that only the first *true* condition encountered is executed, and all the other *statements* and *conditions* are ignored by MATLAB. Figure 8.4 shows the flowchart of the general *if-end* statement with multiple conditions (*elseifs*). Recall that the = sign is used to assign a value to a variable, whereas the == sign is used as a relational operator (is equal to).

R.8.50 Let us illustrate the use of the *if-then* command in the following example:

Create the script file *plot_cos_spikes* that returns the plot of $y(x) = 10 \cos(2\pi x/100)$ over the following ranges $0 \le x \le 24$, $26 \le x \le 49$, $51 \le x \le 74$, and $76 \le x \le 99$, and $y(x)$ takes the following values at the points defined by $x = 25, 50, 75$, and 100, $y(25) = 20$, $y(50) = -30$, $y(75) = 20$, and $y(100) = -30$. The script file *plot_cos_spikes* is executed below and its resulting plot is shown in Figure 8.5.

MATLAB Solution
```
% Script file: plot _ cos _ spikes
for x=1:1:100
    if x==25
        y(x)=20;
    elseif x==50
        y(x)=-30;
    elseif x==75
        y(x)=20;
    elseif x==100
        y(x)=-30;
    else
        y(x)=10*cos(2*pi.*x./100);
    end
end
plot(y), title('[y(x)=10*cos(2pix/100)+ [spikes at x=25,50,75,100]] vs x')
xlabel('x'), ylabel('y(x)'),axis([0 100 -33 23])
```

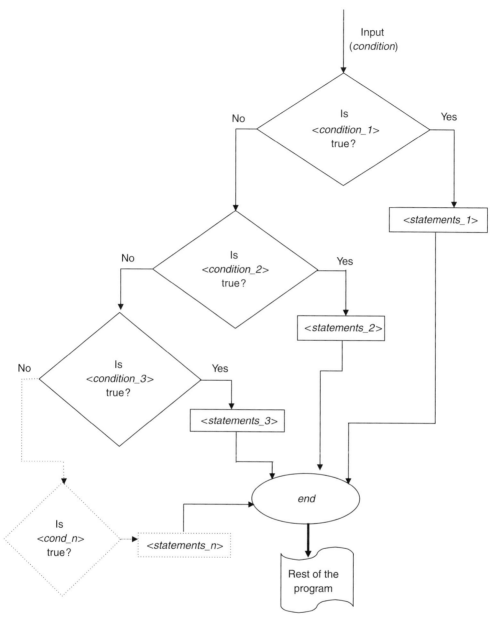

FIGURE 8.4
Flowchart of the *if-end* statement with multiple conditions (*elseif*).

R.8.51 When a loop condition statement is being entered in the command window, the loop is executed and any display is suspended until the execution of the loop. The suspension includes turning off the cursor (within the loop).

Note that the *if* statement does not require semicolons at the end of each line because a line does not represent a command but a partial command, and errors and responses are displayed only when the loop is terminated.

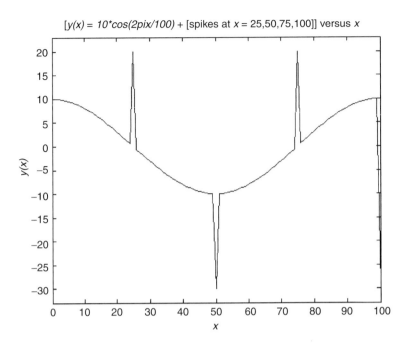

FIGURE 8.5
Resulting plot of executing *plot_cos_spikes* of R.8.50.

R.8.52 For example, analyze the following script file *check_value of_a* given below, which tests the value for a given variable *a* and returns one condition message indicating pass or fail.

The test conditions are

a. *a > 80*, the condition fails

b. *a > 75*, the condition fails

c. *a > 70*, the condition passes

d. *a > 60*, the condition fails

e. *a > 50*, the condition fails

The script file *check_value of_a* is tested for *a = 72*. The reader should follow the logic and observe and verify its response.

MATLAB Solution
```
% Script file: check _ value of _ a
a =72;
if a>80
  disp('*** a>80, the condition fails ***')
elseif a>75
  disp('*** a>75, the condition fails ***')
elseif a>70
  disp('*** a>70, the condition passes ***')
elseif a>60
  disp('*** a>60, the condition fails ***')
elseif a>50
  disp('*** a>50, the condition fails ***')
end
```

The file *check_value of_a* is executed and the result is as follows:

```
>> check _ value of _ a

    *** a>70, the condition passes ***
```

Observe that the script file *check_value of_a* returns only the following message:

```
    *** a>70, the condition passes***
```

The other true conditions such as $a > 60$ and $a > 50$ are ignored because they are never executed by MATLAB.

R.8.53 Let us consider a more sophisticated example.

Create the script file *check_age* that returns one of the following messages: *teenager, adult,* and *senior* given as the input of the variable *age.*

Test the script file *check_age* for the following inputs (age): 34, 13, and 68. Analyze the program's logic, and observe and verify each response.

MATLAB Solution
```
% Script file: check _ age
age = input('Enter the age of the person in question :')
if age<=18
    disp('************************')
    disp('******* teenager**********')
    disp('************************')
elseif age<=62
    disp('**********************')
    disp('****** adult**********')
    disp('**********************')
else
    disp('*********************')
    disp('**** senior **********')
    disp('*********************')
end
```

The script file *check_age* is executed for three *age* inputs (34, 13, and 68), and the results are indicated as follows:

```
>> check _ age

Enter the age of the person in question : 34
age =
      34
**********************
****** adult**********
**********************

>> check _ age

Enter the age of the person in question :13
age =
      13
************************
******* teenager**********
************************
```

```
>> check _ age
```

Enter the age of the person in question : 68
age =
         68
****
****** senior ************
****

R.8.54 Let us now use the *if-end* statement to implement the following function:

$$y(t) = \begin{cases} 0 & \text{for} & t \le 0 \\ 2 & \text{for} & 0 < t \le 3 \\ -3 & \text{for} & t > 3 \end{cases}$$

Draw a flowchart and create the Matlab script file *y_of_t* that returns the value of the function *y(t)* for any given *t*. Test the script file *y_of_t* for the following values of *t* = −5, 1.7, and 6. Trace the logic of the program for each value of *t* and observe and verify their responses.

ANALYTICAL Solution
See Figure 8.6.

MATLAB Solution
```
% Script file: y _ of _ t
format compact
t = input('Enter a numerical value for t=')
if t<=0;
    y = 0
```

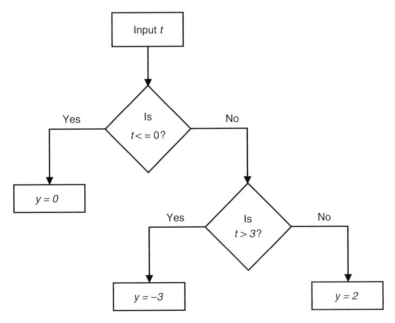

FIGURE 8.6
Flowchart of R.8.54.

```
elseif t>3
    y=-3
else y=2
    end
end
```

The script file *y_of_t* is tested for the following values of $t = -5$, 1.7, and 6, and the results are as follows:

```
>> y _ of _ t

Enter a numerical value for t = -5
t =
    -5
y =
    0

>> y _ of _ t

Enter a numerical value for t = 1.7
t =
    1.7000
y =
    2

>> y _ of _ t

Enter a numerical value for t = 6
t =
    6
y =
    -3
```

R.8.55 The command

for   *<range>*
         <statements>
end

referred as the *for-end* statement is used to create a loop that executes repetitively the *<statements>* a fix number of times based on the spec *<range>*. The spec *<range>* is frequently given by a vector or a matrix. In either case, the commands between the *for* and *end* indicated by *<statements>* are executed once for each column of the *<range>*.

R.8.56 For example, write a program that returns the following sequence:

$C = [1\ 1/2\ 1/3\ 1/4\ 1/5 \dots 1/10]$ as a column vector by using the *for-end* statement.

MATLAB Solution
```
>> format compact
>> for I =1:10;              %  <range>
       C(I) = 1./I;          %  <statements>
       end
```

```
>> disp(C')
        1.0000
        0.5000
        0.3333
        0.2500
        0.2000
        0.1667
        0.1429
        0.1250
        0.1111
        0.1000
```

Observe that the same sequence can be generated by executing the following commands:

```
>> n =1:10;
>> C = (1./n)'
```

R.8.57 Let us consider an additional example.

Let

$$A = \begin{bmatrix} 3 & 7 & 4 \\ 2 & 5 & 7 \\ -1 & -2 & -3 \end{bmatrix}$$

Create the script file *sum_prod* that returns the sum and product of each of the columns of A by using the *for-end* statement and the matrix A as the *<range>*.

Analyze the logic of the script file *sum_prod* given below, and observe and verify each responses.

MATLAB Solution
```
% Script file: sum _ prod
disp('*****************************************')
disp(' This script returns the sum and product ')
disp('of each column of the matrix A defined below')
disp('*****************************************')
format compact
n =1;
A= [3 7 4;2 5 7;-1 -2 -3]
for K= A
    fprintf ('The sum and product of the elements of column
                                                %1.1f\n',n)
    ColumnSum = K(1)+K(2)+K(3)
    ColumnProd = K(1)*K(2)*K(3)
    n = n+1;
 end
```

The script file *sum_prod* is executed and the results are as follows:

```
>> sum _ prod

********************************************
This script returns the sum and product
of each column of the matrix A defined below
********************************************
```

```
A =
         3       7       4
         2       5       7
        -1      -2      -3
The sum and product of the elements of column 1.0
ColumnSum =
                4
ColumnProd =
               -6
The sum and product of the elements of column 2.0
ColumnSum =
               10
ColumnProd =
              -70
The sum and product of the elements of column 3.0
ColumnSum =
                8
ColumnProd =
              -84
```

Observe that the *for-end* statement results in a loop that is executed three times, one for each of the columns of *A*.

R.8.58 If multiple loops are required, the loop structure must be nested, meaning that each loop must be constructed inside another loop. Observe that the resulting nested loop indexes can be used to create matrices where the first index is used to define its rows, whereas the second index is used to define its columns.

R.8.59 For example, write a program that returns a 13 by 3 matrix *A* where the first column of *A* consists of the sequence *1, 2, 3, 4, ..., 13*; the second column of *A* consists of the sequence *2, 4, 6, 8, ..., 24, 26*; and the third column of *A* consists of the sequence *3, 6, 9, 12, ..., 36, 39*.

MATLAB Solution
```
>> format compact
>> for N=1:13;
for M=1:3;
A(N,M)=N*M;
end
end
>> disp(A)
                1       2       3
                2       4       6
                3       6       9
                4       8      12
                5      10      15
                6      12      18
                7      14      21
                8      16      24
                9      18      27
               10      20      30
               11      22      33
               12      24      36
               13      26      39
```

R.8.60 Let us revisit the script file *y_of_t*. Modify this script file and call it *y_of_t_mod*, which now returns the plot of the function *y(t)* over the range $0 \leq t \leq 6$. Recall that *y(t)*, is given by

$$y(t) = \begin{cases} 0 & \text{for} & t \leq 0 \\ 2 & \text{for} & 0 < t \leq 3 \\ -3 & \text{for} & t > 3 \end{cases}$$

MATLAB Solution
```
% Script file: y _ of _ t _ mod
format compact
n =1;
for t =-2:0.01:6;
   if t<=0;
      y(n)  = 0;
      n =n+1;
      elseif t>3
      y(n)  =-3;
      n =n+1;
      else y(n)=2;
      n =n+1;
   end
end
t = -2:0.01:6;
plot(t,y)
xlabel('t')
ylabel('Amplitude')
title('y(t) vs t')
axis([-2 6 -3.3 2.3])
```

The script file *y_of_t_mod* is executed and the resulting plot is shown in Figure 8.7.

R.8.61 Observe that by using the *for-end* command in the previous program, the loop variable *y* is reevaluated over and over; and for each of its new values, a memory expansion is required. It is therefore a good and efficient programming practice to define the final memory size of the affected variables before they are used in a loop. Avoiding multiple memory expansions.

For example, analyze the following program:

```
>> A= ones(1,10);
>> for x = 1:10
            A(x)  = x.^2;
         end
```

Observe that the first instruction *A* = *ones(1, 10)* allocates the total required memory to *A*. If this command is not included, the same *A* is still created; but the computational efficiency of evaluating and storing each element of *A* would be affected.

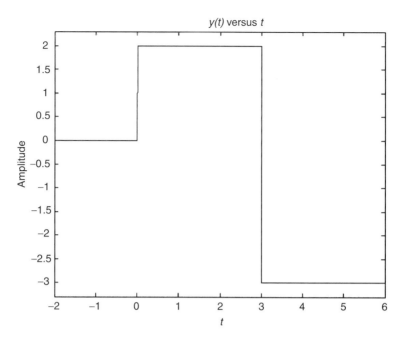

FIGURE 8.7
Plot of R.8.60.

R.8.62 The command

```
while  <condition>
        <statements>
   end,
```

referred as the *while-end* statements is used to create a loop in which the *<statements>* are executed repetitively for an indefinite number of times as long as the specified *<condition>* is and remains *true*. When the *<condition>* is no longer *true*, then the program exits the loop and continues with the normal execution of the remaining of the program by executing the first instruction after the *end* statement.

R.8.63 For example, draw a flowchart and write a program that returns the sequence $C = [1\ 1/2\ 1/3\ 1/4\ 1/5\ ...\ 1/10]$, using the *while-end* statement. Analyze the flowchart (Figure 8.8) and program and observe and verify its output.

ANALYTICAL Solution

See Figure 8.8.

MATLAB Solution
```
>> format compact
>> n = 1;
>> while n<11;                    %  <condition>
        C(n) =1./n;              %  <statements>
        n = n+1;
     end
```

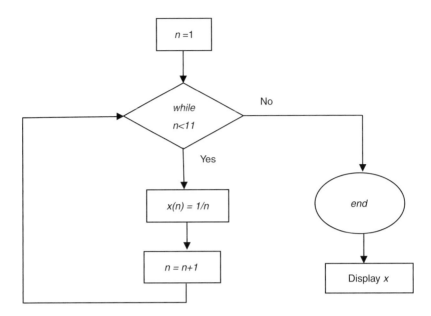

FIGURE 8.8
Flowchart of R.8.63.

```
>> disp(C')
          1.0000
          0.5000
          0.3333
          0.2500
          0.2000
          0.1667
          0.1429
          0.1250
          0.1111
          0.1000
```

The commands between the *while* and *end* referred as *<statements>* are executed 10 times, one for each value of n ($n = 1, 2, 3, ..., 9, 10$) as long as n (*while*) is smaller than *11*.

R.8.64 The *while-end* statement is a powerful command, which can be particularly useful when exploring the behavior of a given equation or relation over a given range.

For example, create the script file *explore* that returns the plot of $y = 5\sqrt{k^{0.5}}$ over the range $1 \le k \le 5$ in discrete increments of $\Delta K = 0.5$, as long as abs(y) < 8.

MATLAB Solution
```
% Script file: explore
figure(1)
k=1;
while k<=5
    y = 5*sqrt(k.^ (0.5));
        k = k+0.5;
            while abs(y)<8
                    plot(k,y,'*'), hold on
                end
        end
```

```
xlabel (' values of k in steps of 0.5')
ylabel ('y')
title('y=5*sqrt(k.^ (0.5)) vs k');
```

The script file *explore* is executed and the resulting plot is shown in Figure 8.9.

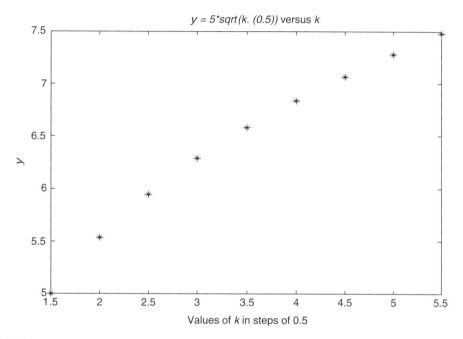

FIGURE 8.9
Plot of R.8.64.

R.8.65 Some general observations about the looping process
   a. The loop variable is updated for each loop pass.
   b. While the loop variable is *true*, the commands inside the *while-end* statement are executed.
   c. Care must be taken not to create an infinite loop. An infinite loop is created when the conditions inside the *while-end* *<statements>* remain always true and the condition for exiting the loop is never met.
   d. Both the *while-end* and *for-end* statements are used to create loops. The main difference between the two commands is that the *for-end* command creates a loop that is executed a specific number of times, whereas the *while-end* command executes a loop for an unspecified implied number of times.
   e. If the number of times of execution of a loop is known in advance, then the *for-end* command is recommended. Otherwise, a decision mechanism to control the looping is required, and the *while-end* or *if-end* commands are better choices.

R.8.66 Observe that the looping condition in a *while-end* and *if-end* statement must be carefully set to avoid the possibility of getting into an infinite loop.

R.8.67 For example, analyze the following sequence of instructions:

```
x = 1
while x>0.5
        y = 2.^x
        x = x + 0.1
end
```

Observe that the condition set for *x* always returns a *true* outcome, resulting in an infinite loop.

R.8.68 Observe that each *if-*, *while-*, and *for* statements must have a corresponding matching *end* statement.

R8.69 Looping construction by means of the *for-end* or *while-end* statements should be avoided whenever possible because of the slow execution time in favor of the *implied* loop (see Chapter 9 for performance analysis).

R.8.70 There are a number of exit or safety mechanisms that can be used when dealing with loops to avoid getting into traps or infinite loops by setting limits on the number of iterations or loop executions.

R.8.71 For example, the following program limits the *while-end* loop to 1000 iterations by using the *if-end* statement and the *break* command (presented in R.8.75 as an exit and safety mechanism).

```
n =1
x = 1
while x>0.5
y = 2.^x
x = x+0.1
n = n+1
        if n=1000
        disp('exit after 1000 iterations')
        break
        end
end
```

R.8.72 Let us review the looping concepts by creating a program that evaluates the *sum_x* defined by the following equation (for the first 10 digits):

$$sum\_x = \sum_{x=1}^{10} x$$

employing the following:
a. The *implied* loop
b. The *for-end* loop
c. The *while-end* loop

MATLAB Solution
```
>> x =1:10;
>> imp_sum_x =sum(x);                    % part (a), the implied
                                          loop solution
```

```
>> imp _ sum _ x
```

imp _ sum _ x =
                    55

```
>> for _ sum _ x = 0;
>> for x =1:10;
       for _ sum _ x = for _ sum _ x+x;        % part (b), the for-end
                                                  loop solution
     end
>> for _ sum _ x
```

for _ sum _ x =
                    55

```
>> while _ sum _ x = 0;
>> i =1;
>> while i <11;                                % part (c), the while-end
                                                  loop solution
       while _ sum _ x = while _ sum _ x + i;
       i = i+1;
     end
>> while _ sum _ x
```

while _ sum _ x =
                    55

R.8.73 The *switch-end* statement is another decision-making command. The general form presents the following format and syntax:

switch flag,
    case flag1
    <statments1>
        case flag2
        <statements2>
            case flag3
            <statement3>

    ....

                otherwise
                <statementn>
                end

where the *flagn* is a string or a variable, which is used to branch when multiple conditions are tested for a common argument.

R.8.74 The commands *break, error,* and *return* are useful when operating inside a loop to stop or control its execution.

R.8.75 The *break* command unconditionally terminates the execution of a loop and the program continues with the first instruction after the *end* command.

R.8.76 The command *error ('text')* stops the execution of a loop, displays the string *text* on the computer screen, and transfers control to the keyboard.

R.8.77 The command *return* produces an unconditional exit from a loop, ignoring the instructions inside the loop. The flowchart illustrating the flow control of the *return* and *break* commands are shown in Figures 8.10 and 8.11, respectively.

R.8.78 A few words of advice—a good programmer must be able to understand the mechanisms and conditions set, and trace the logic used in repeating a block of commands by using loops and nested loops when analyzing a program. Looping and decision making constitutes, in the author's opinion, the main power and capability of most digital computer systems. The analysis of a looping sequence must be followed either mentally by the experienced programmer or by relying on a flowchart or table by the beginner or less experienced programmer. A recommended practice is to assign values to the loop variables tracing its execution for at least two complete cycles to get a good insight of the mechanism used and be able to visualize a pattern of the purpose and nature of the looping algorithm.

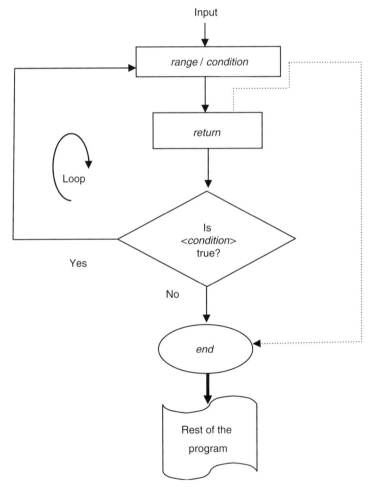

FIGURE 8.10
Flowchart of the *return* command.

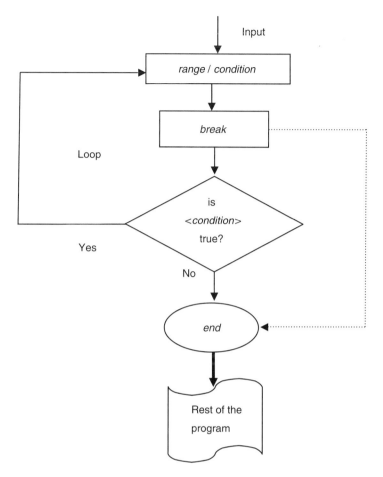

FIGURE 8.11
Flowchart of the *break* command.

8.4 Examples

Example 8.1

Create the script file *max_min* that returns the maximum and minimum values of a given vector *V*, assuming that no two elements in *V* are equal (the commands *sort, max,* and *min* are not allowed). Test the script file *max_min* by using the vector *V* = [1 2 3 4 −3 9].

MATLAB Solution
```
% Script file: max _ min
V = input('Enter the numerical vector V in brackets; V=  ')
n = length(V);
for k = 2:n;
   if V(k) >= V(k-1);
      maxim = V(k);
      minim = V(k-1);
   else
      maxim = V(k-1);
```

```
        minim = V(k);
    end
end
disp('***********************************************************')
disp('          ************  R E S U L T S  *****************')
disp('***********************************************************')
fprintf('The element with the largest value in the array V is
                                        %3.2f\n',maxim);
fprintf('The element with the smallest value in the array V is
                                        %3.2f\n',minim);
disp('***********************************************************')
```

The sript file *max_min* is tested for *V= [1 2 3 4 −3 9]* as follows:

```
>> max _ min

Enter the numerical vector V in brackets; V=   [1 2 3 4 -3 9]
    V =
            1    2    3    4    -3    9
**********************************************************
    ************  R E S U L T S  *****************
**********************************************************
The element with the largest value in the array V is 9.00
The element with the smallest value in the array V is -3.00
**********************************************************
```

Example 8.2

Create the script file *matrix_of_ones* that returns the *n* by *m* matrix *A* that consists of all its elements equal to one, emulating the MATLAB command *ones(n, m)*.
     Test the script file *matrix_of_ones* for *n = 3* and *m = 9*.

MATLAB Solution
```
% Script file: matrix _ of _ ones
% emulates the command ones(n,m)
disp('***************************************************')
disp('This program returns the matrix A consisting of ones. ')
rows = input('Enter the number of rows of the matrix A you want
                                        to create = ')
columns = input('Enter the number of columns of the matrix A you
                                        want to create = ')
for k =1:rows
    for j =1:columns;
        A(k,j) =1;
    end
end
disp('********  The resulting matrix is :  *******************')
A
disp('***************************************************')
```

The sript file *matrix_of_ones* is tested for *n = 3* (rows) and *m = 9* (columns), and the results are shown as follows:

```
>> matrix _ of _ ones
```

```
*****************************************************************
This program returns the matrix A consisting of ones.
Enter the number of rows of the matrix you want to create = 3
rows =
      3
Enter the number of columns of the matrix you want to create = 9
columns =
          9
      ********       The resulting matrix is :       *****************
A =
      1       1       1       1       1       1       1       1       1
      1       1       1       1       1       1       1       1       1
      1       1       1       1       1       1       1       1       1
*****************************************************************
```

Example 8.3

Create the script file *perm_matrix* that returns the $n \times n$ (square) matrix A that consists of the random permutations of the elements of each row of A where the row elements consist of the integers 1 through n. Test the script file *perm_matrix*, for $n = 6$.

MATLAB Solution
```
% Script file: perm _ matrix
n = input('Enter the size of the square matrix A, n =  ')
for k =1:n;
   A(k,:)  =  randperm(n);
end
disp('******************************')
disp(' The permuted row matrix is :  ')
A
disp('******************************')
```

The script file *perm_matrix* is tested for $n = 6$ as follows:

```
>> perm _ matrix


Enter the size of the square matrix A, n =   6
n =
    6
***********************************
The permuted row matrix is :
A =
      3       2       4       1       5       6
      6       5       4       3       1       2
      2       4       3       6       5       1
      4       3       2       5       6       1
      2       1       3       4       5       6
      5       2       3       4       1       6
***********************************
```

Example 8.4

Write a program that returns n versus $\sqrt{n}$, for $n = 1, 2, 3, \ldots, 10$ in a tablelike format by using each of the following commands:

   a. an implied loop
   b. *for-end* statement
   c. *while-end* statement

MATLAB Solution

```
>> % part (a)
>> format compact
>> X=1:10;
>> VA = sqrt(X);
>> Result _ A = [X' VA'];
>> disp('Result part(a)');disp('    n       sqrt(n)');
>> disp('    **************');
>> disp(Result _ A); disp('    **************');
```

```
          Result      part(a)
             n        sqrt(n)
          ******************
          1.0000      1.0000
          2.0000      1.4142
          3.0000      1.7321
          4.0000      2.0000
          5.0000      2.2361
          6.0000      2.4495
          7.0000      2.6458
          8.0000      2.8284
          9.0000      3.0000
         10.0000      3.1623
          ******************
```

```
>> % part (b)
>> for K=1:10;
           VB(K) = sqrt(K);
      end
>> Result _ B = [X' VB'];
>> disp('Result  part(b)'), disp('    n       sqrt(n)');
>> disp('    **************');
>> disp(Result _ B); disp('**************');
```

```
          Result      part(b)
             n        sqrt(n)
          ******************
          1.0000      1.0000
          2.0000      1.4142
          3.0000      1.7321
          4.0000      2.0000
          5.0000      2.2361
          6.0000      2.4495
          7.0000      2.6458
          8.0000      2.8284
          9.0000      3.0000
         10.0000      3.1623
          ******************
```

```
>> % part (c)
>> Y=11;
>> A=1;
>> while A<Y
      VC(A) = sqrt(A);
      A=A+1;
      end
>> Result _ C= [X' VC'];
>> disp('Result  part(c)'), disp('    n       sqrt(n)');
>> disp('    **************');
>> disp(Result _ C)
>> disp(' **************');


       Result   part(c)
         n       sqrt(n)
       ****************
       1.0000   1.0000
       2.0000   1.4142
       3.0000   1.7321
       4.0000   2.0000
       5.0000   2.2361
       6.0000   2.4495
       7.0000   2.6458
       8.0000   2.8284
       9.0000   3.0000
       10.0     3.1623
       ****************
```

Example 8.5

Let the array $x = rand\ (1,\ 20)$.

Draw a flowchart and create the script file *grett_05* that returns the number of elements of x that are smaller than or equal to *0.5*.

ANALYTICAL Solution

The corresponding flowchart is shown in Figure 8.12.

MATLAB Solution
```
% Script file: grett _ 05
format compact
x = rand(1,20);
Addx = 0;
for i =1:20
    if x(i)<=0.5
        Addx = Addx+1;
    else
    Addx = Addx+0;
    end
end
disp(' **************** R E S U L T S ********************* ')
disp('***************************************************************')
disp('The random vector x is given by x = ')
```

```
disp(x)
disp('*********************************************************')
fprintf ('The number of elements of x that are smaller than or
                        equal to 0.5 is % 4.2f\n',Addx)
disp('*********************************************************')
```

The script file grett_05 is executed and the results are as follows:

```
>> grett _ 05

********************* R E S U L T S *********************************
**********************************************************************
The random vector x is given by x =
  Columns 1 through 7
    0.2618    0.5973    0.0493    0.5711    0.7009    0.9623    0.7505
  Columns 8 through 14
    0.7400    0.4319    0.6343    0.8030    0.0839    0.9455    0.9159
  Columns 15 through 20
    0.6020    0.2536    0.8735    0.5134    0.7327    0.4222
**********************************************************************
The number of elements of x that are smaller than or equal to 0.5
is 6.00
**********************************************************************
```

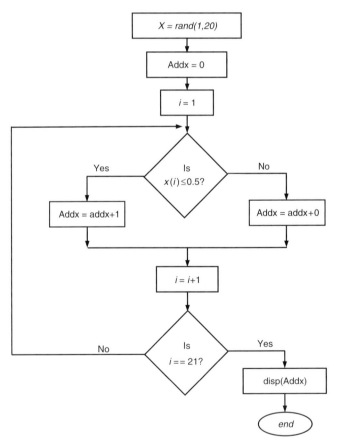

FIGURE 8.12
Flowchart of Example 8.5.

Example 8.6

Create the script file *apprx_exp* that returns the number of terms required to approximate $e = 2.71828182$ by means of the *Maclarin* series with an error of less than 0.000001. Indicate its error as well as the approximation and error plots given by

a. *[error]* versus *[number of terms of the approximation]*

b. *[magnitude of approximation]* versus *[number of terms of the approximation]*

Recall that the *Maclarin* series expansion for e^x is given by

$$e^x = \exp(x) = 1 + x + \frac{x^2}{2!} + \frac{x^3}{3!} + \frac{x^4}{4!} + \cdots + \frac{x^n}{n!}$$

Therefore, for $x = 1$

$$\exp(1) = e = 1 + \frac{1}{1!} + \frac{1}{2!} + \frac{1}{3!} + \frac{1}{4!} + \cdots + \frac{1}{n!}$$

MATLAB Solution

```
% Script file: approx _ exp
format long
exact _ e = exp(1);
error =1;
approx =1;app(1) =1;err(1)  = exact _ e-1;
n=1:100;
x = cumprod(n);i =1;error(1)  = 1.7;
while error>0.000001
   approx=approx+ 1/x(i);
      i=i+1;
               app(i)  = approx
      err(i)  = exact _ e-app(i)
   error = abs(exact _ e-app(i));
end
k=1:i,
subplot(2,1,1)
plot(k,err)
ylabel('[error]');
title('[erros]  vs   [# of terms]')
subplot(2,1,2)
plot(k,app)
ylabel('approximations for e=2.7182...');
xlabel('number of terms')
title('[approximations] vs   [# of terms]')
disp('***************** R E S U L T S ***********************')
disp('*********************************************************')
fprintf('The number of terms required in the approximation is %
4.2f\n',i)
fprintf('The approximation error is % 10.9f\n',error)
disp('*********************************************************')
```

The script file *approx_e* is executed and the results are as follows:

```
>> approx _ exp

****************** R E S U L T S ****************************
**************************************************************
The number of terms required in the approximation is 10.00
The approximation error is 0.000000303
**************************************************************
```

Example 8.7

a. Create the script file *traffic_light_if* that returns one of the following messages based on the conditions stated below:

It is safe to pass, if the traffic light is green

It is not safe to pass, if the traffic light is red

Proceed with caution, if the traffic light is yellow

Light is not functioning, if the traffic light color is neither green, red or yellow.

b. Use the *if-end* statement to check the traffic light specified by the first three letters *gre*, *red*, *yel* and test the script file *traffic_light_if* for the following traffic light colors: *red*, *green*, *yellow*, and *blue*.

c. Repeat part a using the *switch-end* statement by specifying the complete color *green*, *red*, *yellow*.

Create the script file *traffic_light_switch* and test the file for the (same) traffic light colors *red*, *green*, *yellow*, and *blue*.

MATLAB Solution
```
% part(a)
% Script file: traffic _ light _ if
disp('***************************************************************')
disp(' ***************  Traffic light condition  ***************')
disp('***************************************************************')
light = input('Enter the first 3 letters of the following traffic light
color: red, green, yellow,  others:','s')
if light == 'gre'
    disp( 'It is safe to pass')
elseif light == 'red'
    disp('It is not safe to pass')
elseif  light == 'yel'
    disp('Proceed with caution')
else
    disp('Light is not functioning')
end
```

Back in the command window, the script file *traffic_light_if* is executed for each of the traffic light colors *red*, *green*, *yellow*, and *blue*; and the results are as follows:

```
>> traffic _ light _ if

    *******************************************************************
        *************  Traffic light condition  *****************
    *******************************************************************
```

```
Enter the first 3 letters of the following traffic light color: red,
green, yellow, others: red
ligth =
       red
           It is not safe to pass


>> traffic _ light _ if


Enter the first 3 letters of the following traffic light color: red,
green, yellow, others: gre
 ligth =
         gre
             It is safe to pass


>> traffic _ light _ if


Enter the first 3 letters of the following traffic light color: red,
green, yellow, others: yel
 ligth =
        yel
            Proceed with caution


>> traffic _ light _ if


Enter the first 3 letters of the following traffic light color: red,
green, yellow, others: blu
 ligth =
        blu
            Light is not functioning


% part(b)
% Script file: traffic _ light _ switch


disp('****************************************************************')
disp('  ***************  Traffic light condition  ****************')
disp('****************************************************************')
light = input('Enter the following traffic light color: red, green,
                                         yellow, others:','s')
switch light
case'green'
    disp( 'It is safe to pass')
case'red'
    disp('It is not safe to pass')
case'yellow'
    disp('Proceed with caution')
otherwise
    disp('Light is not functioning')
end
```

Back in the command window, the script file *traffic_light_switch* is executed for the traffic light colors—*red, green, yellow,* and *blue;* and the results are as follows:

```
>> traffic _ light _ switch


*****************************************************************
***************  Traffic light condition  *************
*****************************************************************
Enter the following traffic light color: red, green, yellow,
others: red
     light =
               red
     It is not safe to pass


>> traffic _ light _ switch


*****************************************************************
***************  Traffic light condition  ***************
*****************************************************************
Enter the following traffic light color: red, green, yellow,
others: green
     light =
               green
     It is safe to pass


>> traffic _ light _ switch


*****************************************************************
***************  Traffic light condition  ***************
*****************************************************************
Enter the following traffic light color: red, green, yellow, others:
yellow
     light =
               yellow
     Proceed with caution


>> traffic _ light _ switch


*****************************************************************
***************  Traffic light condition  ***************
*****************************************************************
Enter the following traffic light color: red, green, yellow, others:
blue
     light =
               blue
     Light is not functioning
```

See Figure 8.13.

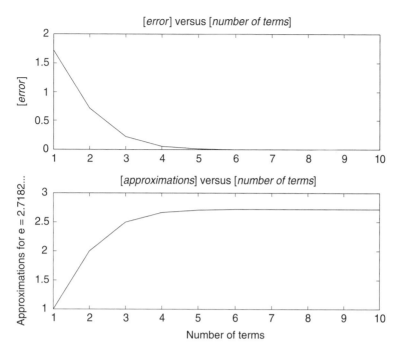

FIGURE 8.13
Plots of Example 8.7.

Example 8.8

Create the script file *voting_age* that, given the name and age of citizen A, returns a message indicating if citizen *A is eligible or not eligible to vote* (the U.S. Constitution states that all its citizens of *age* ≥ 18 are eligible to vote).

Test the script file *voting_age* for the following cases:

Mike Douglas, age 33
Cong Lee, age 43
Carlos Espinosa, age 17

MATLAB Solution
```
% Script file: voting _ age
name = input ('Enter  the citizen's full name :','s');
age = input ('Enter hers/his age in years :');
if age>=18
    disp ('**********************************')
    fprintf ('%s is eligible to vote\n',name)
    disp ('**********************************')
else
    disp ('**********************************')
    fprintf ('%s is not eligible to vote\n',name)
    disp ('**********************************')
end
```

Back in the command window, the script file *voting_age* is tested with the following data: *Mike Douglas, age 33; Cong Lee, age 43; and Carlos Espinosa, age 17.*

The results are indicated as follows:

```
>> voting _ age
```

Enter   the citizen's full name: Mike Douglas
Enter hers/his age in years : 33
****
Mike Douglas is eligible to vote
****

```
>> voting _ age
```

Enter   the citizen's full name: Cong Lee
Enter hers/his age in years : 43
****
Cong Lee is eligible to vote
****

```
>> voting _ age
```

Enter   the citizen's full name: Carlos Espinosa
Enter hers/his age in years : 17
****
Carlos Espinosa is not eligible to vote
****

Example 8.9

Given a person's age, draw a flowchart and create the script file *age_des* that returns a message, which states the person's age status according to the four categories indicated in Table 8.10.

Test the script file *age_des* for the following ages: *34, 12, 77,* and *18*.

ANALYTICAL Solution

The corresponding flow-chart is shown in Figure 8.14.

MATLAB Solution
```
% Script file: age _ des
format compact
disp('* * *AGE   activator "ON" * * * ')
disp('*****************************')
age =input('Enter the persons age:');
```

TABLE 8.10

Person's Age Status

Age	Message
age >= 65	Senior
65 > age >= 20	Adult
20 > age >= 13	Teenager
age < 13	Child

```
disp('*******************************')
 if age>=65
      disp('This person is a «Senior»')
        else
        if age >=20
            disp('This person is an "Adult"')
            else
                if age< 13
                    disp('This person is a "Child"')
                    else
                    disp('This person is a "Teenager"')
                end
          end
    end
disp('*********************************')
```

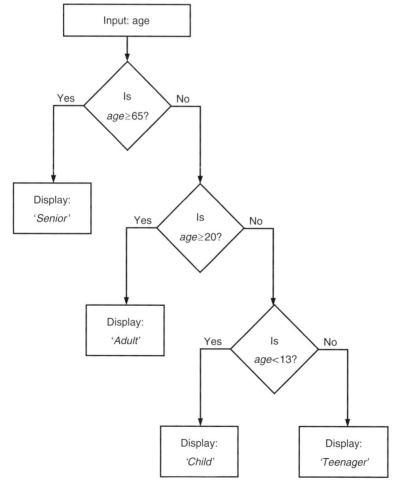

FIGURE 8.14
Flowchart of Example 8.9.

The script file *age_des* is tested for *age*: *34*, *12*, *77*, and *18*. The results are as follows:

```
>> age _ des

* * *AGE    activator "ON" * * *
*********************************
Enter the persons age: 34
*********************************
This person is an "Adult"
*********************************

>> age _ des

* * *AGE    activator "ON" * * *
*********************************
Enter the persons age: 12
*********************************
This person is a "Child"
*********************************

>> age _ des

* * *AGE    activator "ON" * * *
*********************************
Enter the persons age: 77
******************************
This person is a "Senior"
****************************

>> age _ des

* * *AGE    activator "ON" * * *
*********************************
Enter the persons age: 18
*********************************
This person is a "Teenager"
*********************************
```

Example 8.10

Given three unequal numbers *A, B,* and *C,* draw a flowchart and create the script file *order* that returns the three numbers arranged in descending order (the *sort, max,* and *min* commands are not allowed). Test the script file *order* for the following three numbers randomly chosen 10, 5, and 3 and all its possible (six) permutations.

ANALYTICAL Solution
The corresponding flowchart is shown in Figure 8.15.

MATLAB Solution
```
Script file: order
format compact;
A= input ('Enter the value of A =');
B= input ('Enter the value of B =');
C= input ('Enter the value of C =');
disp('*********************************************')
disp('The order of the given inputs: A, B and C is :')
if A>B
      if A>C
```

```
             if B>C
                 disp('A>B>C')
               else disp('A>C>B')
             end
           else disp('C>A>B')
        end
        elseif B>C
           if A>C
               disp('B>A>C')
           else
               disp('B>C>A')
           end
                 else
                     disp('C>B>A')
           end
end
disp('**************************************************')
```

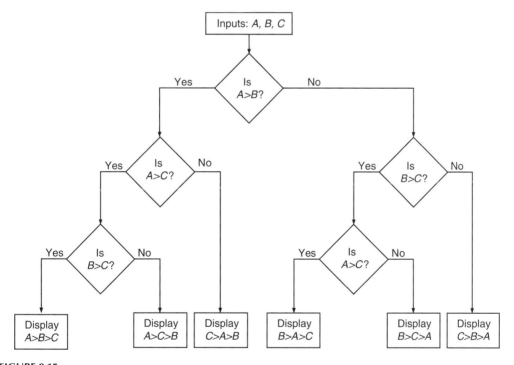

FIGURE 8.15
Flowchart of Example 8.10.

The script file *order* is tested for all possible permutations of the numbers *10, 5,* and *3.* The results are as follows:

```
>> order

Enter the value of A =10
Enter the value of B =5
Enter the value of C =3
```

```
***************************************************
The order of the given inputs: A, B and C is :
A>B>C
***************************************************

>> order

Enter the value of A =10
Enter the value of B =3
Enter the value of C =5
***************************************************
The order of the given inputs: A, B and C is :
A>C>B
***************************************************

>> order

Enter the value of A =5
Enter the value of B =10
Enter the value of C =3
***************************************************
The order of the given inputs: A, B and C is :
B>A>C
***************************************************

>> order

Enter the value of A =5
Enter the value of B =3
Enter the value of C =10
***************************************************
The order of the given inputs: A, B and C is :
C>A>B
***************************************************

>> order

Enter the value of A =3
Enter the value of B =10
Enter the value of C =5
***************************************************
The order of the given inputs: A, B and C is :
B>C>A
***************************************************

>> order

Enter the value of A =3
Enter the value of B =5
Enter the value of C =10
***************************************************
The order of the given inputs: A, B and C is :
C>B>A
***************************************************
```

Example 8.11

Draw a flowchart and create the script file *qua_roots* that, given the coefficients of the quadratic equation of the form $f(x) = ax^2 + bx + c$, returns the following:

1. Its roots
2. The corresponding message, indicating if the roots are real, repeated, or complex conjugate
3. The plot of the roots on the complex plane using zplot and pzmap for the following equation:

$$f(x) = 3x^2 + 9x + 10$$

Test the script file *qua_roots* for the following three equations (each one represents a different case):

1. $f(x) = x^2 + x - 2$
2. $f(x) = x^2 - 4x + 4$
3. $f(x) = 3x^2 + 9x + 10$

ANALYTICAL Solution

See Figure 8.16.

MATLAB Solution
```
% Script file: qua _ roots
format compact;
disp ('**********************************************************')
disp ('This program returns the roots of the quadratic equation')
disp ('of the form f(x)=ax^2+bx+c')
a = input ('Enter the value of the coefficient a=');
b = input ('Enter the value of the coefficient b=');
c = input ('Enter the value of the coefficient c=');
disp('**********************************************************')
p = [a b c];
r = roots(p);
d = b^2-4*a*c;
if   d<0
             disp ('The roots x1 and x2 are complex conjugate')
         elseif   d==0.0
             disp ('The roots x1 and x2 are real and repeated')
         else disp ('The roots x1 and x2 are real and distinct')
      end
disp ('The roots of  x1 and x2 are ='); disp(r)
disp ('**********************************************************')
num =1;
subplot(2,1,1)
zplane (num,p)
title('plot of the roots of the quadratic equation: f(x) = ax^2 +
bx + c')
grid on
subplot(2,1,2)
```

```
pzmap (num,p)
title('plot of the roots of the quadratic equation: f(x) = ax^2 +
bx + c')
grid on
```

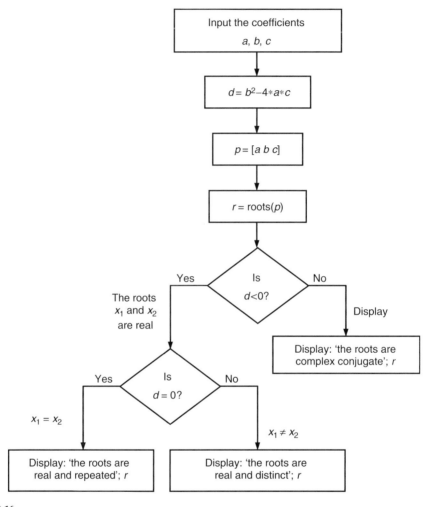

FIGURE 8.16
Flowchart of Example 8.11.

The script file *qua_roots* is tested for the following coefficients:

1. *a=1,b=1,c=−2*
2. *a=1,b=−4,c=4*
3. *a=3,b=9,c=10*

The results are as follows:

```
>> qua _ roots
```

****
This program returns the roots of the quadratic equation
of the form f(x)=ax^2+bx+c

```
Enter the value of the coefficient a=1
Enter the value of the coefficient b=1
Enter the value of the coefficient c=-2
***************************************************************
The roots x1 and x2 are real and distinct
The roots of  x1 and x2 are =
    -2
     1
***************************************************************

>> qua _ roots

***************************************************************
This program returns the roots of the quadratic equation
of the form f(x)=ax^2+bx+c
Enter the value of the coefficient a=1
Enter the value of the coefficient b=-4
Enter the value of the coefficient c=4
***************************************************************
The roots x1 and x2 are real and repeated
The roots of  x1 and x2 are =
     2
     2
***************************************************************

>> qua _ roots

***************************************************************
This program returns the roots of the quadratic equation
of the form f(x)=ax^2+bx+c
Enter the value of the coefficient a=3
Enter the value of the coefficient b=9
Enter the value of the coefficient c=10
***************************************************************
The roots x1 and x2 are complex conjugate
The roots of  x1 and x2 are =
   -1.5000 + 1.0408i
   -1.5000 - 1.0408i
***************************************************************
```

The plot of the roots of $f(x) = 3x^2 + 9x + 10$, using zplot and pzmap on the complex plane, is illustrated in Figure 8.17.

Example 8.12

Create the script file *capital_inter* that returns a tablelike format of the number of years (n) versus the capital (principal plus interest), invested in a bank account, and its corresponding *stem, stairs, bar,* and *plot* (by using the *for-end* command) of its growth, given the principal P (in $), annual interest I, and number of years n of the investment.

Test the script file *capital_inter* for the following case: $P = 1000, $I = 6\%$, and a period of $n = 10$ years.

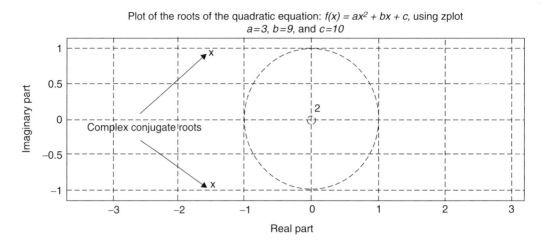

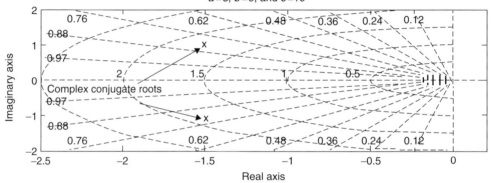

FIGURE 8.17
Plots of Example 8.11 (equation 3).

MATLAB Solution

```
% Script file: capital _ inter
P = input(' Enter the principal in dollars:')
I = input(' Enter the annual percent interest rate: ')
n = input(' Enter the number of years:')
I = I/100;
for k=1:n+1;
    F(k)=P*(1+I)^k;
end
k=0:n;
subplot(2,2,1)
stem(k,F)
title('discrete plot')
ylabel('[principal + interest]');
subplot(2,2,2)
stairs(k,F)
```

```
title('stair plot')
ylabel('[principal + interest]');
subplot(2,2,3)
bar(k,F)
title('bar plot')
ylabel('[principal+ interest]');xlabel ('# of years')
subplot(2,2,4)
plot(k,F,k,F,'s')
title('continuous plot')
ylabel('[principal + interest]');xlabel ('# of years')
disp('            * * * R E S U L T S * * *')
disp('**********************************')
disp('years        amount in $(capital+ interest)')
disp('**********************************')
[k' F']
disp('**********************************')
```

Back in the command window, the script file *capital_inter* is tested for $P = \$1000, I = 6\%,$ and $n = 10$ years. The results are as follows:

```
>> capital _ inter

 Enter the principal in dollars: 1000
    P =
         1000
 Enter the annual percent interest rate: 6
   I =
        6
Enter the number of years: 10
n =
   10
              * * * R E S U L T S * * *
**************************************************
    years        amount in $(capital + interest)

**************************************************


    ans =
          1.0e+003 *
          0.0010          1.0600
          0.0020          1.1236
          0.0030          1.1910
          0.0040          1.2625
          0.0050          1.3382
          0.0060          1.4185
          0.0070          1.5036
          0.0080          1.5938
          0.0090          1.6895
          0.0100          1.7908
**************************************************
```

See Figure 8.18.

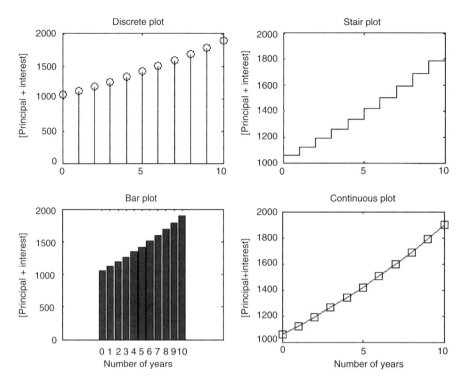

FIGURE 8.18
Plots of Example 8.12.

Example 8.13

Two bank accounts are opened simultaneously: account A with an initial capital of $1535 at an annual interest rate of 8.5% and account B with an initial capital of $2150 at an annual interest rate of 6.3%.

Create the script file *accounts _A_B* that returns the growth of each account in a table-like format and its corresponding plots. Estimate graphically using *ginput*, if and when the two accounts reach the same amount, assuming a time period of $n = 20$ years. Also indicate by means of a plot if account B outperforms account A over the given time period.

MATLAB Solution
```
% Script file: accounts _ A _ B
Account _ A=1535;
Account _ B=2150;
I _ A = 8.5;I _ B = 6.3;
n =20;
I _ A = I _ A/100;I _ B = I _ B/100;
for k=1:n;
    F _ A(k) = Account _ A*(1+I _ A)^k;
    F _ B(k) =Account _ B*(1+I _ B)^k;
    diff(k) =F _ B(k)-F _ A(k);
end
k =1:n;
figure(1)
plot(k,F _ A,'o',k,F _ A,k,F _ B,'h',k,F _ B)
ylabel('[principal + interest] of Accounts A and B');xlabel ('# of
                                                          years')
```

```
title('[growth of Accounts A and B]  vs   [# of years]')
legend('points A', 'plot _ A', 'points B','plot _ B')
disp('*************************************************')
disp('Variables when the two accounts reach the same amount')
[year _ n,amount]=ginput(1)
figure(2)
plot(k,diff,'o',k,diff)
ylabel('[difference between  Accounts B and A');xlabel ('# of years')
title('[growth of Accounts (B - A)] vs   [# of years]')
disp('************************************')
disp('                 Table indicating growth')
disp('years     Ac _ A       Ac _ B      (capital + interest)')
disp('************************************')
[k' F _ A' F _ B']
disp('************************************')
```

The script file *accounts_A_B* is executed and the results are as follows:


```
>> accounts _ A _ B


*************************************************************
Variables when the two accounts reach the same amount
 year _ n =
          16.5438
amount =
           5.9123e+003
*************************************************************
                  Table indicating growth
     years        Ac _ A        Ac _ B        (capital + interest)
*************************************************************
   ans =
   1.0e+003 *
     0.0010    1.6655      2.2855
     0.0020    1.8070      2.4294
     0.0030    1.9606      2.5825
     0.0040    2.1273      2.7452
     0.0050    2.3081      2.9181
     0.0060    2.5043      3.1020
     0.0070    2.7172      3.2974
     0.0080    2.9481      3.5051
     0.0090    3.1987      3.7260
     0.0100    3.4706      3.9607
     0.0110    3.7656      4.2102
     0.0120    4.0857      4.4755
     0.0130    4.4330      4.7574
     0.0140    4.8098      5.0571
     0.0150    5.2186      5.3757
     0.0160    5.6622      5.7144
     0.0170    6.1435      6.0744
     0.0180    6.6657      6.4571
     0.0190    7.2322      6.8639
     0.0200    7.8470      7.2963
*************************************************************
```

Observe from the table and plots that account B outperforms account A during the first *16.54* years; but over the 20-year period, account A outperforms account B by *$550*.
See Figures 8.19 and 8.20.

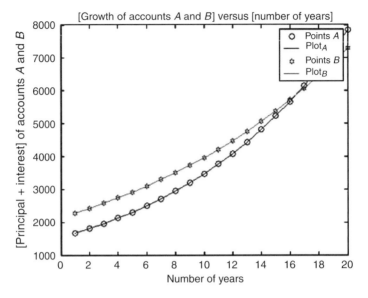

FIGURE 8.19
Plots of growth of accounts A and B of Example 8.13.

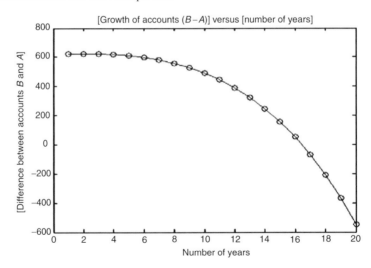

FIGURE 8.20
Plots of growth of accounts $(B - A)$ of Example 8.13.

Example 8.14

The stock value of a company XYZ taken during the closing of *15* consecutive trading days of two time periods A and B are given. Analyze the performance of the company's stock by comparing the two time periods. Draw a flowchart and create the script file *stock* that returns the following:

1. The performance plot of the stock during each period A and B.

2. Let A be the array that represents the stock values during period A and B the array that represents the stock values during period B (with 15 entries each). Determine the number of days that $\{A(n) \geq B(n)\}$ the stocks during period A outperformed or showed equal performance with respect to the stocks during period B.

3. Evaluate the number of days that the stocks during period B outperformed the stocks during period A.

4. Evaluate which period of time shows better performance. Let $C = A - B$ and $D = B - A$. Implement a simple performance mechanism by comparing the $sum(C)$ with $sum(D)$.

5. An equal valid model of performance prediction is the evaluation of the area under the curve of $magnitude[A(n) - B(n)]$ versus n, or $magnitude[B(n) - A(n)]$ versus n, for $n = 1, 2, 3, \ldots, 15$.

6. Note that the magnitude plots and areas in part 5 are the same, but with opposite sign.

7. Observe that the sign in part 6 can be used to indicate performance and the *area* may be used to indicate the level of performance.

See Figure 8.21.

ANALYTICAL Solution

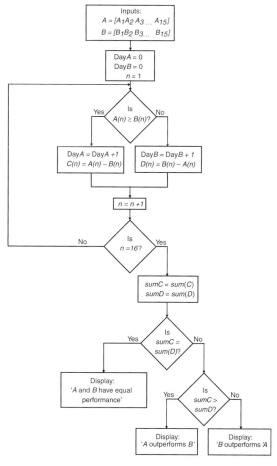

FIGURE 8.21
Flowchart of Example 8.14.

MATLAB Solution

```
% Script file: stock
format compact
A= input('Enter the 15 stock values in brackets at the closing
during period a = ');
B= input('Enter the 15 stock values in brackets at the closing
during period b =');
    if length(A) == length(B)
    else disp('A mistake was made!!!')
       disp('The input data does not have the same number of
       days closings')
    end
    if length(A) ==15
    else disp('A mistake was made!!!')
       disp('The input data does not have 15 days closings')
    end
    DayA =0;
    DayB =0;
    for n =1:15
       if A(n)>=B(n);
          DayA=DayA+1;
             else DayB=DayB+1;
       end
       C(n)  =A(n)-B(n);
       D(n)  =B(n)-A(n);
    end
    sumc = sum(C);
    sumd = sum(D);
    disp('****************************************************')
    disp('********************R E S U L T ********************')
    disp('****************************************************')
    if sumc == sumd
       disp('The stock performance during the periods A and B
                                                are equal')
    elseif sumc>sumd;
       disp('The stock during the period A outperforms the
                                                period B')
       fprintf ('Period A outperformed period B during %4.2f
                                                days\n',DayA)
       disp('from 15 days')
    else disp('***The stock during the period B outperforms
    the period A***')
       fprintf ('Period B outperformed period A during %4.2f
                                                days\n ',DayB)
       disp('from 15 days')
    end
    disp('****************************************************')
    disp(' ')
    n =1:15;
```

```
figure (1)
plot (n,A,'o',n,A,n,B,'s',n,B)
ylabel ('[Stocks values during periods A and B]');xlabel
                                          ('n in days ')
title ('[Stocks values during periods A and B] vs  [days]')
legend ('points A','stock _ A','points B', 'stock _ B')

figure(2)
subplot(2,1,1)
plot (n,C,'o',n,C)
fill(n,C,'k')
ylabel ('[period A-period B]');xlabel ('n in days ')
title ('[Difference between periods A and B] vs.  [days]')
subplot (2,1,2)
plot (n,D,'o',n,D)
fill(n,D,'k')
ylabel('[period B-period A]');xlabel ('n in days ')
title('[Difference between periods B and A] vs.  [days]')
area _ A = trapz(n,C);
area _ B = trapz(n,D);
disp ('****************************************************')
fprintf ('The area under [period A- period B] is =
                          %4.2f\n',area _ A)
fprintf ('The area under [period B- period A] is =
                          %4.2f\n',area _ B)
disp ('****************************************************')
disp ('****************************************************')
```

The script file *stock* is executed and the results are as follows:

```
>> stock

Enter the 15 stock values in brackets at the closing during
period A =
[1.8 2 2.3 3 4 5 6.7 4.5 5.6 5 3.5 4.6 2.3 3.6 4.2]
Enter the 15 stock values in brackets at the closing during
period B =
[1.3 2.5 3.3 3.9 4 5.4 4.7 3.5 3.3 5 3.25 4.36 3.3 3.86 4.4]

*********************************************************************
****************** R E S U L T ************************************
*********************************************************************
*********************************************************************
    The stock during the period A outperformed the period B
    Period A outperformed period B during 8.00 days
    from 15 days
*********************************************************************
    The area under [period A- period B] is = 1.94
    The area under [period A- period B] is = -1.94
*********************************************************************
```

See Figures 8.22 and 8.23.

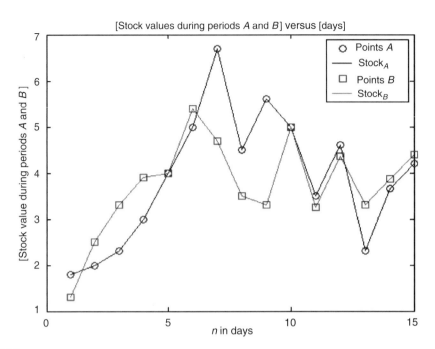

FIGURE 8.22
Stock performance plots of Example 8.14.

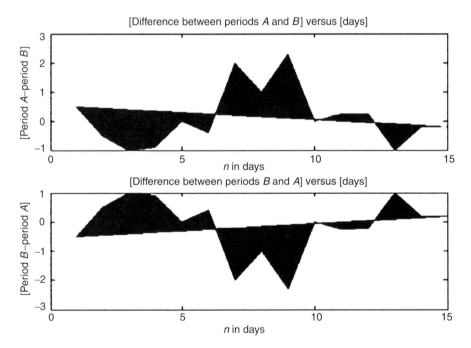

FIGURE 8.23
Performance plots of the differences between *A* and *B* of Example 8.14.

Example 8.15

Let a system polynomial equation be $y(x) = 3x^5 + (4 + k)x^4 + (5 - k)x^3 - 2kx^2 - kx + (2 + k)$, over the range $-2 \leq k \leq 2$. Evaluated k in linear increments of 0.5 represents a system, where x is its input, y its output, and k may represent a disturbance.

Create the script file *plots_roots* that returns the following plots:

a. [the real and imaginary part of the roots of $y(t)$] versus k
b. {abs[the roots of $y(x)$]} versus k
c. [the real part of the roots of $y(x)$] versus [the imaginary part of the roots of $y(x)$]

MATLAB Solution
```
% Script file: plots _ roots
figure(1)
for K=-2:0.5:2;
      Y = [3 4+K 5-K -2*K -K 2+K];
      K,r = roots(Y);
      K = K*ones(1,5);
      plot(K,real(r),'*',K,imag(r),'o' )
      hold on
end
grid on
title('Plot of [ roots of y(x) vs [k],   for-2<k<2')
xlabel('disturbance-k');ylabel('roots of y(x)')
legend('real part','imag. part')
figure(2)
for K=-2:0.5:2;
      Y = [3 4+K 5-K -2*K -K 2+K];
      K,r = roots(Y);
      k=K*ones(1,5);
      plot(k,abs(r),'o' )
      hold on
      end
      grid on
      title('Plot of [ abs(roots of y(x))] vs [k],   for-2<k<2')
      xlabel('k');ylabel('magnitude')
      figure(3)
      for K =-2:0.5:2;
          Y= [3 4+K 5-K -2*K -K 2+K];
          R= roots(Y);
          imagR=imag(R);
          realR=real(R);
          plot(realR,imagR,'*',realR,imagR)
          hold on
      end
      grid on
      title('Plot of the roots of y(x) on the complex plane
                                  for -2<k<2')
      xlabel('real axis');ylabel('imaginary axis')
```

See Figures 8.24 through 8.26.

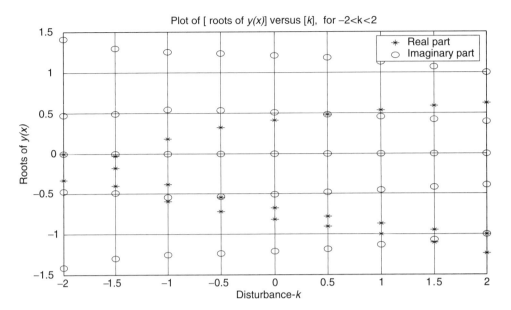

FIGURE 8.24
Plots of roots of Example 8.15.

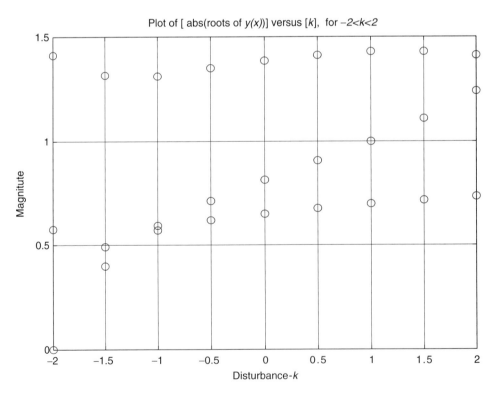

FIGURE 8.25
Plots of the magnitude of the roots of Example 8.15.

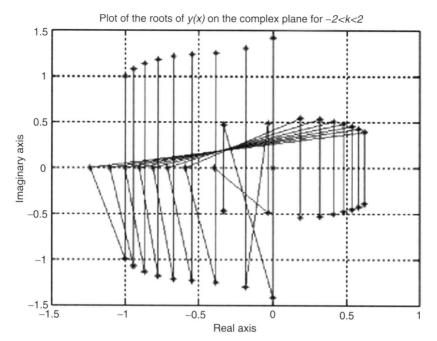

FIGURE 8.26
Plots on the complex plane of the roots of Example 8.15.

Example 8.16

Let $y_1(t) = sin(2 \pi t)$. Create the script file *waves* that returns the following plots by properly transforming $y_1(t)$:

1. $y_1(t)$ versus t
2. $y_2(t)$ versus t
3. $y_3(t)$ versus t
4. $y_4(t)$ versus t
5. $y_5(t)$ versus t
6. $y_6(t)$ versus t

over the range $0 \leq t \leq 2$ by converting $y_1(t)$ into the periodic functions $y_n(t)$, for $n = 2, 3, 4, 5, 6$, defined as follows:

1. $y_2(t) = y_1(t) = sin(2 \pi t)$, for $0 \leq t \leq 0.5$, which repeats periodically with $T = 0$ ($y_2(t)$ is known as a fully rectified wave)

2. $y_3(t) = \begin{cases} sin(2\pi t) & \text{for} \quad 0 \leq t \leq 0.5 \\ 0 & \text{for} \quad 0.5 < t \leq 1 \end{cases}$

   ($y_3(t)$ is known as a half rectified wave)

3. $y_4(t) = \begin{cases} 0.5 & \text{for} \quad y_1(t) \geq 0.5 \\ -0.5 & \text{for} \quad y_1(t) \leq -0.5 \\ y_1(t) & \text{otherwise} \end{cases}$

   ($y_4(t)$ is known as a clipped wave)

4. $y_5(t) = \begin{cases} 1 & \text{for} \quad 0 \le t \le 0.5 \\ 0 & \text{for} \quad 0.5 < t \le 1 \end{cases}$

        ($y_5(t)$ becomes a square wave with unity peak-to-peak value)

5. $y_6(t) = \begin{cases} 10 & \text{for} \quad 0 \le t \le 0.5 \\ -5 & \text{for} \quad 0.5 < t \le 1 \end{cases}$

        ($y_6(t)$ becomes a rectangular wave with a peak-to-peak value = *15*)

MATLAB Solution
```
% Script file: waves
format compact
figure(1)
t = 0:0.01:2;
Y1 = sin(2*pi.*t);
subplot (2,1,1);
plot (t,Y1)
title('[y1(t)=sin(2*pi*t)] vs. t')
ylabel('Amplitude'); axis([0 2 -1.2 1.2 ])
n = 1;
for   n =1:201;
    if Y1(n)<0;
        Y2(n)  = -1*Y1(n);
        n = n+1;
    else Y2(n)=Y1(n);
        n = n+1;
    end
end
subplot (2,1,2)
plot (t,Y2)
title ('[y2(t)] vs. t (full rectified wave)')
ylabel ('Amplitude')
xlabel ('t. in sec.') ;axis([0 2 -0.2 1.2 ])
K=1;
figure(2)
while K<202;
    if Y1(K)<0;
        Y3(K)=0;
        K= K+1;
    else Y3(K) =Y1(K);
        K= K+1;
    end
end
subplot(2,1,1);
plot (t,Y3)
title('[y3(t)] vs. t (half rectified wave)')
ylabel('Amplitude')
axis([0 2 -0.2 1.2 ])
for M =1:201
    if Y1(M)>=0.5;
        Y4(M)=0.5;
    elseif Y1(M)<=-0.5
        Y4(M)=-0.5;
    else Y4(M)=Y1(M);
    end
```

```
end
subplot (2,1,2);
plot (t,Y4)
title ('[y4(t)] vs. t   (clipped wave)')
ylabel ('Amplitude')
xlabel ('t in sec.');axis([0 2 -1 1 ])
figure(3)
for K= 1:201;
     if Y1(K)<=0;
        Y5(K) =0;
     else Y5(K) =1;
     end
end
subplot (2,1,1);
plot (t,Y5)
title ('[y5(t)] vs. t ( square wave with a unity peak to peak value)')
ylabel ('Amplitude ')
axis ([-0.1 2 -0.2 1.2 ])
for K=1:201;
     if Y1(K)<=0
        Y6(K) =-5;;
     else Y6(K) =10;
     end
end
subplot (2,1,2);
plot (t,Y6)
xlabel ('t in sec.'); ylabel('Amplitude')
title ('[y6(t)] vs. t (rectangular wave with a peak to peak value=15)')
axis ([-0.1 2 -6 11 ])
```

The script file *waves* is executed and the resulting plots are shown in Figures 8.27 through 8.29.

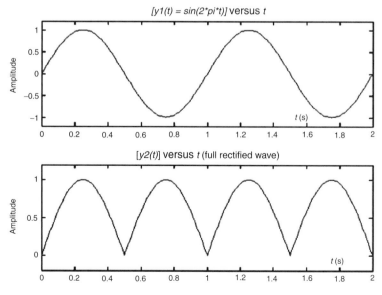

FIGURE 8.27
Plots of $y_1(t)$ and $y_2(t)$ of Example 8.16.

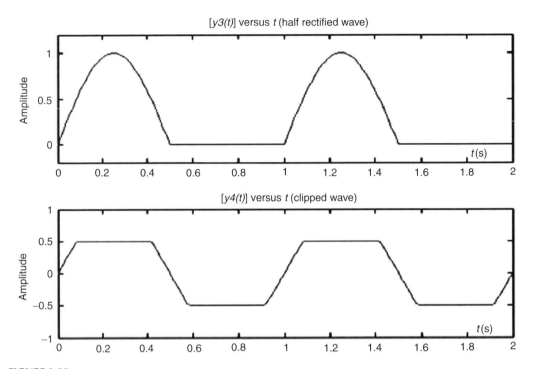

FIGURE 8.28
Plots of $y_3(t)$ and $y_4(t)$ of Example 8.16.

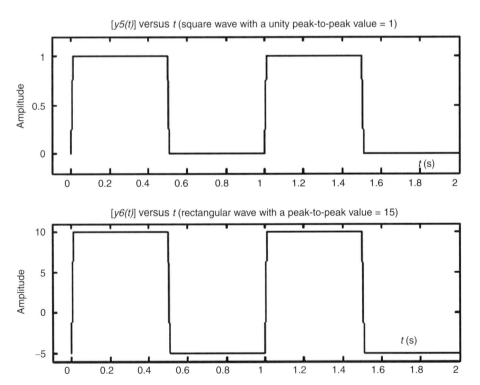

FIGURE 8.29
Plots of $y_5(t)$ and $y_6(t)$ of Example 8.16.

8.5   Further Analysis

Q.8.1   Load and run the script file *max_min* of Example 8.1.

Q.8.2   Draw a clear flowchart of the script file *max_min*.

Q.8.3   Estimate the number of times that the program is in a loop.

Q.8.4   Modify the script *max_min* in which every time a loop is executed, MATLAB returns a message indicating the loop number, starting with *1 (2, 3, 4, ...).*

Q.8.5   Modify the program so that the maximum and minimum is displayed for each loop iteration.

Q.8.6   Discuss the case when two elements in *V* are equal.

Q.8.7   Modify the script *max_min* to take into account the case in which two or more equal elements are part of *V*.

Q.8.8   Modify the script to perform the search by using the absolute values of *max* and *min*.

Q.8.9   Load and run the script file *matrix_of_ones* of Example 8.2.

Q.8.10 Draw a flowchart of the script file *matrix_of_ones*.

Q.8.11 Estimate the number of loops in the script program.

Q.8.12 Define the purpose of each loop.

Q.8.13 Modify the script to return a matrix consisting of zeros (similar to *zeros*).

Q.8.14 Modify the script to return the *eye* (square) matrix.

Q.8.15 Modify the script to return a diagonal (square) matrix with the element values of *1, 2, 3, ..., n* in the main diagonal (the *diag* command is not allowed).

Q.8.16 Load and run the script file *perm_matrix* of Example 8.3.

Q.8.17 Verify that each row sum returns the same result.

Q.8.18 Modify the script file so that the permutations of the elements are done columnwise.

Q.8.19 Load and run the program of Example 8.4.

Q.8.20 Draw a flowchart for each of the solutions (*a*, *b*, and *c*).

Q.8.21 Which solution in your opinion is the *best*. Discuss and define the term *best*.

Q.8.22 Can you implement Example 8.4 using the *if-end* statement? If possible indicate how.

Q.8.23 Load and run the script file *grett_05* of Example 8.5.

Q.8.24 Define the variable *Addx* and indicate if the command *Addx = Addx + 0* can be omitted.

Q.8.25 Redraw the flowchart for the case of estimating the number of elements in *x* greater than *0.5*.

Q.8.26 Rerun the modified script for Q.8.25 and test the program with a new array *x*.

Q.8.27 Redraw the flowchart to return the number of elements whose values are between 0.5 and 0.7 using the *if-end* statement.

Q.8.28 Code and rerun the program for Q.8.27 and test the modified program with a new sequence.

Q.8.29 How many *if-end* statements are required in Q.8.28?

Q.8.30 Modify the script file of Example 8.5 to obtain the sum of all the elements smaller than 0.5 in addition to the number of elements.

Q.8.31 Load and run the script file *approx_exp* of Example 8.6.

Q.8.32 Draw a clear flowchart of *approx_exp*.

Q.8.33 Define the variables *app* and *approx*.

Q.8.34 If the desired number of terms of the approximation is 10, what is the total number of iterations executed?

Q.8.35 Modify the script to evaluate the number of terms required to obtain an approximation with an error of less than 1%.

Q.8.36 From the plots (Figure 8.13) it seems that a five term approximation is appropriate. If that is the case, then estimate the error in terms of the absolute and percentage values.

Q.8.37 Load and run each of the script files of Example 8.7, *traffic_light_if* and *traffic_light_switch* {parts (a) and (b)}.

Q.8.38 Draw a flowchart for each of the preceding script files.

Q.8.39 Modify the script files to accept inputs that are not case sensitive. Test the modified files with your own input strings.

Q.8.40 Modify the script file *traffic_light_if* to include the full traffic light colors as inputs.

Q.8.41 Compare the two script file solutions {part (a) with part (b)}.

Q.8.42 Modify each script file so that the output message is repeated 10 times when the traffic light is red.

Q.8.43 Load and run the script file *voting_age* of Example 8.8.

Q.8.44 Draw a clear flowchart of *voting_age*.

Q.8.45 Modify the program so that the input includes gender (male or female).

Q.8.46 Modify this program to return (display) also the voting place, such as females in booth #A and males in booth #B.

Q.8.47 Load and run the script file *age_des* of Example 8.9.

Q.8.48 Modify the flowchart and script to include the additional category of *toddler* that corresponds to the age range $age \le 2$. Test the modified script for $age = 1$.

Q.8.49 Modify the program to include the party's name as an additional input.

Q.8.50 Modify the display to replace *This person* by the person's name (Q.8.49).

Q.8.51 Load and run the script file *order* of Example 8.10.

Q.8.52 Redraw the flowchart and modify the program that would allow for equal values.

Q.8.53 Rerun the program with the modifications and test it for $A = 1$, $B = 2$, and $C = 2$.

Q.8.54 Modify the flowchart and scripts of Example 8.10 with the objective of returning A, B, and C in descending order.

Q.8.55 Modify and redraw the flowchart of Example 8.10 for the case of four unequal numbers.

Q.8.56 Load and run the script file *qua_roots* of Example 8.11 for the case $y(x) = x^2 - x - 6$.

Q.8.57 Test the script file *qua_roots* of Example 8.11 for the following cases:

a. $y_1(t) = x^2 - 6x - 7$

b. $y_2(t) = x^2 - 6x + 9$

c. $y_3(t) = x^2 - 6 + 10$

Analyze and verify the results for each of the above cases.

Q.8.58 Verify the values obtained in Q.8.57 by hand.

Q.8.59 Verify Q.8.57 by obtaining the plots for each equation.

Q.8.60 Load and run the script file *capital_inter* of Example 8.12 for $P = \$2500$, $I = 7.15\%$, and $n = 12$ years.

Q.8.61 Assuming that monies are withdrawn at the rate of $60 annually at the end of every year, modify the program that returns a table and plots of the balance in the account over a 10-year period.

Q.8.62 Load and run the script file *accounts_A_B* of Example 8.13.

Q.8.63 Modify the script to return automatically the number of years when the two accounts reach the same value (the command *ginpu* is not allowed).

Q.8.64 Load and run the script file *stock* of Example 8.14.

Q.8.65 Analyze the given flowchart. Does the script *stock* agree with the given flowchart?

Q.8.66 Define the purpose of the variables *DayA*, *DayB*, and *n*.

Q.8.67 Define and discuss how the performance of the stock is evaluated. Discuss the model and state if you agree.

Q.8.68 Redraw the flowchart and include the *area* to measure performance.

Q.8.69 Modify the program so that the average value of the stocks over each period (*A* and *B*) is used to evaluate their performance.

Q.8.70 Define performance now in terms of areas (divided by 15). Modify the flowchart and script to include this new approach.

Q.8.71 Compare the performance models.

Q.8.72 Load and run the script file *plots_roots* of Example 8.15.

Q.8.73 Estimate the number of times the command *roots(y)* is executed.

Q.8.74 Estimate the number of times the plot command is executed.

Q.8.75 Draw a clear flowchart of the script of Example 8.15.

Q.8.76 Load and run the script file *waves* of Example 8.16.

Q.8.77 Draw a flowchart for the script file of Example 8.16.

Q.8.78 Replace the *while-end* command by the equivalent *for-end* command and rerun the script *waves*.

Q.8.79 Modify the script file of Example 8.16 to return $y_7(t)$ defined below as a plot:

$$y_7(t) = \begin{cases} y_1(t) & 0 < t \le 0.5 \\ -1 & 0.5 < t \le 1 \\ 1 & 1 < t \le 1.5 \\ y_1(t) + 2 & 1.5 < t \le 2 \end{cases}$$

8.6 Application Problems

P.8.1 Let $A = [-2\ 0\ 4\ 8\ 1]$ and $B = [-1\ 1\ 0\ 7\ 1]$.
Evaluate the following expressions by hand and using MATLAB:

a. $C = \sim A \,\&\, B$

b. $D = A \,\&\, \sim B$

    c. $E = \sim(A \ \& \ B)$

    d. $F = \sim A/B$

    e. $G = \sim X \ or \ (A, \sim B)$

    f. $H = (A > B) \ \& \ (A = = B)$

    g. $J = \sim(A = = B)$

P.8.2   Given

$$A = \begin{bmatrix} 0 & 1 \\ 2 & 3 \end{bmatrix}, B = \text{'a b c'}, C = 0, \quad \text{and} \quad D = [1\ 2\ 3\ 4]$$

Anticipate the display and verify your answers after executing the following commands:

    a. *all(D)*

    b. *find(A)*

    c. *isinf(A)*

    d. *ischar(B)*

    e. *isnumeric(B)*

    f. *exist(C)*

    g. *any(D)*

    h. *isreal(A)*

    i. *isempty(A)*

    j. *isempty(C)*

    k. *isnan(A./C)*

P.8.3   Evaluate and determine if the following pair of expressions are equivalent:

    a. $(a==b)|(c==d)$

       $\sim(a==b)|\sim(c==d)$

    b. $(a==b) \ \& \ (c==d)$

       $\sim(a==b) \ \& \ \sim(c==d)$

    c. $(a==b) \ \& \ \sim(c==d)|(d==a)$

       $(a==b)|(c==d) \ \& \ (d==a)$

    d. $(a==b) \ xor \ (c==d) \ \& \ (d==a)$

       $(a==b) \ \& \ (c==d) \ xor \ (d==a)$

Assign arbitrary values to *a*, *b*, *c*, and *d*; then test and verify your answers using MATLAB.

P.8.4   Indicate for what ranges of *x* are the following expressions true:

    a. $a = (x < 15) \ \& \ \sim(x >= 25)$

    b. $b = (x >= 2315) \ \& \ (x < 33)$

    c. $c = (x >= 0)|(x <- 3)$

P.8.5   For what ranges of *x* are the expressions given in P.8.4 false.

P.8.6   Evaluate how many times each loop is executed and what are the resulting loop variables after executing each of the following segments of programs:

```
a. for i = 1:2:25
        y = i + 2
    end

b. for j   = -1:2:26.
        k = (j) ^ 2
    end

c. y = 2; z = 25.
    while  y > z
            x = y - 25
            y = y + 2
        end

d. for i = [3,7,9,13]
        d = i.^2
          end

e. for i = 1:4
        for j = 1:3
                A(i,j) = i*3/j
        end
    end

f. n = 2, y(1) = 1,
    while n<4
            y(n) = n*2
                n = n+1
        end

g. n =2,m=5,
    if n>2
        y = n*3
    elseif n<2
            z=m*n
    end

h. n =0
    for x= linspace(0,2*pi,100)
            if x>pi
                n=n+1
                y(n)=sin(x)
            elseif x>2*pi
                    n=n+1
                    y(n)=-1
            else n=n+1
                    y(n)=1
    end
```

P.8.7   Draw clear flowcharts for each segment of P.8.6.

P.8.8   Anticipate the display for the following short programs and verify your answers using MATLAB:

```
a. for  x = 1: 10
        disp('This is a loop')
    end
```

```
b. y = 0; z = 10;
   if y < z
        disp('Loop =', y)
        y = y+1
   end
```

```
c. for i = 1:5
      for j = 4
          A(i,j) = (i*j)^2
          disp(A(i,j))
      end
   end
```

```
d. for k = 0:pi/10: 2*pi
      a(k) = k*sin (k)
      b(k) = -k*cos(k)
   end
   disp(a(k)' b(k)')
```

P.8.9   Draw clear flowcharts for each program of P.8.8.

P.8.10 Create a script file that returns the message *Matlab is fun*, 10 times.

P.8.11 Create a script file that, given a number *N*, returns a message indicating if *N* is a multiple of 7 and 11 or both.

P.8.12 Draw a flowchart for the program of P.8.11.

P.8.13 Draw a flowchart and write a program (using conditional statements where the commands, not allowed *cumprod* or *prod*, are not allowed) that returns n! (factorial). Test your program for *n* = 6, 15, and 34.

P.8.14 Evaluate the following series:

$$\sin(x) = x - \frac{x^3}{3!} + \frac{x^5}{5!} - \frac{x^7}{7!} + \frac{x^9}{9!} - \cdots \frac{x^n}{n!} \quad \text{for} \quad n = 1, 3, 5, 7, \ldots, \text{odd}$$

and

$$\exp(x) = 1 + x + \frac{x^2}{2!} + \frac{x^3}{3!} + \cdots + \frac{x^n}{n!} \quad \text{for} \quad n = 1, 2, 3\ldots$$

for *x* = 0:5 and *n* = 25 in a tablelike format using the *for-end* and *while-end* commands.

P.8.15 Draw a flowchart, test and verify if the results obtained in P.8.14 are correct. Estimate in each case the error.

P.8.16 Draw a flowchart and write a MATLAB program that returns *log(x)* for *x* = 1:10 in a tablelike format, using
   a. *implied looping*
   b. *for-end command*
   c. *while-end command*

P.8.17 Write a MATLAB program that uses the *for-end* and *while-end* commands that return the plots of the following functions:
   a. $f(t) = \frac{1}{\pi} t + 1$, over the domain $-\pi < t < 0$
   b. $f(t) = \cos(t)$, over the domain $0 < t < \pi$
   c. $f(t) = \frac{1}{\pi} t - 2$, over the domain $\pi < t < 2\pi$

P.8.18 The *Fibonacci* sequence is defined by the following recursive equation: $F_n = F_{n-1} + F_n$, for $n = 2, 3, 4, \ldots$, where $F_o = 0$ and $F_1 = 1$. Write a program that returns the nth *Fibonacci* coefficient given n.

P.8.19 Write a program that returns the sum of the *Fibonacci* sequence for any given n.

P.8.20 Draw a flowchart and write a program that returns the number of required terms resulting in the sum of the *Fibonacci* sequence closest to *1000*.

P.8.21 Let $A(i, j) = (i.j)$ for the case of a *3 by 5 matrix*. Draw a flowchart and write a program that returns the matrix A, using the *for-end-* and *while-end* commands.

P.8.22 Draw a flowchart and write a program (using conditional branching) that, given four unequal numbers, returns the smallest- and largest number. The *sort, max,* and *min* commands are not allowed.

P.8.23 Draw a flowchart and create a script file that, given the number of hours worked during a week and the hourly pay for a given employee, returns the weekly gross salary. The salary is computed in the following way:

a. The first 35 hours are paid according to the agreed hourly salary

b. Overtime is paid for the hours worked in excess of the first 35 h, but less than 50 hours at a rate of one and a half the base pay

c. The hourly rate is twice the base rate when the weekly number of hours worked exceeds 50 h

Test the script for the following case:

Base hourly rate at $13.45, and the weekly number of hours worked 27, 38, 44, 55, and 62.

P.8.24 Draw a flowchart and write a program that, given a (telegram) message, returns its cost C. The cost C is calculated in the following way:

a. A flat $4.15 for the first *100* characters or less

b. $0.05 for any additional character

P.8.25 Evaluate and plot the absolute value of the roots of the equation $y = x^3 - x + (2 + k)$, over the range $k = -1{:}0.1{:}1$, using

a. The *while-end* loop

b. The *for-end* loop

c. The *implied* loop

P.8.26 Given a positive number A and an integer B, draw a flowchart and write a program that evaluates

a. $A{*}B$ without using *

b. A/B without using /

c. A^B without using ^

P.8.27 The variance *VAR* for n samples is defined as

$$VAR = \frac{1}{n-1}\left[\sum_{k=1}^{n} x_k^2 - n\bar{x}_i^2\right] \quad \text{for } n > 1$$

$$\text{where } \bar{x}_n = \frac{1}{n}\sum_{k=1}^{n} x_k$$

Let $x = [1\ 3\ 2.1\ 3\ 4\ 6\ 8\ 9.1\ 4\ 5\ 9.2\ 2]$

Draw a flowchart and write a program that returns the plots of *VAR* versus n over $3 \leq n \leq 12$.

P.8.28   The equations of the geometric mean (GM), RMS, and harmonic mean (HM) for n samples are defined as

a. $GM = \sqrt[n]{\prod_{i=1}^{n}(x_i)}$

b. $RMS = \sqrt{\dfrac{1}{n}\sum_{i=1}^{n}(x_i)^n}$

c. $HM = \dfrac{n}{\sum_{i=1}^{n}(1/x_i)}$

For the case of x given in P.8.27, draw a flowchart and write a program that returns GM, RMS, and HM using the *for-end* and *while-end* statements.

P.8.29   Implement the MATLAB commands *sum(x)* and *cumsum(x)* using the *for-end* and *while-end* commands. Draw a flowchart and test the program using the array x defined in P.8.27.

P.8.30   Implement the MATLAB command *diag(A)*, for a given square matrix A, using the *for-end* and *while-end* statements.

P.8.31   Implement the MATLAB command *find(B)*, where B is an arbitrary array using the *for-end* and *while-end* statements.

P.8.32   Given $g(t) = 5\,sin(2\pi t)$, draw a flowchart and write a program that returns the plot of $f(t)$ versus t over the range $0 \leq t \leq 5$, where

$$f(t) = \begin{cases} 5sin(2\pi t) & \text{for } |g(t)| < 2.7 \\ 2.7 & \text{for } |g(t)| > 2.7 \end{cases}$$

P.8.33   Repeat P.8.32 for the following cases:

a. $f(t) = \begin{cases} 5\,sin(2\pi t) + 2\,cos(\pi t) & \text{for } |g(t)| < 2.7 \\ 2.7 & \text{for } |g(t)| > 2.7 \end{cases}$

b. $f(t) = \begin{cases} tsin(2\pi t) & \text{for } |g(t)| < 2.7 \\ 2.7 & \text{for } |g(t)| > 2.7 \end{cases}$

P.8.34   Two bank accounts A and B are opened simultaneously: account A with an initial deposit of $1300 at an annual interest rate of 8.25% and account B with an initial deposit of $2200 at an annual interest rate of 6.50%. Write a program that returns, when and if the two accounts reach, the same amount (in term of years, months, and days), assuming that no withdrawals or additional deposits are ever made.

P.8.35   Mr. X opens a bank account with $5000 at an annual interest rate of 5.50%. Mr. X plans to deposit at the end of every year an additional $1000. How many years are required for the account to reach $250,000 if no withdrawals are ever made. Write a program and draw a flowchart that returns the plot showing the growth of the account over time.

9

Files, Statistics, and Performance Analysis

Man's mind, once stretched by a new idea, never regains its original dimensions.

Oliver Wendell Holmes

9.1 Introduction

The command window uses the interactive mode. In this mode, each instruction once entered is processed and a response is returned before a new instruction is processed.

Another way to process MATLAB® instructions or run a program is to save a program in a file, and then execute the file. The file that contains a program can be called by typing the filename and pressing the <enter> key, while in the command window. The instructions in the file are then executed, line by line, in sequential order, just as they would be by entering them at the command window. This approach is particularly useful when a set of instructions in a program is used repeatedly in many places of the program. In this case, the interactive mode is not adequate or efficient.

A more efficient way is to store a block of frequently used instructions in a file, called either a script or function file, and execute each file as a command when needed. Of course, these files are stored in a permanent place and can be accessed when desired. This is not a new topic. Recall that script and function files were introduced in Chapter 2 and used in different chapters of this book. In this chapter, M-files are revisited and further discussed.

Script files are used to store data and/or instructions, whereas function files are designed by the user to return a calculated function similar to the built-in MATLAB functions—*sort(x), max(x), abs(x), cos(x), log(x)*, etc.

Recall that script files as well as function files are M-files. These files (M-files) are similar to subroutines in the old traditional computer languages such as Fortran or Basic. Let us review and summarize the main purpose, goal, and objectives of the M-file.

1. A long complex program can be made more manageable by breaking up the program into smaller and simpler modular segments. Each of these segments can constitute an M-file.

2. Before starting the coding of a program, an experienced programmer draws a flowchart (Chapter 1) indicating and identifying the steps and variables involved and grouping some commands into possible M-files. Sometimes, it is also useful and helpful to describe the M-files using simple English sentences before starting the process of program coding.

3. It is always recommended to choose variables and filenames that best describe and identify their purpose. Also, comments should be included to clarify, document, and explain the step(s) involved in a program.

4. The partitioning or modular formatting of a program makes the program in general more readable, logical, easier to follow, and better-structured and organized. The segmentation as well as the planning of a program involves some level of experience (recall the heuristics from Chapter 1), a task learned best by doing and not necessarily by just reading.

5. Error detection, error correction, maintenance, and upgrading of programs can be made easier by testing the individual modules with simple inputs and by providing warning and error checks at appropriate points.

6. By constructing a modular program in terms of blocks of commands, the program can easily be updated, modified, maintained, and can grow and be improved by incorporating new blocks and changing others.

7. By using blocks or M-files, the end user or programmer can build up a library of useful modular functions that can be used and reused in other applications.

8. The segments, blocks, or modules can make up a good portion of a program. These segments can be written and tested separately by different programmers, located in different geographic locations, since each segment is completely independent from one another.

9. In summary, by breaking up the program into segments, the following benefits are evident:

   a. The program can be shorter, user-friendly, readable, logical, and compact.

   b. The segment may result in an over all economy of codes and labor, promoting code reuse.

   c. In general, the overall program would be more efficient. MATLAB M-files are similar to the source code in C or Fortran. When a MATLAB function is used for the first time, each instruction is interpreted or compiled into internal pseudocodes, an action called parsing. MATLAB saves the parsed version of a function in a p-file for later use, saving valuable processing time especially when a function is repeatedly called.

   d. The modular M-file structure and organization is also used by MATLAB for its own management and control. For example, when MATLAB is activated, two files are called and executed. They are *matlabrc.m* and *startup.m*, and when a session is terminated, MATLAB calls *finish.m* to execute the *exit* or *quit* commands. The main objective of these files *(matlabrc.m* and *startup.m)* is to set the default features like format, color, and access to the figure window, whereas the *finish.m* file confirms the quitting action by way of a dialog menu.

This chapter deals with files; file organization and addresses; the MATLAB search file path (to access a given file); recommendations about file structure; and in general the many commands associated with file creation, modification, deleting, existence checking, storing, loading, etc.

The creation and usage of special MATLAB files are also introduced and discussed. Statistical performance analysis, performance techniques, and tools to improve the execution efficiency, error detection, and dependencies are also introduced and discussed. Many examples solved in previous chapters are revisited with different objectives in mind—to present an alternate solution in some cases; a more efficient solution in other cases; show the techniques and quantize the efficiency of a solution in some cases; just to compare the structure, presentation, or a new approach to an old problem in others. Many cases are used to develop the analytical tools, techniques, equations, graphs, and models to evaluate, measure, estimate, and represent a variety of statistical data.

Some of the revisited problems are

a. The solution of a second-order equation
b. The capital–interest problem is generalized (to include annual, semiannual, weekly, daily, and continuous interest rates)
c. The solution of a set of linear equations with a disturbance (over a range)
d. Area estimation
e. Statistical data analysis (HM, variance, deviation, etc.)
f. Performance analysis of different programming implementations (estimating *cputime*, *elapse-time*, etc.)
g. Function file versus script file, comparisons, and recommendations
h. Function file analysis (using the profiler)
i. Histogram and pie representation of data
j. Numerical and computation performance efficiencies of sparse versus full matrices

Since this is the last chapter of *Practical MATLAB® Basics for Engineers*, an effort was made to use this chapter to review and revisit the type (or classes) of problems most often encountered by students and professionals, when they are first exposed to MATLAB.

9.2 Objectives

After reading this chapter, the reader should be able to

- Define the different file types
- Know the meaning and application of the different file types
- Know the difference between ASCII and binary files
- Learn the computer terminology and meaning such as loading, saving, and storing
- Know the file format and syntax
- Review the steps involved to create, save, and run a script file
- Know how to use the MATLAB editor
- Know how to copy and delete a file
- Know how to open and modify a file
- Know how to add and delete a file
- Know how to use the *diary* file to record the workspace activity
- Know what is a mat file
- Revisit script and function files
- Know in what directory a file is saved
- Know the path file search
- Determine or alter the path file search

- Create, add, and remove directories from the file search path
- Know the difference between global and local variables
- Set and test a global variable
- Understand how files are executed and converted into pseudocodes (parsing)
- Understand the meaning of a p-file
- Know the directories and the current directory
- Organize the directories into subdirectories
- Know how the information search is done when executing the *lookfor* and *help* commands
- Revisit the control commands used in files (*error, warning, and keyboard*)
- Know the statistical methods, techniques, and models used in measuring performance
- Know the performance commands
- Estimate the time required by MATLAB to evaluate standard and non-standard operations
- Estimate the best techniques and practices in the construction of efficient programs

9.3   Background

R.9.1   A file may contain a program, a set of instructions, or data that are stored in a storage computer media such as a hard disk, floppy disk, or CD-ROM.

R.9.2   Files can be classified according to their contents as

  a. Data files

  b. Program files

  c. Executable files

  d. Source files

  e. Batch files

R.9.3   Data files contain numbers, graphs, words, or pictures. These files can be viewed, edited, saved, or printed, and they may also constitute the input to other programs or files.

R.9.4   Executable files are files that contain sequence of instructions that tell the computer how to perform a particular task. Examples of executable files are the programs that make up the operating system, application software such as MATLAB, and icons that accessed applications in the Microsoft (MS) Window environment, by a click of the mouse.

R.9.5   Source files are computer programs that are translated into computer code instructions that can run under an application software.

R.9.6   Batch files are source files that must be translated before they can be executed by the application software. Frequently, batch files are used by the operating system to customize the user's computer system during the initialization process. For example, the action in an IBM-compatible computer when executing the batch file *autoexec.bat*.

R.9.7   Files are stored or saved in storage devices identified by the computer software with a letter such as

   a. *A* or *B*, which identify a 3.5 in. floppy disk

   b. *C*, which identifies the hard drive

   c. *D, E, ..., Z*, which identify additional storage devices

R.9.8   The process of storing a file in a computer storage facility is referred as *saving*.

R.9.9   The process of accessing a file with the purpose of executing, or reading its contents is referred as *loading*.

R.9.10   The process of reading a file is commonly referred as *opening* a file.

R.9.11   Files are identified by a format consisting of a filename followed by a dot (.) and suffix (.*m*), referred as its extension.

R.9.12   The filename consists of characters that begin with a letter, uppercase or lowercase, followed by any letter, number, or underscore (similar to a MATLAB variable name).

R.9.13   The filename's length is restricted by

   a. The MATLAB software allows up to 31 characters

   b. The MATLAB software allows 19 characters for the case of a script file

   c. The user's operating system imposes restrictions, usually choosing the shortest length of the restriction imposed by the various software applications used

R.9.14   A filename must be unique, different from any MATLAB function, another M-file, or a variable name.

R.9.15   Since MATLAB searches first for variables in the workspace (and if a file is named the same way as a variable), once the variable is found, the search is stopped and MATLAB never accesses the intended file.

R.9.16   Recall that the command *exist('filename','file')* checks for the existence of *filename* in the *file* or *dir* and returns a 0 if the file *filename* does not exist and a 2 if it does exist.

R.9.17   In general, a filename is followed by a dot (.) and one to three characters that constitute its extension. The extension of a filename further serves to identify and describe its content. Table 9.1 illustrates frequently used file applications and their extensions.

R.9.18   Files are stored in directories, also referred as folders.

R.9.19   The location of a file is given by an address referred as its path, consisting of *drive:\ application*. For example, *C:\MATLAB* (drive C, application MATLAB).

TABLE 9.1

File Types and Extensions

File Type or Application	Extension
MATLAB (script or function)(ASCII)	.m
MATLAB (binary)	.mat
MATLAB (C or Fortran)	.mex
MATLAB (parsed)	.p
Text	.txt
MS Word document	.doc
Compressed	.zip
Executable	.exe
Sound recording	.wav
Batch	.bat

R.9.20 To access a file, MATLAB must know its precise location. The path tells MATLAB what directories to search for accessing a particular file.

R.9.21 Directories may have subdirectories. The path to a subdirectory is the path to a directory followed by \ and the name of the subdirectory. The objective of dividing a directory into subdirectories is to create a better-organized environment. For example, a directory can be *courses* and the files in the subdirectory may be academic subjects such as Mathematics, English, and Physics. For example, the file path to *math.txt* (mathematics) would then be C:\courses\math.txt. The preceding command (path) states that the subdirectory *math.txt* is in the directory *courses*, which is located in drive C.

R.9.22 The word *path* is a MATLAB-reserved filename, which contains the path of all the directories that are automatically included in the MATLAB's search. The *path* provides the operating system with the routing information required to find a desired file. The command *path* returns the MATLAB directory path. The MATLAB search path can be modified by clicking file from the command window and by choosing *set path ...* to add a path.

R.9.23 Some MATLAB versions automatically create the directory *work* as the default directory where all the M-files created by the MATLAB are automatically saved.

R.9.24 The current directory and its path is displayed at the top of the command window's toolbar in a small window clearly identified by

$$current\ directory:C\backslash MATLAB\backslash\ ...$$

R.9.25 The MATLAB command *addpath dir_A dir_B dir_C* adds the directories *dir_A*, *dir_B*, and *dir_C* automatically to the MATLAB search path.

R.9.26 The MATLAB command *rmpath dir_A dir_B dir_C* removes the directories *dir_A*, *dir_B*, and *dir_C* from the MATLAB search path.

R.9.27 The MATLAB command *mkdir('dir_new')* sets the directory *dir_new* as the current directory.

R.9.28 The MATLAB command *cd* returns the current directory. The general command *cd A* changes the current directory to *A*.

R.9.29 The MATLAB command *pwd* returns the current working directory.

R.9.30 The commands *dir* or *ls* return a list of all files in the current directory. The more general command *dir C*, for example, returns a list of all the files in the C directory.

R.9.31 The command *what* returns a list of all MATLAB files in the current directory. The more general command *what C*, for example, returns a list of all the files in the C directory.

R.9.32 The command *type filename* returns the content of the file *filename*. For example,

$$>> type\ filename.m\ \%\ returns\ the\ file\ called\ filename$$

R.9.33 Recall that the sequence of instructions to access a file from the command window are given by

File → *Open* → *filename* (opens the file *filename* for editing purposes).

The same sequence of instructions *File* → *New* → *M-file* (opens a blank edit window and a new file can be created).

Information can then be entered and stored by following the sequence: *File →
Save as* (type a *filename*) → <enter> key.

Recall that this process is not new and should be familiar to the reader since
it was introduced in Chapter 2 and used frequently throughout this book. It is
stated only for completeness and review.

R.9.34 For example, to create the test file *test_file*, the following actions are taken (while
in the command window): *file → new → M-file →* opens the edit window, and the
following MATLAB statements are entered (just to illustrate the process):

```
% Script file: file:test_file
disp('*********************************************')
disp('This is a script test file called test_file')
disp('It contains no specific program or data ')
disp('*********************************************')
```

To save the file *test_file*, follow the sequence *File → Save as →* type *test_file*
(in the dialog box displayed) followed by the <enter> key.

R.9.35 The command *edit* activates (opens) a blank edit window while in the command
window.

R.9.36 The command *edit filename.m* opens the file *filename* while in the command
window.

R.9.37 The command *which filename.ext* returns the path to the file *filename*, if this file is in
the current working directory or on the MATLAB path.

For example, the path to the just created file *test_file* is returned by executing

```
>> which test_file
```

```
C:\MATLAB6p1\work\test_file.m
```

R.9.38 The command *inmem* returns a list of the M-function currently in memory.

For example,

```
>> inmem

ans =
    'javachk'
    'iscellstr'
    'edit'
    .........
    'pathdef'
    'matlabrc'
```

R.9.39 The command *copyfile (filename_old, filename_new)* copies the contents of *filename_
old* into the new file *filename_new*.

R.9.40 The command *copyfile (filename, X)* copies the content of the file *filename* into directory *X*, where *X* can be any storage devices such as *A, B, C, ..., Z*.

R.9.41 The command *movefile (source, destination)* is used to move a file or directory to a
new destination.

R.9.42 The command *delete filename.m* deletes (erases) the file *filename.m* from the
computer memory.

R.9.43 A file is created when instructions or data are typed while in the editing mode. The
typed information is stored in memory after a valid filename or a default untitled
name is assigned, after the saving process is executed.

R.9.44 The command *fopen (filename)* opens the existing file *filename* for reading purposes. *Filename* can be opened in binary mode (the default) or in text mode. The more general form is *fopen (filename, mode, format)*, where *mode* and *format* are defined as follows:

The *mode* can include one of the following arguments:

'r'	Read
'w'	Write
'a'	Append
'r+'	Read and write
'w+'	Truncate or create for read and write
'a+'	Read and append (create if necessary)
'W'	Write without automatic flushing
'A'	Append without automatic flushing

To open a file in text mode, add *'t'* to the mode, for example, *'rt'* and *'wt+.'*
The *format* can include one of the following arguments:

'native' or *'n'*—local machine format—the default

'ieee-le' or *'ieee-be'*—IEEE floating point

'vaxd' or *'vaxdg'*—VAX floating point

'cray' or *'c'*—Cray floating point

'ieee-le.l64' or *'a'* or *'s'*—IEEE floating point with 64 bits

R.9.45 The command *[A, num_ele] = fread (fid)* reads binary data from the opened file (*fid = fopen filename*) into matrix *A*. The returning argument *num_ele* represents the number of elements read successfully.

R.9.46 The work performed in the command window during a MATLAB session, consisting of commands as well as responses (excluding the figure window), can be recorded by typing the command *diary* followed by pressing the <enter> key. The interactive session is then stored in an ASCII file named *matlab.mat*.

R.9.47 The recording information into a *diary file* can be controlled by the commands *diary off* or *diary on* to stop or start recording.

R.9.48 The *diary* file can be retrieved by typing the command *load*, while in the command window. MATLAB returns a message indicating if the file *matlab.mat* is not found.

R.9.49 The command *diary* creates the default file *matlab.mat* during each session, meaning that the same filename will be reused and rewritten during each session.

R.9.50 The command *get (o, 'diary')* returns the status of the *diary* file.

R.9.51 A standard and unique file can be created with all the characteristics of the diary file by entering the command *diary diaryname*. The file *diaryname* will record all the commands and responses during an interactive session, while at the command window.

R.9.52 The interactive session can be saved with the filename *diaryname* by entering the command *save diaryname* (a prefix can be included to indicate the data type).

R.9.53 The filename *diaryname* is saved as a *mat-file* (extension.mat) in binary format.

R.9.54 The command *get (o, 'dairyfile')* returns the status of *diaryfile*.

R.9.55 The file created by *diary* has similarities and differences with respect to the file created by *dairy diaryname* (both files record a MATLAB session but using different data type).

R.9.56 The command *load diaryname* retrieves the file *diaryname.mat* and all the workplace variables of the previous MATLAB session are restored in the active workspace.

R.9.57 The command *save filename A, B, C* saves the selected variables *A, B,* and *C* from an interactive session (after pressing the <enter> key).

R.9.58 The command *load filename* (without specifying any variable) restores the variables *A, B,* and *C* previously saved in *filename.*

R.9.59 The command *save* saves all the workspace variables in the default file *matlab.mat.*

R.9.60 The command *load* loads or restores all the workplace variables previously saved by the default file *matlab.mat.*

R.9.61 If a MATLAB session is interrupted and then resumed at a later date, select *save workplace* in the file menu followed by a destination and filename. For example, *save workplace A: session.mat.*

   The file menu is illustrated in Figure 9.1. The file *session.mat* saves all the workspace variables in binary format and cannot be opened, but it can be loaded back into the workspace while in MATLAB by using the command *load A: session.mat.*

R.9.62 The MATLAB software distinguishes three types of files:

a. M-files (with extension.m)

b. Mat files (with extension.mat)

c. Mex-files (with extension.mex)

R.9.63 The general characteristics of M-files are summarized below:

a. The information is stored in an ASCII format.

b. The filename extension consists of *.m* (dot follow by *m*).

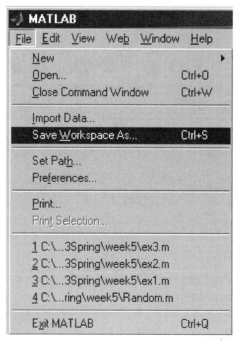

FIGURE 9.1
(See color insert following page 342.) File menu.

    c. It contains either a script- or function file.

    d. A function file is equivalent to a built-in function (inputs are given to produce outputs).

    e. A script file is just a collection of MATLAB instructions.

    f. Can be created by any word processor software from any machine.

    g. Can be used to store data or programs.

R.9.64 The general characteristics of mat files are summarized below:

    a. The information is stored in binary format.

    b. The filename extension is *.mat* (dot followed by *mat*).

    c. They are generally used to save variable and/or data.

    d. They can only be processed by the machine that generates them.

R.9.65 Mex files are C or Fortran files that may be called while in MATLAB. Mex files are not used or discussed in this book, since this topic is beyond the scope and objectives of a simple, introductory, practical MATLAB textbook.

R.9.66 Binary files are, in general, more compact and efficient than ASCII files. ASCII variables can be saved by using either single precision (one byte) or double precision (two bytes), by typing save *filename-ASCII* for the case of one (byte) or *save filename-double* for the case of two bytes, or double precision. MATLAB uses efficiently the disk space and stores every variable by using full precision, and no precision is lost due to the conversion to or from ASCII.

R.9.67 Let us revisit script files. Recall that a script file consists of a sequence of MATLAB instructions and/or data. Script files are generally created for a particular application and are then discarded after used.

R.9.68 A script file is executed at the command window by typing its filename without the extension (*.m*) followed by the <enter> key. The instructions stored in the file are then executed, one instruction after the other in the same sequence as they were typed. The effect of executing this type of file is the same as manually entering each instruction of the file, while at the command window.

R.9.69 The steps involved in the creation of a script file in an MS Window or Mac environment are summarized as follows:

    a. Access the command window

    b. Once in the command window, select (click) *File → New → M-file*

    c. A blank edit window, label editor, or debugger is then opened

    d. Create as an example, a script by typing the following instructions:

```
t = 0:.01:10;
infor = 5*cos(t);
carr = cos(10.*t);
modul = infor.*carr;
plot(t,mod);
xlabel('time(sec)');
ylabel('magnitude');
title('AM Signal');
```

See Figure 9.2.

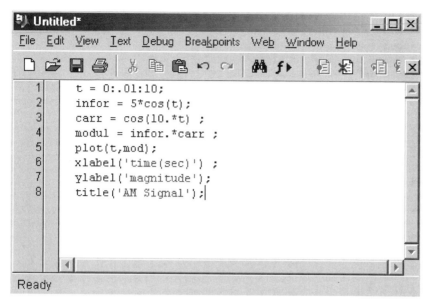

```
t = 0:.01:10;
infor = 5*cos(t);
carr = cos(10.*t) ;
modul = infor.*carr ;
plot(t,mod);
xlabel('time(sec)') ;
ylabel('magnitude');
title('AM Signal');
```

FIGURE 9.2
(See color insert following page 342.) The edit window with the program just entered R.9.69.

R.9.70 Recall that the file's name must be unique. The command *exist ('filename')* can be used to test for uniqueness. For this case, test the file existence by executing *exist ('AM')*, while at the command window, and if the response is 0, that would indicate that *AM* is a valid filename (not previously used). Any other response would indicate that the name was already assigned to another file or variable. Observe that the existence of the filename search is done generally before the editing process is started.

R.9.71 Recall that the MATLAB file search is done as follows:

a. MATLAB looks first for a variable defined in the workplace by the filename.

b. If the given filename is not a variable, then MATLAB looks for a built-in filename function.

c. If the filename is not a variable or a built-in function, then MATLAB looks in the C directory (C:\) for the filename where the M-files are usually stored.

R.9.72 When creating a file, it is highly recommended to dedicate the first line to define and describe the file's purpose, goals, and objectives by using key words as comments (%).

Recall that the contents of the first comment line are returned when a search is performed by either the *help* or *lookfor* commands.

R.9.73 It is essential to use the subsequent (%) comment lines to include additional comments that describe the purpose of a file in question, logic and algorithms used, author of the file (program), date it was written, and subsequent modifications and updates, syntax, inputs and outputs, and in general, any additional information regarding the file to make a given file readable, identifiable, and clear (user-friendly) for any possible user.

R.9.74 The following file illustrates the format and content, of a well-structured and organized file, using the script file *AM* (R.9.69) as example.

MATLAB Solution
```
% Script file name: AM.m
% Returns the plot of the amplitude modulated signal
% consisting of the carrier: cos(10t), and the signal = 5cos (t)
% Created by: M.K.                    date: Dec. 2005
% New York City College of Technology, Brooklyn, NY 11201.
% ******************************************************
t= 0:.01:10;            % creates a time vector t
infor = 5*cos(t);       % creates the cosine vector name infor with
                          w =1
carr = cos(10.*t);      % creates the cosine vector name carr with
                          w =10
mod = infor.*carr;      % creates the vector mod, cos(t).cos(10t)
plot(t,mod);            % plots t vs. mod
xlabel('time/sec')
ylabel('Magnitude')
title('AM Signal with w_c = 10, and w_i = 1')
grid on;
```

R.9.75 Recall that the preceding file can now be saved by selecting (clicking) *File → Save
        as... →* and a dialog box will be displayed. Type the filename *AM.m* in the reserved
        field followed by (clicking) *Save.* Figures 9.3 and 9.4 illustrate the menus followed in
        the saving process.

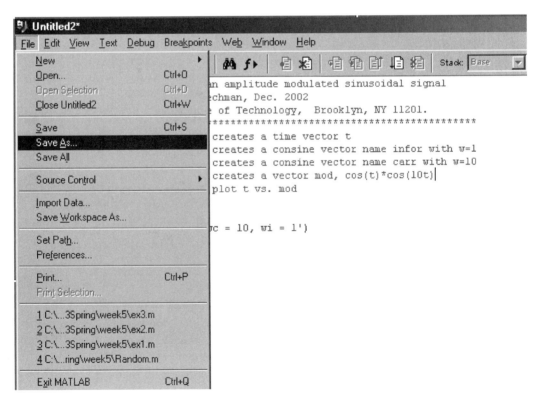

FIGURE 9.3
(See color insert following page 342.) Saving the file.

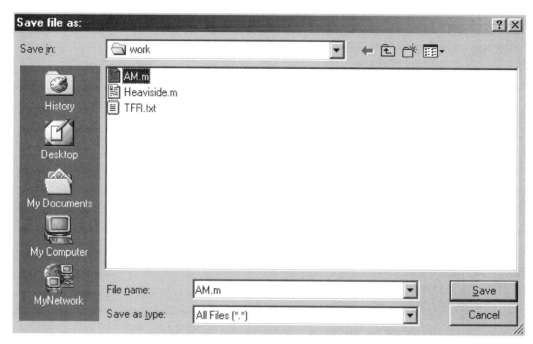

FIGURE 9.4
(See color insert following page 342.) Saving the script file as *AM.m*.

R.9.76 Once the file *AM.m* is saved, *exit* the edit window and return to the command window to test (execute) the file just saved (*AM.m*).

Observe that *AM.m* is stored in the default directory *work*, as indicated in Figure 9.4.

R.9.77 Let us test the file *AM.m*, while in the command window, by executing the following commands and observing their responses:

a. *exist('AM')*

b. *help AM*

c. *lookfor AM*

d. *whos*

e. *type AM.m*

f. *AM*

g. *whos*

MATLAB Solution
```
>> exist('AM')                    % part (a)

    ans =
            2

>> help AM                        % part (b)
```

```
% Script file name: AM.m
% Returns the plot of the amplitude modulated signal
% consisting of  the carrier: cos(10t), and the signal = 5cos(t)
% Created by: M.K.                         date: Dec. 2005
% New York City College of Technology, Brooklyn, NY 11201.
% ************************************************************
```

```
>> lookfor AM                            % part(c)
```

```
    AM.m: % Script file name: AM.m
```

```
>> whos                                  % returns the variables used,
                                            part (d)
```

```
    Name      Size        Bytes    Class
    ans       1x1         8        double array
    x         1x1         8        double array
    Grand total is 2 elements using 16 bytes
```

```
>> type AM.m       % returns the AM file without opening the file,
                      part (e)
```

```
    Script file name: AM.m
    % Returns the plot of the amplitude modulated signal
    % consisting of  the carrier: cos(10t),  and the signal =
      5cos(t)
    % Created by: M.K.                         date: Dec. 2005
    % New York City College of Technology,  Brooklyn, NY 11201.
    % ************************************************************
    t= 0:.01:10;             % creates a time vector t
    infor =5*cos(t);         % creates a cosine vector name
                                infor with w = 1
    carr = cos(10.*t);       % creates a cosine vector name carr
                                with w = 10
    mod = infor.*carr;       % creates a vector mod,
                                cos(t)*cos(10t)
    plot(t,mod);             % plots t vs. mod
    xlabel('time/sec')
    ylabel('Magnitude')
    title('AM signal with wc = 10 and wi = 1')
    grid on;
```

```
>> AM       % executes the AM.m file and returns the plot shown in
               Figure 9.5
>> whos     % returns all the variables used, part(g)
```

```
    Name      Size        Bytes    Class
    ans       1x1         8        double array
    carr      1x1001      8008     double array
    infor     1x1001      8008     double array
    mod       1x1001      8008     double array
    t         1x1001      8008     double array

    Grand total is 4005 elements using 32040 bytes
```

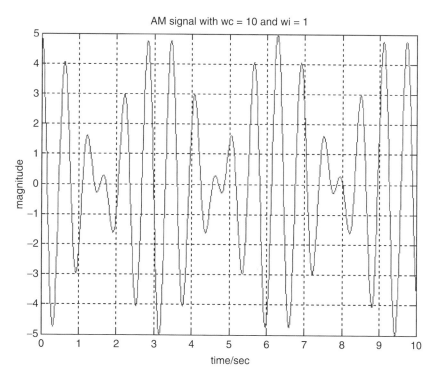

FIGURE 9.5
Plot of the script file *AM.m*.

R.9.78 If the file *AM.m* was saved in a directory other than C (e.g., in a floppy), MATLAB must be directed to include this directory (the floppy) in its search path inorder to find and execute the referred file.

R.9.79 Recall that script files can be used also as data files. For example, let us assume that, measured data, the result of an experiment is stored as a vector *V* in the file *data. m*. This data can be made available while in the command window by typing *data* followed by the <enter > key, assuming that *data.m* is in the search MATLAB path. Let us also assume that the vector *V* in *data.m* is the input to a program stored as another script file named *analysis_V.m*, which has as its objectives to analyze *V* and return the following information of interest:

a. The number of elements in *V*

b. The maximum value in *V*

c. The minimum value in *V*

d. The average and median values in *V*

e. Rearrange the elements in *V* in ascending order

f. The standard deviation of the elements in *V*

g. Plot [sample value] versus [sample number]

The following solution consists of two files *data.m* and *analysis_V.m*:

MATLAB Solution
```
% Script file: analysis _ V.m
format compact
disp('*********** RESULTS****************')
length _ of _ V= length(V)
max _ value _ in _ V = max(V)
min _ value _ in _ V = min(V)
average _ of _ V = mean(V)
median _ of _ V = median(V)
sort _ samples _ V = sort(V)'
std _ in _ V = std(V)
disp ('**********************************')
plot (x,V,x,V,'s')
title (' [sample value] vs [sample num.]')
xlabel ('sample num.')
ylabel ('sample value')
disp ('**********************************')
```

For testing and illustrative purposes, let us assume that the vector *V* located in the file *data.m* consists of a sequence of 10 random numbers. The *data.m* file is as follows:

```
% Script file: data.m
V = (rand(1,10).*3); x=1:10;
disp ('The data is given by V (below)'), disp ('sample num.value');
disp ('^^^^^^^^^^^^^^^^^^^^^^^^^^^^')
[x' V']
disp ('^^^^^^^^^^^^^^^^^^^^^^^^^^^^')
```

Observe that the process of analyzing the data (vector *V*) is accomplished by executing two commands while at the command window.

```
>> data                    % loads the data represented by V
>> analysis _ m            % returns the analysis using vector V
```

The process is illustrated as follows:

MATLAB Solution
```
>> data

The data is given by  V  (below)
 sample num.      value
^^^^^^^^^^^^^^^^^^^^^^^^^^^^^^^^^^^^

ans =
    1.0000     0.0458
    2.0000     2.2404
    3.0000     1.3353
    4.0000     2.7954
    5.0000     1.3980
    6.0000     1.2559
    7.0000     2.5387
    8.0000     1.5755
    9.0000     0.6079
   10.0000     2.0164
^^^^^^^^^^^^^^^^^^^^^^^^^^^^^^^^^^^^
```

```
>> analysis _ V

*********** RESULTS****************
length _ of _ V =
                10
max _ value _ in _ V =
                2.7954
min _ value _ in _ V =
                0.0458
average _ of _ V =
                1.5809
median _ of _ V =
                1.4867
sort _ samples _ V =
                0.0458
                0.6079
                1.2559
                1.3353
                1.3980
                1.5755
                2.0164
                2.2404
                2.5387
                2.7954
std _ in _ V =
                0.8511
**********************************
```

See Figure 9.6.

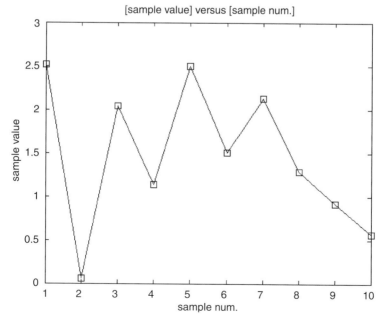

FIGURE 9.6
Plot of the 10 samples in *V* of R.9.79.

R.9.80 Observe from the last example that the script file variables operate globally on the workspace and become part of the overall MATLAB session. Note that the variable *x* defined in *data.m* file is used in the *analysis_V* file (global).

R.9.81 Recall that when an M-file is executed, its commands are not displayed on the screen unless the *echo* command is included. Observe that the punctuation used in the construction (and display) of a file follows the standard MATLAB syntax rules. Recall also that the effect of the command *echo* is to display each command of the file as they are executed. The *echo on* command is particularly useful when the intermediate variable values created during the execution of a file are of particular interest to the programmer.

R.9.82 Recall that function files are also M-files whose purpose is to create new MATLAB functions. The variables used in the creation of a function file are local, defined, and used only inside the file and have no incidence on the global workspace. These variables are erased once the file is executed unless they are declared *global*.

R.9.83 The command *global A B C* declares the variables *A*, *B*, and *C* as global. The workspace then shares a single copy of the variables *A*, *B*, and *C*, and any assignment to *A*, *B*, and *C* in any function used in the workplace is available to all the other functions used.

R.9.84 The command *isglobal A* returns a 1 if *A* has been declared or is global, and 0 otherwise.

R.9.85 The command *help funfun* returns a list of the MATLAB's functions of functions.

R.9.86 Recall that the first line in a function file is a function definition line that defines the function name, input variables (*I1, I2, …, In*), and corresponding output variables [*O1, O2, …, Om*].

R.9.87 The general format and syntax of a function file is as follows:

function [O1 O2 … Om] = filename (I1, I2, …, In)

Observe that the keyword *function* must be typed in lowercase letters. Note also that the output variables *O1, O2, …, Om* are enclosed in square brackets, whereas the input variables *I1, I2, …, In* are enclosed in parentheses.

R.9.88 The comment lines (%) defining the purpose of a function file should use key (descriptive) words and be placed immediately after the line that contains the file function definition. Recall that the comment lines are accessed by the *help* and *lookfor* commands when a search is done.

Recall also that the first character of the comment line must be % with no preceding spaces, and the first comment line after the function definition line is the only line searched by the *lookfor* command.

R.9.89 If the function file has a single output variable *O1*, then the square brackets are not required, and the syntax is as follows:

function O1 = filename (I1, I2, …, In)

R.9.90 If the function file has no output variables, then the square brackets as well as the equal sign are omitted, and the syntax is as follows:

function filename (I1, I2, …, In)

R.9.91 Recall that the objective and purpose of a function file is to take the input variables *I1, I2, …, In* and transform them into the output variables *O1, O2, …, Om*.

The process of transforming the inputs into outputs is usually hidden from the end user (unless the *echo* command is included in the file). Function files use a temporary workspace, and the variables defined are deleted after the execution of the (function) file.

The first noncomment line constitutes the beginning of the program and blank lines are ignored.

R 9.92 The steps involved in the creation of a function file are summarized as follows (the steps are similar to the case of script files in MS Window or Mac environments):

a. Start at the command window.

b. Use the *exist* command to check if the function name to be chosen is valid (should be unique).

c. While at the command window, select the *file* menu.

d. Click *New → M-file*.

e. A blank edit window label edit or debugger is then opened.

f. Create the function file's definition line.

g. The next line is used to describe and document the file to be created as a comment line using key words.

h. Include additional comments regarding the file such as inputs, outputs, syntax, date, author.

i. Type in the function program (output variables in term of input variables).

j. Once the program is completed, the file is saved by selecting from the menu *File → Save as … →* type in the *filename.m* chosen in the reserved field followed by (clicking) *Save*.

k. Click *File →* Exit (the edit or debugger window), and MATLAB returns to the command window where the file just created can now be tested.

l. Test the file with simple verifiable data.

R.9.93 For example, the following file *function [A, B] = fn(I1, I2, I3)*, where *fn* is its function filename, can be executed using the following format:

a. *[A, B] = fn(I1, I2, …, I3)*, where *I1, I2, I3* have been already defined.

b. *[A, B] = fn(0, 5, pi)*, where *I1 = 0, I2 = 5*, and *I3 = π*.

c. *fn(2, –2.5, 3.89)*, where the inputs are *I1 = 2, I2 = –2.5*, and *I3 = 3.89*; but the output variables *A* and *B* are assigned no values.

R.9.94 For example, create the function *sine_fn* that returns *y(t) = A sin(wt + ph)*, over the range *0 ≤ t ≤ 5*, where *A* represents its amplitude, *w* frequency, and *ph* phase angle. The function's input variables are *A, w*, and *ph*, and its output variables are *t* and *y*.

MATLAB Solution
```
function [t, y] = sine _ fn(A, w, ph)
% sine _ fn : returns a table and plot of   [y(t) = A sin(wt + ph)]
                                          vs. t,   for 0 < t < 5
% Variables:  A= Amplitude, w = angular frequency in radians/sec,
                                   ph = phase in radians
% y(t)= Asin(wt+ph)
% Author: M. K                           date : Dec 2006
% Call syntax: [t,y] = sine _ fn(A, w, ph)
```

```
% Inputs: A, w, ph
% Output: [t, y]  for 0 ≤ t ≤ 5,
% ***********************************************************************
t = linspace(0, 5, 15);    %  creates a 15 elements vector t over
                                the range 0 to 5
y = A* sin(w.*t + ph) ;
plot(t, y, t,y,'s');         % plots [Asin (wt+ph)] vs. t
title([ '[A sin (wt+ph)] vs t, for 0 < t < 5, where A'= num2str(A),
                    ', w =', num2str(w), 'and ph= ', num2str(ph),])
xlabel('t (time in sec.)')
ylabel(' y(t) =A sin (wt+ph)')
```

R.9.95   The function file *sine_fn*, created in R.9.94, is tested by executing the following commands and observing their responses:

a. *exist (sine_fn)*

b. *help sine_fn*

c. *lookfor sine_fn*

MATLAB Solution
```
>> format compact
>> exist sine _ fn                          % part  (a)

        ans =
             2

>> help sine _ fn                           % part  (b)


sine _ fn : returns a table and plot of  [y(t) = A sin(wt + ph)]  vs.
                                          t,  for 0 ≤ t ≤ 5
Variables:  A= Amplitude, w = angular frequency in radians/sec, ph =
phase in radians
y(t)= A sin(wt+ph)
 Author: M. K                     Date : Dec 2006
 Call syntax: [t,y] = sine _ fn(A, w, ph)
 Inputs: A, w, ph
 Output : [t, y]  for 0 < t < 5
***********************************************************************

>> lookfor sine _ fn                        % part  (c)

sine _ fn : returns a table and plot of  [y(t) = A sin(wt + ph)]  vs.
t, for 0 ≤ t ≤ 5
```

R.9.96   The function file *sine_fn* is now tested by executing the following instructions:

a. *[t, y] = sine_fn(3, pi, pi/4)*

b. *sine_fn(3, pi, pi/4)*

c. *A = 3.5*sqrt(2), w = pi − 2/3, ph = pi/7),[t, y] = sine_fn(A, w, ph)*

d. *which sine_fn*

e. *isglobal t*

```
MATLAB Solution
>> [t,y] = sine _ fn(3,pi,pi/4)                    % part (a)
   t =
   Columns 1 through 8
    0    0.3571    0.7143    1.0714    1.4286    1.7857    2.1429    2.5000
   Columns 9 through 15
      2.8571    3.2143    3.5714    3.9286    4.2857    4.6429    5.0000
   y =
   Columns 1 through 8
    2.1213    2.8316    0.3359    -2.5402    -2.5402    0.3359    2.8316
    2.1213
   Columns 9 through 15
    -0.9908    -2.9811    -1.5961    1.5961    2.9811    0.9908    -2.1213


>> sine _ fn(3,pi,pi/4)                    % part (b)


>> A = 3.5*sqrt(2);                        % part (c )
>> w = pi-2/3;
>> ph = i/7;
>> [t,y] = sine _ fn(A,w,ph)


   t =
   Columns 1 through 8
    0    0.3571    0.7143    1.0714    1.4286    1.7857    2.1429    2.5000
   Columns 9 through 15
    2.8571    3.2143    3.5714    3.9286    4.2857    4.6429    5.0000
   y =
   Columns 1 through 8
    2.1476    4.8101    3.9529    0.2033    -3.6951    -4.8897    -2.5064
    1.7109
   Columns 9 through 15
    4.6763    4.2199    0.6758    -3.3629    -4.9408    -2.9035    1.2584


>> which sine _ fn                         % part (d)


   C:\MATLAB6p1\work\sine _ fn.p


>> isglobal t                              % part (e)


   ans =
          0
```

See Figures 9.7 and 9.8.

R.9.97 Any changes, upgrades, revisions, and debugging of an M-file is done using the edit or debugger window.

R.9.98 The execution of an M-file terminates or stops with the execution of the last instruction of the file, or it is interrupted when it encounters either one of the commands—*return, error, warning, input, keyboard, pause,* or *waitforbuttonpress.*

R.9.99 Recall that the *return* command causes a return to the invoking function or keyboard.

R.9.100 Recall that the command *error ('string')* in an M-file is used to stop the execution of the file and returns control to the command window displaying the message *string.*

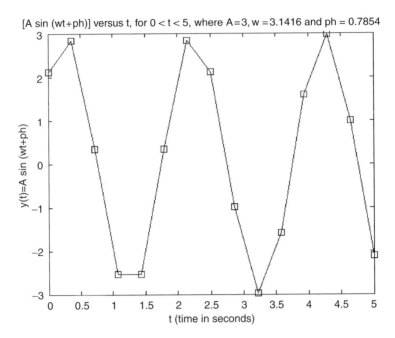

FIGURE 9.7
Plot of the function file *sine_fn* of R.9.96.

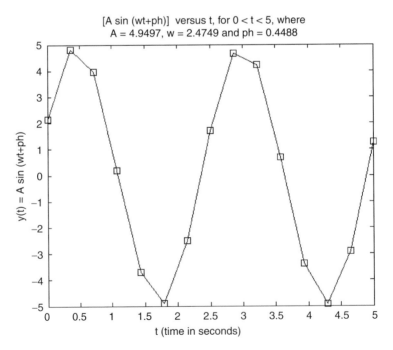

FIGURE 9.8
Second plot of the function file *sine_fn* of R.9.96.

R.9.101 Recall that the command *warning ('string')* returns the message ('string') when certain programmed conditions are not satisfied. The *warning* command can be switched by using *warning on* or *warning off*.

R.9.102 Recall that the command *A = input ('string')* introduced and used in early chapters causes the execution of a file to stop and returns the message 'string' in the command window and waits for the user to enter data that is assigned to *A*, after the <enter> key is executed.

R.9.103 Recall that the command *keyboard* inside a file stops the execution of the file and transfers control to the keyboard while in the command window where the prompt becomes *k >>*. At this point, the user can execute any instruction or return to complete the execution of the M-file by typing the instruction *return* followed by the <enter> key. This instruction is particularly useful when dealing with a long file, and it is a good strategy in many cases to stop the execution of a file at key points, and to check the partial results, if indeed they make sense before continuing with the execution of the rest of the file.

R.9.104 Recall that the *pause* command stops the execution of a file, transfers control to the *command* window, and waits for the user to enter any character to resume the execution of the file. The command *pause (n)* stops the execution of the file during *n* second (see Chapter 2, Table 2.1).

R.9.105 The command *waitforbuttonpress* stops the execution of the file until a character is entered either by the keyboard or mouse.

R.9.106 MATLAB function files can call other function files including themselves during their execution.

R.9.107 When a function file is called for the first time, MATLAB compiles the file and any other file called during the execution, where each function uses its own independent workplace. The executed function files are stored in precompiled format. This process saves valuable time when a function is called several times during the execution of a program.

R.9.108 Strictly speaking, MATLAB does not compile each input statement, what it does is something very similar, called parsing. Parsing is the process of converting the MATLAB instructions into a lower-level language, similar to assembly or machine language. The parsing process also checks for errors and inconsistencies.

R.9.109 The performance of a MATLAB function file can be evaluated by using the command *profile*, which evaluates the performance of a function file (no script files) in terms of the time spent on each line (in terms of 0.01 s).

R.9.110 The output of the command *profile* is in the form of a report (table) or a plot. The profile report returns an HTML file, displayed by (clicking)

view → desktop layout → default → current directory → 0.html

The *profiler* is a powerful, simple, and useful tool that consists of a family of commands, some of which are defined in the following discussion.

R.9.111 The command *profile on* activates the *profiler* immediately followed by the command *profile fn*, where *fn* is the function's name.

R.9.112 The profile function also checks optimization and efficiency of the codes used in the file *fn*. To get reliable results, the function *fn* must be executed a number of times to accumulate sufficient statistical data about its performance.

R.9.113 The command *profile off* deactivates the *profiler.*

R.9.114 The command *profile resume* restarts the *profiler* without clearing the previous function statistics.

R.9.115 The command *profile clear* clears all the previous statistical data.

R.9.116 The command *profile reset* resets the statistical compiled data used to evaluate the performance of the function file *fn* and restarts the *profiler.*

R.9.117 The command *profile report* returns a report consisting of the line codes of the function file *fn* that consumes the greatest amount of time.

R.9.118 The command *profile report n* returns the *n* code lines of the function file *fn* that consumes the greatest amount of time.

R.9.119 The command *profile report m*, where *m* has a range over $0 \leq m \leq 1.0$, returns a report consisting of the code lines of the function file *fn,* which consumes a time greater than $m*100\%$ of the total execution time of *fn.*

R.9.120 The command *profile plot* returns a bar graph of the function *fn* with the commands that consume the greatest amount of time.

R.9.121 The command *profile('status')* returns the information about the current profiler state such as *profiler status (on or off), detail level,* and *history tracking (on or off).*

R.9.122 The command *profreport(basename)* returns a report using the current profiler statistics that are automatically saved by the profiler in file *basename.*

R.9.123 For example, let the script file *performance* be used to accumulate performance data and return the profiler's status and *profile report 2* and *0.4* of the function file *sine_fn* (R.9.94). To get reliable statistical data, the function file is executed 150 times as illustrated by the following script file *performance:*

```
MATLAB Solution
% Script file: performance
profile sine_fn
profile on
profile clear;
n = 150;
m = 1;
while m<n
  A=3;
  w = m/150;
  ph = 2*pi*m/150;
  [t,y] = sine _ fn(A,w,ph);
  m = m+1;
end
profile report
profile plot
disp('**************************************')
disp('********profile status**************')
profile('status')
disp('**************************************')
profile report 2
profile report 0.4
```

The script file *performance* is executed and the results are indicated below:

```
>> performance

**********************************
********profile status***************
ans =
      ProfilerStatus: 'off'
        DetailLevel: 'mmex'
     HistoryTracking: 'off'
*****************************************************************************
              MATLAB Profile Report: Summary
             Report generated 20-Jan-2007 13:01:04
                   Total recorded time:
                        2.17 s
                  Number of M-functions:
                        21
                 Number of M-subfunctions:
                        4
                   Clock precision:
                        0.010 s
                   Function List
                        Name
```

	Time		Calls	Time/call	Self time		Location
sine_fn	2.173	100.0%	149	0.014584	0.191	8.8%	C:\MATLAB6p1 \work\sine_ fn.p
newplot	0.561	25.8%	149	0.003765	0.050	2.3%	
....							
title	0.530	24.4%	149	0.003557	0.450	20.7%	
xlabel	0.401	18.5%	149	0.002691	0.291	13.4%	
xlabel	0.401	18.5%	149	0.002691	0.291	13.4%	
....							
num2str	0.300	13.8%	447	0.000671	0.170	7.8%	
newplot	0.291	13.4%	149	0.001953	0.020	0.9%	
clo	0.271	12.5%	149	0.001819	0.191	8.8%	
gcf	0.240	11.0%	596	0.000403	0.240	11.0%	
ylabel	0.190	8.7%	149	0.001275	0.170	7.8%	
gca	0.110	5.1%	596	0.000185	0.090	4.1%	
isappd	0.100	4.6%	447	0.000224	0.070	3.2%	
int2str	0.060	2.8%	894	0.000067	0.060	2.8%	
allchild	0.060	2.8%	149	0.000403	0.060	2.8%	
isfield	0.030	1.4%	447	0.000067	0.030	1.4%	
gca	0.110	5.1%	596	0.000185	0.090	4.1%	
log10	0.030	1.4%	447	0.000067	0.030	1.4%	
deblank	0.020	0.9%	447	0.000045	0.020	0.9%	
strvcat	0.020	0.9%	447	0.000045	0.020	0.9%	
setdiff	0.020	0.9%	149	0.000134	0.010	0.5%	
unique	0.010	0.5%	298	0.000034	0.010	0.5%	

See Figure 9.9.

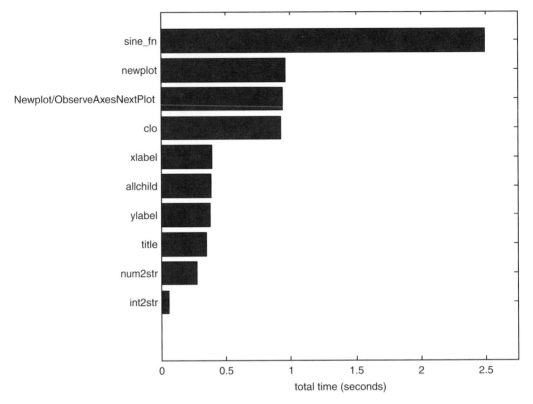

FIGURE 9.9
Profile plot of the performance of the file *sine_fn* of R.9.94.

R.9.124 A given function file *fn.m* can be parsed (compiled) by executing the command *pcode fn*. This command creates the parsed (compiled) function, which is saved in the new created file *fn.p*, for later use. This file is placed in the current MATLAB directory. Note that the *fn.m* file may be anywhere on the MATLAB search path.

R.9.125 For example, let us parse the function file *sine_fn* and check its parsed existence and location.

MATLAB Solution
```
>> pcode sine_fn
>> exist('sine_fn.p')

    ans =
          6

>> which sine _ fn.p

    C:\MATLAB6p1\work\sine _ fn.p
```

R.9.126 The command *pcode fn1 fn2 ... fnn* creates and stores the parse p-files of the functions *fn1, fn2, ..., fnn* in the current directory.

R.9.127 The command *pcode *.m* creates the parse files of all the M-files in the current directory (stored as p-files).

R.9.128 The command *pcode fn-inplace* creates and then stores the p-file of *fn* in the same directory where the *fn* M-file is located.

R.9.129 The command *nlint fn* parses the M-file *fn.m* in MATLAB 7. Recall that the objective of parsing also includes searching for errors, inconsistencies, and inefficiencies.

R.9.130 The command *mlock('functionname')* locks the parse function, and the *clear* command does not clear the lock function. A function can be unlocked by using the command *numlock('functionname')*.

R.9.131 The command *mislocked(functionname')* returns a 1 (true) if the function is locked.

R.9.132 Some commands that may be of interest in the performance analysis process are *etime, clock, tic-tac, flops,* and *cputime,* introduced in early chapters (Chapter 2, Table 2.1). These commands are revisited in the discussion below.

R.9.133 Recall that the command *etime(t1, t0)* returns the elapsed time in seconds between *t1* and *t0,* where *t1* and *t0* are the times defined by six fields given by *year, month, day, hour, minute,* and *second.*

R.9.134 The command *clock* returns the current date as a six-field vector consisting of *year, month, day, hour, minute,* and *seconds.* For example, the current date is given by

```
>> clock

    ans =
        1.0e+003 *
        2.0070    0.0010    0.0200    0.0140    0.0110    0.0597
```

R.9.135 The command *flops* returns the number of floating-point operations. With the incorporation of LAPACK in MATLAB 6, the *flops* command is not supported by MATLAB 6.0 and subsequent releases (2000).

R.9.136 The command *tic, <statements>, toc* where *tic* starts a stopwatch timer and *toc* stops it. The sequence *tic <statements> toc* returns the time in seconds elapsed between the activation of *tic* and execution of *toc.*

R.9.137 The command *cputime* returns the cpu time in seconds that has been used by the MATLAB processor. For example, the sequence *t0 = cputime; <statements>, t1 = cputime,* returns *cpu_time = t1 − t0* representing the processing and execution time for *<statements>.*

R.9.138 By taking advantage of the sparsity of a matrix, a substantial saving in operational time as well as storing facilities is attained. Recall that a matrix is sparsed if it contains a high number of zero elements (see Chapter 3).

R.9.139 The following example, given by the script file *sparse_versus_full* shows how to generate a large 125 × 125 sparse matrix A and evaluate the number of floating-point operations required to evaluate $A^3 = A\wedge3$, for both the sparse and full matrix versions.

Observe that matrix A is created for a range of different densities (0.01 to 0.2) and returns the following plots:

a. *[densities]* versus *matrix*

b. *[inefficiency]* versus *[densities]*

c. *[# of operations]* versus *[densities]*

Recall that the density of the matrix A is given by

$$[\text{density of the matrix } A] = \frac{[\text{number of nonzero elements in } A]}{[\text{total number of elements in } A]}$$

MATLAB Solution
```
% Script file: sparse _ vs _ full
n = 125;k=1;
disp('            * * * R E S U L T S * * * ')
disp ('****Performance results for sparse vs. full matrix oper. *****');
disp (' dens    # oper. sparse       # oper. full');
disp ('****************************************************');

figure(1)
for dens = 0.01:0.02:0.2;
A = sprand (n,n,dens);
flops(0);
prodsp=A^3;
opersp = flops; sp(k) = opersp;
subplot (5,2,k)
spy(A)
B = full(A);
flops(0);
prodfull=B^3;
operfull=flops;fu(k)=operfull;k=k+1;
fprintf('%10.3f %6.1f %6.1f\n',dens,opersp,operfull);
end
disp ('****************************************************');
dens = 0.01:0.02:0.2;
ineffic= (fu-sp).*100./fu;
figure(2)
plot(dens,sp)
xlabel('density of matrix A')
title(' # of operations of A^3 vs density')
ylabel('# of oper.for sparse A^3')

figure(3)
plot(dens,ineffic)
xlabel('density of matrix')
ylabel('percentage inefficiency')
title('inefficiency vs density')
```

The script file *sparse*_versus_*full* is executed and the results are as follows:

```
>> sparse_vs_full
```

```
                * * * R E S U L T S * * *
****Performance results for sparse vs. full matrix oper.  ******
        dens       # oper. sparse         # oper. Full
****************************************************
        0.010            1082.0            11718750.0
        0.030           16336.0            11718750.0
        0.050           63706.0            11718750.0
```

0.070	134646.0	11718750.0
0.090	235714.0	11718750.0
0.110	341024.0	11718750.0
0.130	473312.0	11718750.0
0.150	575486.0	11718750.0
0.170	679488.0	11718750.0
0.190	774470.0	11718750.0

****

Observe that the preceding script uses the command *flops* (floating-point operation count), which is no longer supported by MATLAB 6.0, release 12 (November 2000), and subsequent releases. This program clearly illustrates the computational efficiency of the sparse matrices over the full matrices. The reader can still verify the results of this program by replacing the flops and its dependent instructions by an equivalent set of instructions that count the number of operations performed.

Note that as the sparsity decreases, the density and number of floating-point operations increases, whereas the number associated with the full matrices operations remains constant at 11718750.0, independent of its sparsity.

See Figures 9.10 through 9.12.

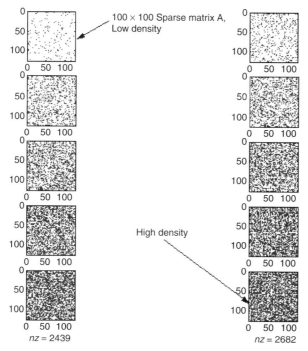

FIGURE 9.10
Plots of sparse matrix *A* over a range of densities.

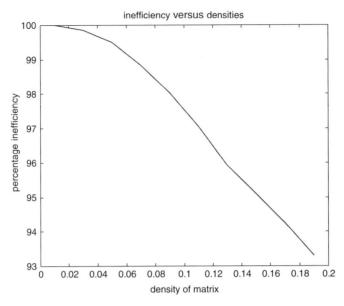

FIGURE 9.11
Inefficiency plot of A^3 over a range of densities of R.9.139.

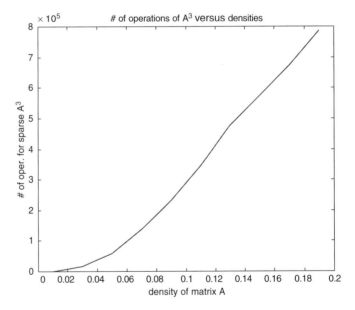

FIGURE 9.12
Plot of $\left[\frac{full\text{-}sparse}{full} * 1000\right]$ versus *density* of R.9.139 of A^3.

9.4   Examples

Example 9.1

Create and execute the script file *redrose* that returns a four-petal red rose. Illustrate and discuss the steps in the creation of this script file and show

a. The edit window with the script file *redrose*

b. The file path

MATLAB Solution

From the command windows click

Edit → New M File → and enter the following instructions

```
% Script file: redrose.m
% The script file redrose returns a
% red rose, if  a color printer is available, otherwise returns a
  black rose.
% Inputs: none
% Outputs: plots of a red rose
% Call syntax: :redrose
% Author: M.K ....................date: Jan,2004
%*****************************************
clf
n = 2;
t = linspace(0,2*pi,200);
r = 5*cos(n*t);
x = abs(r).*cos(t);
y = abs(r).*sin(t);
fill(x,y,'r');axis('equal');
title('red rose')
grid on;
```

Once the preceding program is entered, (click) *File,* → *Save as ...,* → and type in the reserved field the filename *redrose.m* and (click) *Save*.

The edit window with the created file *redrose.m* is shown in Figure 9.13.

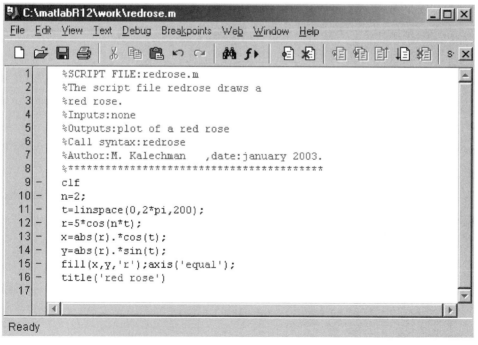

FIGURE 9.13
(See color insert following page 342.) Edit window with the script file *redrose*.

See Figure 9.14.

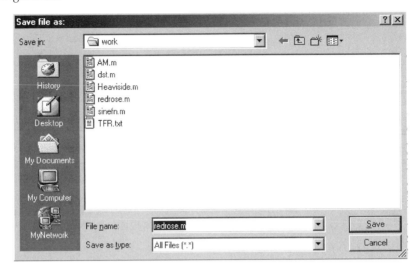

FIGURE 9.14
(See color insert following page 342.) The script file *redrose* is stored in the folder *work*.

Once the file *redrose* is saved in the directory *work*, the directory is checked, as indicated in Figure 9.14, to confirm that the file is stored there.

Back in the command window, the script file *redrose* is tested, and the resulting plot is shown in Figure 9.15.

```
>> redrose;
```

See Figure 9.15.

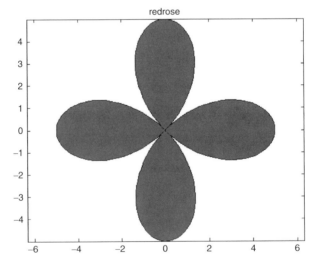

FIGURE 9.15
(See color insert following page 342.) Plot of *redrose* of Example 9.1.

Example 9.2

Create and test the script file *sinc33.m* that returns the plot consisting of threading of two lines with a frequency of $w = \pi$, over the range $0 \le t \le 6$, with an initial magnitude of 1 at $t = 0$.

ANALYTICAL Solution

The following two super impose equations return the desired plot:

$f1 = abs\ (sin(\pi\ t)/(\pi t),$

$f2 = -abs(sin(\pi t)/(\pi t)$

MATLAB Solution

```
% Script  file :  sinc33.m
% Returns the  plot of thread of two lines, given by the function
% abs (sin (pi.*t)./(pi.*t)) and -abs(sin(pi.*t)./(pi.*t)) vs t
% for 0<t<6 in increments of 0.25, and thread frequency w=pi
% Inputs:none
% Outputs: plots of abs(sin(pi.*t)./(pi.*t)) , plus
%                   -abs(sin(pi.*t)./(pi.*t)) vs. t
% Call syntax: sinc33
% Author: M. K.......................      date: Jan. 2007
%%%%%%%%%%%%%%%%%%%%%%%%%%%%%%%%%%%%%%%%%%%%%%%%%%%%%%

t = 0:0.25:6;
f = abs(sin(pi.*t)./(pi.*t));
plot (t,f)
axis ([0.25 6 -1 1.25]);
hold on;
plot (t, -f)
title ('[abs(sin(pi.*t)./(pi.*t))] vs. t, and [-abs(sin(pi.*t)./(pi.*t))]
vs. t');
xlabel ('time');
ylabel ('Amplitude')
grid on;
```

The script file *sinc33.m* is executed as follows, and the resulting plot is shown in Figure 9.16.

```
>> sinc33
```

See Figure 9.16.

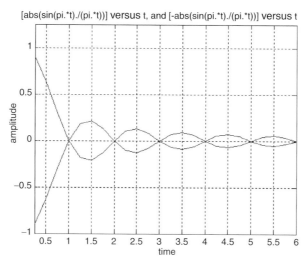

FIGURE 9.16
Plot of *sinc33* of Example 9.2.

Example 9.3

The objective of this example is to explore the execution time of a script versus an equivalent function file, as well as its parse and nonparse responses.

Let us revisit the solution of a quadratic equation of the form $ax^2 + bx + c = 0$ by

a. Creating the script file *quadratic*

b. Test the script file *quadratic* by creating the new script file *test_quadratic*, which returns the execution time of the parse and nonparse version for the following equation, used for testing purposes:

$$3x^2 + 7x + 13 = 0$$

c. Creating the equivalent function file and label it *quadraticf* (that returns the roots of a quadratic equation)

d. Test the function file *quadraticf* by creating the new script file *test_quadraticf*, which returns the execution time of the parse and nonparse version, by using the same equation $3x^2 + 7x + 13 = 0$ (twice, parsed and nonparsed)

e. Compare and discuss the performance of the script and function solutions for the parse and nonparse cases

MATLAB Solution

```
%      ( part (a) )
% Script file: quadratic
% This script file solves
% the quadratic equation
% of the form,a*x^2+b*x+c = 0
% Inputs : coeficient,a,b,c.
% Outputs: roots x1,x2
% Call syntax: quadratic
% Author: M. K.                    date: Jan, 2007
%************************************************************
clc;
disp('*************************************************')
disp('This program solves the quadratic equation:')
disp('    a*x^2+b*x+c=0    ')
disp('**************************************')
disp('Provide the values of the coefficients ')
disp('      of the quadratic equation')
disp('***                        ****')
disp('          ')
a = input('Enter the value of the coefficient for a = ');
b = input('Enter the value of the coefficient for b = ');
c = input('Enter the value of coefficient for  c = ');
x1=(-b-sqrt(b^2-4*a*c))/(2*a);
x2=(-b+sqrt(b^2-4*a*c))/(2*a);
disp('^^^^^^^^^^^^^^^^^^^^^^^^^^^^^^^^^^^^^^^^^^^^^^^^^^^')
disp('The solutions of the given quadratic equation are :   ');
disp(['x1=',num2str(x1),' and x2=',num2str(x2)])
disp('^^^^^^^^^^^^^^^^^^^^^^^^^^^^^^^^^^^^^^^^^^^^^^^^^^^')

%(part(b))
% Script file: test _ quadratic
tic;
```

```
quadratic
t1 = toc;
time_not_parsed = t1;
tic;
quadratic
toc;
time_parsed =toc;
disp('***********Time Results*****************')
disp('First calculations of the roots of the quadratic equation')
fprintf('non-parse time (in sec)=%f\n',time_not_parsed')
disp('Second calculations of the roots of the quadratic equation')
fprintf('parse time(in sec)=%f\n',time_parsed')
disp('********************************************')
```

The script file *test_quadratic* is executed and the results are as follows:

```
>> test_quadratic

*****************************************
This program solves the quadratic equation:
   a*x^2+b*x+c=0
*****************************************
Provide the values of the coefficients
     of the quadratic equation
***                              ****
Enter the value of the coefficient for a=3
Enter the value of the coefficient for b=7
Enter the value of the coefficient for c=13

^^^^^^^^^^^^^^^^^^^^^^^^^^^^^^^^^^^^^^^^^^^^^^^^^
The solutions of the given quadratic equation are :
x1=-1.1667-1.724i and x2=-1.1667+1.724i
^^^^^^^^^^^^^^^^^^^^^^^^^^^^^^^^^^^^^^^^^^^^^^^^^

*****************************************
This program solves the quadratic equation:
   a*x^2+b*x+c=0
*****************************************
Provide the values of the coefficients
     of the quadratic equation
***                              ****
Enter the value of the coefficient for a=3
Enter the value of the coefficient for b=7
Enter the value of the coefficient for c=13
^^^^^^^^^^^^^^^^^^^^^^^^^^^^^^^^^^^^^^^^^^^^^^^^^
The solutions of the given quadratic equation are :
x1=-1.1667-1.724i and x2=-1.1667+1.724i
^^^^^^^^^^^^^^^^^^^^^^^^^^^^^^^^^^^^^^^^^^^^^^^^^

*****************Time Results*****************
First calculations of the roots of the quadratic equation
non-parse time(in sec) = 6.048000
Second calculations of the roots of the quadratic equation
parse time(in sec)= 5.659000
*****************************************************
```

Observe that the time difference for the identical executions of the script file *quadratic* for the same test equation, given by $3x^2 + 7x + 13 = 0$, between the parse- (compiled script stored in computer memory) and nonparse script (executed for the first time) is 5.659000 and 6.048000 s, respectively.

MATLAB Solution
```
(part (c))
function quadraticf(a,b,c)
%This function file solves
%the quadratic equation
%of the form,a*x^2+b*x+c=0
%Inputs:a,b,c.
%Outputs:x1,x2
%Call syntax: quadraticf(a,b,c)
% Author:M.K...............   date: Jan, 2007
%****************************************
x1= (-b-sqrt(b^2-4*a*c))/(2*a);
x2 = (-b+sqrt(b^2-4*a*c))/(2*a);
disp ('       ')
disp ('*****************************************************')
disp ('The solutions of the given quadratic equation are :  ');
disp (['x1=',num2str(x1),' and x2=',num2str(x2)])
disp ('*****************************************************')

% Script file:test _ quadraticf
tic;
quadraticf (3,7,13);
t1=toc;
time _ not _ parsed=t1;
tic;
quadraticf(3,7,13);
toc;
time _ parsed = toc;
disp('***********Time Results*****************')
disp ('First calculations of the roots of the quadratic equation')
fprintf ('non-parsed time(in sec)=%f\n',time _ not _ parsed')
disp ('Second calculations of the roots of the quadratic equation')
fprintf ('parsed time(in sec)=%f\n',time _ parsed')
```

The script file *test _quadraticf* is executed and the results are as follows:

```
>> test _ quadraticf

********************************************************************
The solutions of the given quadratic equation are:
x1 = -1.1667-1.724i and x2=-1.1667 + 1.724i
********************************************************************


********************************************************************
The solutions of the given quadratic equation are:
x1 = -1.1667-1.724i and x2 = -1.1667+1.724i
********************************************************************

********************Time Results*****************************
First calculations of the roots of the quadratic equation
non-parse time (in sec) = 0.101000
Second calculations of the roots of the quadratic equation
parse time (in sec) = 0.020000
********************************************************************
```

Observe the significant time difference (from 0.101 to 0.02 s) for the identical two responses of the function file *quadraticf*, for the same arguments ($a = 3$, $b = 7$, and $c = 13$), between the parse (compiled script stored in computer memory) and the nonparse versions.

Also observe the significant time improvement between the function solution (0.02 s) and equivalent script solution (5.659 s).

Example 9.4

Create the function file *[dist]* = dst(x1, y1, x2, y2) that, given two Cartesian coordinate points $P1 < x1, y1>$, and $P2 <x2, y2>$, returns the calculated distance between the points and a plot of their distance.

Test the function file *dst*, and observe the respective responses for the following instructions:

a. *dst* for $<x1 = 2, y1 = 2>$, $<x2 = 5, y2 = 6>$ and $<x1 = -2, y1 = 3>$, $<x2 = 5, y2 = -6>$
b. *help dst* (description of the function file *dst*)
c. *which dst* (the file path)
d. *what* (the files in the current directory)
e. *inmem* (list of functions in memory)
f. *exist('dst')* (checks the existence of *dst*)
g. *whos* (list of the variables in the workplace)
h. *profile dst* for the points $P1 <-2, 3>$ and $P2 <5, -6>$ See Figure 9.17.

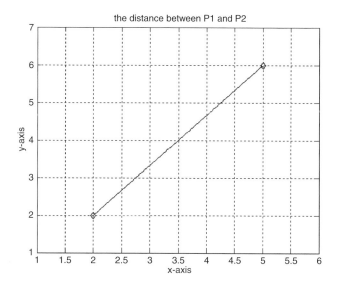

FIGURE 9.17
Plot of the function file *dst* of the Example 9.4(a).

    i. *profile report*
    j. *profile report2* for the two lines of *dst* that consume the longest amount of line
    k. *profile report* for the lines of *dst* that consume more than 50% of the total execution time

MATLAB Solution
```
function [dist] = dst(x1,y1,x2,y2)
% This function file computes the distance
% between points P1(x1,y1) and P2(x2,y2) ,
```

```
% and returns the  plot of the shortest distance
% between P1 and P2 as well as the points:
% P1(x1,y1) and P2(x2,y2).
% Inputs : x1,y1,x2,y2
% Outputs: dist and plot of the dist
% Author : M.K  .....................date: Jan,2007
%*******************************************************
clc,clf;
xd = x2-x1;
yd = y2-y1;
dist = sqrt(xd^2+yd^2);
disp('          * * *  R E S U L T * * *')
disp('*************************************************')
fprintf(' The calculated distance between points P1 and P2
is=%f\n',dist);
disp('*************************************************')
disp('The graphic distance between points')
disp('P1 and P2 is shown in the figure window.....');
disp('              ');
disp('*********************************************');
fprintf('The approximate distance is......');
x = [x1 x2];
y = [y1 y2];
maxx = max(x);minx = min(x);
maxy = max(y);miny = min(y);
MAXX = maxx+1;MINX = minx-1;
MAXY = maxy+1;MINY= miny-1;
x = linspace(x1,x2,100);
y = linspace(y1,y2,100);
plot(x1,y1,'d',x2,y2,'d',x,y)
title(' The distance between P1 and P2')
xlabel('x-axis')
ylabel('y-axis')
axis([MINX MAXX MINY MAXY]);
grid on
```

The function file *dst* is tested for the following Cartesian points:

$<x1 = 2, y1 = 2>$ and $<x2 = 5, y2 = -6>$, and the results are indicated as follows:

```
>> % part(a)
>> dst(2,2,5,6)
                * * *  R E S U L T * * *
********************************************************************
The calculated distance between points P1 and P2 is = 5.000000
********************************************************************
The graphic distance between points
P1 and P2 is shown in the figure window.....
********************************************************************
The approximate distance is......
ans =
     5

>> help dst                              % part(b)

  This function file computes the distance
  between points P1(x1,y1) and P2(x2,y2) ,
```

```
    and returns the  plot of the shortest distance
    between P1 and P2 as well as the points:
    P1(x1,y1) and P2(x2,y2).
    Inputs : x1,y1,x2,y2
    Outputs: dist and plot of the dist
    Author : M.K  .....................date: Jan,2007
    *****************************************************
```

```
>> which dst % part(c)

    C:\MATLAB6p1\work\dst.m
```

```
>> what                             % part (d)

    M-files in the current directory C:\MATLAB6p1\work
    Impfun          quadraticf
    diana           sine_fn
    dst             test_file
    f               test_quadratic
    func _ quad _ sol
    perf
    pz
    quadratic
    P-Files in the current directory C:\MATLAB6p1\work
```

```
>> inmem % part(e)
    ans =
    'dst'
    'getappdata'
    .........
    'closereq'
    'grid'
    'axis'
    'ylabel'
    v'xlabel'
    'title'
    'gcf'
    'C:\MATLAB6p1\toolbox\matlab\graphics\private\clo'
    'C:\MATLAB6p1\toolbox\matlab\general\private\openm'
```

```
>> exist('dst')                        % part (f)

    ans =
          2
```

```
>> whos                               % part(g)

    Name                Size              Bytes Class
    ans                 1x1               8 double array
    distance _ P1 _ P2  1x1               8 double array
        Grand total is 2 elements using 16 bytes
```

```
>> profile dst                          % part (h)
>> profile on
>> dst (-2, 3, 5, -6)
```

```
                  * * *  R E S U L T * * *
*******************************************************************
The calculated distance between points P1 and P2 is =11.401754
*******************************************************************
```

```
The graphic distance between points
P1 and P2 is shown in the figure window.....
******************************************************************
The approximate distance is ......
 ans =
        11.4018
```

See Figure 9.18.

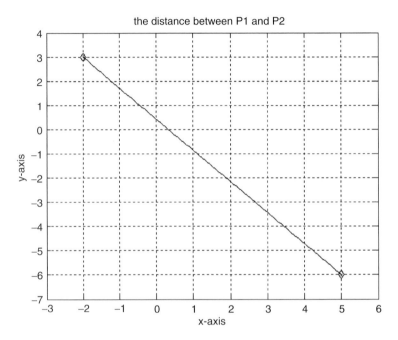

FIGURE 9.18
Plot of the function file *dst* (−2, 3, 5,−6) of the Example 9.4(a).

```
>> profile report                              %  part (i)

Total time in "c:\matlab\work\dst.m": 1.07 seconds
100% of the total time was spent on lines: [11 38 42 39 16 36 14
                                            41 40]
        10: %**************************************************
0.51s, 48%   11: clc;clf;
             12: xd=x2-x1;
             13: yd=y2-y1;
0.02s, 2%    14: dist=sqrt(xd^2+yd^2);
             15:disp('*****************************************')
0.06s, 6%    16: fprintf('The calc distance bet P1 and P2
                 is=%f\n',dist);
             17: disp('*****************************************')
             35: MAXY=maxy+1;MINY=miny-1;
0.02s, 2%    36: x=linspace(x1,x2,100);
             37: y=linspace(y1,y2,100);
0.19s, 18%   38: plot(x1,y1,'d',x2,y2,'d',x,y)
0.10s, 9%    39: title(' The distance between P1 and P2')
0.01s, 1%    40: xlabel('x')
```

```
0.01s,  1%    41: ylabel('y')
0.15s, 14%    42: axis([MINX MAXX MINY MAXY]);
              43: grid on

>> profile report 2                           % part (j)

Total time in "c:\matlab\work\dst.m": 1.07 seconds
65% of the total time was spent on lines: [11 38]
              10: %**********
0.51s, 48%    11: clc;clf;
              12: xd= x2-x1;
..........................................

..........................................
              37: y=linspace(y1,y2,100);
0.19s, 18%    38: plot(x1,y1,'d',x2,y2,'d',x,y)
              39: title(' The distance between P1 and P2')


>> profile report 0.5                         % part (k)


Total time in "c:\matlab\work\dst.m": 1.07 seconds
No line took more time than 0.5*total_time.
```

Observe that the total execution time is 1.07 s, and 0.5 ∗ 1.07 = 0.535 s.
Indeed there is no line that takes more than 0.535 s. The closest line is *clc,clf* with an execution time of 0.51 s.

Example 9.5

Create the script file *calc_area* that returns the calculated area indicated by the shaded plot for any arbitrary cubic polynomial of the form

$$f(x) = ax^3 + bx^2 + cx + d \quad \text{over the range} \quad x_1 \le x \le x_2$$

Test the script file *calc_area* for the following cases:

a. $f_1(x) = 3x^3 + 5x^2 - 7x + 13$, over the range $-3 \le x \le 4$
b. $f_2(x) = -3x^3 + 2x^2 - x + 4$, over the range $-2.5 \le x \le 3$
c. $f_3(x) = 10x^3 + 3x^2 - 5x + 28$, over the range $-4 \le x \le 5$

MATLAB Solution
```
% Script file : calc_area.m
% This program-file returns the area under
% a cubic polynomial of the form:
% y = a*x^3 + b*x^2 +c *x + d ,
% over the range: x1 ≤ x ≤x2
% Input variables: a, b, c, d, x1, x2.
% Call syntax: calc_area
% Author: M.K............................ date: June, 2007
disp('*****************************************************')
disp('This program-file calculates and plots  ')
```

```
disp ('the area under a cubic polynomial defined by: ')
disp ('       y = a*x^3+b*x^2+c*x+d ')
disp('      over the range: x1 ≤ x ≤ x 2')
disp ('*****************************************************')
disp ('                                ')
disp ('                                ')
disp ('Provide the following data about the polynomial')
a = input('Enter the coefficient a = ');
b = input('Enter the coefficient b = ');
c = input('Enter the coefficient c = ');
d = input('Enter the coefficient d = ');
x1 = input('Enter the lower limit x1 = ');
x2 = input('Enter the upper limit x2 = ');
x = linspace(x1,x2,100);
y = a.*x.^3+b.*x.^2+c.*x+d;
trarea = trapz(x,y);
disp ('******************RESULTS**************************')
fprintf ('The area is=%f\n',trarea)
disp (['of y=a*x^3+b*x^2+c*x+d ,between ',num2str(x2),'<x<',num2str(x1)]);
disp ('The shaded plot is displayed in the figure window')
disp ('*****************************************************')
area (x,y)
xlabel ('x');
ylabel ('y');
grid on
title('Area of y=a*x^3+b*x^2+c*x+d vs x')
```

The script file *calc_area.m* is tested as follows:

```
>> calc _ area                          % part (a) for f1(x)


***************************************************
This program-file calculates and plots
the area under a cubic polynomial defined by:
    y=a*x^3+b*x^2+c*x+d
    over the range: x1 ≤ x ≤ x 2
***************************************************
Provide the following data about the polynomial
Enter the coefficient a= 3
Enter the coefficient b= 5
Enter the coefficient c= -7
Enter the coefficient d= 13
Enter the lower limit x1= -3
Enter the upper limit x2= 4
******************RESULTS**************************
The area is=349.472078
of y=a*x^3+b*x^2+c*x+d ,between -3<x<4
The shaded plot is displayed in the figure window
***************************************************
```

See Figure 9.19.

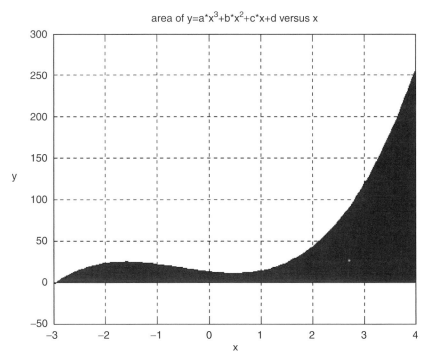

FIGURE 9.19
Plot of *calc_area* of Example 9.5(a).

```
>> calc _ area              % part(b) for f2(x)

*******************************************************
This program-file calculates and plots
the area under a cubic polynomial defined by:
    y=a*x^3+b*x^2+c*x+d
    over the range: x1 ≤ x ≤ x2
*******************************************************
Provide the following data about the polynomial
Enter the coefficient a= -3
Enter the coefficient b= 2
Enter the coefficient c= -1
Enter the coefficient d= 4
Enter the lower limit x1= -2.5
Enter the upper limit x2= 3
******************RESULTS*************************
The area is=17.587834
of y=a *x^3+b*x^2+c*x+d ,between   -2.5<x<3
The shaded plot is displayed in the figure window
*******************************************************
```

See Figure 9.20.

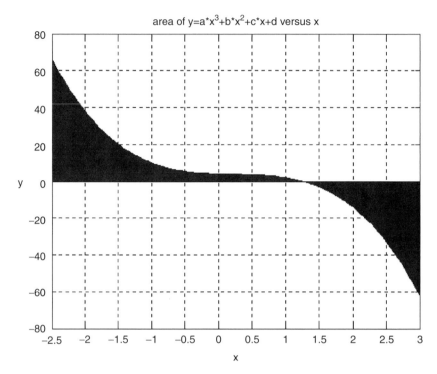

FIGURE 9.20
Plot of *calc_area* of Example 9.5(b).

```
>> calc _ area                    % part (c) for f3(x)


*****************************************************
This program-file calculates and plots
 the area under a cubic polynomial defined by:
     y =a*x^3+b*x^2+c*x+d
     over the range: x1 ≤ x ≤ x2
*****************************************************
Provide the following data about the polynomial
Enter the coefficient a= 10
Enter the coefficient b= 3
Enter the coefficient c= -5
Enter the coefficient d= 28
Enter the lower limit x1= -4
Enter the upper limit x2= 5

******************RESULTS*************************
The area is=1341.223140
of y=a*x^3+b*x^2+c*x+d ,between -4<x<5
The shaded plot is displayed in the figure window
*****************************************************
```

See Figure 9.21.

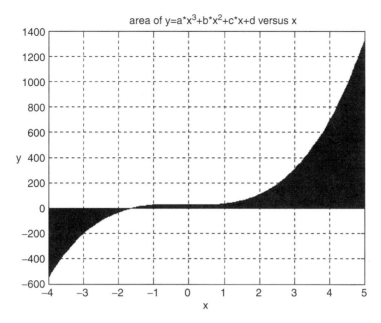

FIGURE 9.21
Plot of *calc_area* of Example 9.5(c).

Example 9.6

Create the function file *prodz1z2* that returns the product of two complex numbers *z1* and *z2* (in polar form) as magnitude and angle where angle is expressed in radians and degrees, given *z1* and *z2* in rectangular form.

Test the function file *prodz1z2* for the following case *z1 = 2 + 3i* and *z2 = 4 + 5i*.

MATLAB Solution
```
function[mag,phase _ rad,phase _ deg] = prodz1z2(z1,z2)
% This function file performs the product of
% two complex numbers z1 and z2 given in
% rectangular format, and returns their product
% in polar format with the angle expressed
% in radians and degrees.
% Inputs : z1, z2 (rectangular format)
% Output: magnitude and angle in radians and degrees
% Call syntax : [mag,angle _ rad,angle _ deg]= prodz1z2 (z1,z2)
% Author:M.K......................................... date: june 2007
%****************************************************
clc;
disp ('            ')
disp ('******************************************')
disp('**********R  E  S  U  L  T  S  *****************')
disp ('******************************************')
disp ('    The mag. and angle (in rad and degrees)');
disp ('    of the product of z1*z2 is given by:')
```

```
mag _ z1= abs(z1);
mag _ z2 = abs(z2);
mag = mag _ z1*mag _ z2;
angle _ z1 = angle(z1);
angle _ z2 = angle(z2);
phase _ rad = angle _ z1+angle _ z2;
phase _ deg = phaserad*(180/pi);
disp ('*************************************')
```

The function file *prodz1z2.m* is tested and the results are as follows:

MATLAB Solution
```
>> z1 = 2+3i;
>> z2 = 4+5i;
>> [mag,angle _ radians,angle _ degrees] = prodz1z2(z1, z2)

*********************************************
**********R E S U L T S *******************
*********************************************
   The mag. and angle (in rad and degrees)
   of the product of z1*z2 is given by :
   mag =
            23.0868
   angle _ radians =
               1.8788
   angle _ degrees =
              107.6501
*******************************************
```

Example 9.7

Let us revisit the capital–interest problem.

a. Create the script file *growth.m* that returns the growth of an initial capital of $1000 in a tablelike format that earns an interest rate of 6, 8, and 10% per annum during 10 years.
b. Change part a for the case of the interest rates (6, 8, and 10%) compounded quarterly. Call this new script file *growth_quot.m* that returns in addition to the table the plot indicating the nonlinearity growth, over the range $1 \le n$ (years) ≤ 10.

ANALYTICAL Solution

Part a. Let P be the present value ($1000), F the principal in the account at the expiration of n interest periods and referred as future worth, I the interest rate, and n (1, 2, ..., 9, 10) the number of years. Then the principal at the end of n periods is given by (Kurtz, 1985)

$$F = P(1 + I)^n \quad \text{(from Chapter 1)}$$

Part b. The preceding equation ($F = P * (1 + I)^n$) is modified for smaller intervals such as quarters as follows:

$$F = P * (1 + I/4)^{n*4}$$

MATLAB Solution
```
% part(a)
```

```
% Script file : growth.m
% The script file growth displays
% the in a tabe like format
% the growth of an initial capital
% of $1000, that earns 6%,8% and 10%
% per annum, during a period of 10 years.
% Inputs: none
% Outputs: growth of $1000 at 6%, 8% and 10%
% Call syntax: growth
% Author: M. K  .......................date: January, 2007
%*******************************************
clc
format compact
P = 1000;
n = 1:10;
one = ones(1,10);
one6 = one+.06;
one8 = one+.08;
one10 = one+.1;
G6 = (one6.^n).*P;
G8 = (one8.^n).*P;
G10 = (one10.^n).*P;
% Display table
disp ('*******************************************')
disp('         TABLE OF  GROWTH       ')
disp('     years        Fat6%      Fat8%     Fat10% ')
disp([n' G6' G8' G10'])
disp ('*******************************************')
```

The script file *growth.m* is tested and the results are as follows:

```
>> growth

*******************************************
            TABLE OF GROWTH
    years       Fat6%      Fat8%      Fat10%
   1.0e+003  *
     0.001      1.0600     1.0800     1.1000
     0.002      1.1236     1.1664     1.2100
     0.003      1.1910     1.2597     1.3310
     0.004      1.2625     1.3605     1.4641
     0.005      1.3382     1.4693     1.6105
     0.006      1.4185     1.5869     1.7716
     0.007      1.5036     1.7138     1.9487
     0.008      1.5938     1.8509     2.1436
     0.009      1.6895     1.9990     2.3579
     0.010      1.7908     2.1589     2.5937
*******************************************
```

MATLAB Solution
```
% part(b)
% Script file : growth_quot.m
% The script file growthquot.m
% traces the growth of an initial
% capital  of $1000 at an annual
```

```
% interest rate of 6%, 8% and 10%
% compounded quarterly during a
% period of 10 years.
% Inputs: none
% Outputs: growth of $10,000 at 6%,8%
%          and 10% compounded quarterly.
% Call syntax: growth _ quot
% Author: M.K.........................date: January 2007
%*****************************************
clc;clf;
format compact;
P =1000;
n =1:10;one=ones(1,10);
one6q = one+.02;
one10q = one+.025;
G6q= (one6q.^(n.*4)).*P;
G8q = (one8q.^(n.*4)).*P;
G10q = (one10q.^(n.*4)).*P;
disp ('***********************************************')
disp ('       TABLE OF GROWTH COMPOUNDED QUARTERLY');
disp ('*********************************************** ')
disp ('                                             ');
disp ('    YEARS       Fat6%       Fat8%       Fat10% ');
disp ('***********************************************');
disp ([n' G6q' G8q' G10q']);
disp ('***********************************************')
plot (n,G6q,':o',n,G8q,'*-',n,G10q,'+',n,G10q)
xlabel ('years');
ylabel ('growth($)');
title ('Quarterly growth of an initial $10,000');
legend ('6%','8%','10%');
grid on
```

The script file *growth_qout.m* is executed and the results are as follows:

```
>> growth _ quot

*******************************************************
TABLE OF GROWTH COMPOUNDED QUARTERLY
*******************************************************
YEARS        Fat6%       Fat8%       Fat10%
*******************************************************
  1.0e+003 *
     0.001    1.0614      1.0824      1.1038
     0.002    1.1265      1.1717      1.2184
     0.003    1.1956      1.2682      1.3449
     0.004    1.2690      1.3728      1.4845
     0.005    1.3469      1.4859      1.6386
     0.006    1.4295      1.6084      1.8087
     0.007    1.5172      1.7410      1.9965
     0.008    1.6103      1.8845      2.2038
     0.009    1.7091      2.0399      2.4325
     0.010    1.8140      2.2080      2.6851
*******************************************************
```

See Figure 9.22.

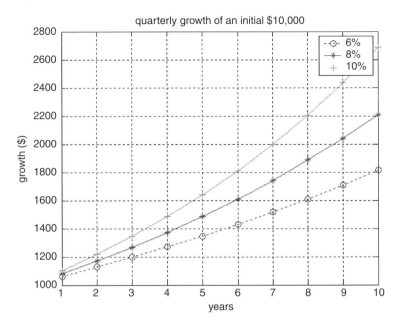

FIGURE 9.22
Growth plots of Example 9.7(b).

Example 9.8

Create the script file *effective_int.m*. that returns the effective interest rate of 10% per annum and its effect on the growth of an initial capital of $10,000, during 10 years, compounded in the following ways:

a Annually

b. Semi annually

c. Quarterly

d. Monthly

e. Weekly

f. Daily

g. Continuously

Indicate the different growth plots over the range $1 \leq n$ (years) ≤ 10. Also indicate, in a different plot, the last 8.4 months of growth of the different modes, over the range $9.3 \leq n$ (years) ≤ 10.

MATLAB Solution
```
% Script file: effective_int.m
% returns the different effective
% interest rates and display the plot of
% the growth of $10,000 at an annual interest
% rate of 10 % during 10 years, and the last 8.4 months
% with the following effective interest rates compounded:
%     (a) annually
%     (b) semi-annually
```

```
%      (c) quarterly
%      (d) monthly
%      (e) weekly
%      (f) daily
%      (g) continuously
% Inputs: none
% Outputs:  (a) plot of the   growth of $10,000 at 10%
%               during a period of 10 years, and the last 8.4 months
%           (b) Effective interest of 10% compounded:
%                   (1) annually
%                   (2) semi annually
%                   (3) quarterly
%                   (4) monthly
%                   (5) weekly
%                   (6) daily
%                   (7) continuously
% Call syntax: effective _ int
% Author: M.K ...................date: January, 2007
%*******************************************
clc; clf;
format compact
I =.10;
P = 10000;
I6month = (1.0+(10/2)*.01)^2-1;
I3month = (1.0+(10/4)*.01)^4-1;
I1month = (1.0+(10/12)*.01)^12-1;
I1week = (1.0+(10/52)*.01)^52-1;
I1day = (1.0+(10/365)*.01)^365-1;
I _ cont = exp(.1)-1;
disp('************************************************')
disp('******* EFFECTIVE INTEREST RATES**************');
disp('************************************************');
disp(['EFFECTIVE ANNUAL INTEREST=',num2str(I)]);
disp(['EFFECTIVE SEMMI-ANNUAL INTEREST=',num2str(I6month)]);
disp(['EFFECTIVE QUARTERLY INTEREST=',num2str(I3month)]);
disp(['EFFECTIVE MONTHLY INTEREST=',num2str(I1month)]);
disp(['EFFECTIVE WEEKLY INTEREST=',num2str(I1week)]);
disp(['EFFECTIVE DAILY INTEREST=',num2str(I1day)]);
    disp(['EFFECTIVE CONTINOUS COMPOUNDING INT=',num2str (I _ cont)]);
disp('*********************************************')
n=1:10;
Fyear=P.*(1+I).^n;
F6month=P.*(1+I6month).^n;
F3month=P.*(1+I3month).^n;
F1month=P.*(1+I1month).^n;
F1week=P.*(1+I1week).^n;
F1day=P.*(1+I1day).^n;
F _ cont=P.*(1+I _ cont).^n;
clf;
subplot(2,1,1);
grid on
plot(n,Fyear,n,F6month,n,F3month,n,F1month,n,F1week,n,F1day,n,F_cont);
title('years vs growth');
ylabel('Growth')
```

```
subplot(2,1,2);
plot(n,Fyear,n,F6month,n,F3month,n,F1month,n,F1week,n,F1day,n,F_cont);
title('years(9.3 through 10) vs growth');
ylabel('Growth')
xlabel('Years')
axis([9.3 10 25000 28000])
grid on
```

The script file *effective_int.m* is executed and the results are as follows:

```
>> effective_int

**********************************************************
******* EFFECTIVE INTEREST RATES********************
**********************************************************
EFFECTIVE ANNUAL INTEREST = 0.1
EFFECTIVE SEMMI-ANNUAL INTEREST = 0.1025
EFFECTIVE QUARTERLY INTEREST = 0.10381
EFFECTIVE MONTHLY INTEREST = 0.10471
EFFECTIVE WEEKLY INTEREST = 0.10506
EFFECTIVE DAILY INTEREST = 0.10516
EFFECTIVE CONTINOUS COMPOUNDING INT = 0.10517
**********************************************************
```

See Figure 9.23.

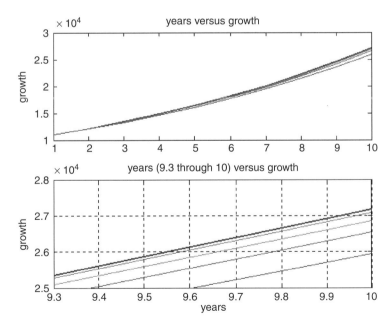

FIGURE 9.23
Growth plots of Example 9.8.

Example 9.9

Create the function file *growthf1* that returns the growth of an initial capital P in a table-like format, given the three interest rates $I1$, $I2$, and $I3$ per annum during a time period of n years.

Test the function file *growthf1* for the following cases:

   a. *P = 10,000, I1 = 3%, I2 = 6%, I3 = 9%,* and *n = 10*
   b. *P = 10,000, I1 = 5%, I2 = 7%, I3 = 9%,* and *n = 11*

MATLAB Solution
```
function growthf1(P,I1,I2,I3,n)
% This function file computes the growth
% rates I1,I2,I3 and duration of n years.
% Inputs :P,I1,I2,I3,n.
% Call syntax:growthf (P,I1,I2,I3,n)
% Author :M. Kalechman ,date:june 2002.
%*******************************************
one = ones(1,n);
oneI1 = one+one.*(I1*(.01));
oneI2 = one+one.*(I2*(.01));
oneI3 = one+one.*(I3*(.01));
m =1:1:n;
F1= (oneI1.^m).*P;
F2=(oneI2.^m).*P;
F3=(oneI3.^m).*P;
disp('*********************************************');
disp('   Years n    Fut.%I1   Fut.%I2    Fut.%I3');
disp('*********************************************');
disp([m' F1' F2' F3']);
disp('*********************************************');
```

The function file *growthf1.m* is tested for the following cases:

   a. *P = 10,000, I1 = 3%, I2 = 6%, I3 = 9%,* and *n = 10*
   b. *P = 10,000, I1 = 5%, I2 = 7%, I3 = 9%,* and *n = 11*

and the results are as follows:

```
>> growthf1(10000, 3, 6, 9, 10)        %***** part (a)

*********************************************
    Years n     Fut.%I1   Fut.%I2    Fut.%I3
*********************************************
    1.0e+004 *
      0.0001    1.0300    1.0600    1.0900
      0.0002    1.0609    1.1236    1.1881
      0.0003    1.0927    1.1910    1.2950
      0.0004    1.1255    1.2625    1.4116
      0.0005    1.1593    1.3382    1.5386
      0.0006    1.1941    1.4185    1.6771
      0.0007    1.2299    1.5036    1.8280
      0.0008    1.2668    1.5938    1.9926
      0.0009    1.3048    1.6895    2.1719
      0.0010    1.3439    1.7908    2.3674
*********************************************

>> growthf1(10000, 5, 7, 9, 11)        %***** part(b)
```

```
***************************************************
   Years n      Fut.%I1   Fut.%I2    Fut.%I3
***************************************************
   1.0e+004  *
     0.0001     1.0500    1.0700     1.0900
     0.0002     1.1025    1.1449     1.1881
     0.0003     1.1576    1.2250     1.2950
     0.0004     1.2155    1.3108     1.4116
     0.0005     1.2763    1.4026     1.5386
     0.0006     1.3401    1.5007     1.6771
     0.0007     1.4071    1.6058     1.8280
     0.0008     1.4775    1.7182     1.9926
     0.0009     1.5513    1.8385     2.1719
     0.0010     1.6289    1.9672     2.3674
     0.0011     1.7103    2.1049     2.5804
***************************************************
```

Example 9.10

Create the function file *solv* that returns the solution of the following set of linear equations:

$$(3 - u)x + 5y + 2z = 2$$

$$(-2u)x + 2y - 6z = -1$$

$$(4/u)x - 3y + 5z = 4$$

where *u* may represent a disturbance (over a range). Also explore and evaluate the conditions for the existence of the solutions.

a. Use the help command to obtain information about the file *solv*.
b. Test the function file *solv*, by creating the script file *test_solv*, that returns the solution of the given set of equations for $u = -3.61$.
c. Repeat part b over the range $a \le u \le b$ with linear increments given by the variable $c(u = a{:}c{:}b)$.
d. Test part c for $1 \le u \le 3$ with unit increments $(u = 1{:}1{:}3)$.

ANALYTICAL Solution

The preceding set of equations can be expressed in a matrix format as

$$\begin{bmatrix} (3-u) & 5 & 2 \\ (-2u) & 2 & -6 \\ (4/u) & -3 & 5 \end{bmatrix} \begin{bmatrix} x \\ y \\ z \end{bmatrix} = \begin{bmatrix} 2 \\ -1 \\ 4 \end{bmatrix}$$

Let

$$A = \begin{bmatrix} (3-u) & 5 & 2 \\ (-2u) & 2 & -6 \\ (4/u) & -3 & 5 \end{bmatrix}, \quad B = \begin{bmatrix} 2 \\ -1 \\ 4 \end{bmatrix}, \text{ and } XYZ = \begin{bmatrix} x \\ y \\ z \end{bmatrix}$$

then

$$[A] * [XYZ] = [B]$$

Recall that the existence of a solution of a system of equations is associated to the value of the *det(A)* and condition number *cond(A)* (recall that the condition number indicates the singularity condition of *A*).

MATLAB Solution

```
function[x,y,z,detA,condA]=solv(u)
%*******************************************
% This function file solves the following
% set of linear equations:
%         (3-u)*x+5*y+2*z= 2   ;
%         (-2*u)*x+2*y-6*z=-1;
%         (4/u)*x-3*y+5*z= 4   ;
% for the variables x, y, z, det(A(u)), cond(A(u))
% where A=[(3-u) 5 2 ;(-2*u) 2 6;(4/u) -3 5] ;
% Inputs : u
% Outputs:x, y, z, detA, condA
%          where detA= det(A) and condA = cond(A)
% Author: M.K.                 date:june 2002.
% Call syntax:[x, y, z, detA, condA] = solv(u)
%***********************************************
clc;
A = [(3-u) 5 2 ;(-2*u) 2 6;(4/u) -3 5] ;
B = [2;-1;4];
DetA = det(A);
condA = cond(A);
xyz = inv(A)*B;
disp('************RESULTS*********************')
disp('The values of  x, y, and z are:');
disp([xyz]);
disp('The values of det(A) and the cond(A) are:');
disp([detA condA])
disp('*******************************************')

% Script file = test_solv
disp(' This file solves the following  set equations:')
disp('         (3-u)*x+5*y+2*z= 2 ') ;
disp('         (-2*u)*x+2*y-6*z=-1');
disp('         (4/u)*x-3*y+5*z= 4 ') ;
disp(' for the variables x, y, z, det(A(u)), cond(A(u))');
disp('  where A= [(3-u) 5 2 ;(-2*u) 2 6;(4/u) -3 5]');
disp('over the range a ≤u ≤b')
disp('*******************************************')
disp(' Provide the following information about the range of the
                                    disturbance (u)')
a=input('Enter the lower value of the disturbance =')
b=input('Enter the higher  value of the disturbance =')
c=input('Enter the increment size=')
for u =a:c:b
fprintf('The result is for perturbation u=%f\n',u)
solv(u)
end
```

The file *solv* and *test_ solv* are tested and the results are as follows:

```
>> help solv                    % part(a)
```

```
*********************************************
This function file solves the following
set of linear equations:
        (3-u)*x+5*y+2*z= 2   ;
        (-2*u)*x+2*y-6*z=-1;
        (4/u)*x-3*y+5*z= 4   ;
for the variables x, y, z, det(A(u)), cond(A(u))
where A=[(3-u) 5 2 ;(-2*u) 2 6;(4/u) -3 5] ;
Inputs : u
Outputs:x,y,z,detA,condA
        where detA=det(A) and condA=cond(A)
Author: M. K.            date: june 2002.
Call syntax :[x,y,z,detA,condA]=solv(u)
*********************************************
```

The function *solv.m* is executed first for $u = -3.61$ and then over the range $1 \le u \le 3$ by calling the file *test_solv*.

```
>> solv(-3.61)                  % part (b)

************RESULTS**********************
The values of x, y, and z are:
   -2.8276
    3.2812
    2.1421

The values of det(A) and the cond(A) are:
  -67.5489    15.4333

*********************************************

>> test_solv                    % part(c/d)

This file solves the following set of equations:
            (3-u)*x+5*y+2*z = 2
            (-2*u)*x+2*y-6*z =-1
            (4/u)*x-3*y+5*z = 4
for the variables x, y, z, det(A(u)), cond(A(u))
where A=[(3-u) 5 2 ;(-2*u) 2 6;(4/u) -3 5]
over the range a ≤u ≤b')
Enter the lower value of the disturbance =1
Enter the higher  value of the disturbance =3
Enter the increment size=1

*********************************************
The result is  for perturbation u=1.000000
************RESULTS**********************
The values of x, y, and z are:
    0.8604
    0.0090
    0.1171

The values of det(A) and the cond(A) are:
  222.0000     1.9714

*********************************************
The result is for perturbation u=2.000000
************RESULTS**********************
```

```
The values of x, y, and z are:
    0.9363
    0.0343
    0.4461
The values of det(A) and the cond(A) are:
   204.0000    2.1279

***********************************************
The results is for perturbation u=3.000000
*************RESULTS************************
The values of x, y, and z are:
    0.8656
    0.1390
    0.6526

The values of det(A) and the cond(A) are:
   220.6667    2.4688

***********************************************
```

Example 9.11

Evaluate the performance of each of the three script files, define below where each script returns the sum of the square roots of all the positive integers smaller than 501 by creating

  a. The script file *vector_loop* that employs the *implied* loop
  b. The script file *forend_loop* that employs the *for-end* command
    i. Without preallocation memory for the loop variable
    ii. With preallocation memory for the loop variable
  c. The script file *whileend_loop* that employs the *while-end* command

For each of the preceding files, estimate their performance by evaluating the following timing parameters:

  1. Estimated time
  2. Cpu time
  3. Elapsed time

ANALYTICAL Solution

The objective of each one of the three files defined earlier is to evaluate the y variable defined by

$$y = \sum_{i=1}^{500} (i)^{1/2}$$

MATLAB Solution

```
% Script file: vector _ loop.m
% This program-file returns
% the sum of the square
% roots of all the numbers
% between 0 and 500,
% using the implied loop method.
% Input: none
```

```
% Output:  (1)  sum(sqrt(1,2,...500))
%          (2)  statistics: a. estimated time
%                           b. cpu time
%                           c. elapse time
% Call syntax: vector _ loop
% Author: M.K..................date: January 2007
%*****************************************
% solution (a)
tstart = clock;tic;
ta = cputime;
xa =1:500;
solution _ a = sum(sqrt(xa));
estimea = etime(clock,tstart);
tictoca = toc;
t1= cputime-ta;
disp ('***********RESULTS USING THE IMPLIED LOOP*******')
disp (['Sum of  sqrt of 1 _ 500 =',num2str(solution _ a)]);
disp ('****************************************************')
disp ('******TIME STATICS USING THE IMPLIED LOOP*******')
fprintf ('est.time =   %6.6f\n',estimea);
fprintf ('cpu.time =   %6.6f\n',t1);
fprintf ('time elapsed =%6.6f\n',tictoca);
disp ('****************************************************')

% Script file: forend _ loop.m
%*****************************************
% This program-file returns
% the sum of the square
% roots of all the numbers
% between 0 and 500,
% using the implied loop method.
% Input: none
% Output:   (1)  sum(sqrt(1,2,...500))
%           (2)  statistics:  a. estimated time
%                             b. cpu time
%                             c. elapse time
% Call syntax: vector _ loop
% Author: M.K  ............   date: January 2007
%*****************************************
% solution (b)
tstartb = clock;tic;
tb = cputime;
for xb=1:500;
   b(xb)=sqrt(xb);
end
solution _ b=sum(b);
estimeb=etime(clock,tstartb);
tictocb=toc;
t2=cputime-tb;
disp('***********RESULTS USING THE FOR-END**********')
disp(['Sum of  sqrt of 1 _ 500 =',num2str(solution _ b)]);
disp('****************************************************')
disp('*******TIME STATISTICS USING THE FOR-END********')
fprintf('est.time =   %6.6f\n',estimeb);
fprintf('cpu.time =   %6.6f\n',t2);
```

```
fprintf('time elapsed =%6.6f\n',tictocb);
disp('*************************************************')

% Script file: whileend_loop.m
% This program-file returns
% the sum of the square
% roots of all the numbers
% between 0 and 500,
% using the implied loop method.
% Input: none
% Output: (1) sum(sqrt(1,2,...500))
%          (2) statistics: a. estimated time
%                          b. cpu time
%                          c. elapse time
% Call syntax: vector _ loop
% Author: M.K.................date: January 2007
%***************************************
% solution (c)
tstartc = clock;tic;
tc= cputime;
xc =1;y =501;c =0;
while xc<y;
c = c+sqrt(xc);
xc = xc+1;
end
solution_c= c;
estimec = etime(clock,tstartc);
tictocc = toc;
t3 = cputime-tc;
disp ('***********RESULTS USING THE WHILE-END***********')
disp (['Sum of  sqrt of 1_500 =',num2str(solution _ c)]  );
disp ('*************************************************')
disp ('*******TIME STATISTICS USING THE WHILE-END******')
fprintf ('est.time =%6.6f\n',estimec);
fprintf ('cpu.time =%6.6f\n',tc);
fprintf ('time elapsed =%6.6f\n',tictocc);
disp ('*************************************************')
```

Each of the preceding script files are executed, and the results are indicated as follows:

```
>> vector _ loop                    % part(a)

***********RESULTS USING THE IMPLIED LOOP******
Sum of  sqrt of 1 _ 500 =7464.5342
*************************************************
******* TIME STATICS USING THE IMPLIED LOOP****
est.time =      0.010000
cpu.time =      0.030000
time elapsed = 0.031000
*************************************************


>> forend _ loop                    % part (b1)

***********RESULTS USING THE FOR-END************
```

```
Sum of  sqrt of 1 _ 500 = 7464.5342
**************************************************
****** TIME STATISTICS USING THE FOR-END*******
est.time =    0.010000
cpu.time =    0.030000
time elapsed = 0.030000
**************************************************

>> b= zeros(1,500);
>> forend _ loop                    % part (b2)

***********RESULTS USING THE FOR-END************
Sum of  sqrt of 1 _ 500 =7464.5342
**************************************************
*******TIME STATISTICS USING THE FOR-END*******
est.time =    0.011000
cpu.time =    0.020000
time elapsed =0.020000
**************************************************


>> whileend _ loop                  % part (c)

***********RESULTS USING THE WHILE-END*********
Sum of  sqrt of 1_500 =7464.5342
**************************************************
****TIME STATISTICS USING THE WHILE-END***
est.time = 0.020000
cpu.time = 11.686000
time elapsed = 0.040000
**************************************************
```

Observe that the sum of the square roots of all positive numbers smaller than 501 is exactly the same (7464.5342) in all the four cases; but the computational times are quite different, with the *implied* loop vector approach as the most efficient and the *while-end* as the least efficient. Observe that the *for-end* solution shows good time performance that is significantly improved by preallocating the memory size of the loop variable (part b).

Example 9.12

Create the function file *statis_perf* that returns the following statistics of a given row array *x*:

- The average
- The standard deviation
- The variance
- The GM
- The RMS
- The HM
- The timing statistics
  - Estimated execution time
  - Cpu time
  - Elapse time

Test the function file *statis_perf* for the following input arrays:

a. *x1 = rand(1, 33)*

b. *x2 = randn(1, 99)*

c. Get information about *statis_perf*

d. Discuss and compare the timing statistics

ANALYTICAL Solution

The variables to be used in *statis_perf* are defined as follows:

$$n = \text{length}(x) \qquad (n \text{ is the number of elements})$$

$$\text{ave} = \frac{1}{n}\sum_{i=1}^{n} x(i) \qquad (\text{average})$$

$$\text{dev} = \sqrt{\sum_{i=1}^{n} (x(i))^2/n - \text{ave}} \qquad (\text{standard deviation})$$

$$\text{geomean} = \sqrt{\prod_{i=1}^{n} (x(i))} \qquad (\text{GM})$$

$$\text{rms} = \sqrt{\frac{1}{n}\sum_{i=1}^{n} x(i)^2} \qquad (\text{RMS})$$

$$\text{harmean} = \frac{n}{\sum_{i=1}^{n} \frac{1}{xi}} \qquad (\text{HM})$$

MATLAB Solution

```
function statis_perf(x)
% This function file returns the
% statistical values of a given row
% array of data x such as :
%    * average (ave)
%    * standard deviation(dev)
%    * variance (var)
%    * geometric mean (geomean)
%    * root mean square (rms)
%    * harmonic mean (harmean)
%    * timing statistics:
%            • estimated time
%            • cpu time
%            • elapse time
% Input : x (row array of data)
% Outputs: ave, dev, var, geomean, rms, harmean, performing
%          statistics
% Call syntax: [sta,perf] = statis_perf(x)
% Author : M.K ....................date: January, 2007
%***********************************************
tstart = clock;
tic;
t0 = cputime;;
n = length(x);
ave = sum(x)/n;
dev = sqrt(sum(x.^2)/n-ave^2);
one = ones(1,n);
var = sum(x-one.*ave)/(n-1);
```

```
geomean = sqrt(prod(x));
rms = sqrt(sum(x.^2)/n);
harmean = n/sum(one./x);
sta = [ave dev var geo_mean rms har_mean];
disp ('**************STATISTICAL  RESULTS*******************');
disp ('   ave    dev    var    geo_mean   rms    har_mean');
disp (sta)
disp ('********************************************************');
timeelaps = etime(clock,tstart);
ter = toc;
tf = cputime-t0;
disp ('****************TIMING  RESULTS**********************');
fprintf ('est.time =......%6.6f\n',timeelaps);
fprintf ('cpu.time =......%6.6f\n',tf);
fprintf ('time elapsed =...%6.6f\n',ter);
disp('********************************************************');
```

```
>> help statis _ perf                          % part(c)
```

This function file returns the
statistical values of a given row
array of data x such as :
    * average (ave)
    * standard deviation(dev)
    * variance (var)
    * geometric mean (geomean)
    * root mean square (rms)
    * harmonic mean (harmean)
    * timing statistics:
        • estimated time
        • cpu time
        • elapse time
 Input : x (row array of data)
 Outputs: ave, dev, var, geomean, rms, harmean, performing
       statistics
 Call syntax: [sta,perf] =statis _ perf(x)
 Author : M.K ....date: January, 2007

```
**********************************************
>> x1 = rand(1, 33);                           % part(a)
>> statis _ perf(x1)
*************STATISTICAL RESULTS***********************************
      ave      dev      var      geo _ mean    rms     har _ mean
     0.5089   0.2710   0.0000      0.0000     0.5766    0.1867
******************************************************************
*****************TIMING   RESULTS*********************************
est.time =......   0.090000
cpu.time =.....  .0.091000
time elapsed =...0.090000
******************************************************************
>> x2 = randn(1,99);                          % part (b)
>> statis _ perf(x2)
***************STATISTICAL  RESULTS*******************************
      ave      dev      var      geo _ mean    rms     har _ mean
    -0.1317   0.9302  -0.0000      0.0000     0.9394     1.0871
```

```
**********************************************************************
********************TIMING   RESULTS*********************************
est.time =......   0.100000
cpu.time =......   0.110000
time elapsed =...0.110000
**********************************************************************
```

Observe that *x2* consists of an array of elements that is three times greater than *x1*, but the execution time increases by less than 12%.

Example 9.13

Create the function file *graph_perf* that, given a (data) row array *x*, returns the following plots:

1. The [number of elements of *x*] versus [magnitude of the elements of *x*]
2. The histogram of *x* consisting of five bins
3. Pie graph of the histogram
4. Profile plot and report

Test this function file by executing the following:

a. The *help graph_perf* command
b. Let the input array *x* be given by *x1* = *rand(1, 33)*
c. Let the input array *x* be given by *x2* = *rand(1, 99)*, and display
    1. The sample and pie graphs
    2. The profile report and profile plot
d. Discuss and compare the results

MATLAB Solution
```
function graph _ perf(x)
% This function file returns the plots of the
% statistical behavior of the row data
% array  x, , and the profile
% (timing) performance of this file.
% This function returns:
%     * plot of   [Amplitude of x] vs [#of samples]
%     * histogram of x consisting of 5 bins
%     * pie graph
%     * profile plot
%     * profile report
% Inputs : x (row array of data)
% Outputs:     * plot of [Amplitude of  x] vs. [# of samples of x]
%                * histogram of x consisting of 5 bins
%                * profile plot
%                * profile report
% Call syntax: graph _ perf(x)
% Author: M.K.............................      date: January 2007
%***********************************************************
profile graph _ perf
profile on
n = length(x);
m = 1:n;
subplot (2,1,1);
plot(m,x)
```

```
title ('Amplitude vs # of SAMPLES');
xlabel ('# of samples');
ylabel ('Amplitude');
grid on;
subplot (2,1,2);
[N,X] = hist(x,5);                          % histogram with 5 bins
hist (x,5);                                 % histogram plot
title ('Histogram with 5 bins')
disp ('****************************************************');
disp ('If you wish , print the plots'.);
disp ('Enter «return» to continue the execution');
disp ('of this function ; next figure is the pie plot.');
disp ('****************************************************');
keyboard;                                   % pause
pie(N)
title ('Pie plot of the histogram')
legend ('bin#1','bin#2','bin#3','bin#4','bin#5')
disp('****************************************************')
disp ('If you wish , print the pie plots.');
disp ('Enter «return» to continue the execution');
disp ('of this function ; next is the profile report.')
disp ('****************************************************')
keyboard                                    % pause
profile report
disp ('****************************************************')
disp ('Enter «return» to continue the execution')
disp ('of this function ; next figure is the profile plot')
disp ('****************************************************')
keyboard;                                   % pause
profile plot

>> help graph _ perf                        % part(a)

  This function file returns the plots of the
  statistical behavior of the row data
  array x, , and the profile
  (timing) performance of this file.
  This function returns:
     * plot of [Amplitude of x] vs [# of samples of x]
     * histogram of x consisting of 5 bins
     * pie graph
     * profile plot
     * profile report
  Inputs: x (row array of data)
  Outputs: * plot of [Amplitude of  x] vs. [# of samples of x]
       * histogram of x consisting of 5 bins
       * profile plot
       * profile report
  Call syntax: graph_perf(x)
  Author: M.K..........................        date: January 2007

**************************************************

>> x1= rand(1,33);                          % part (b)
>> graph _ perf(x1);
**************************************************
```

```
If you wish, print the plots.
Enter "return" to continue the execution
of this function ; next figure is the pie plot.
****************************************************
```

See Figure 9.24.

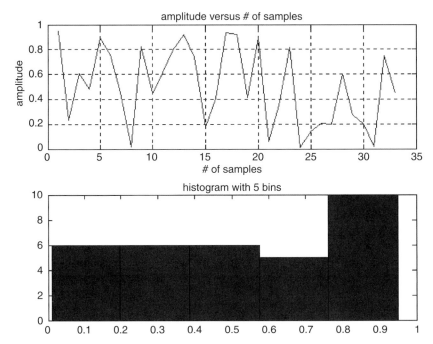

FIGURE 9.24
Sample plot and histogram of Example 9.13.

```
K>> return
****************************************************
If you wish , print the pie plots.
Enter "return" to continue the execution
of this function ; next is the profile report.
****************************************************
```

See Figure 9.25.

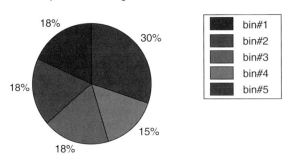

FIGURE 9.25
(See color insert following page 342.) Pie plots of x_1 of Example 9.13.

```
K>> return

Total time in "C:\MATLAB\work\graph _ perf.m": 204.58 seconds
100% of the total time was spent on lines: [40 49 33 43 41 25 26
                                                         32 31 30]
                    24: m=1:n;
  0.07s,    0%     25: subplot(2,1,1);
  0.06s,    0%     26: plot(m,x)
                    27: title('MAG. vs # of SAMPLES');
                    28: xlabel('# of samples');
                    29: ylabel('Magnitude');
  0.02s,    0%     30: grid on;
  0.02s,    0%     31: subplot(2,1,2);
  0.04s,    0%     32: [N,X]=hist(x,5);%histogram with 5 bins
  0.31s,    0%     33: hist(x,5);%histogram plot
                    34: title('Histogram with 5 bins')

                    39: disp('**************************************');
173.57s,   85%     40: keyboard;%pause;
  0.24s,    0%     41: pie(N)
                    42: title('Pie plot of the histogram')
  0.25s,    0%     43: legend('bin#1','bin#2','bin#3','bin#4','bin#5')
                    44: disp('**************************************')

                    48: disp('**************************************')
 29.98s,   15%     49: keyboard; % pause;
                    50: profile report


**********************************************************
Enter "return" to continue the execution
of this function ; next figure is the profile plot.
**********************************************************
```

See Figure 9.26.

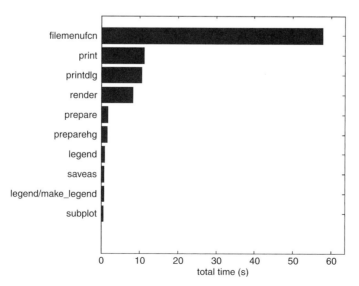

FIGURE 9.26
Profile plot of x_1 of Example 9.13.

```
K>> return

>> x2 = rand (1, 99);
>> graph_perf (x2);                                % part(c)


***********************************************************
If you wish , print the plots.
Enter "return" to continue the execution
of this function ; next figure is the pie plot.
***********************************************************
K>> return


***********************************************************
If you wish , print the pie plots.
Enter "return" to continue the execution
of this function ; next is the profile report
***********************************************************
```

See Figure 9.27.

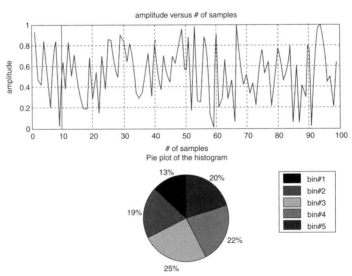

FIGURE 9.27
(See color insert following page 342.) Pie plots for x_2 of Example 9.13.

```
K>> return

Total time in "C:\MATLAB\work\graph_perf.m": 132.88 seconds
100% of the total time was spent on lines: [40 49 33 41 26 43 25
                                                    48 42 35]
                    24: m = 1:n;
    0.02s,  0%      25: subplot(2,1,1);
    0.04s,  0%      26: plot(m,x)
                    27: title('MAG. vs # of SAMPLES');
                    32: [N,X] = hist(x,5);%histogram with 5 bins
    0.11s,  0%      33: hist(x,5);%histogram plot
                    34: title('Histogram with 5 bins')
    0.01s,  0%      35: disp('***************************************');
                    36: disp('If you wish ,print the plots');
```

```
                         39: disp('****************************************');
104.61s,  79%            40: keyboard;%pause;
  0.09s,   0%            41: pie(N)
  0.01s,   0%            42: title('Pie plot of the histogram')
  0.03s,   0%            43: legend('bin#1','bin#2','bin#3','bin#4','bin#5')
                         44: disp('************************************')
                         47: disp('of this function ;next is the profile
                             report')
  0.01s,   0%            48: disp('************************************')
 27.93s,  21%            49: keyboard;%pause;
                         50: profile report

****************************************************
Enter "return" to continue the execution
of this function ;next figure is the profile plot.
****************************************************

K>> return
```

See Figure 9.28.

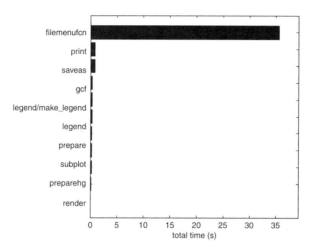

FIGURE 9.28
Profile plot of x_2 of Example 9.13.

Observe from the file reports that the total time spent in the execution of the program lines decreased from 204.58 s (array $x1$) to 132.88 s (array $x2$), and the plots indicate, for example, that the print command consumes approximately 12 s (for array $x1$) and approximately 1 s (for array $x2$), when the number of elements of $x2$ is three times greater (99) than the elements of $x1$ (33). Note that the *profile* results were obtained by executing the same function file. Observe also that the better performance is a consequence of the parsed or precompiled attribute of MATLAB.

Example 9.14

Let the 100 × 100 sparse matrix A be defined by the following MATLAB command:

$$A = sprand(n, n, dens) + speye(n), \quad \text{for } n = 100,$$

over the following range of densities $0.02 \leq dens \leq 0.2$, with linear increments of 0.02.

Create the script file *sparse_full* that compares the timing statistics of performing sparse matrix operations versus full matrix operations, returning the following:

a. A table with the densities of *A* over the indicated range and timing required to perform *[A * inv(A)]* and *[B * inv(B)]*, where *B = full(A)*

b. The cumulative cpu and estimated times to perform *[A * inv(A)]* and *[B * inv(B)]* over the indicated density range

c. Graphs of the sparse matrix *A* over the indicated density range

d. Plots of the cumulative cpu times of sparse *[A * inv(A)]* versus *[densities]* and the cumulative cpu times of full *[B * inv(B)]* versus *[densities]*

e. Plots of the cumulative estimated times of sparse *[A * inv(A)]* versus *[densities]* and cumulative estimated times of full *[B * inv(B)]* versus *[densities]*

MATLAB Solution

```
% Script file = sparse_full
n =100;k=1;
disp ('********** Performance Results ***************')
disp ('********* Sparse vs  full matrix *************')
disp (' ************* [A*inv(A)]   ******************');
disp (' dens    cpu.time _ sp  est.time _ sp cpu.time _ full est.
                                              time _ full');
tstart = clock; totaltimeA = 0; totaltimeB =0;
totalcpuA = 0; totalcpuB = 0;


figure(1)
for dens = 0.02:0.02:0.2;
   A= sprand(n,n,dens)+speye(n);            % 100x 100 sparse matrix
   tstartA = clock; t0A = cputime;
   prodsp = A*inv(A);
   tfA = cputime-t0A;
   totalcpuA = tfA+totalcpuA;tcpu _ A(k)  = totalcpuA;
   timeelapsA = etime(clock,tstartA);
   totaltimeA = totaltimeA+timeelapsA; telapse _ A(k)=totaltimeA;
   subplot(5,2,k)
   spy(A)
   B = full(A); tstartB = clock; t0B = cputime;
   prodfull =B*inv(B);
   tfB =cputime-t0B; totalcpuB = tfB + totalcpuB;
   tcpu_B(k) = totalcpuB;
timeelapsB = etime(clock,tstartB);
totaltimeB = totaltimeB+timeelapsB; telapse _ B(k) = totaltimeB;
fprintf('%4.3f %6.5f %6.5f %10.8f  %10.8f\n',dens,tfA,timeelapsA,tfB,
                                              timeelapsB);


k = k+1;
end
dens = 0.02:0.02:0.2;


figure(2)
```

```
plot (dens,tcpu _ A,dens,tcpu _ A,'s',dens,tcpu _ B,dens,tcpu _ B,'h');
xlabel ('density of matrix A');
title (' [time elapsed in sec.] vs [density of A]');
ylabel (' time elapsed in sec');


figure (3)
plot (dens,telapse_A,dens,telapse _ A,'*',dens,telapse _ B,dens,
                                      telapse _ B,'+');
xlabel ('density of matrix A');
ylabel ('cpu time in sec. ');
title (' [cpu time of (A*inv(A)]]] vs [density of A]');
disp ('************************************************************');
disp ('*******Perf. results for sparse[ A*inv(A)]****************');
fprintf ('total exec. time =......%6.6f\n',totaltimeA);
fprintf ('total cpu.time =......%6.6f\n',totalcpuA);
disp ('************************************************************');
disp ('***********Perf. results for full[ A*inv(A)]***************');
fprintf ('total exec.time =......%6.6f\n',totaltimeB);
fprintf ('totaL cpu.time =......%6.6f\n',totalcpuB);
disp ('************************************************************')
```

See Figure 9.29.

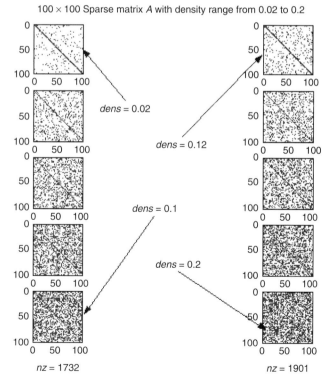

FIGURE 9.29
Spy diagrams of the 100×100 sparse matrix A, over the density range 0.02 to 0.2.

The script file *sparse_full* is executed and the results are as follows:

```
>> sparse _ full

A********** Performance Results **********************************
********* Sparse vs full matrix **********************************
************[A*inv(A) ]*******************************************
dens   cpu.time _ sp   est.time _ sp   cpu.time _ full   est.time _ full
0.020  0.01000         0.01000         0.00000000        0.00000000
0.040  0.01000         0.01000         0.01000000        0.01000000
0.060  0.02000         0.02000         0.01000000        0.01000000
0.080  0.02000         0.02100         0.01000000        0.01000000
0.100  0.02000         0.02000         0.01000000        0.01000000
0.120  0.02000         0.02000         0.00000000        0.00000000
0.140  0.02000         0.02000         0.01000000        0.01000000
0.160  0.02000         0.02000         0.00000000        0.00000000
0.180  0.02000         0.02000         0.01000000        0.01000000
0.200  0.02000         0.02000         0.01000000        0.01000000
*****************************************************************
******* Perf. results for sparse[ A*inv(A)] ********************
total exec. time =....... 0.181000
total cpu. time =......  0.180000
*****************************************************************
************ Perf. results for full[ A*inv(A)] ****************
total exec.time =......0.070000
totaL cpu.time  =......0.070000
**************************************************************
```

Observe the unexpected results. The operations using the sparse matrix *A* consume more time than the equivalent operations using the full matrix *A* (Figures 9.30 and 9.31).

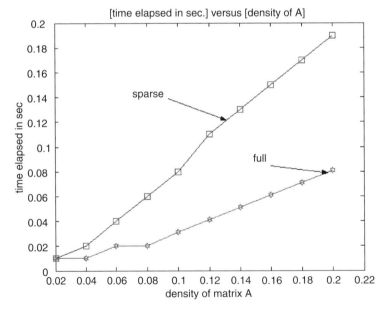

FIGURE 9.30
Evaluations of sparse and full matrix operations of Example 9.14.

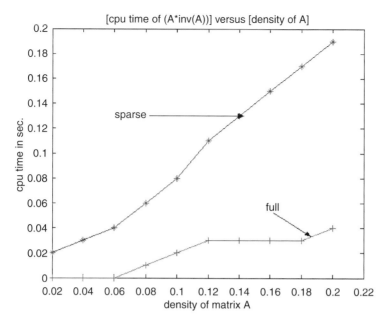

FIGURE 9.31
Computational time of sparse and full matrix operations of Example 9.14.

Example 9.15

Create the script file *timing_oper* that returns the average required time (computed by executing 10,000 times the same operation) to perform the following simple MATLAB operations in a tablelike format and corresponding plot for

1. Addition
2. Multiplication
3. Division
4. Squaring
5. Square root
6. Exponentiation

MATLAB Solution
```
% Script file= timing_oper
n = 1;
disp('^^^^^^^^^^^^^^^^^^^^^^^^^^^^^^^^^^^^^^^^^^^^^^^^^^^^^^')
disp('**********Performance Timing Results ***********')
disp(' ^^^^^^^^^^^^^^^^^^^^^^^^^^^^^^^^^^^^^^^^^^^^^^^^^^^^^');
% addition
tic
for I=1:10000;
    add(I)=I+1;
end
Average_add = toc/10000;
A(n) = Average _ add;
N = n+1;
% product
```

```
tic
for J=1:10000;
    pro(J)=J*2;
end
Average_pro = toc/10000;
A(n) =Average _ pro;
n = n+1;
% division
tic
for K=1:10000;
    div(K)=K/2;
end
Average _ div=toc/10000;
A(n) =Average _ div;
n =n+1;
% squaring
tic
for L=1:10000;
    sqr(L)=L^2;
end
Average _ sqr = toc/10000;
A(n) = Average _ sqr;
n = n+1;
% square root
tic
for M =1:10000;
    srt(M) = sqrt(M);
end
Average _ srt = toc/10000;
A(n) =Average _ srt;
n = n+1;
% exponentiation
tic
for N =1:10000;
    ext(J) =exp(N);
end
Average _ ext = toc/10000;
A(n) =Average _ ext;
X =1:6;
stem(x,A)
title('Average time per operation')
xlabel('operations')
ylabel(' time in seconds')
axis([0 7 0 0.00009]);
disp('********************************************************');
fprintf('The ave. time of a simple addition in sec. is =......
                                  %10.9f\n',Average _ add);
fprintf('The ave. time   of a simple multiplication in sec. is=......
                                  %10.9f\n',Average _ pro);
fprintf('The ave. time of a simple division in sec. =......
                                  %10.9f\n',Average _ div);
```

```
fprintf('The ave. time of squaring in sec. =......
                       %10.9f\n',Average _ sqr);
fprintf('The ave. time of a square root in sec. =......
                       %10.9f\n',Average _ srt);
fprintf('The ave. time of an exponentiation in sec. =......%
                       %10.9f\n',Average _ ext);
disp('********************************************************');
```

The script file timing _ oper is executed and the results are as follows:

```
>> timing _ oper
^^^^^^^^^^^^^^^^^^^^^^^^^^^^^^^^^^^^^^^^^^^^^^^^^^^^^^
**********Performance Timing Results ************
^^^^^^^^^^^^^^^^^^^^^^^^^^^^^^^^^^^^^^^^^^^^^^^^^^^^^^
*********************************************************************
The ave. time of a simple addition in sec. is =......0.000055100
The ave. time  of a simple multiplication in sec. is =......
                                         0.000063100
The ave. time of a simple division in sec. =......0.000061100
The ave. time of squaring in sec. =......0.000055100
The ave. time of a square root in sec. =......0.000071100
The ave. time of an exponentiation in sec. =......0.000004000
*********************************************************************
```

See Figure 9.32.

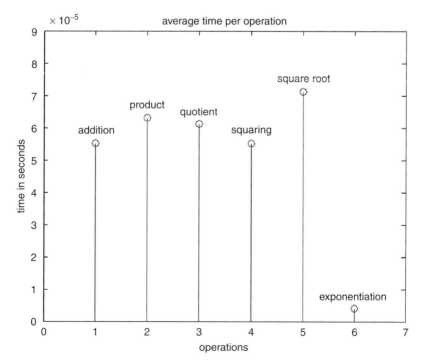

FIGURE 9.32
Computational time of basic operations of Example 9.15.

9.5   Further Analysis

Q.9.1   Load and run the script file *redrose* of Example 9.1.

Q.9.2   Transform the script file *redrose* to a function file *blackrose(n)* that returns a black rose of n petals, where n is an even integer (see Example 5.13).

Q.9.3   Transform the function file *blackrose.m* of Q.9.2 into a file that returns a rose of any color with any number of petals (see Examples 5.1 and 5.12).

Q.9.4   Load and run the script file *sinc33.m* of Example 9.2.

Q.9.5   Transform the script file *sinc33.m* to the function file *sinc33f.m* where the magnitude and frequency of the *sinc* function constitute the input variables.

Q.9.6   Modify the function file *sinc33f.m* to return the area inside the times.

Q.9.7   Load and run the script file *quadratic.m*, of Example 9.3, for the equation $y(x) = 2x^2 + 7x + 6$. Verify that the roots are −2 and −1.5.

Q.9.8   Modify Example 9.3 so that the roots $x1$ and $x2$ as well as the messages indicated in the following table are displayed when the conditions apply.

See Table 9.2.

TABLE 9.2

Characteristics of the Roots of a 2^d Order Equation

Condition	Message
$b^2 > 4 \cdot a \cdot c$	Real and distinct roots
$b^2 = 4 \cdot a \cdot c$	Repeated real roots
$b^2 < 4 \cdot a \cdot c$	Complex roots

Q.9.9   Test the script file of Q.9.8 with examples that satisfy the following conditions:

$$a = 0, \quad b \neq 0, \quad c \neq 0$$

$$a = 0, \quad b = 0, \quad c \neq 0$$

$$a \neq 0, \quad b = 0, \quad c \neq 0$$

$$a \neq 0, \quad b = 0, \quad c = 0$$

and print the roots $x1$ and $x2$ displaying the appropriate message.

Q.9.10   Repeat Q.9.7, Q.9.8, and Q.9.9 for the function file *quadraticf.m* of Example 9.3.

Q.9.11   Load and run the function file *dst.m* of Example 9.4 for the Cartesian points given by *P1(2, −1)* and *P2(3, 2)*.

Q.9.12   Define and describe the purpose of each line of the function *dst.m* as comments (%).

Q.9.13   Modify the function file *dst.m* to output the distance and slope defined by the line that passes through points *P1* and *P2*.

Q.9.14   Load and run the modified file of Q.9.13.

Q.9.15   Evaluate by hand Q.9.14 and compare with the results obtained.

Q.9.16   The profile report indicates the timing used for each line of the file tested. Indicate and discuss which lines of *dst* consume 75% of the total time.

Q.9.17 Load and run the script file *calc_area.m* of Example 9.5 for the following cases:

a. $f_1(x) = 8x^3 + 2x^2 - 6x - 35$, over the range $-2 \le x \le 5$

b. $f_2(x) = -3x^3 + 2x^2 - x + 4$, over the range $-2.5 \le x \le 3$

c. $f_3(x) = 10x^3 + 3x^2 - 5x + 28$, over the range $-4 \le x \le 5$

Q.9.18 Modify Example 9.5 into the equivalent function file *calc_areaf.m*.

Q.9.19 Evaluate, compare, and discuss the timing performance of the files *calc_areaf.m* and *calc_area.m*.

Q.9.20 Load and run the function file *prodz1z2.m* of Example 9.6 for the following case:

$$z1 = 3 - 8i \quad \text{and} \quad z2 = -5 + 7i$$

Q.9.21 Modify the function file *prodz1z2* to output the product of *z1* and *z2* in rectangular form.

Q.9.22 Load and run the script file *growth.m* of Example 9.7, part a.

Q.9.23 Modify the script file *growth* to include two additional output columns: Fat 12% and Fat 14%.

Q.9.24 Load and run the script file *growthquot.m* of Example 9.7, part b.

Q.9.25 Compare the growth of an initial capital of $10,000 at 10% compounded annually (Example 9.7) versus the same initial capital compounded quarterly.

Q.9.26 Modify Example 9.7 to include an additional plot that displays the extra benefits of the interest when compounded quarterly instead of annually versus years for the cases of 6, 8, and 10%.

Q.9.27 Load and run the script file *effective_int.m* of Example 9.8.

Q.9.28 Identify the variables that define the effective interest compounded annually, semi-annually, quarterly, monthly, weekly, daily, and continuously and define these variables in the form of equations.

Q.9.29 Modify Example 9.8 to display in a tablelike format the growth of $10,000 by using the different effective interest rates per year during a time period of 10 years.

Q.9.30 Load and run the function file *growthf1.m* of Example 9.9 for the following cases:

a. $P = 25,000$, $I1 = 5\%$, $I2 = 10\%$, $I3 = 15\%$, and $n = 15$ years

b. $P = 50,000$, $I1 = 6\%$, $I2 = 12\%$, $I3 = 18\%$, and $n = 18$ years

Q.9.31 Load and run the function file *solv.m* of Example 9.10 for the following cases:

a. $u = -2$

b. $u = 6$

c. $-2 < u < 6$ in linear increments of 0.5

Q.9.32 Load and run each of the script files of Example 9.11 (*vec_loop*, *forend_loop*, and *whileend_loop*) and compare the execution time in each case. Discuss the results. Which program is the most efficient? Which program is the least efficient?

Q.9.33 Load and run the function file *statis_perf.m* of Example 9.12 for the data array defined as follows:

$$n = 1:33$$

$$x(n) = \frac{5 \cos(n\pi\ 3/2)}{0.01n}$$

Q.9.34 Load and run the function file *graph_perf.m* of Example 9.13 for the data defined in Q.9.33.

Q.9.35 Load and run the script file *sparse_full* of Example 9.14.

Q.9.36 Run the file *sparse_full* for the case $AA = A^3 + A^2 + A$ twice for A sparse and A full.

Q.9.37 Evaluate and estimate the efficiencies of using the sparse versus the full version.

Q.9.38 Compare the execution time for the sparse and full version.

Q.9.39 Compare the efficiencies for the operation $A*inv(A)$ for the two cases, sparse and full.

Q.9.40 Why are the efficiencies so different? Discuss.

Q.9.41 Load and run the script file *timing_oper* of Example 9.15.

Q.9.42 Explore the execution time of A^n, for $n = 3, 4, 5, 6$ where A is the magic matrix, returning the results in the form of the plot n versus *execution time*.

Q.9.43 Obtain the equivalent of Figure 9.32 for the case of matrices.

9.6 Application Problems

P.9.1 Create a script file that returns the square matrices $X = magic(3)$ and $Y = hilb(3)$ and evaluate the following matrices:

$$A = X + Y$$
$$B = X - Y$$
$$C = X * Y$$
$$D = X \cdot * Y$$
$$E = A^B$$
$$F = A^3$$

Estimate also the execution time for each operation.

P.9.2 Create a script file that computes and displays a conversion table between temperature scales of Celsius, Fahrenheit, and Kelvin over the range 0–100°C with liner increments of 10°C (use the conversion shown in Table 2.5).

P.9.3 Create a script file that computes and displays a conversion table between miles, meters, and feet over the range of 0–1000 m (use the conversions shown in P.2.29) in steps of 100 m.

P.9.4 Create a script file that returns a conversion table between gallon, liters, and cubic feet over the range of 0–100 L, in steps (increments) of 5 L.

P.9.5 Create a script file that traces the amount accumulated during 5 years, of an initial investment of $5000 at an interest of 6.5% annually, compounded quarterly.

P.9.6 Create a script file that returns a conversion table between the following currencies:

U.S. dollars, euros, Denmark kronens, British pounds, Jordan dinars, and Venezuelan bolivars over a range of $0–$100 (use the conversion shown in Table 2.6), in increments of $5.

P.9.7 Create a script file that returns the plots of the functions $y1$ versus x and $y2$ versus x on the same graph, where

$$y1 = x \cos(x)$$

$$y2 = 1/3 * x^2 \sin(x)$$

over the range $-\pi \le x \le +\pi$.

P.9.8 Given an input vector x, create a script and its equivalent function file that returns the following:

a. Sum of all the elements of x

b. Product of all the elements of x

c. Sort the elements of x in ascending order

d. Sort the elements of x in descending order

P.9.9 Create a function file where the inputs $<x_1, y_1>$, $<x_2, y_2>$, and $<x_3, y_3>$ represent the vertices of a triangle in Cartesian coordinates and return the plot of the triangle as well as its area and perimeter.

P.9.10 Create a function file that draws a circle of radius r centered at $<x_0, y_0>$ in the rectangular coordinate system.

P.9.11 Create a function file that, given the coefficients of a set of three linear equations, returns its solution in terms of the three unknown x, y, and z.

P.9.12 Given a square matrix A, create a function file that returns the following:

• The value 1 if A is square and 0 if it is not

• The $cond(A)$

• The $det(A)$

• The $inv(A)$

• The $eigenvalues$ and $eigenvectors$ of A

P.9.13 Draw a clear flowchart of the function file that returns a rose inside a circle with radius r and n petals, where $n \le 18$.

P.9.14 Create the script file *matrix_1_n* that returns the matrix A defined as

$$A = \begin{bmatrix} 1 & 2 & 3 & 4 & . & . & n \\ 2 & 3 & 4 & . & . & n & 1 \\ 3 & 4 & 5 & . & n & 1 & 2 \\ 4 & . & . & n & . & . & . \\ . & . & n & . & . & . & . \\ . & n & . & . & . & . & . \\ n & 1 & 2 & 3 & . & . & n-1 \end{bmatrix}$$

for any given integer $n \le 10$.

P.9.15   Convert the script file of P.9.14 into a function file with n as input, for any integer $n \leq 10$.

P.9.16   Create a function file that converts the time t, given as a vector $t = [hh\ mm\ ss]$, into hours.

P.9.17   Create a function file with input $f(t)$ that returns the function $f(-t)$, over a given range $t_1 \leq t \leq t_2$.

P.9.18   Create a function file with input $f(t)$ that returns the following functions:

- $f(t - t_0)$
- $-f(t)$
- $f(t + t_0)$
- $-f(-t + t_0)$

over a given range $t_1 \leq t \leq t_2$.

P.9.19. Create a function file with input $y = f(x)$ that returns

a. dy/dx

b. d^2y/dx^2

c. $\int y dx$

d. $\int\int y dx dx$

P.9.20   Repeat P.9.19 that returns the function plots.

P.9.21   For the script file of P.9.1, estimate the execution and cpu time for each operation.

P.9.22   Let $x = rand(1, 1000) * 5$. Repeat P.9.8 and estimate in each case the number of operations as well as the total execution time.

P.9.23   For the vector x defined in P.9.22, create a program that returns the following statistical data:

a. Average (ave)

b. Standard deviation (dev)

c. Variance (var)

d. GM (geomean)

e. RMS (rms)

f. HM (harmean)

P.9.24   Estimate the following statistical data for P.9.23:

a. Number of operations

b. Estimated time

c. Cpu time

d. Elapsed time

P.9.25   Using the data of P.9.22, create a program that returns an 8-bins histogram and its corresponding pie graph.

P.9.26   Evaluate the performance of the function file of P.9.9 by executing the profile report and profile plot.

P.9.27   Create a function file that returns the effective interest rate compounded annually, monthly, and weekly, given an initial investment P, interest I, and number of years n.

P.9.28   Repeat P.9.27 for the case of effective interest rate compounded quarterly, weekly, and daily.

P.9.29   Test the function file of P.9.13 for $r = 3$ and $n = 9$, and estimate the cpu time.

P.9.30   Run P.9.29 twice and estimate and discuss the improved efficiency.

P.9.31   Create the sparse matrix $A = \text{eye}(3)$ and full matrix $B = \text{eye}(3)$, and perform the following operations:

  a. $C = A^2 + B^2$

  b. $D = B^2 * A^2$

  In each case, estimate the execution time and the resulting matrix.

P.9.32   Repeat P.9.31 for the following cases:

  a. $E = A^2 + A^2$

  b. $F = B^2 * B^2$

P.9.33   Compare and discuss the execution times obtained in P.9.31 and P.9.32.

Bibliography

Adamson, T., *Structured Basic*, 2nd Edition, Merrill Publishing Co., New York, 1993.

Ana, B., The state of the research isn't all that grand, *The New York Times*, September 3, 2006.

Attia, J.O., *Electronics and Circuits, Analysis Using Matlab*, CRC Press, Boca Raton FL, 1999.

Austin, M. and Chancogne, D., *Engineering Programming C, Matlab, Java*, Wiley, New York, 1999.

Ayres, A., *Theory and Problems of Differential Equations*, Schaum's Outlines Series, McGraw-Hill, New York, 1952.

Balador, A., *Algebra*, Deama Octava Re-impression, Publications Cultural, Mexico, 2000.

Biran, A. and Breiner, M., *Matlab for Engineers*, Cambridge, Great Britain, 1995.

Bogard, T.F., Beasley, J.S., and Rico, G., *Electronic Devices and Circuits*, 5th Edition, Prentice-Hall, New York, 1997.

Borse, G.J., *Numerical Methods with Matlab*, P.W.S. Publishing Co., Boston, 1997.

Brooks, D., The populist myths on income inequality, *The New York Times*, September 7, 2006.

Buck, J., Daniel, M., and Singer, A., *Computer Explorations in Signals and Systems Using Matlab*, 2nd Edition, Prentice-Hall, Upper Saddle River, NJ, 2002.

Caputo, D.A., The world's best education; remade in America, *The New York Times*, A29, December 6, 2006.

Carlson, A.B., *Circuits*, Brooks/Cole, Pacific Grove, CA, 2000.

Cartinhour, J., *Digital Signal Processing*, Prentice-Hall, New York, 2000.

Chapman, S., *Matlab for Engineers*, Brooks/Cole, Pacific Grove, CA, 2000.

Clawson, C.C., *Mathematical Mysteries*, Perseus Books, Cambridge, MA, 1996.

Connor, F.R., *Circuits*, Calabria, Barcelona, Spain, 1976.

Cyganski, D., Orr, J.A., and Vaz, R.F., *Information Technology Inside and Outside*, Prentice-Hall, New York, 2001.

Deziel, P.J., *Applied Digital Signal Processing*, Prentice-Hall, New York, 2001.

Dwyer, D. and Gruenwald, M., *Precalculus*, Thomson-Brookes/Cole, Pacific Grove, CA, 2004.

Edminister, J., *Theory and Problems of Electric Circuits*, Schaum's Outlines Series, McGraw-Hill, New York, 1965.

Etter, D.M., *Introduction to Matlab for Engineers and Scientist*, Prentice-Hall, Upper Saddle River, NJ, 1996.

Etter, D.M. and Kuncicky, D.C., *Introduction to Matlab*, Prentice-Hall, New York, 1999.

Exxon Mobil, Multiplier effects, *The New York Times*, A33, December 19, 2006.

Friedman, T.L., Learning to keep learning, *The New York Times*, A33, December 13, 2006.

Gabel, R. and Roberts, R., *Signals and Linear Systems*, Wiley, New York, 1973.

Gawell, K., report available at http:/www.geo-energy.org/publications/reports.asp. 2007.

Grover, D. and Deller, J.R., *Digital Signal Processing and the Microcontroller*, Prentice-Hall, New York, 1999.

Grunwald, M., *The Clean Energy Scan*, Time Magazine, 2008.

Gustafsson, F. and Bergman, N., *Matlab for Engineers Explained*, The Cromwell Press, Springer-Verlag, London, U.K., 2004.

Hanselman, D. and Littlefield, D., *The Student Edition of Matlab*, Version 5, User's Guide, Prentice-Hall, Upper Saddle River, NJ, 1997.

Hanselman, D. and Littlefield, D., *Mastering Matlab 7*, Prentice-Hall, Upper Saddle River, NJ, 2005.

Harman, T.L., Dabney, J., and Richer, N., *Advanced Engineering Mathematics Using Matlab*, Brooks/Cole, Pacific Grove, CA, 2000.

Hayt, W. and Kemmerly, J.E., *Engineering Circuit Analysis*, McGraw-Hill, New York, 1962.

Hill, D. and David, E.Z., *Linear Algebra Labs with Matlab*, Second Edition, Prentice-Hall, Upper Saddle River, NJ, 1996.

Hodge, N., Solar energy stock, *Wealth Daily*, Anger Publishing LLC, 2007.

Hsu, H.P., *Analysis De Fourier*, Fondo Education Interamericano, S.A., Bogata, Colombia, 1973.

Ingle, V. and Proakis, J., *Digital Signal Processing*, Brooks/Cole, Pacific Grove, CA, 2000a.

Ingle, V. and Proakis, J., *Digital Signal Processing Using Matlab*, Brooks/Cole, Pacific Grove, CA, 2000b.

Jack, K., *Engineering Circuit Analysis*, McGraw-Hill, New York, 1962.

Jairam, A., *Companion in Alternating Current Technology*, Prentice-Hall, Upper Saddle River, NJ, 1999.

Jairam, A., *Companion in Direct Current Technology*, Prentice-Hall, Upper Saddle River, NJ, 2000.

Jensen, G., *Using Matlab in Calculus*, Prentice-Hall, Upper Saddle River, NJ, 2000.

Joseph, J.D. III, Allen, R.S., and Ivan, J.W., *Feedback and Control Systems*, Schaum Publishing Co., New York, 1967.

Judith, M. and Muschle, G.R., *The Math Teacher's Book of Lists*, Prentice-Hall, Upper Saddle River, NJ, 1995.

Kamen, E.W. and Heck, B.S., *Fundamentals and Systems Using the Web and Matlab*, 2nd Edition, Prentice-Hall, Upper Saddle River, NJ, 2000.

Kay, D.A., *Trigonometry*, Cliffs Quick Review, First Edition, Cliffs Notes, Lincoln, NE, 1994.

Keedy, M., Bittinger, M.L., and Rudolph, W.B., *Essential Mathematics for Long Island University*, Brooklyn Campus, Pearson-Addison-Wesley, Custom Publishing, Boston, 1992.

Keedy, M., Griswold, A., Schacht, J., and Mamary, A., *Algebra and Trigonometry*, Holt, Rinehart and Winston Inc., New York, 1967.

Krauss, C., Move over, Oil. There's Money in Texas Wind, *The New York Times* (pp. A1, A15), February 23, 2008.

Kurtz, M., *Engineering Economics for Professional Engineer's Examinations*, Third Edition, McGraw-Hill, New York, 1985.

Lathi, B.P., *Modern Digital and Analog Communication Systems*, 3rd Edition, Oxford University Press, New York, 1998.

Leon, S., Eugene, H., and Richard, F., *ATLAST, Computer Exercises for Linear Algebra*, Prentice-Hall, Upper Saddle River, NJ, 1996.

Linderburg, M., *Engineer in Training Review*, 6th Edition, Belmont, CA, 1982.

Lindfield, G. and Penny, J., *Numerical Methods Using Matlab*, Prentice-Hall, Upper Saddle River, NJ, 2000.

Lipschutz, S., *Theory and Problems of Linear Algebra*, Schaum's Outlines Series, McGraw-Hill, New York, 1968.

Lutovac, M.D., Tosic, D.V., and Evans, B.L., *Filter Design for Signal Processing (Using Matlab and Mathematics)*, Prentice-Hall, Upper Saddle River, NJ, 2001.

Lynch, W.A. and Truxal, J.G., *Signals and Systems in Electrical Engineering*, McGraw-Hill, The Maple Press Company, York, PA, 1962.

Magrad, E.B., Azarm, S., Balachandran, B., Duncan, J.H., Herold, K.E., Walsh, G., *An Engineer's Guide to Matlab*, Prentice-Hall, Upper Saddle River, NJ, 2000.

Maloney, T.J., *Electric Circuits: Principles and Adaption*, Prentice-Hall, Upper Saddle River, NJ, 1984.

Markoff, J., Smaller than a pushpin, more power than a PC, *The New York Times*, C3, February 7, 2005.

Markoff, J., A chip that can move data at the speed of laser light, *The New York Times*, C1, September 18, 2006.

Markoff, J., Intel prototype may herald a new age of processing, *The New York Times*, C9, February 12, 2007.

Matt, R., Start up fever shift to energy in Silicon Valley, *The New York Times*, A1/C4, March 14, 2007.

McClellan, J.H., Schafer, R.W., and Yoder, M.A., *DSP First: A Multimedia Approach*, Prentice-Hall, Upper Saddle River, NJ, 1998.

McMenamin, S.M. and John, F.P., *Essentials Systems Analysis*, Yourdon Press Computing Series, Prentice-Hall, Englewood Cliffs, NJ, 1984.

Meador, D., *Analog Signal Processing with Laplace Transforms and Active Filter Design*, Belmont, CA, 2002.

Miller, M.L., *Introduction to Digital and Data Communications*, West Publishing Company, St. Paul, MN, 1992.

Minister, J.A., *Electric Circuits*, Schaum Publishing Co., New York, 1965.

Miroslav, D., Tosic, D., and Evan, B., *Filtering Design for Signal Processing*, Prentice-Hall, Upper Saddle River, NJ, 2001.

Mitra, S.K., *Digital Signal Processing Laboratory Using Matlab*, McGraw-Hill, New York, 1999.

Mitra, S.K., *Digital Signal Processing*, McGraw-Hill, New York, 2001.

Nashelsky, L. and Boylestad, R.L., *Basic Applied to Circuit Analysis*, Merrill Publishing Co., Columbus, OH, 1984.

Navarro, H., *Instrumentacion Electronica Moderna*, Univsidad Central de Venezuela, 1995.

New York City Board of Education, *Sequential Mathematics*, 1989 (Revision).

Newman, J., *The World of Mathematics*, Volumes 1, 2, 3 and 4, Simon and Schuster, New York, 1956 (commentaries).

Novelli, A., *Lecciones De Analisis I*, Impresiones Avellaneda, Buenos Aires, Argentina, 1998.

Novelli, A., *Algebra Lineal Y Geometria*, 2nd Edition, Impresiones Avellaneda, Buenos Aires, Argentina, 2004a.

Novelli, A., *Lecciones De Analisis II*, Impresiones Avellaneda, Buenos Aires, Argentina, 2004b.

O'Brien, M.J. and Larry, S., *Profit from Experience*, Bard & Stephen, Austin, TX, 1995.

Oppenheim, A.V., Schafer, R.W., and Buck, J.R., *Discrete-Time Signal Processing*, 2nd Edition, Prentice-Hall, Upper Saddle River, NJ, 1999.

Oppenheim, A., Willsky, A. and Young, I., *Signals and Systems*, Prentice-Hall, Upper Saddle River, NJ, 1983.

Palm, W.J., III, *Introduction to Matlab for Engineers*, McGraw-Hill, Natick, MA, 1998.

Parson, J.J., and Oja, D., *Computer Concepts*, Thomson Publishing Company, 1996.

Patrick, D.R. and Fardo, S.W., *Electricity and Electronics: A Survey*, 4th Edition, Prentice-Hall, Upper Saddle River, NJ, 1999.

Petr, B., *A History of TT*, St. Martin's Press (Golem), New York, 1971.

Petruzella, F., *Essentials of Electronics*, 2nd Edition, McGraw-Hill, New York, 2001.

Polking, J. and Arnold, D., *Ordinary Differential Equations Using Matlab*, 3rd Edition, Prentice-Hall/ Pearson, Upper Saddle River, NJ, 2004.

Pratap, R., *Getting Started with Matlab*, Saunders College Publishing, Orlando, FL, 1996.

Proakis, J., and Salekis, M., *Contemporary Communication Systems Using Matlab*, PWS Publishing Co., Boston, 1998.

Randall, S., It's not who you know. It's where you are, *The New York Times*, B3, December 22, 2006.

Recktenwald, G., *Numerical Methods with Matlab*, Prentice-Hall, Upper Saddle River, NJ, 2000.

Rich, B., *Elementary Algebra*, Schaum's Outline Series, McGraw-Hill, New York, 1960.

Rich, B., *Modern Elementary Algebra*, Schaum's Outline Series, McGraw-Hill, New York, 1973.

Robert, J.P., *Introduction to Engineering Technology*, Third Edition, Prentice-Hall, Englewood Cliffs, NJ, 1996.

Robinson, D., *Fundamentals of Structured Program Design*, Prentice-Hall, Upper Saddle River, NJ, 2000.

Russel, M.M., and Mark, J.T.S., *Digital Filtering: A Computer Laboratory Textbook*, John Wiley, New York, 1994.

Ruston, H. and Bordogna, J., *Electric Networks: Functions, Filters Analysis*, McGraw-Hill, New York, 1966.

Sarachik, P., *Principles of Linear Systems*, Cambridge University Press, New York, 1997.

Schilling, R. and Harris, S., *Applied Numerical Methods for Engineers, Using Matlab and C*, Brooks/Cole, Pacific Grove, CA, 2000.

Schuller, C.A., *Electronics Principles and Applications*, McGraw-Hill, New York, 1989.

Shenoi, K., *Digital Signal Processing in Telecommunications*, Prentice-Hall, Upper Saddle River, NJ, 1995.

Sherrick, J.D., *Concepts in Systems and Signals*, Prentice-Hall, Upper Saddle River, NJ, 2001.

Silverman, G. and Tukiew, D.B., *Computers and Computer Languages*, McGraw-Hill, New York, 1988.

Smith, D.M., *Engineering Computation with Matlab*, Pearson, Boston, 2008.

Smith, M.J.T. and Mersereau, R.M., *Introduction to Digital Signal Processing*, Wiley, New York, 1992.

Spasov, P., *Programming for Technology Students Using Visual Basic*, 2nd Edition, Prentice-Hall, Upper Saddle River, NJ, 2002.

Spiegel, M.R., *Laplace Transforms*, Schaum's Outline Series, McGraw-Hill, New York, 1965.

Sprankle, M., *Problem Solving and Programming Concepts*, 5th Edition, Prentice-Hall, Upper Saddle River, NJ, 2001.

Sprankle, M., *Problem Solving for Information Processing*, Prentice-Hall, Upper Saddle River, NJ, 2002.

Stanley, W.D., *Network Analysis with Applications*, 4th Edition, Prentice-Hall, Upper Saddle River, NJ, 2003.

Stanley, W.D., *Technical Analysis and Applications With Matlab*, Thomson V Delmar Learning, New York, 2005.

Stearns, S. and David, R., *Signal Processing Algorithms in Matlab*, Prentice-Hall, Upper Saddle River, NJ, 1996.

Stein, E.I., *Fundamentals of Mathematics*, Modern Edition, Allyn and Bacon Inc., Boston, 1964.

Steve, L., Parsing the truths about visas for tech workers, *The New York Times*, April 15, 2007.

Sticklen, J. and Taner, E.M., *An Introduction to Technical Problem Solving with Matlab*, Volume 7, Second Edition, Great Lakes Press Inc., Wildwood, MO, 2006.

Strum, R. and Kirk, D., *Contemporary Linear Systems Using Matlab*, Brooks/Cole, Pacific Grove, CA, 2000.

Tagliabue, J., The eastern bloc of outsourcing, *The New York Times*, C1/5, April 19, 2007.

Tahan, M., *EL Hombre the Calculaba Segunola*, Edition Ampliaoa, Buenos Aires, Argentina (Traduudo por Mario Cappetti), 1938.

The Math Work Inc, *The Student Edition of Matlab*, Version 4, User's Guide, Prentice-Hall, Englewood Cliffs, NJ, 1995.

Theodore, F.B., *Basic Programs for Electrical Circuit Analysis*, Reston Publishing Company, Inc., Reston, VA, 1985.

Van de Vegte, J., *Fundamentals of Digital Signal Processing*, Prentice-Hall, Upper Saddle River, NJ, 2002.

Van Valkenbung, M.E., *Network Analysis*, Third Edition, Prentice-Hall, Englewood Cliffs, NJ, 1974.

Von Seggern, D., *Standard Curves and Surfaces*, CRC Press, Boca Raton, FL, 1993.

White, S., *Digital Signal Processing*, Thomson Learning, Albany, NY, 2000.

Young, P.H., *Electronic Communication Techniques*, 4th Edition, Prentice-Hall, Upper Saddle River, NJ, 1999.

Zabinski, M.P., *Introduction to TRS-80 Level II Basic*, Prentice-Hall, Upper Saddle River, NJ, 1980.

Index